D1599142

About Island Press

Island Press is the only nonprofit organization in the United States whose principal purpose is the publication of books on environmental issues and natural resource management. We provide solutions-oriented information to professionals, public officials, business and community leaders, and concerned citizens who are shaping responses to environmental problems.

In 1998, Island Press celebrates its fourteenth anniversary as the leading provider of timely and practical books that take a multidisciplinary approach to critical environmental concerns. Our growing list of titles reflects our commitment to bringing the best of an expanding body of literature to the environmental community throughout North America and the world.

Support for Island Press is provided by The Jenifer Altman Foundation, The Bullitt Foundation, The Mary Flagler Cary Charitable Trust, The Nathan Cummings Foundation, The Geraldine R. Dodge Foundation, The Ford Foundation, The Vira I. Heinz Endowment, The W. Alton Jones Foundation, The John D. and Catherine T. MacArthur Foundation, The Andrew W. Mellon Foundation, The Curtis and Edith Munson Foundation, The National Fish and Wildlife Foundation, The National Science Foundation, The New-Land Foundation, The David and Lucile Packard Foundation, The Surdna Foundation, The Winslow Foundation, The Pew Charitable Trusts, and individual donors.

Avian Conservation

Avian
Conservation
Research and Management

Edited by
John M. Marzluff and Rex Sallabanks

ISLAND PRESS

WASHINGTON, D.C. • COVELO, CALIFORNIA

No copyright claim is made in work by the following employees of the federal government: Robert C. Fleischer, Martin G. Raphael, Beth M. Galleher, Richard L. Hutto, Scott R. Derrickson, Steven R. Beissinger, Noel F. R. Snyder, John R. Squires, James F. Gore, Frank R. Thompson III, Therese M. Donovan, Fritz L. Knopf, Gordon H. Rodda, Michael N. Kochert, Michael W. Collopy, Sallie J. Hejl, Kathleen Milne Granillo, Joseph L. Ganey, Cecelia M. Dargan.

Grateful appreciation is acknowledged for permission to use the following copyrighted material: Figure 10.1 is reprinted from "Using Animal Behavior for Conservation: Case Studies in Seabird Restoration from the Maine Coast, USA," by S. W. Kress, *Journal of Yamashina Institute for Ornithology,* vol. 29, no. 1 (1997). Figure 14.1 is reprinted from "Energy and Large-Scale Patterns of Animal- and Plant-Species Richness," by David J. Currie, *American Naturalist,* vol. 137 (1991). Tables 9.2 and 9.3 are reprinted with permisssion of the British Ornithologists' Union.

Library of Congress Cataloging-in-Publication Data

Avian conservation : research and management / edited by John M.
   Marzluff and Rex Sallabanks.
      p.   cm.
   Some papers presented at a symposium held at the 1996 joint
meeting of the American Ornithologists' Union and the Raptor
Research Foundation.
   Includes bibliographical references and index.
   ISBN 1-55963-569-X (cloth : alk. paper)
   1. Birds, Protection of—Congresses.   I. Marzluff, John M.
II. Sallabanks, Rex.
QL676.5.A96   1998
333.95'816—dc21   98-10506
                                                     CIP

Contents

Part IV: Conservation in Forested Landscapes

Part V: Conservation in Nonforested and Urban Landscapes

Part VI: Global Variation in Conservation Needs

Part VII: Relevance of Conservation Research to Land Managers

Preface

This book by avian conservationists reviews current research and identifies information gaps that need to be filled if we are to effectively conserve birds in the next several decades. Approximately half of the chapters were presented orally in a symposium at the 1996 joint meeting of the American Ornithologists' Union and the Raptor Research Foundation organized by ourselves. The remaining chapters were solicited to provide a broad review of avian conservation, fill important gaps in the coverage of this expanding field, and allow working managers and conservation pioneers to interject their thoughts. We strived to obtain chapters that not only discussed research needs, but also suggested how to most effectively implement research results for avian conservation.

The book is organized into seven parts. First, we reflect on recent approaches, growing pains, and future courses of action of the growing field of avian conservation biology. We then discuss important new techniques for conserving and monitoring birds. Genetics, spatial modeling, and methods for monitoring abundance and demography, important tools that subsequent chapters rely on, are discussed. Captive breeding, reintroduction, translocation, population viability modeling, and proactive conservation are discussed in part III on approaches for conserving endangered and sensitive species. Specific case studies reviewing research and future research needs in forested and nonforested landscapes are the focus of parts IV and V. Most of these chapters rely on research conducted in North America. We broaden this perspective in part VI by presenting discussions of research results and needs from a variety of European, island, neo- and paleotropical projects. Part VII includes discussions about how to make research applicable to management needs. These chapters are written by land managers and researchers working for management agencies. We provide short introductions to each section to provide cohesion needed with a multiauthored volume such as this.

Several threads weave the book together. First, an expanding human population has caused widespread changes in bird habitats. All habitats are affected to varying degrees, and birds respond quickly to these changes. Second, most avian conservation is currently based on short-term, correlative studies of bird abundance, which may be misleading. Instead, conservation needs to be rooted on long-term, experimental studies that measure avian population viability (productivity, survival, and dispersal). Third, research has often proceeded independent of consultation with actual managers. A closer relationship needs to be developed between researchers and managers if effective conservation strategies are to be developed.

The decline of many bird species throughout the world is alarming and the cause of much research and management. We hope this book makes an important contribution to the conservation of birds by: (1) highlighting the peril many species experience; (2) showcasing some of the best research projects designed to understand why birds are declining and reverse those trends; and (3) presenting the advice of practicing managers to underscore important research approaches often not considered by research scientists. Our intent is to provide a recipe for successful conservation during the next ten to fifty years. We hope conservation biologists, managers, and students interested in the conservation of birds will take the messages in the book to heart. Use the ideas herein to: (1) gauge the quality and relevance of your research; (2) stimulate new research areas; and (3) improve your communication of needs to researchers and results to managers. Above all else, use these thoughts to improve the conservation of birds as we enter a new century with ever increasing human demands on our natural resources.

We offer our sincere thanks to all who encouraged, reviewed, and guided our development of this book. Each chapter was peer reviewed, and we thank the following colleagues for their comments: Yoram Ayal, Paul Banko, Steve Beissinger, Carl Bock, Jeff Brawn, Henry Campa, Andy Carey, Dick DeGraff, John Faaborg, Hugh Ford, Mark Fuller, Rob Fuller, Joe Ganey, Fred Gehlbach, Lauren Gilson, Alejandro Grajal, Sue Haig, Andy Hansen, Sallie Hejl, Chuck Hunter, Dick Hutto, Fran James, Scott Johnston, Robert Kenward, Steve Knick, Rick Knight, Mike Kochert, Stephen Kress, Dave Manuwal, Bruce Marcot, Bill McComb, Mike Morrison, Raymond O'Connor, Berry Pinshow, Richard Porter, Martin Raphael, Harry Recher, Scott Robinson, John Rotenberry, Ian Rowley, Vickie Saab, Tom Sherry, Bill Sydeman, Jeff Walters, John Wiens, and Amotz Zahavi. Financial support for the project was provided by Sustainable Ecosystems Institute and the AOU/RRF 1996 Local Committee. Barbara Dean and her staff at Island Press showed great patience and guidance in preparing the final copy. Lastly, we would like to thank all the authors for contributing and revising in a timely (or less than timely!) fashion.

J. M., Clearview, Washington
R. S., Meridian, Idaho

PART I

Introduction: The Current Status of Avian Conservation Biology and Prescriptions for the Future

We start the book with two reflective chapters that look at the past approaches and future directions of avian conservation biology. Marzluff and Sallabanks (chapter 1) sampled the scientific literature to quantify recent approaches to avian conservation biology. From this benchmark they suggest the general direction future research should follow. Their review found that current avian conservation research most often addresses the effects of habitat change due to timber harvest, urban development, and agricultural intensification. Most studies: (1) produce correlative, rather than causal relationships; (2) measure avian abundance or behavior, rather than demographics; (3) are proactive, rather than reactive; and (4) are conducted at short (one- to two-year) temporal and small spatial scales.

To improve research on avian conservation in the future, chapter 1 suggests that we need to simultaneously increase the scale and the resolution of studies. Adaptive management will be necessary to extrapolate results obtained at small spatial, temporal, or ecological scales to larger scales relevant to management of avian biodiversity. Research is also needed on ways to accurately extrapolate experimental, demographic analyses conducted on a single species or population to the scale of entire avifaunas. We need to identify the mechanisms responsible for causal relationships between avian population viability and human disturbance. This will require research on predators, brood parasites, food availability, and abiotic factors. Quality research of the scope called for is probably beyond the ability of a single researcher. Therefore, cooperative projects that last beyond the usual funding and graduate career cycles will be necessary if we are to provide managers with the information needed to successfully conserve our avifauna.

Fran James, a leading ornithologist, recently changed the direction of her research to address problems in the burgeoning field of avian conservation biology. In chapter 2, she reflects upon her career, pointing out that the two main jobs of avian conservation biologists are to provide the best possible scientific information about bird populations that are in trouble and to participate in their management. She notes that this assignment is much more difficult to do well than it sounds. It requires knowledge of the basic biology of birds, expertise in standardized methods of monitoring populations, ability to diagnose causal relationships, and skill in working with policy makers and managers. Only if all these efforts are successful will ornithologists be able to conserve their favored class of organisms and contribute to solving resource problems that include but are not limited to birds. She uses three examples to illustrate what she thinks are some of the growing pains that the field is having as it learns from experience.

James argues that with the endangered Red-cockaded Woodpecker on both public and private land, managers are putting too much emphasis on the short-term benefits of providing nest boxes and transporting birds among sites and too little on the long-term objective of restoring the all-age pine ecosystem of the southeastern states, of which the woodpecker is one component. With long-distance migrant songbird populations, the continuing tendency to make unjustified generalizations about population trends and weak arguments about causes is standing in the way of the formulation of sound policies. Lastly, she notes that the field of environmental contamination is simply not getting enough attention.

She uses the nonornithological case of the reduction of chlorofluorocarbons (CFCs) in the atmosphere as a good example of the fact that cooperation among scientists, government policy makers, and even industrial interests can achieve major conservation objectives. At the time of the signing of the Montreal Protocol in 1987, there was no proof that chlorofluorocarbons in the atmosphere were allowing more ultraviolet radiation to reach the earth, but policy makers were convinced that reductions were justified. A decade later the levels were 76 percent lower. Similar cooperation among scientists and policy makers on biodiversity issues is needed. But first conservation biologists have to assure all the stakeholders that they are getting the science right and doing a good job of identifying priorities.

The need to identify priorities for conservation efforts is echoed in both chapters. Given the limited funding available for conservation research, we must devise some way to get the money where it is most urgently needed, not just where the political will is strongest. Marzluff and Sallabanks urge the ornithological community to establish a committee to set major research priorities, solicit cooperative proposals to address priorities, and lobby for funding to support such proposals. James argues that we must first identify problems, then analyze their causes. She suggests we need to move beyond wide-scale claims of endangerment based upon a group's life history (e.g., migrants to the Neotropics) and root our claims of endangerment in solid science. We need to identify which species are in the most trouble, regardless of migratory status or other attributes, determine where in their geographic range the problems are most severe, analyze the potential causes of the problems, and then study the feasibility of management. Objective priority setting, based upon a species' conservation needs should be a "no brainer." However, until we agree on criteria of endangerment, meet as a group to discuss relative endangerment of all birds, and form a consensus as to which species are in need of help, we will continue to see independent research on species of political interest, rather than coordinated efforts to reduce threats to avian diversity.

CHAPTER 1

Past Approaches and Future Directions for Avian Conservation Biology

John M. Marzluff and Rex Sallabanks

In Utopia, research into the few existing conservation problems is well designed, longterm, sensitive to scale, and manipulative so that causal relationships between human activity and avian population viability can be developed across spatial, temporal, and ecological scales. The effect of each human action on each bird's fitness is experimentally tested and verified, and the importance of such an effect on ecosystem functioning is fully understood. Research results are then quickly translated into management recommendations, which are successfully implemented by managers to improve the conservation of avifauna. Conservation biologists in the real world rarely have it so good. Instead, limited time and money reduce our ability to execute the perfect research program before biodiversity declines. As practitioners of a "crisis science" (Soulé 1985) we often forgo complete understanding for quick and dirty assessments. We all want Utopian research because understanding causal relationships between our activity and avian diversity would remove the guesswork from converting research results into management prescriptions. But the pressing need to do something, anything, now before a species declines further pushes us to accept the uncertainty of implementing management prescriptions derived from correlational associations. The reality is that researchers cannot wait until they have Utopian knowledge; instead, they must share their results with managers as soon as they can (Hejl and Granillo, this volume).

As a way to assess the proximity of current avian conservation research to Utopian research we reviewed a sample of the recent (1990–96) ornithological, wildlife management, and conservation literature (*Auk, Condor, Journal of Wildlife Management, Wildlife Society Bulletin, Wildlife Monographs, Ecology, Ecological Monographs, Ecological Applications,* and *Conservation Biology*). We found 218 articles that addressed avian conservation research. Review articles and those dealing with policy, model development, simulation, or genetic description were excluded. We scored each article for its: (1) study duration; (2) scale of analysis (spatial and ecological); (3) type of human disturbance investigated; (4) use of an experimental (replicated treatments at a minimum) or a correlative approach; (5) measures of demographic components of fitness (survival, fecundity, nest success, dispersal) or proximate indicators of fitness (abundance, behavior); and (6) birds investigated and their status (endangered, threatened, sensitive, agent of decline). We use this review to quantify the recent approach to avian conservation research. From this benchmark we suggest the directions future research should follow. These directions are echoed and amplified for specific topics in the chapters of this book.

Recent Approaches to Avian Conservation Research

What Factors Do We Study?

The most commonly studied form of human disturbance was timber harvest (figure 1.1A). Studies of bird responses to timber harvest were typically conducted in western North America, the tropics, and northern Europe. Habitat disruption resulting from construction and expansion of human population centers, and effects of agricultural practices also were studied often. Few studies investigated other sources of human disturbance such as recreation, introduction of exotics, and power development. Two types of potentially important factors were rarely studied: the effects of our own research activities and the effects of harvesting birds for falconry.

If we assume that understanding causal relationships between bird populations and types of human disturbance requires experimental manipulation of disturbance and is best when demographic responses to such manipulations are measured, then we have some understanding of only a subset of the disturbances we studied (figure 1.1B, C). These disturbances are primarily logging, urbanization, and agricultural intensification. Only 28 of the 218 studies (13 percent) used experiments to assess avian demographic responses to human disturbance (figure 1.1C). Even for timber harvest (the most common form of human disturbance studied experimentally), the majority of published studies used artificial nests to assess rates of predation in various types of managed forests. Rarely were harvest treatments randomly allocated across a landscape or avian demographics measured before and after harvest. Urbanization and agricultural practices were the only other disturbances with more than two experimental studies that measured demography.

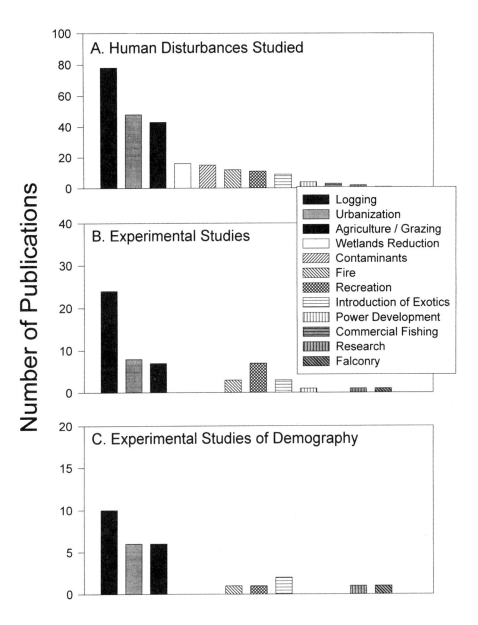

Figure 1.1. Types of human disturbances researched by avian conservation biologists 1990–96. (A) The relative abundance of disturbances investigated in the 218 research articles sampled. (B) The relative abundances of disturbances investigated in the 55 articles that conducted experiments (did manipulations or replicated control and disturbance treatments in a designed manner). (C) The relative abundance of disturbances investigated in the 28 articles that conducted experiments and measured demographic (survival, reproduction, or dispersal) responses of birds to human disturbances.

Are We Detecting Causal Relationships?

Principles of experimental design (randomization, replication, control) must be employed to document the causal relationship between a type of disturbance and an avian response (Caughley and Gunn 1996; James, Hess, and Kufrin 1997). Most studies relating birds to human disturbance (n = 171; 78 percent) did not have a rigorous experimental design. Treatments were rarely randomly allocated across the environment, and controls were uncommon. Instead, most studies measured a gradient of disturbance and correlated the degree of disturbance with avian abundance, population size, or demography. Such studies are useful at suggesting experiments that may produce causal relationships but are insufficient for establishing causal relationships.

Are We Measuring the Influence of Disturbance on Avian Fitness?

Measuring a component of fitness (fecundity, survival, dispersal) in a free-ranging population of birds is time consuming, expensive, and rarely done in either basic or applied research. Only 74 studies (34 percent) measured some component of fitness (nearly always reproductive success). A relatively complete understanding of how disturbance affects a population's demography requires a comparison of life tables, growth matrices, population viability models, or species-centered environmental analyses in disturbed and undisturbed areas. This type of study is possible for only a few well-studied (and typically threatened or endangered) birds (Northern Spotted Owl [*Strix occidentalis*], Florida Scrub Jay [*Aphelocoma coerulescens*], Red-cockaded Woodpecker [*Picoides borealis*]; Thomas et al. 1990; MaGuire, Wilhere, and Dong 1995; Breininger et al. 1996; James, Hess, and Kufrin 1997).

Are We Reactive or Proactive?

Despite the ability of sexy, endangered species to capture the public's eye and pocketbook, the majority of studies we reviewed (n = 147; 67 percent) were conducted on sensitive species, not threatened or endangered species. Thus, research tends to be proactive because birds that might respond negatively to human activities are identified before they become severely threatened, and management actions are initiated to reduce impacts (e.g., reducing snag removal for cavity nesters). The most commonly studied sensitive species were cavity nesters, neotropical migrant songbirds, and shorebirds. This proactive research is fueled by concerns over reductions in standing dead trees, breeding and wintering area habitat fragmentation, and wetland draining, respectively. Most reactive research was conducted on an elite group of threatened and endangered birds (Piping Plover [*Charadrius melodus*], Northern Spotted Owl, Red-cockaded Woodpecker, Bald Eagle [*Haliaeetus leucocephalus*]).

Although we are proactive in addressing the needs of many sensitive species, we rarely study the factors that directly threaten these populations. Only five studies (3 percent) focused on the predators (usually corvids) or brood parasites (Brown-headed Cowbird [*Molothrus ater*]) thought to be responsible for songbird declines. None experimentally assessed predator, competitor, or brood parasite demographic responses to human disturbance.

At What Scales Do We Conduct Research?

Avian conservation research is done at a very small temporal scale, one that is equal to the generation time of our typical grants and graduate degrees but rarely to that of our subjects. Seventy-three percent ($n = 159$) of the studies we reviewed lasted < 3 years; only 5 percent ($n = 11$) lasted a decade or more.

Long-term studies that include experimental assessments of avian demography are rare (figure 1.2). The most common avian conservation research study lasted 2–3 years and correlated bird abundance (not demography) with a disturbance factor. Only four studies (2 percent) investigated causal relationships between

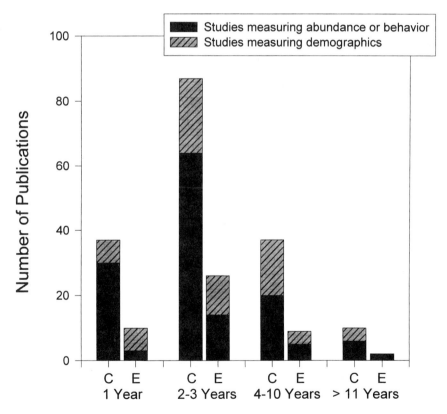

Figure 1.2. The combination of study duration (years), experimental approach (C = correlative study, E = experimental study), and measurement of demographics (indicated by shading) in a sample ($n = 218$) of avian conservation research studies conducted 1990–96. The most common study lasted two to three years, was correlative, and did not measure demographics.

avian demography (usually nesting success inferred from artificial nest experiments) and disturbance for > 4 years.

Our review suggested that we study avian conservation at both the species (n = 114; 52 percent) and community (n = 104; 48 percent) scale. We may study interactions between competitors or predators and prey, or investigate the role of abiotic factors on population persistence, but we found no instances where avian researchers truly studied the role of ecosystem functioning (energy flow, nutrient cycling) on species of concern (one example appeared shortly after our survey; James, Hess, and Kufrin 1997).

Although most research on avian conservation related bird abundance to the attributes in a few, relatively small study areas, a growing number of recent studies (n = 54; 25 percent) related landscape attributes to birds. This has been made possible by recent technology that allows us to measure habitat and model bird responses to human disturbance at large spatial scales (Villard, Schmidt, and Maurer, this volume) and a resurgent interest in metapopulation theories that remind us that a particular deme's stability may depend upon dispersal from distant demes (Pulliam 1988).

Few studies conducted at the landscape scale are experimental (n = 8), last > 4 years (n = 14), or measure avian demography (n = 17). None of the studies we reviewed fit into all three categories, and only seven were both experimental and measured demography.

Future Directions for Avian Conservation Research

Our survey suggests that the avian conservation research arena suffers from the same problems identified for ecological research in general. The major problems relate to the scale of investigation, which is usually too small spatially and too short temporally, and the experimental design of the investigation, which all too often seeks correlative rather than causal relationships (Weatherhead 1986; Kareiva and Anderson 1989; Tilman 1989). These problems limit inference from our studies. In basic ecology, this may waste careers in pursuit of incorrect hypotheses. In conservation biology, incorrect inference from poorly designed studies can result in the extinction of species. Here we suggest a general research framework that we hope will reduce the likelihood of incorrect inference. We do not advocate experimentation for elegance's sake. Understanding a species and its important limiting factors including anthropogenic ones, through literature and field study, should precede experimentation so that *relevant* experiments are conducted.

The scale at which an investigation is conducted has profound effects on the interpretation of results (Wiens 1989; May 1994). Thus, scale should be considered carefully before a study is executed. Rather than choosing a scale that is convenient for conducting elegant experiments, conservation biologists must choose scales that are relevant for addressing pressing environmental problems (May 1994). In general, we need to increase the scale of our research to include more long-term studies of landscape processes. Rather than dropping a research project at the end of a graduate career, academic researchers should encourage new students to take up a project where the last student left off so that long-term assessments can be built from a series of studies with shorter duration. Research is needed at a variety of scales to determine: (1) if accurate assessments of conservation problems can be obtained with short-term data (Hutto, this volume); (2) if human effects on birds vary from year to year; and (3) the geographic scale at which birds are sensitive to various human disturbances.

Management actions based on research that identifies cause–effect relationships between avian responses and human disturbances are more likely to succeed than actions based on correlative relationships. However, causal relationships are extremely difficult to determine as we increase the scale of our research. How can we resolve these important, but conflicting, needs for avian conservation research? Adaptive management (Walters 1986) can be used to progress from correlative relationships (which function as testable hypotheses) to causal ones so that we use our best guesses to manage a species initially, but modify these as causal relationships emerge. In other words, progression from a small scale, where causal relationships are identified, to the large scale necessary for conservation, can be done through a series of experimental management actions. In a similar vein, detailed knowledge about the causal response of a single species to disturbance can be used to identify important community- and ecosystem-level responses to disturbance (James, Hess, and Kufrin 1997).

Ideally, we will continue to increase our understanding of how human disturbance affects individual fitness and population viability of bird species. This is critical to the assessment of human disturbance, but more research is needed to determine if reliable, but more easily obtained, predictors of fitness can be found. Abundance may be a poor correlate of productivity (Van Horne 1983; Vickery, Hunter, and Wells 1992b), but it does indicate something about the effect of a disturbance or the importance of a particular habitat to a species (Haufler, this volume; Hutto, this volume). The necessity to contrast landscape-level changes on birds leaves us with the daunting task of collecting demographic data in a variety of landscapes to fully understand a disturbance. Studies from the midwestern United States show the power of this approach (Faaborg et al., this volume) but also the limitation (only a few species are adequately understood). Research is critically needed on ways to accurately extrapolate experimental, demographic analyses conducted on a single species or population to the scale of entire avifaunas. Defining the relationship between avian demography and measures of relative abundance across a variety of habitat types would be a good starting point.

Understanding the influence of a single form of human disturbance at the scales required to conserve avifaunas is not adequate. Human disturbances do not occur in isolation. They are repeated over time, and many types of disturbance usually affect any given location. We need research on the cumulative and interactive effects of disturbance (Riffell, Gutzwiller, and Anderson 1996; Rotenberry, this volume; Herkert and Knopf, this volume).

Proactive research may enable us to conserve more birds for less money because expensive restoration projects and curtailment of resource extraction can be avoided. However, proactive research is difficult to sell to funding agencies because its societal benefits are unproven. Research is needed to quantify the benefits of proactive versus reactive research. Without good rationale, proactive research is likely to become rarer in the future than it is at present, especially given the current move within federal agencies to remove formal listings of sensitive species (Squires, Hayward, and Gore, this volume).

Well-designed experiments conducted at appropriate scales can establish causal relationships between human disturbance and avian population viability. Such experiments, however, do not necessarily identify the mechanism that is responsible for reducing viability. For example, we may know that habitat fragmentation reduces population viability, but we do not necessarily know if this is due to habitat loss, increased predation, or alteration of microclimate. Managers have few options to stabilize or increase population viability if they do not understand the mechanism linking disturbance to viability. Therefore, research on potential mechanisms, notably nest predators and brood parasites, is critical. Only 3 percent of the studies we surveyed conducted such research and none of these were long-term, experimental investigations. Another important mechanism, food availability, has been essentially ignored by avian researchers. None of the articles we reviewed attempted to measure the response of prey (e.g., insects) to human disturbance and relate these changes to observed patterns of avian reproductive success (Gehlbach 1994 is a good example where prey was measured).

While we strongly advocate the use of well-designed experiments, we recognize that in some cases, notably with rare birds, this is not possible. In such cases, studies of responses to human activities over periods of time sufficient to cover population and environmental cycles are needed for the development of management recommendations (Gehlbach 1994). Although replicated experiments may be difficult when dealing with rare birds, measurement of demography can, and should, be done.

Those of us who conduct research on birds must strive to deliver our results to land managers so that effective conservation can ensue. While we could not specifically evaluate whether results from recent studies have been utilized by managers, we are doubtful this occurred in most cases. Future research on avian conservation needs to include an element of "technology transfer" to the management community. Currently, the gap between research and management is large (Finch and Patton-Mallory 1993), and most managers feel that researchers are not effectively communicating their results in a management context (Arnett and Sallabanks, this volume; Hejl and Granillo, this volume). Making research meaningful to the manager should be an important component of any future study of avian conservation (Young and Varland, this volume).

Will collaborative, long-term, experimental research in which demography is emphasized and managers are included be adequate to conserve avian diversity? We believe it will, but success is critically dependent upon allocation of available research money. Funding decisions are often made by executives with little knowledge of research priorities or by local managers with specific but narrow needs. This creates a diverse array of funding opportunities, allowing us to market our research ideas in a variety of places and often secure some funding. Such a system allows researchers to address local problems effectively, but funding for large research programs is driven more by public attraction to charismatic species (e.g., raptors, Neotropical migrants) or litigation over endangered species than by objective determination of research needed to manage avian diversity. For example, why did policy makers in the 1970s decide to spend millions of dollars to restore Peregrine Falcons (*Falco peregrinus*) in the United States (endangered there but distributed around the world) rather than invest in the proactive conservation of entire endemic avifaunas such as found on the Hawaiian archipelago? We are unaware of formal discussions about such decisions and suggest that such discussion should occur before decisions are made.

The large-scale research projects we suggest are necessary for avian conservation will require initial prioritization of research needs so that cooperatives can tailor research to specific problems and obtain sufficient funding to adequately respond to them. Researchers are inherently creative and independent, but to avoid redundancy and assure complete coverage of important research questions, direction from a central committee would be helpful. The current system of fragmented funding sources with relatively small budgets precludes decisions about funding peregrines versus archipelagoes. No one seems to have the budget or the coordination to solicit and review such major proposals simultaneously. We suggest that current ornithological organizations field such a committee (in the United States, perhaps through the National Science Foundation, Ornithological Council, or proposed National Institute for the Environment) and charge it with: (1) polling researchers and managers to determine research priorities; (2) lobbying for funding; (3) soliciting cooperative proposals; and (4) supporting research aimed at addressing top priorities. The following chapters in this book start us on the path toward identifying critical research needs. We hope that the ornithological community will discuss these needs and establish a committee to evaluate them and synthesize them into specific research initiatives for which proposals and funding are sought.

We have a difficult road to travel before avian conservation research approaches research in Utopia. We need to increase the temporal and spatial scales at which we work, while simultaneously increasing the rigor and depth of our experiments. This means we must increase the extent and resolution of our studies, two features of research traditionally at odds with each other (Wiens 1989). Quality research of this type is probably beyond the ability of a single researcher. Instead, cooperative projects in which individuals from a variety of disciplines and geographic regions collect and combine data will be necessary (Faaborg et al., this volume). Developing cooperative research groups is one of our biggest challenges for the future. Making them last beyond the usual funding and graduate career cycles may be an even larger challenge. Accomplishing both is crucial for the conservation of our avifauna.

Acknowledgments

This research was supported by Sustainable Ecosystems Institute. We thank Fred Gehlbach and Fran James for constructive review of our ideas.

CHAPTER 2

The Growing Pains of Avian Conservation Biology

Frances C. James

Trained primarily as ecologists, avian conservation biologists are better prepared to contribute to the field's basic knowledge base than to the formulation of policy. Nevertheless, environmental problems abound, and biologists of all stripes are scrambling to rise to the occasion. Like many others I have been caught up in this expanding effort to conduct primary research that is relevant to solving environmental problems and also to contribute in some way to the process of policy formulation. In spite of the advances that are being made, I am frustrated by how slowly they come.

I have seen many changes in the file of ornithology since I attended my first American Ornithologists' Union meeting in Baton Rouge, Louisiana. At that meeting, Grace Murphy, the wife of Robert Cushman Murphy, persistently suggested that the AOU establish a committee on the conservation of birds. Her suggestion was discussed in the halls in disparaging terms and dismissed. The main view was that ornithologists should focus on basic science and leave advocacy to conservation-minded organizations like the National Audubon Society. As I recall, George Sutton, Sewall Pettingill, and Roger Tory Peterson were more sympathetic with Grace's view, but theirs was definitely a minority position. I was just the graduate student who worked the tape recorder in the council meeting, awestruck at being in the same room with the top ornithologists of the day. Now, of course, there is an active conservation program in the AOU, and new journals like *Conservation Biology* and *Ecological Applications* have been established by other societies.

Today, as a professor at Florida State University, I teach a course in conservation biology, conduct research on an endangered species (the Red-cockaded Woodpecker, *Picoides borealis*), and try to analyze long-term population trends in North American land birds (from Breeding Bird Survey data). In my spare time I work with organizations that speak at the national level with policy makers. My research on determinants of character variation in birds has been sitting in filing cabinets for more than a decade.

I have heard biologists argue that, because the environmental problems are so urgent and countervailing economic and political pressures so great, they are justified in overstating reports about the deteriorated state of the environment. I disagree with this position because it taints technical judgments with value judgments. One can argue that all science is value laden, in the sense that it contains underlying models that have a subjective component, but that is not what I mean here. My own view is that, within the paradigms we use, we should never compromise the science. Science demands that we get the most reliable knowledge and state the level of its uncertainty. Value judgments can be stated and opinions expressed, so long as they are identified as such. In a crisis situation, it is more important than ever to set priorities carefully, allocate management efforts efficiently, and be alert to previously undetected problems. It is a truism that management and research should proceed together to reduce the uncertainties, but in practice there are few examples (Walters and Holling 1990; Underwood 1995). In the next sections I will discuss the two research topics mentioned above and then comment on the emerging field of environmental contamination. Finally, I will use the nonornithological example of the chlorofluorocarbons to illustrate a good case in which cooperation among scientists and policy makers has achieved a major conservation objective.

Biologists have been trying to help evaluate the science underlying the current version of the Endangered Species Act including its amendments. I was disappointed that the two major recent attempts (Carroll et al. 1996; NRC 1995) were not stronger in their call for a reevaluation of how priorities are set. It seems to me that biologists who are advocates for protecting ecologically important habitats, hotspots of endemism, rare populations (evolutionarily significant units), and full endangered species need to sit down together and develop a consensus. Then they should negotiate with sociologists and economists about how to plan for the future. One current proposal would transfer some federal responsibilities for overseeing the management of endangered species on private lands to states (Kennedy, Costa, and Smathers 1996), probably further hindering future collaboration between scientists and managers. Some authors feel that problems with the act are not as great as the problems with its implementation (Clark, Reading, and Clarke 1994).

The Endangered Red-cockaded Woodpecker

Like the Northern Spotted Owl (*Strix occidentalis caurina*) in the Pacific Northwest, the Red-cockaded Woodpecker in the southeastern states occurs only in commercially valuable coniferous forests. Both birds have been used to focus conservation efforts on entire ecosystems under the Endangered Species Act, and both have been a source of conflict between timber interests and conservationists. To respond to legal action to halt environmental degradation in old-growth forests on federal land, the Clinton administration sponsored the preparation of a comprehensive ecosystem-level plan for the management of forests in the Pacific Northwest (FEMAT 1993). That plan eclipsed a new recovery plan for the Northern Spotted Owl (USDOI 1992). Similarly, legal pressure halted timber harvest by the U.S. Department of Agriculture (USDA) Forest Service in Red-cockaded Woodpecker habitat for several years, until in 1995 the agency produced new management guidelines. But management of the Red-cockaded Woodpecker is still being cast largely as single-species management, under the assumption that what is good for the woodpecker is also good for its ecosystem. A new recovery plan is being drafted. Most remaining Red-cockaded Woodpeckers occur on federal land, primarily national forests and military reservations, but about one-fourth of them are on private land. Although there is virtually no old-growth forest left in the southeastern states, Red-cockaded Woodpeckers persist in widely scattered areas of open, reasonably mature pine forests (about forty to sixty years old) that retain some relict trees (usually over one hundred years old) and often other elements of the original, highly diverse, mature, natural pine ecosystem.

Unlike most other woodpeckers, which roost at night in cavities in dead trees, the Red-cockaded Woodpecker roosts in cavities in living trees, especially in the longleaf pine (*Pinus palustris*) ecosystem with its wiregrass (*Aristida stricta*) ground cover. This system has an exceptionally high diversity of endemic animals and plants (Costa 1995). Originally, frequent natural fires swept through the "piney woods" burning the snags, promoting the grassy component of the ground cover, and discouraging the hardwoods that invade when fire is excluded.

Until recently, conservation efforts for the Red-cockaded Woodpecker have emphasized retention of trees with cavities and protection of foraging habitat and prescribed fire, often in combination with clearing hardwood midstory vegetation. Many losses of small local populations have been attributed to hardwood encroachment. Since 1990 managers have been using a new tool, provision of nest boxes inserted into living pine trees (Allen 1991; Watson et al. 1995). The woodpeckers readily accept such boxes and will establish new social groups when they are available in proper habitat. After the major destruction of cavity trees in the Francis Marion National Forest in South Carolina in September 1989, the provision of inserted nest boxes allowed the retention of many birds that surely would have been lost otherwise (Watson et al. 1995).

Inserts have to be put in the large old trees that have high resin production, so the birds can continue to wound the area around the cavity and maintain a protective layer of sticky sap around its entrance. What is worrisome is that in the long term these successes may be illusory, if the ecosystem as a whole is being allowed to decline in quality. Also, it is not clear yet how many years an insert is acceptable to the Red-cockaded Woodpecker or whether having an insert shortens the life of a pine tree. An alternative to the inserted boxes, drilled artificial cavities, causes less damage to trees and would probably be a preferable management tool. Before managers started to use inserts, Copeyon (1990) in collaboration with Walters (Copeyon, Walters, and Carter 1991; Walters et al. 1995) showed that the drilling technique also induces birds to establish new social groups. Thus far, this option, which is technically more difficult, has not been popular with most managers.

I think the conservation program for the Red-cockaded Woodpecker should be broadened to include the maintenance of the full biological diversity of its mature, pine, fire-maintained ecosystem. Such a program should include timber harvest, and some of the profits from timber sales should go directly into management. There should be incentives for managers of both public and private lands to participate in this program. Old trees should be protected, and natural regeneration of pines in openings in forests that have all ages of trees should be encouraged. There should be no plowing of the land associated with silviculture. Incentives should be designed to assure that the habitat receives prescribed fire on an average of every three years. The new USDA Forest Service Guidelines (USDA 1995a) take major steps in these directions (Escano 1995). Another document, the draft management plan for national forests in Florida (USDA 1997a) calls for an uneven-age type of timber management called "group selection," whereby small patches of trees are harvested approximately every ten years. Experiments are badly needed to compare this method with other methods. The issue of the management of the Red-cockaded Woodpecker and other endangered species on private land (Peters 1996; Bean, Fitzgerald, and O'Connell 1991) is too complex to treat here, except to say that the subject needs more input from biologists (Kaiser 1997).

In collaboration with the Forest Service, Paul Hendrix, Walter Tschinkel, Stewart Reed, and an enthusiastic group of students, I am trying to test several management-related hypotheses in the Apalachicola National Forest. The Forest Service is implementing a five-year program of prescribed burning in which sixty management compartments (averaging more than 400 ha each) have been randomly assigned to one of three burning regimes. We are monitoring one family group of woodpeckers in each compartment. In twenty-four of the sixty cases we are tracking the movement of nutrients through the soil, the understory, the trees, and their arthropod fauna. In some compartments, we are removing Red-bellied Woodpeckers *(Centurus carolinus)* to test the impact of their competition with Red-cockaded Woodpeckers for nest cavities. In a separate experiment we are fertilizing 20 x 20-meter plots with calcium, nitrogen, and phosphorous, in order to track the passage of nutrients through the system. From previous work we suspect that trophic relationships associated with fire history are more important to the ecosystem than has been appreciated in the past (James, Hess, and

Kufrin 1997). We hope to be able to make recommendations about how to use fire efficiently and how much timber can be removed without degrading the system. One counterintuitive finding thus far is that the selective removal of more trees than current guidelines suggest would probably be beneficial. We also want to know more about the effects of differing fire regimes and to be able to predict what would happen if interspecific competition for cavities were reduced.

In conclusion, the Red-cockaded Woodpecker story has several messages for avian conservation biology. At present conservation of the Red-cockaded Woodpecker does not put sufficient priority on the restoration of the full diversity of the fire-maintained pine ecosystems of the southeastern states, especially the longleaf pine–wiregrass ecosystem. A partnership between conservation and timber management is essential. Without management the Red-cockaded Woodpecker and what remains of its ecosystem will be lost. Too much reliance on the provision of artificial cavities may risk long-term damage to the ecosystem. Single-species management can probably benefit the full ecosystem, but its effects should be documented by studies of population regulation in key species in the system other than the Red-cockaded Woodpecker. As the Department of Interior urges policy changes that emphasize ecosystem management (Babbitt 1995), it should acknowledge that the units being affected are populations of the component species and that research at the population level will be necessary to understand its effects. The case of the Red-cockaded Woodpecker is just one example of the kinds of difficulties that typify implementation of the Endangered Species Act (Clark, Reading, and Clarke 1994). The act was not really designed as an overall policy act for the preservation of the nation's ecosystems (NRC 1995b).

Trends in Populations of Migratory Songbirds

A tenet of avian conservation biology for more than the last fifteen years has been that long-distance migratory songbirds in North America, especially those that nest in forests, need special attention because they are in decline. This conclusion has been based on the vague subjective opinion that because there is environmental degradation of various sorts and long-distance migrants are presumably subjected to more hazards than are short-distance migrants or resident birds, they ought to be declining. There are three lines of evidence to support this opinion, all of which are very weak. First, some long-distance migratory species have declined in small suburban parks and nature reserves in the eastern United States (Askins, Lynch, and Greenberg 1990). Second, radar detection of birds migrating northward in spring past Lake Charles, Louisiana, in 1987–89 indicated activity on fewer suitable days than in 1965–67 (Gauthreaux 1992). Third, some analyses of data from the broad-scale government-sponsored Breeding Bird

Survey (BBS) indicate large increases in the numbers of long-distance migrants in the 1970s and decreases in the 1980s (Robbins et al. 1989; Sauer and Droege 1992; Peterjohn, Sauer, and Robbins 1995). Long-distance migrants may in fact be declining but none of these lines of evidence is sufficient to support that generalization. In both the analysis of trends in local reserves and the report of radar observations, the areas sampled were unlikely to be representative of population-level phenomena that occur across thousands of square miles (Hutto 1988; Finch 1991). Given an adequate sampling regime, a reasonable null hypothesis would be that population sizes of long-distant migrant species have been fluctuating randomly around stable values, so one could look for evidence of declines in more than 50 percent of species. The most recent analysis of BBS data for all Neotropical migrant landbirds (Peterjohn, Sauer, and Robbins 1995) and our own analyses of the same data for warblers (Parulinae) (James, McCulloch, and Wiedenfeld 1996; James, Wiedenfeld, and McCulloch 1992) both estimate that the proportion of species with increasing trends in the full BBS period since 1966 is higher than 50 percent.

The tendency of the conservation community has been to elevate the subject of declining migrants to the status of a crisis, ignore the evidence of increases and exaggerate the evidence of declines, and generalize beyond what is justified. The BBS staff have been careful to state repeatedly that, although some species are clearly in trouble, over the entire BBS period (since 1966) there is no evidence of general declines in Neotropical migrant songbirds (Sauer and Droege 1992; Peterjohn, Sauer, and Robbins 1995).

With my collaborators, David Wiedenfeld, Charles McCulloch, and others (James, McCulloch, and Wiedenfeld 1996 James, Wiedenfeld, and McCulloch 1992; James and McCulloch 1995;) I have been studying geographic variation in population trends, mostly using nonlinear trajectories to compare patterns among regions. The results show that the numbers of a few species are declining generally, some are declining in parts of their ranges and increasing in other parts, and some are stable and increasing throughout their breeding ranges. We do not see the increases in the 1970s and declines in the 1980s reported in the papers mentioned above. The one multispecies phenomenon we have identified thus far is a statistically significant pattern of decreasing populations of both migrants and nonmigrants in highland areas like the Adirondack Mountains, the Blue Ridge Mountains, the Cumberland Plateau, and the Ozark Mountains, even in species that are stable or increasing in other parts of their geographic ranges (James, Wiedenfeld, and McCulloch 1992; James, Wiedenfeld, and McCulloch 1996).

Robinson et al. (1995) documented that in Illinois and adjacent states nests of migrant songbirds in small patches of forest in agricultural landscapes suffer low reproductive success because of high levels of predation on nests and high nest parasitism by Brown-headed Cowbirds (*Molothrus ater*). Because the numbers of migrants there are apparently not declining, Robinson's hypothesis is that central

Illinois is a sink for migrants. Testing this hypothesis would be difficult. It would require demonstrating with marked birds that the populations in fragments are being renewed from source populations elsewhere to a greater extent than is the case with populations of the same species in areas of continuous forest.

Askins (1995) in an accompanying perspective entitled "Hostile Landscapes and the Decline of Migratory Songbirds" says that Robinson's findings explain why migrants have declined in isolated forests in suburban landscapes in the eastern states. Testing Askins's hypothesis that predation and cowbird parasitism are causing declines in populations of migrant (but not resident) species in fragments of forest in the eastern states would require the systematic elimination of alternative hypotheses. One problem with this hypothesis is that levels of cowbird parasitism have not been increasing in the eastern states during the period in question (Hoover and Brittingham 1993). Both Robinson's and Askins's hypotheses confuse processes that affect nesting productivity with processes that affect recruitment, and they ignore alternatives like variation in the resource base. Their emphasis on predation and cowbird parasitism would be weakened for example if the arthropod food supply were affected by the isolation of forests in either agricultural or suburban landscapes.

A recent review by Robinson (1997) acknowledges that BBS data do not indicate general declines in migratory songbirds, but gives mixed messages. The article is entitled "The Case of the Missing Songbirds." His recommendation for the management of long-distance migrant songbirds is a program to lessen the degree of forest fragmentation in areas that already have fairly continuous forests in the north-central states. Otherwise, migrants might start declining in small fragments. No one is going to argue against restoration of habitats. But, if money is limited and there is concern about continental populations of native birds, shouldn't priority be placed on the identification of which species are in the most trouble regardless of their migratory status, where in their geographic ranges the problems are most severe, the analysis of possible causes of these phenomena, and study of the feasibility of management? Figure out, for example, whether and where it is feasible to help birds like the Prairie (*Dendroica discolor*) and Cerulean (*D. cerulea*) Warblers, which happen to be long-distance migrants and are known to have been in serious decline for decades.

When you think about how predisposed people are to thinking that bird populations must be declining, it is not hard to see how such weak causal analysis gets started. Once started, it seems to perpetuate itself. Even when there is good evidence of declines, ecologists have a tendency to attribute them to some combination of their favorite causes (Hutto 1988). Causal analysis is a poorly developed area of population biology (Murdoch 1994; Caughley and Gunn 1996). The solution here is to work harder on the identification of problems in the first place and then on the analysis of causes, always being careful to state the uncertainty of the cause.

A Note about Environmental Pollution

A new book entitled *Our Stolen Future* (Colburn, Dumanoski, and Myers 1996) summarizes recent literature showing that wildlife problems associated with the severe pollution of the Great Lakes in the 1960s and 1970s (Fry and Toone 1981; Fry, Peard, Speich, and Toone 1987; Geisy, Ludwig, and Tillit 1994) have not ended, even though the production of DDT was restricted starting in 1972 and manufacture of PCBs has been banned in the United States since 1976. As recently as 1993, Bald Eagle (*Haliaeetus leucocephalus*) chicks have been reported with deformities like crossed bills (Bowerman et al. 1994). This more recent evidence shows that synthetic chemicals, which now can be detected in the tissues of virtually all vertebrate animals on the earth (including human beings), are affecting their development and reproduction (Colburn, vom Saal, and Soto 1993). It reviews the wildlife studies and the laboratory experiments about developmental and reproductive abnormalities, making the case that a large number of synthetic chemicals can mimic natural hormones in the vertebrate body, where they act as "endocrine disruptors." These compounds accumulate in body fat and produce transgenerational effects, which can alter sexual development, thyroid function, and fertility. These and other problems have all been demonstrated in birds and other wildlife (Facemire, Gross, and Guillette 1995). Conservation biologists have been paying very little attention to this problem (Rosemarin 1988), which seems likely to get much worse in the future.

Chlorofluorocarbons

The ozone experience is only indirectly relevant to avian conservation biology, but it contains an important lesson about the science-policy interface, an area where avian conservation biology is weak. By now everyone knows that there is a layer of ozone in the stratosphere (ten to fifty kilometers above the earth's surface) that protects animals and plants from harmful ultraviolet (UV-B) radiation. Some of the problems expected with its depletion would be more cases of skin cancer, lower agricultural yields, and damage to aquatic life. The story began with Nobel prize–winning work in the early 1970s presciently suggesting that industrial chemicals called chlorofluorocarbons (CFCs) might reach the stratosphere, where solar radiation would break them apart, releasing reactive chlorine atoms that would destroy ozone on a massive scale. Public pressure to ban the use of CFCs, which are used in aerosol cans and as solvents and coolants, led to an international meeting on the subject in 1977 and the establishment of a committee of scientists, plus government, industry, and nongovernmental organizations, to evaluate the situation. Then, in the fall of 1984, members of the British Antarctica Survey detected a 40 percent loss of stratospheric ozone over Antarctica

(Farman, Gardiner, and Shanklin 1985). In 1987 a landmark international cooperative agreement was signed, called the Montreal Protocol on Substances That Deplete the Ozone Layer. That agreement has now been ratified by more than 150 countries. At the time of the signing there were no data showing increases in the ultraviolet radiation reaching the earth. But because delayed action might result in serious damage, policy makers were willing to take action based on the precautionary principle (Cameron and Abouchar 1996). The agreement put restrictions on the production and use of CFCs and halons, and subsequent modifications added restrictions on methyl chloroform, methyl bromide, and carbon tetrachloride. Several industries (not including coal and oil) that originally complained that the development of substitutes would be too costly, came to agree to successively more restrictive accords. Some governments used excise taxes to discourage the use of CFCs, while developing countries were allowed longer phase-out periods. The United Nations Environmental Program played a leading role in these policy matters. By 1995 the global production of CFCs was down 76 percent from its peak in 1988. Because of the lag time, scientists predict that the soonest expected time for detection of a healing of the ozone shield is the year 2000 (French 1997).

What this example shows is that even major international environmental problems involving multibillion dollar industries can be addressed effectively in the absence of complete knowledge. It requires continual updates of good scientific information that is independent of government influences, public interest, joint agreements among the stakeholders, and international coordination. In other words, with good estimates of the uncertainties involved, international environmental negotiation can proceed and even big business may be persuaded to cooperate. Negotiation where uncertainty is involved is now being explored in other areas closer to avian conservation biology, like loss of biodiversity.

Techniques for Conserving and Monitoring Birds

Most research on avian conservation does not require specialized techniques. However, the technological boom of the last few decades has given several important tools to avian conservationists. Three such tools (genetical analysis, geographic information systems, and spatially explicit models) are discussed in the first three chapters of part II.

Conserving genetic diversity is a fundamental tenet of conservation biology. Recent developments have allowed us to better quantify avian genetic diversity and gain some understanding about factors that endanger birds by reducing genetic diversity. In chapter 3, Fleischer reviews standard techniques for measuring genetic diversity and goes beyond this traditional use of conservation genetics to suggest a variety of ways genetic analyses can aid in the conservation of birds. An important topic he discusses is the determination of "significant units" for evolution and management. This is central to conservation, as we must know when we are dealing with populations significantly divergent enough to warrant special protection. We must not waste limited resources or demand sacrifices from the public to conserve evolutionarily insignificant variants.

The development of geographical information systems (GIS) has revolutionized the way avian conservationists work. It has allowed us to quantify landscape patterns and changes and relate these to patterns of bird abundance and population viability. Focus on the importance of landscapes has been fueled by a "rediscovery" of the metapopulation concept of avian population structure. This has served to emphasize the critical importance of dispersal to avian population viability, a point expanded on later in the book by Walters (chapter 12). Coupled with temporal projections, spatial modeling allows the avian conservationist to simultaneously assess the impact of human activity on birds at various spatial and temporal scales. Investigation of human impacts on birds at varying scales is essential to a full understanding of how we affect birds. Villard, Schmidt, and Maurer (chapter 4) discuss how GIS and statistical analyses can be used to develop spatial models of avian abundance, distribution, and demography at the local, landscape, and geographic scales. Raphael, McKelvey, and Galleher (chapter 5) provide details of a case study in which spatial modeling was used to simulate the effects of timber harvest on the viability of Northern Spotted Owl populations.

One of the most basic tools needed by the avian conservationist is a good technique for assessing bird populations. A variety of programs exists in Europe and North America for monitoring the abundance and productivity of birds. Long-standing programs such as the Breeding Bird Survey, the Christmas Bird Count, the Mettnau-Reit-Illmitz Program, and the British Common Bird Census have traditionally measured abundance in consistent locations at yearly intervals. Hutto (chapter 6) describes a similar extensive approach to monitoring songbirds during the breeding season. He argues that important relationships between birds and habitat can be ascertained by sampling the entire landbird community throughout a large area (the northern Rocky Mountains in his example). Moreover, reliable bird-habitat relationships might be determined in as little as one year. The uncertainty with which abundance correlates with avian population viability and the need to understand how and why land management activities reduce bird abundance have recently prompted the development of intensive monitoring schemes designed to measure avian demographics. DeSante and Rosenberg (chapter 7) review these efforts and detail a scheme they have developed and implemented throughout North America (MAPS, Monitoring Avian Productivity and Survivorship). This type of monitoring approach should enable us to determine if the models relating habitat to bird abundance also correctly identify habitat features associated with productive (source) versus unproductive (sink) populations. The challenge with such programs is to provide unbiased estimates of demographics at large scales and maintain precise estimates at small scales so that local issues can be addressed (chapters 1, 7).

In order to monitor birds most effectively, a combined approach that uses elements from intensive and extensive schemes would be optimal. DeSante and Rosenberg point out that this is starting to occur in Europe and North America, but it needs to become the norm, not the exception. Various monitoring schemes currently in place need to be conducted at overlapping sites. This combined approach would also allow researchers to determine biases of different monitoring protocols when such programs are used independently. Studies that combine the use of point counts, winter surveys, constant-effort mistnetting, and nest searching and monitoring would not only assess the status of bird populations most thoroughly but would also provide opportunities to calibrate each monitoring technique with respect to each other.

The diverse topics in this section are intended to acquaint the avian conservationist with important techniques, but studies that integrate these techniques are most likely to lead the march of avian conservation into the next century. Spatial models of avian abundance and productivity determined by extensive point counts and intensive mistnetting and nest-searching stations are critically needed. As pointed out in chapter 4, such models are in their infancy. Especially important here is a spatially explicit comparison of change in abundance, distribution, genetic composition, productivity, and survivorship through time. Simultaneous measures of habitat change and human activity need to accompany such studies. Ideally, spatial models of temporal change will be compared among replicated control and manipulated (disturbed by some type of human activity) study areas.

Two important human societal issues are touched upon in this introductory section and echoed in later chapters. First, Fleischer reminds us that unchecked human population growth and increasingly industrialized society are the ultimate reasons so many birds are in apparent peril. The human population of the earth has grown and expanded at an explosive exponential rate since the last ice age. It has become especially large during the past few decades, and there is no hopeful sign that the growth is leveling off (Horiuchi 1992; Cohen 1995). The recent, anomalous, increased rates of avian extinction and endangerment result from direct and indirect environmental impacts caused by this multitudinous human population. Anyone that doubts the ability of humans to have devastating impacts on avian diversity need only be shown the clearly elucidated history of Pacific Island avifaunas following their contacts with colonizing humans (Steadman 1996). Anyone concerned with biodiversity must work to convince the world how important it is to decrease our own species' birth rate and resource use. Second, Hutto points out that although we focus on important research needs in this book, what we really need is the "widest possible variety of quality research." Moreover, he suggests that the limiting factor in avian conservation is not a shortage of quality data, but a lack of willingness to *use* it. Thus, important tasks for avian conservationists, outside of the collection of our data, are the general education of the public to be more sensitive to the perils of overpopulation and the determination of the types of data most likely to influence policy makers.

CHAPTER 3

Genetics and Avian Conservation

Robert C. Fleischer

Anthropogenic causes of avian endangerment and extinction may vary among taxa, geographic regions, and habitats (Dobson et al. 1997). While habitat destruction and fragmentation are clearly the most important factors, other agents, such as overhunting, pesticides, pollution, increased UV-B, global warming, salinization, and introduced predators, competitors, parasites, and diseases can also be significant, if not primary, causes of population decline for some bird species (Pulliam and Babbitt 1997). In the face of this variety of threats to biodiversity, genetic theory and technologies have been increasingly used to assist in the management of taxa in peril. Methods of conservation genetics have been applied in many case studies, especially during the past decade, and there have been many reviews that discuss these applications in detail (e.g., Soulé 1980, Lande and Barrowclough 1987, Hedrick and Miller 1992, Avise 1994, Moritz 1994a, O'Brien 1994, Frankham 1995a, Lynch 1996, Hedrick et al. 1996). Most conclude, of course, that genetic methods are not by themselves a panacea for the conservation problems that organisms face. However, most also conclude that such methods can provide useful information for biodiversity management.

In general, conservation genetic methods have not been applied to birds as much as to other vertebrate taxa, especially mammals (see volume with Haig and Avise 1996, or O'Brien 1994 for evidence of this). In addition, most previous studies in avian conservation genetics assessed genetic diversity within and among populations or species, and then recommended ways to conserve these levels and patterns (Haig and Avise 1996). There are alternative uses for genetic methods that have not been greatly explored. Here I take a functional approach to how these applications can be categorized and divide the primary uses into:

1. forensic applications (i.e., identification of individuals, populations, or species to solve "crimes against nature");
2. assessments of inbreeding and associated fitness depression;

3. assessments of current (along with past and future) levels of genetic variability within and among populations;

4. use of population genetic models with molecular data to estimate long-term gene flow rates, effective population sizes, and population fluctuations; and

5. identification of evolutionarily significant and management units (ESUs and MUs), hybrids, and determining if such units are "worth" saving.

These categories are not mutually exclusive: some applications involve conceptual bases and methodologies that overlap to some degree. In addition to those listed there are several more-novel ones. In particular, because of recent conceptual and technical advances in developmental genetics, immunogenetics, and applied genetic engineering, a number of applications may before long be possible that will assist management efforts in ways undreamed of before.

In this chapter I review the uses (and what I view as some potential misuses) of genetics for the conservation of avian diversity. I begin with a brief caveat, followed by categorization and description of the primary quantitative and molecular tools in conservation genetics. I then characterize and illustrate (with mostly avian examples) the five types of conservation applications noted above, and provide rationales for and problems with each approach. I conclude with some predictions concerning possible future uses for genetic theory and molecular genetic methods in avian conservation biology.

As in most applied fields of study, there are controversies within conservation genetics about how different methods should be applied, and about the relative merits of particular applications for management decisions (Hedrick et al. 1996). For example, one controversy revolves around whether inbreeding depression and genetic variation have important consequences for short- or long-term survival of a population or species. Another concerns how we should preserve levels and patterns of genetic variation—as they presently occur, at maximum variability, or how they occurred prior to human impacts? Or perhaps we should be concerned only with restoring or maintaining healthy species so that an ecosystem can function properly, regardless of the genetic makeup of its populations. These questions reflect the types of uncertainty that often exist for other aspects of conservation biology. In many cases we may be required to make management decisions based on conflicting or flawed theory or empirical databases, or when we cannot properly experiment and consider alternatives because of considerations of time, impact to endangered populations, or financial cost. These controversies may also reflect a tendency of us as competitive social organisms to take extreme positions partly out of concern for our standing in dominance hierarchies ("political" reasons). Lastly, as much as we biologists like to think we have all of the solutions, we must realize that most issues cannot be resolved entirely with biological or even other scientific information, but also often involve philosophical, legal, economic, or social concerns.

Methods of Conservation Genetics

Pedigree Analyses

The analysis of pedigrees is useful for the detection of inbreeding and its associated depression of fitness (Ralls, Ballou, and Templeton 1988), and for designing breeding strategies to avoid future losses of genetic variation or adaptation to captivity (Foose and Ballou 1988; Lacy 1989; Haig, Ballou, and Derrickson 1990). Such analyses are usually conducted with captive populations and involve maintaining records of parentage for all individuals in large databases called studbooks. Coefficients of inbreeding or coancestry (i.e., the probability that two alleles sampled from an individual are identical by descent; Ballou 1983) are calculated and then used to design breeding strategies that maintain particular levels or patterns of genetic variability (Foose and Ballou 1988; Lacy 1989; Hedrick and Miller 1992). Probabilities of loss of genetic variation also are often estimated via simulation experiments ("gene drops"; MacCluer et al. 1986). Alternatively, measures of fitness can be regressed onto these coefficients to assess inbreeding depression (Ralls, Ballou, and Templeton 1988).

Models of Genetic Structure

A second set of quantitative methods involves the use of demographic data within population genetics models to make predictive estimates of genetic variability and structure (e.g., for birds: Barrowclough 1980; Fleischer 1983; Rockwell and Barrowclough 1987). For example, Barrowclough (1980) used field-based estimates of bird dispersal and effective population sizes in equations of isolation-by-distance and stepping-stone models to estimate expected levels of among population genetic differentiation. Two concerns about this approach are that these models rely on accurate input values (that can often be very difficult to obtain without long-term field studies) and on assumptions that need to be met to varying degrees in order to have any predictive power.

Quantitative Genetics

A group of methods receiving increased attention in conservation genetics is that of the more traditional field of quantitative genetics (Falconer 1981). In fact, a growing cadre (e.g., Lande and Barrowclough 1987; Hamilton 1994; Lynch 1996) believes that such methods reveal variation that is of greater conservation importance than the putatively "neutral" variation revealed by molecular or other methods (but see Hughes 1991). They reason that traits that are most likely to affect fitness are polygenic or quantitative trait loci (QTLs) as opposed to single-locus traits. Effective mutation rates at QTLs are very high ($\sim 10^{-3}$ mutations/gamete/generation; Lande and Barrowclough 1987) in comparison to more typical coding loci ($\sim 10^{-5}$ to 10^{-7}; Nei 1987) but are similar to rates found for variable number tandem repeat loci, VNTRs (e.g., Jeffreys et al. 1988; see Fleischer, Fuller, and Ledig 1995).

Quantitative genetic methods may be useful for conservation because they estimate genetic variability in potentially selected and important traits, like body size or clutch size. However, estimation of heritabilities (not to mention nonadditive genetic variances and covariances) requires that relatively large numbers of family groups be measured for such traits. This may not be a simple matter for some avian species, especially endangered ones that are sparsely distributed or for which nest disturbance could lead to abandonment. In addition, heritabilities could be biased if much extra-pair mating occurs, which is normal for many passerine and some nonpasserine birds (Fleischer 1996). Lastly, the work of James (1983) suggests that growth in different environments may impact heritabilities in ways that are not straightforward, although a recent review (Weigensberg and Roff 1996) revealed a high correlation between heritabilities estimated in captive animal studies with those measured in nature.

Biochemical Methods

The majority of genetic applications in avian conservation have used electrophoretic analyses of allozymes and other proteins (e.g., Haig and Oring 1988; Barrowclough and Gutiérrez 1990; Fleischer, Conant, and Morin 1991). Proteins can be assayed from soft tissues, blood, and feather pulp, and the latter two, less invasive approaches have obviously been favored in conservation applications. Unfortunately, blood and feather pulp yield fewer protein loci than soft tissues. When this drawback is combined with their relatively low mutation rate ($\mu = 10^{-5}$ to 10^{-7}) and variability (mean heterozygosity for birds is ~5.8 percent; Barrowclough, Johnson, and Zink 1985), allozymes generally appear less useful than other methods for assessment and comparison of genetic variability. The low proportion of variable loci makes it difficult to resolve differences in heterozygosity and assess population structure, especially in bottlenecked populations. Also, if a population has low allozyme variability, it is difficult to estimate the timing of a bottleneck, as it would take about $1/\mu$ generations (Nei, Maruyama, and Chakraborty 1975) for mutation to restore to equilibrium the neutral variation that had been lost. Some advantages of protein electrophoresis include its relatively simple extraction, electrophoresis, and staining protocols; the universality of its application; and its low cost compared to most molecular methods. See Hillis, Moritz, and Mable (1996) for a recent review of allozyme methods.

Molecular Genetic Methods

There has been a virtual explosion of new molecular genetic (nucleic acid) methods over the past decade, many of which have application to avian conservation biology. Some of the latest techniques are extremely powerful, and with rapid advances in technology, many of the methods are becoming easier and, to some extent, less expensive than before. There is a tendency among researchers, however, to use newer methods even when some of the older, less expensive methods may produce adequate results.

The methods can be categorized by the type of analysis that is performed and the type of marker that is characterized. In figures 3.1 to 3.3 and below, I briefly summarize some of the common molecular methods. For detailed treatments of methods and explicit protocols, see Avise (1994); Hoelzel (1992); Hillis, Moritz, and Mable (1996); or Ferraris and Palumbi (1996). An older method, the analysis of Restriction Fragment Length Polymorphism (RFLP; figure 3.1) uses enzymes to cleave DNA at particular short sequences. A more recent and powerful method, the Polymerase Chain Reaction (PCR; figure 3.2A, 3.2B) copies, or amplifies, specific DNA sequences using a DNA replicating enzyme (polymerase) and short, synthesized DNA primers. PCR is very sensitive and can amplify sequences from very small starting quantities of DNA, including partly degraded DNA from feathers and avian museum specimens (Ellegren 1991a; Leeton, Christidis, and Westerman 1993; figure 3.2C), mammalian feces (e.g., Paxinos et al. 1997), and avian subfossil bones (Cooper et al. 1996). Its extreme sensitivity demands careful execution to avoid accidental contamination with foreign DNA. DNA sequencing (Figure 3.3A) determines exact nucleotide sequences of PCR amplified (or cloned) genes. There are many novel alternatives to DNA sequencing that also reveal variation (see Avise 1994; Ferraris and Palumbi 1996 for such methods).

Mitochondrial DNA

The only organellar genome identified in birds is mitochondrial DNA (mtDNA), which has proven to be one of the most useful (and used) marker systems (Moritz 1994a; Avise 1994). As in nearly all vertebrates, mtDNA has been shown to be maternally inherited, nonrecombining, and relatively rapidly evolving in birds. Thus, it is a very effective marker for detecting genealogical structure within and among avian populations and for reconstructing phylogenetic relationships among taxa (e.g., Avise 1994; Moore 1995). The primary method of assaying variation in mtDNA is PCR amplification of all or part of a particular gene, followed by direct sequencing. Different regions of the molecule appear to have different levels of selective constraints and thus exhibit different levels of variability (with the slowest-evolving regions being the ribosomal and transfer RNAs and the fastest usually being the noncoding control region). Because mtDNA is a single linkage group, any force (e.g., selection) that impacts one part of the molecule also impacts others (Degnan 1993; Rand 1996). In addition, copies of mtDNA genes transposed to the nuclear genome have been found in birds (e.g., Quinn 1992; Sorenson and Fleischer 1996), and these result in a variety of problems (as well as some opportunities).

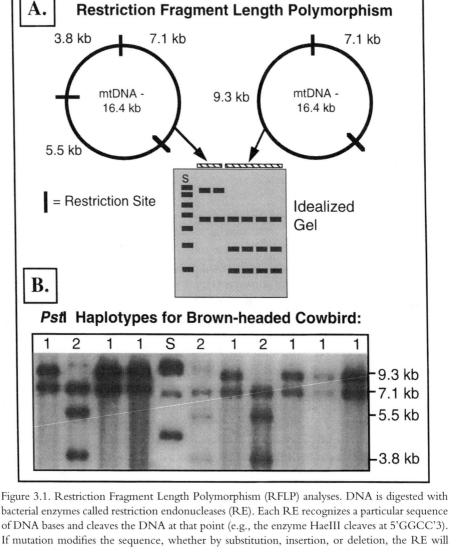

Figure 3.1. Restriction Fragment Length Polymorphism (RFLP) analyses. DNA is digested with bacterial enzymes called restriction endonucleases (RE). Each RE recognizes a particular sequence of DNA bases and cleaves the DNA at that point (e.g., the enzyme HaeIII cleaves at 5'GGCC'3). If mutation modifies the sequence, whether by substitution, insertion, or deletion, the RE will not cleave the DNA. By comparing the distributions of fragments on gels following digestion, one can usually determine the presence (or absence) of a restriction site and tally the minimum changes required for the pattern. Because most mutations involve substitutions, and usually only a single base change per site, models of restriction site evolution can be used to estimate the proportional change in nucleotide sequence (Nei 1987). (A) Schematic diagram of RFLP analysis. The molecule on the right has two recognition sites, and digestion by RE results in two fragments. The molecule on the left has three restriction sites, and digestion produces three fragments. Fragment sizes add up to (approximately) the same size, based on size standards run in adjacent lanes (S). (B) Example of an actual RFLP analysis of mtDNA in the Brown-headed Cowbird (*Molothrus ater*) using the RE PstI. The number above each lane indicates the haplotype (1 = 1 cut, 2 = 2 cuts). S = size marker. Sizes of the mtDNA fragments are noted in kilobases to the right of the gel.

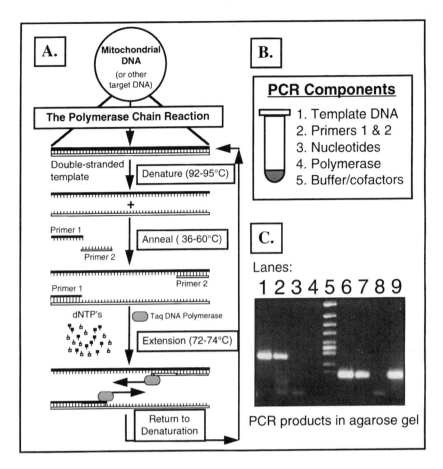

Figure 3.2. Polymerase Chain Reaction (PCR). PCR copies (amplifies) specific sequences of DNA using a DNA polymerase and synthetic primers. (A) Schematic of PCR showing the three major steps: *denaturation* of double-stranded template DNA to single-stranded by heating; *annealing* of complementary flanking primers (at temperatures that depend on the degree of match between primer and template and the proportion of G:C versus A:T bonds); and *extension* from the 3' end of the primer as the polymerase covalently binds deoxynucleotides from solution. The result is an accurately synthesized complementary strand of sequence. The process is repeated thirty or more times, each time as much as doubling the number of copies of the sequence. (B) Components of the PCR mix: "Template DNA" is from the organism of study, nucleotides = dNTPs. The polymerase (Taq) is thermally stable, obtained originally from bacteria capable of living in water near boiling temperature. (C) Examples of PCR products for the control region of mtDNA amplified from DNA isolated from museum specimens of Hawaiian Honeycreepers (Tarr et al., unpublished). A size marker is in lane 5. The lighter bands near the bottoms of lanes 2, 3, and 8 are PCR artifacts called primer dimers. Lanes 3 and 8 contain amplifications of extract controls (all components but template DNA) and serve to identify possible contamination. Amplification products of the correct size were obtained in lanes 1, 2, 6, 7, and 9. Amplified products can be run in a gel, visualized by one of a variety of methods, cleaned of primers and nucleotides, and then cloned into plasmid or viral vectors in bacteria or directly analyzed for variability by RFLP, DNA sequencing, or other methods.

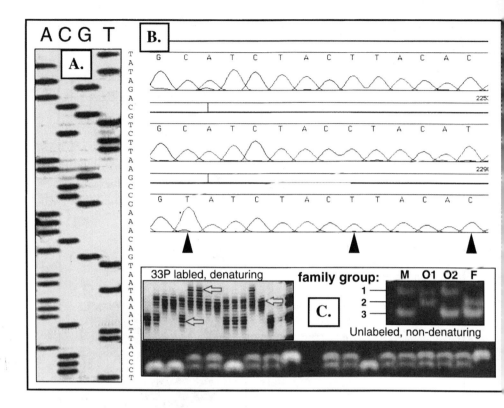

Figure 3.3. DNA Sequencing and Microsatellite VNTR analysis. DNA sequencing involves reactions similar to PCR: complementary strand synthesis from a primer annealed to a template DNA. However, DNA sequencing involves primers or free deoxynucleotides that are covalently labeled (radioactively in "manual" sequencing and with fluorescent dyes in "automated" sequencing) and other nucleotides that lack a hydroxyl group needed for strand synthesis (i.e., dideoxynucleotides). The dideoxynucleotides "randomly" terminate sequence extension, thus leaving a nested series of labeled, single-stranded products in size increments of a single base pair. (A) Manual DNA sequence. Extension reactions for each base (A, C, G, and T) are completed in separate tubes, and the denatured (single-stranded) products are loaded into adjacent lanes of a denaturing polyacrylamide gel. The sequence is read up the ladder as shown to its right. (B) Automated sequence. Cytochrome b mtDNA sequence of three Corvus species. In automated sequencing, each dideoxynucleotide (A, C, G, or T) is labeled with a different color of fluorescing dye. The sequencing reactions are completed in one tube and loaded into a single lane of a denaturing gel. The labeled, synthesized strands are run past a scanning laser at the bottom of the gel. The laser-excited dye fluoresces, and each color and its intensity are recorded into a computer and displayed as continuous spectrum in a "chromatogram." The bases are called according to the colors of the peaks (σ marks variable sites). (C) Microsatellites PCR amplified from drepanidine DNA. Upper left shows microsatellites PCR amplifed with a ^{33}P end-labeled primer followed by electrophoresis in a denaturing polyacrylamide gel. Note that "slippage" bands (a polymerase artifact) make it difficult to score the genotypes consisting of three alleles (marked with dots). Example below and on the right are of unlabeled amplifications of microsatellites followed by electrophoresis on nondenaturing gels. No slippage is evident, and genotypes and alleles are easy to score. The family group shows microsatellites for a male (M), female (F), and two offspring (O1 and O2) and matches Mendelian expectation for inheritance of variants.

Nuclear Genes

These are diverse in their rates of evolution and usefulness in conservation applications. The most used are the variable number of tandem repeat loci, random genomic markers, protein coding regions (exons), noncoding or spacer regions (introns), and immune system multigene families such as the major histocompatibility complex (MHC).

One of the best molecular marker systems for conservation genetics developed to date is the variable number of tandem repeat loci (see Fleischer 1996 for literature summarized here), which include minisatellite (7–25 bp repeat length) and microsatellite (1–6 bp repeat length) sequences in a tandem repetitive array or "core." These loci are highly variable, primarily because of their high mutation rates (10^{-2}–10^{-4} mutations per gene per generation). The mode of mutation also differs from other types of DNA markers. Mutation generally involves replication slippage or unequal sister chromatid exchange, rather than point substitutions, and results in a loss or gain in the number of repeats within a repeat array. Thus, array size varies, causing variation in the size (and thus position) of a fragment or product in a gel (figure 3.3C).

Multilocus DNA fingerprinting involves nearly "universal" minisatellite and microsatellite probes in an RFLP analysis. Multilocus probes can assay for as many as 15–25 highly variable loci on a single gel with one probe, and filters can be reprobed with several independent probes. Some problems with the method are that a large amount of DNA (> 2 μg per individual) is required (not so great a problem for birds because of their nucleated erythrocytes), specific loci and alleles cannot usually be determined, and individuals can usually not be reliably compared among gels (Burke et al. 1991; Fleischer 1996). Some minisatellite probes take advantage of single-copy sequences that flank minisatellite repeat arrays and thus reveal variation at only one locus (Hanotte et al. 1991).

PCR allowed the development of highly variable single-locus genetic markers (Tautz 1989). This method usually involves *microsatellites* as opposed to minisatellites. Microsatellite probes (e.g., CAn, CACn) are initially used to screen random genomic or microsatellite-enriched libraries of a species of interest in order to locate clones containing microsatellite sequences (Fleischer 1996). Regions flanking the repeat region are sequenced, and synthetic oligonucleotide primers are designed so that they will PCR amplify across the microsatellite to produce small products (<300 bp) that can be resolved on a polyacrylamide gel. The products can be sized exactly in multiples of the repeat length, and are highly variable, sometimes with more than 10–20 alleles and heterozygosities above 80–90 percent (figure 3.3C).

Microsatellites are less common (perhaps by as much as an order of magnitude) and variable in birds than in other vertebrates (Ellegren 1992; Hanotte et al. 1994; Fleischer 1996; Glenn 1997). Microsatellites can be difficult and time-consuming to develop, and primers are not universal and may reveal variable and/or interpretable patterns only with related species (Hanotte et al. 1994). In addition, problems, such as nonamplifying or null alleles (e.g., Pemberton et al. 1995), slippage or other artifacts, and uncertainty about the modes of evolution (e.g., Valdes, Slatkin, and Friemer 1993), can confound their interpretation and analysis. Nonetheless, amplifiable microsatellites are the most powerful markers in avian conservation genetics and have a wide range of applications (e.g., estimation of parentage, dispersal, genetic variation and structure, subspecies systematics, and genome mapping).

Some studies use markers that are, in a sense, randomly sampled from the genome (e.g., in birds: Haig, Rhymer, and Heckel 1994; Nusser et al. 1996). Anonymous single-copy markers can be specifically amplified by PCR from sequences identified from random clones (Avise 1994, 72), but the easiest markers to develop and use, though not necessarily the most trouble free, are Randomly Amplified Polymorphic DNAs (RAPDs; Grossberg, Levitan, and Cameron 1996). In this method, one or two random sequence primers usually $\geq$ 10 bp in length are used in a PCR reaction. In regions of the genome where the primers can anneal in opposite orientations while flanking a small (<2,000 bp) sequence, amplification of a product will occur. Sometimes amplification will not occur if priming site sequences are polymorphic. Thus, RAPD fragments usually show complete genetic dominance (i.e., presence of a fragment indicates homozygote dominant or heterozygote, while absence indicates homozygous recessive). Fragment profiles often vary among individuals (but see Nusser et al. 1996) and, with certain assumptions (Lynch and Milligan 1994), can provide useful markers for estimation of population heterozygosity and structure. They are perhaps most useful for development of simple markers to assess gene flow, hybridization among differentiated taxa, or population of origin (e.g., Haig et al. 1997).

Other potentially important single-copy markers are introns, exons (coding sequences), and pseudogenes (exons no longer translated to protein and thought to be free from selective constraints). Variation in neutral markers such as introns and pseudogenes presumably reflects processes (μ, m and N_e) that occur in the absence of selection (assuming no strong linkage to selected loci). Some feel that coding or regulatory regions may be preferable for conservation studies because they code for traits possibly related to fitness. Variation in one such group of genes, the MHC (highly polymorphic proteins that recognize foreign antigens), is putatively maintained by balancing selection caused perhaps by differential resistance to infectious disease (Hughes 1991), but the perils of relying on only one gene system for conservation management, especially one as difficult to characterize in birds as the MHC (Edwards, Grahn, and Potts 1995; Jarvi et al. 1995), have been noted (e.g., Miller and Hedrick 1991; Haig, Ballou, and Derrickson 1990).

Applications in Avian Conservation

Forensics

There has been a virtual explosion in the use of molecular genetic methods in human forensics over the past ten years, and these same methods have been introduced to wildlife conservation biology with only a minor lag period. In most cases some crime has been committed and the molecular data are used as evidence in a trial. The primary use of such methods is to identify or exclude at a variety of levels (i.e., that of the individual, relative, population, or species). Most such applications involve identification of particular wildlife species using mitochondrial DNA (e.g., Paxinos et al. 1997) or of individuals using VNTRs. Another important type of "forensic" method involves the diagnosis and identification of parasites or diseases in tissue samples (e.g., *Plasmodium* in the blood of Hawaiian birds at different elevations; Feldman, Freed, and Cann 1995).

Relatedness, Inbreeding, and Inbreeding Depression

Inbreeding, as defined as the mating between close relatives ($r > 0.125$, sometimes called consanguineous mating), is relatively rare in birds and mammals in nature (Ralls, Harvey, and Lyles 1986), although the preference to mate with relatives has been found in some captive bird studies (e.g., Bateson 1982). Interestingly, in the one case (the Splendid Wren, *Malurus splendens*) in which Ralls et al. note a very high rate of inbreeding in birds in nature (19.4 percent of pairings based on observed pedigree), the species has turned out to have a high rate of extra-pair fertilization that effectively cancels out the putatively high degree of inbreeding (Brooker et al. 1990). This is an example of molecular data invalidating an incorrect claim of high inbreeding without inbreeding depression in a wild population. There are also a number of examples in which calibrations of molecular data with relatedness of individuals (e.g., Rave et al. 1994; Haig and Avise 1996) have been applied to determine inbreeding levels.

The usually negative fitness consequences of inbreeding (known as inbreeding depression) have long been known by people who conduct captive breeding of livestock, laboratory animals, and zoo animals (Falconer 1981; Ralls, Ballou, and Templeton 1988; Lacy, Petric, and Warneke 1993). Inbreeding effects can be even more pronounced in studies in which outbred animals are brought in from wild populations and bred in captivity (Lynch 1977; Lacy, Petric, and Warneke 1993). There is also considerable variability in the incidence and impact of inbreeding depression among different taxa in captivity (Ralls, Ballou, and Templeton 1988; Frankham 1995a,b; Lacy, Alaks, and Walsh 1996), and some of this variability may be related to the degree of prior inbreeding and thus the "purging" of deleterious recessive alleles (but see Frankham 1995a). Molecular markers such as VNTRs have also been used in studies of captive birds to show relationships between the genetic similarity of pairmates and their reproductive success (e.g., in parrots and Hawaiian geese: Brock and White 1992; Rave et al. submitted).

There are few data that directly reveal inbreeding depression in wild populations (Lacy, Petric, and Warneke 1993). One recent exception is from a study of song sparrows (Keller et al. 1994) in which outbred individuals had greater survivorship during a natural bottleneck than inbred ones. A rather ingenious study by Jiminez et al. (1994) demonstrated that inbred field mice (*Peromyscus polionotus*) survived less well following release than inbred lines remaining in captivity and all outbred controls. The general prediction is that inbreeding depression should be more severe in nature (Frankham 1995b), where conditions are generally more stressful than in captive environments (in which food, medical care, and mates are provided more or less *ad libitum*)) and where inbreeding may generally be rare because of intrinsic inbreeding avoidance mechanisms (Blouin and Blouin 1988), and thus little or no purging of deleterious alleles occurs.

Can inbreeding impact fitness to the extent that it causes the extinction of populations? Contrasting theoretical results have been obtained from Lande (1988) on the one hand and Mills and Smouse (1994) and Frankham (1995b) on the other; however, recent studies by Lande (1994) and Lynch, Conery, and Burger (1995) have indicated that "mutational meltdowns" are likely to have a very strong impact on fitness and may have greater impacts on population survival than demographic or even environmental stochasticity. There are few direct empirical assessments of this question, and none that I know of in birds. Frankham's study (1995b) of inbred strains of *Drosophila* and mice provides the clearest direct demonstration that inbreeding can directly affect rates of extinction of lines, but there are a huge number of indirect studies in the agricultural and laboratory genetics literatures that strongly suggest such a relationship. In particular, because the theory is so complex and contentious, this is one area of research where additional direct empirical tests are of critical importance (Frankham 1995a).

Genetic Variability and Structure

Genetic variation is thought to be important to the probability of population extinction at both proximate and ultimate levels (Soulé 1980). In the former, variability is thought to be important because of mechanisms that increase individual fitness, such as heterozygote advantage or frequency-dependent selection (e.g., in disease resistance genes). In the latter, standing genetic variation is important because it allows a population to respond more rapidly to selection than if mutation is the only source of new variability; i.e., it is required for adaptation to occur. Lastly, in order for higher levels of evolutionary change to have an opportunity to occur (i.e., cladogenesis) the structuring of genetic variation should also be conserved.

The evidence for the importance of heterozygosity to proximate fitness is mixed, with a few studies showing a strong, clear relationship, and perhaps most showing little or no direct impact (see Mitton 1994 for review). In addition, a recent meta-analysis of the available data relating fitness to heterozygosity revealed only weak support for such a relationship (Britten 1996).

Particular criticisms have been raised (Lande 1988; Caughley 1994) and countered (Lande 1994; Lynch, Conery, and Burger 1995; O'Brien 1994; Hedrick et al. 1996) about theory concerned with the genetic impacts within the "small population paradigm": that is, do small or endangered populations have reduced genetic variation, and does this reduction significantly impact short- or long-term survival of populations? There are clearly cases in which apparently healthy avian populations have extremely low allozyme variability (e.g., see review of allozyme studies in Haig and Avise 1996). Given the low mutation rate in allozymes, it might not be surprising that these populations could harbor significant amounts of quantitative genetic variation for fitness in spite of low allozyme variability. Perhaps better markers to reflect polygenic variation may be hypermutable VNTRs (Fleischer, Fuller, and Ledig 1995; Hedrick et al. 1996), which have mutation rates nearer to those of quantitative trait loci. However, several studies have revealed high similarity among individuals and thus low variability for VNTRs (e.g., Blue Duck [*Hymenolaimus malacorhynchos*], Triggs et al. 1992; Nene [*Branta sandvicensis*], Rave et al. 1994; Light-footed Clapper Rail [*Rallus longirostris*], Fleischer, Fuller, and Ledig 1995; Black Robin [*Petroica traversi*], Ardern and Lambert 1997), and yet these populations are currently surviving if not thriving. One should note that there is often an unrealistic expectation that when populations get very small (bottlenecks to the tens or hundreds of individuals), they will go extinct. We need only to look at the current population sizes and fitnesses of species for which small founder populations were introduced, such as House Sparrows (*Passer domesticus*) or Red-whiskered Bulbuls (*Pycnonotus jocosus*) in Hawaii (Long 1981), to see that this is not the case. In fact, reduced population size and reduced genetic variation only imply a greater probability of reduced fitness and population survival, rather than guaranteeing it (Hedrick and Miller 1992; Frankham 1995a).

Estimation of Demographic Variables

Because the levels and patterns of variability in neutral genes (i.e., those not subject to selection) are determined by the population processes of mutation (μ), migration (m), and genetic drift, the analysis of molecular genetic or quantitative trait variation can provide "indirect" estimates of demographic parameters such as m, and N_e and their direction of change (Slatkin and Maddison 1989; Tajima 1989; Hudson 1990; Donnelly and Tavaré 1995; Wakeley and Hey 1997; but see Moritz 1994a and Rand 1996 for caveats). Models can be directly analytical or can use simulations to obtain parameter estimates (e.g., Neigel, Ball, and Avise 1991; Slatkin and Barton 1989). Coalescent, or genealogical, methods (Donnelly and Tavaré 1995) are particularly powerful in that they take advantage of all of the information in a set of molecular data (i.e., both the frequency of an allele

and its divergence from other alleles), and they can provide historical information on population change and migration largely independent of mutation rates.
Such models combined with better methods of tree construction are revolution-
izing population genetics and have considerable application in conservation.

Applying these models invariably requires assumptions, some of which need
to be met more stringently than others. One assumption of most models is that
populations have reached genetic equilibrium (with regard to mutation, drift, and
migration). Assuming equilibrium conditions can be problematic for populations
that fluctuate greatly, are recently founded, or that have been recently impacted
by humans (like many endangered ones). A second assumption of such models is
the mode of mutation acting on the markers under study (e.g., infinite alleles,
stepwise mutation; Nei 1987).

A third assumption for which violation can confound estimates of demographic variables is neutrality of alternative variants (Rand 1996). For example,
gene diversity can be calculated from pairwise sequence divergence (π) or from
the number of variable (segregating) sites under a stochastic model (θ)(Nei 1987).
These should be equivalent (and equal to $4N_e\mu$) under equilibrium conditions,
but a value of π significantly less than θ may indicate either a recent population
decline or an episode of directional selection (Tajima 1989). Determining the
liable alternative may be resolved by comparing π and θ for more than one
unlinked marker (e.g., Tajima 1989; Rand 1996): it would be very unlikely for
unlinked loci to show the same signature of selection, but all loci should reveal
the impact of a population bottleneck. Anomalous genes may be under selection
and can be excluded from estimating demographic parameters.

A variety of models have been used to assess effective population sizes and gene
flow rates in birds. For example, Fleischer, Fuller, and Ledig (1995) used minisatellite data and the analytical models of Lynch (1991) to estimate N_e and the
effective number of migrants ($N_e m$) among endangered populations and subspecies of the Clapper Rail in California. A coalescent approach was used with
an mtDNA control region sequence by Edwards (1993) to estimate and provide
evidence for long-distance gene flow among populations of the Grey-crowned
Babbler (*Pomatostomus temporalis*).

Evolutionarily Significant Units, Hybridization, and Phylogenetic Triage

Sometimes decisions have to be made about what constitutes a unit to be conserved. In essence, this results in two problems: how to define a unit for evolutionary or management significance, and how to determine whether such a unit
is "worth" the investment of resources required to recover or conserve it.

Several definitions of "evolutionarily significant unit" have been proposed
since the term was hesitantly coined by Ryder (1986). In general, an ESU is a
population or group of populations that has had an independent evolutionary trajectory through time, which should be evidenced by some degree of genetic

differentiation from other such populations. The keys to designating ESUs are in the analytical methods used to hierarchically order individuals into such units and in how much differentiation should exist among units.

A number of approaches to making ESU determinations have been developed (e.g., Ryder 1986; Barrowclough 1992; Vogler and DeSalle 1994; Moritz 1994b). The most stringent essentially applies the phylogenetic species concept (Vogler and DeSalle 1994): that is, an ESU is "delimited by characters that diagnose clusters of individuals to the exclusion of other such clusters." Thus, any fixed, alternative difference between taxa makes them "diagnosable" as a distinct unit. This cladistic approach is the most straightforward and perhaps logical, but it is not without difficulties. For example, under this definition a "single nucleotide change in a DNA sequence" defines an ESU. From models of genetic drift we know that small populations can diverge and fix variants quite rapidly (the number of generations required to fix a new mutant is directly proportional to the effective population size, i.e., $4N_e$). Thus if populations become fragmented to small N_e by recent human impacts, some alternate genetic variants may "fix" in a very short period of time. We would then be delineating an "unnatural" ESU, one based on an evolutionarily unimportant and anthropogenically originated event (e.g., Fleischer, Tarr, and Pratt 1994).

Moritz (1994b) defined ESUs as requiring reciprocal monophyly of taxa in an mtDNA phylogenetic tree along with evidence of genetic differentiation at nuclear loci. This, like Vogler and DeSalle's method, uses phylogenetic methods to avoid the question of how much divergence is enough? But should we avoid that question? Some differences that define ESUs could be trivial, either because they do not truly reveal prehistoric patterns of diversity and levels of gene flow, or because they would not result in any biological problems for the taxa should they be managed as a single unit (e.g., Zink and Kale 1995). One way to address the how much question is to assay highly variable markers in nonfragmented populations of the same species or closely related members of the same genus (e.g., Tarr and Fleischer 1995). Divergence between putative ESUs may not be important if it is significantly less, for the same markers, than variability within the relative. Examination of genetic markers from museum or subfossil material may also reveal that genetic change has resulted from recent, anthropogenically induced, population change. Moritz (1994b) also defined management units as "populations with significant divergence of allele frequencies . . . regardless of the phylogenetic distinctiveness" and suggested that they be considered equivalent to "stocks" in wildlife biology.

Another problem with ESU and MU approaches, especially for species that view the world as a fine-grained entity (like most birds), are the rationales for an ESU approach and its management implications. These include:

1. Mixing of ESUs or MUs could result in reduced fitness of the hybrids (outbreeding depression) because of local adaptation and/or co-adapted gene

complexes. There is little evidence for either phenomenon in avian or other vertebrate species (Hedrick and Miller 1992; Ballou 1995). For example, both the survival of many bird species introduced to new ranges and environments (Long 1981) and the relatively high levels of gene flow found in most avian taxa (e.g., Barrowclough 1980; Rockwell and Barrowclough 1987; Fleischer and Rothstein 1988) argue against local adaptation. Even hybridization between fairly divergent populations or subspecies rarely appears to result in significant outbreeding depression, both in the few bird species assessed (e.g., Rhymer and Simberloff 1996) and in a sizeable sample of subspecies or population crosses of mammals in captivity (Ballou 1995; Frankham 1995a). These factors may be more problematic for plants, invertebrates, and lower vertebrates, which appear to have greater potential for local adaptation.

2. Several small populations can maintain more variability than a single large one with the same number of individuals because drift in small populations will fix alternative alleles (Varvio, Chakraborty, and Nei 1986; Nei and Takahata 1993). However, if fitness is impacted by small size and low heterozygosity in the subpopulations, this mode of management may be detrimental.

3. Behavioral isolating mechanisms alone may be insufficient to maintain distinct species. The high level of hybridization and introgression among fairly differentiated avian taxa (e.g., summarized in Rhymer and Simberloff 1996) suggests that most avian subspecific taxa may not have behaviors different enough to cause problems with reproductive isolation.

4. Preservation of "natural" patterns of variation and processes of evolution may have far-reaching consequences. Perhaps a "philosophical" rationale— but do patterns we find today reflect structure after human impacts? What if preservation of ESUs impacts the probability of survival of one or both ESUs (Ryder 1986)? On the other hand, what if ESUs are weakly defined and protected under the endangered species act principally to protect vulnerable habitats? We must be careful not to abuse this approach until the ESA changes to allow it, as trivial designations, or ones that can be easily challenged, may make us appear biased or fickle in the eyes of management agencies, congress, and the public.

Another problem for defining conservation units and survival of avian endangered species is hybridization (Rhymer and Simberloff 1996). How should hybrid individuals or populations be managed, especially if only one of the hybridizing taxa is endangered (e.g., O'Brien and Mayr 1991)? Each case may require independent evaluation as to the level of introgression reached. Habitat changes and species introductions have greatly modified ranges, resulting in greater opportunity for hybridization. Molecular genetic methods are proving useful for measuring the extent of hybridization (Rhymer and Simberloff 1996).

Phylogenetic information can also help in allocating effort and resources to endangered species recovery. This approach is sometime called phylogenetic triage after the method of culling hopeless cases from wartime casualties (Barrowclough 1992), and decisions depend on how phylogenetically unique a taxon is relative to other surviving taxa. Several quantitative measures of uniqueness have been devised and applied (e.g., Vane-Wright et al. 1991; Krajewski 1994). Avian examples of triage ask whether more resources should go into preserving barely differentiated, but endangered, populations of a common species—e.g., the Common Amakihi (*Hemignathus virens*) on Molokai (Tarr and Fleischer 1993); Mississippi populations of Sandhill Crane (*Grus canadensis*) (Krajewski 1994); and San Clemente Island Loggerhead Shrikes (*Lanius ludovicianus mearnsi*) (Mundy, Winchell, and Woodruff 1997)—or into preservation of phylogenetically (not to mention morphologically and ecologically) unique species on the brink of extinction (e.g., the California Condor [*Gymnogyps californianus*], Poouli [*Melamprosops phaeosoma*], and Spix Macaw [*Cyanopsitta spixii*]). Clearly other factors enter into triage decisions, including, how likely is recovery of the taxon, how attractive or important it is to the public, and how might preserving a species result in umbrella protection for other organisms.

The Future?

With fears about incorrect prognostication thrown aside, I attempt below to discuss advances in theory and technology that may, in the future, be applied to particular cases in avian conservation biology.

Better Models and Faster Molecular Methods

As noted above, population genetics is undergoing a revolution as new theories and analytical methods are being applied to large databases derived from faster molecular techniques. In particular, coalescent models and simulations that estimate age of most-recent common ancestor, effective number of migrants, and effective population sizes are able to maximize information from DNA sequence data (Slatkin and Maddison 1989; Donnelly and Tavaré 1995; Wakeley and Hey 1997). Population viability models also need improvement, with impacts of low genetic variability and inbreeding on extinction probabilities more accurately incorporated (Lande 1994; Lynch, Conery, and Burger 1995; Frankham 1995b).

Molecular methods of analysis have become incredibly rapid and efficient during the past ten years (primarily with the advent of PCR), and there is no doubt that the methods will become even faster during the next ten years. In fact, technologies have already changed to the point that many complex protocols are automated via robotics, and "digitally" assayed human genetic markers (e.g., DNA coated "chips"; Schena et al. 1996) are in development. We may soon be able to electronically "type" thousands of DNA variants per individual in very short periods (approaching the tricorder of *Star Trek* fame!).

Another recent technological advance with potential application to conservation of birds is the analysis of "ancient DNA." For example, mtDNA sequences from crane museum specimens were used to directly assess loss of variability associated with a population bottleneck (Glenn 1997), and DNA sequences from subfossil bones have been used to confirm that Laysan Ducks (*Anas laysanensis*) recently existed on islands on which they are now extinct (Cooper et al. 1996). Lastly, hypervariable markers such as VNTRs appear very useful in assays of the mutational effects of environmental pollutants (e.g., in Herring Gulls, *Larus argentatus*; Yauk and Quinn 1996).

Genetic Engineering and Immunogenetics

For a relatively small number of cases there may be genetically based problems that impact individual fitness and population survival. In these cases, advances in gene therapy and genetic engineering may prove useful. Although their routine use is probably more than five or ten years away, I speculate here about how such methods may eventually be applied.

Some cases may involve genetic disease. VNTR and RAPD markers are proving extremely valuable for gene mapping and they are also useful, along with other approaches, for gene identification (e.g., Georges et al. 1993; Ferraris and Palumbi 1996; Ghosh and Collins 1996). One avian disease that may have a genetic basis is hemochromatosis (iron storage disease), evidenced by the circulation of excess iron in the blood and in the liver and most commonly observed in frugivorous species in captivity (Kincaid and Stoskopf 1987). If a simple genetic basis for absorption of iron can be isolated, it may be possible to clone it and conduct gene therapies to correct the problem. Molecular "fixes" perhaps should initially be engineered into somatic cells, so they can be easily reversed if problems later arise.

Another future application of genetic engineering methods is to modify genetic systems that confer resistance to infectious disease. Drepanidines (Hawaiian Honeycreepers) have been greatly impacted by introduced disease, along with other factors (van Riper et al. 1986). Currently, only twenty-four (of fewer than fifty) species exist, and of these only eight are not endangered. Infectious disease is considered to be a major limiting factor in drepanidine distribution, especially the introduced mosquito-transmitted disease, malaria, caused by *Plasmodium relictum*. Many honeycreeper species do not occur, or occur in very small numbers, at lower elevations where mosquito density is high. Most species at higher elevations appear to lack natural resistance to malaria (van Riper et al. 1986; Atkinson et al. 1995), and mosquitoes may be gradually moving upslope. Common Amakihi occur at low and high elevations on several islands, and half or more survive malarial infection in controlled challenge experiments (including presumably "unchallenged" individuals from high elevations; Atkinson et al.

1995). Such resistance appears to be genetically based, and Jarvi, Fleischer, and Atkinson (unpublished) are attempting to identify this basis by searching for linkage disequilibrium between malaria resistance and variants of: (1) MHC proteins (which do appear to provide some resistance to infectious disease in chickens and to malaria in humans [Hill 1996]); and (2) microsatellite markers (Ghosh and Collins 1996). If we can locate the genetic complexes that confer resistance in the Common Amakihi, it may be possible in the future to develop gene therapies or even gene transfers to susceptible, endangered drepanidines. Another fanciful application is to engineer genes for rejection of parasitic eggs (if these exist in a simple, clonable form) from rejector species (e.g., Eastern Warbling Vireos, *Vireo g. gilvus*) to closely related acceptor species impacted by brood parasitism (e.g., Western Warbling Vireos, *V. g. swainsoni;* Sealy 1996).

The recently described method of cloning sheep from adult cell lines (Wilmut et al. 1997) may also present some opportunities for conservation biology. For example, founders of a captive population that are senescent or ill and will not breed in captivity could be cloned and maintained for future breeding into the population. An individual that shows resistance to an epidemic disease could be cloned and the clones used to "swamp" the population with disease-resistant genes. In this light, it may have been remarkably prescient of zoo geneticists to initiate cell lines from different endangered species and store them in "frozen zoos" since the late 1970s (Benirschke, Lasley, and Ryder 1980).

The biotechnology applications described above should enhance conservation efforts for only a small number of endangered species. They will also genetically alter the species, if only by a small amount. Numerous questions have been raised about the ethics and safety of genetic manipulations of humans and their agricultural products. Such issues have not yet been considered intensively for the prospect of genetic engineering in wildlife that are then released into nature, but I imagine this will change in the near future.

Acknowledgments

I would like to thank Cheryl Tarr, Sabine Loew, Jon Ballou, Kathy Ralls, Matt Hamilton, Judith Rhymer, Susan Jarvi, Carl McIntosh, Bob Zink, and Dina Fonseca for enlightening discussions on many of the topics included in this chapter, and Cheryl Tarr, Carl McIntosh, and Pedro Cordero for examples of their lab work for display in figures. Sue Haig and John Marzluff provided feedback on an earlier draft. The Smithson Society, Friends of the National Zoo, National Biological Service, National Geographic Society, and National Science Foundation funded original research illustrated in this chapter.

CHAPTER 4

Contribution of Spatial Modeling to Avian Conservation

Marc-André Villard, Elise V. Schmidt, and Brian A. Maurer

The conservation of vagile organisms like birds poses challenges that are intrinsically spatial in nature. Serious conservation strategies must consider a hierarchy of scales over which bird movements take place and the effects of the amount and distribution of habitat at every scale on these movements. A wide variety of movement types have to be considered (figure 4.1): (1) local movements related to foraging, territory defense, search for mates, etc.; (2) dispersal movements within or between years; (3) migratory movements; and (4) long-term, large-scale movements among centers of abundance and lower-density peripheral populations (Maurer and Villard 1994). To add to this already complex task, behavioral responses to the spatial distribution of habitat may vary substantially among species (Villard and Taylor 1994; Machtans, Villard, and Hannon 1996; Desrochers and Hannon, in press). In this context, spatial models are not only useful, they may represent the only approach to developing realistic conservation strategies. However, in spite of the rapid progress in the technology and software available to model the spatiotemporal dynamics of living organisms, the usefulness of spatial models ultimately depends on the quality of empirical data available to parameterize these models (Wennergren, Ruckelshaus, and Kareiva 1995).

Environments are not homogeneous at any spatial scale, and organisms cannot be expected to perceive them as such (Wiens 1976). Although the assumption of habitat homogeneity greatly facilitates model computation, studies or models relying on it often lack biological realism (Wiens 1995). The potential of a model to predict individual or species response to habitat alteration increases if it incorporates spatial information. Spatial information should include: (1) habitat distribution; (2) characteristics of movements of individuals within and among habitat patches; and (3) reproductive success of individuals in different habitat patches.

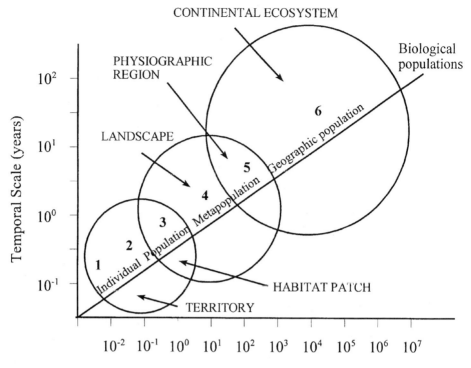

Figure 4.1. Temporal and spatial scales of spatially explicit habitat models used in avian conservation. The circles represent the three scales discussed here: the scale of an individual animal or territory, the landscape scale, and the continental (geographic) scale. Note that these arbitrary scales often overlap for a given modeling effort. Two significant regions of overlap are the habitat patch and physiographic region (see definitions of these scales below). The diagonal line represents a continuum of levels of organization. Lower levels are implied to be contained within larger ones. The bold numbers on the graph denote:

1. Models of individual movements. These models attempt to simulate the use of space by individuals for foraging and other activities related to fitness. 2. Models of territory selection and placement and microhabitat use. These models seek to describe and predict how individual organisms will select territories and other locations used for breeding or other important life-history activities. 3. Population dynamics models. The interest of these models is to describe the dynamics of populations that result from spatial arrangements of individuals within continuous patches of habitat. Some spatially explicit population models describe such dynamics within patches and tie them into the dynamics of populations in other patches. 4. Metapopulation dynamics models. These models describe how a species persists in a heterogeneous landscape, where habitat quality varies spatially. Populations within patches are explicitly connected through the dispersal of individuals. Extinction-recolonization dynamics are used to describe metapopulation dynamics and persistence. 5. Models of persistence within a geographic region. Interest at this scale is the description and prediction of whether a species is declining or will persist within a relatively uniform geographic region (identified as a physiographic region in the BBS) or arbitrary political region (e.g., state or province). 6. Models of global persistence. These models describe how population trends and processes at lower scales determine the ultimate fate of a species (over ecological time). Conservation concern at this scale is in identifying species at risk prior to extended periods of widespread population decline.

In this chapter, we review conservation issues that are spatial in nature and applications of spatial models to these issues. We use the term "model" in its broad sense to include statistical, analytical, and simulation models. For clarity, we divided the chapter into sections corresponding to three scales of investigation, although some movement types clearly overlap between scales (figure 4.1).

The Local Scale

The *local* scale is loosely defined here as the scale of an individual's home range. Spatial modeling of movements of individual birds within their home range is an important tool to improve understanding of resource use. It is particularly relevant to species with large area requirements or relatively specialized habitat use since resources tend to be patchily distributed within their home ranges, and human alteration of habitat tends to increase this patchiness for most species.

This section summarizes spatially explicit approaches that have been used to assess the ways in which individuals and populations use their local habitat. Attributes of individuals or populations that can provide researchers and managers with an indication of habitat suitability from the perspective of the organism include movements related to foraging and territory defense, as well as measures of reproductive success.

Conservation Issues and Previous Research

Technological advances in equipment and computer hardware and software have greatly facilitated studies of animal movement and distribution. Remote tracking, predominantly using radiotelemetry, allows for tracking of individually marked animals as they move through their environment (White and Garrott 1990). Satellite technology has greatly expanded the applications of remote tracking by (1) eliminating the requirement of active field monitoring, and (2) allowing the monitoring of animals under previously intractable conditions. Finally, geographical information systems greatly facilitate the exploration and analysis of large data sets containing spatial attributes (Aronoff 1989). Information about a particular area, at local, regional, or geographic scales, can be organized in an explicitly spatial manner. At the local scale, important habitat parameters may include attributes such as breeding territories, movement of individuals, and food resources patches. GIS provide the capability to integrate spatial analyses of habitat with life-history characteristics of individuals as well as with population or species parameters.

Models of animal movement have also been refined in recent years, and movement can be investigated from perspectives ranging from simple diffusion models (Johnson, Milne, and Wiens 1992) to more complicated fractal models (Wiens et al. 1995). Diffusion models describe movement of individuals as random events. These models have been modified to incorporate correlated movement (Boone and Hunter 1996) or movement through suitable and unsuitable habitat patches at local scales (Johnson, Milne, and Wiens 1992). Inclusion of a fractal component provides a scale-independent measure of movement for comparison among species of different size, or species that move over vastly different scales (Wiens and Milne 1989; With 1994).

Features of Research That Improved Conservation

Studies Using Radiotelemetry

Radiotelemetry has proven to be a very important tool for gathering spatial data relevant to conservation issues. For example, a radiotelemetry study of the Palila (*Loxioides bailleui*), an endangered Hawaiian bird, indicated that individuals tend to exhibit strong site fidelity (Fancy et al. 1993). Home-range sizes and movement distances were small compared to the potential mobility of the species. Fancy et al. suspect that the strong site fidelity of Palila may prevent them from repopulating favorable habitat in their former range. However, this study provides some hope that translocations of Palilas within their former range will be successful.

Radiotelemetry can also be useful to investigate detailed habitat use patterns. In Capercaillie (*Tetrao urogallus*) populations of central Europe, Storch (1995) used radiotelemetry to compare home-range characteristics of males and females. He found that males tended to include intermediate-aged forest in their home ranges, while females tended to restrict their activities to older stands. Thus, females are more likely to be affected by fragmentation of old forest than males. Radiotelemetry also helped determine the size of forest patches and the type of forest necessary to conserve leks and maintain viable populations.

Applications of Geographic Information Systems

Geographic information systems are useful to visualize spatial parameters of habitat use and to analyze relationships among these parameters. For example, movements and population densities of two species of endangered Hawaiian birds, the Akepa (*Loxops coccineus coccineus*) and the Hawaii Creeper (*Oreomystis mana*), were modeled using GIS to examine their conservation needs (Ralph and Fancy 1994). Individual locations were mapped on a habitat grid along with other attributes such as date, sex, and age. Home range was determined, and overlap between individual territories was computed. Overlap of individual home ranges within species was high, and the occurrence of unpaired males within home ranges of other males was also high. Because these birds do not defend large feeding territories, models indicate that both species should be able to maintain high-density populations. Analyses of nest distribution also indicated that current populations may be limited to high-elevation native forests above the zone of avian malaria-carrying mosquitoes.

Other studies have used GIS to assess the spatial characteristics of Spotted Owl (*Strix occidentalis caurina*) home ranges to help elaborate conservation strategies for this subspecies (Raphael, McKelvey, and Galleher, this volume) and to investigate selection of nesting habitat by Sandhill Cranes (*Grus canadensis tabida*) (B. W. Baker et al. 1995). The latter study demonstrated the importance of habitat characteristics within two hundred meters of the nest, whereas there was no apparent selection for habitat parameters on a larger scale.

Research Needed to Further Conservation

Studies of avian populations have often used data on frequency of occurrence, density, and reproductive success to characterize habitat quality. However, when applied to landscape or regional scales, models based on such data often do not accurately predict population or species persistence, or even the distribution of these species (see below). Suitable habitat may be patchily distributed, and the ability of the organism to use this habitat may depend on the area, connectivity (Taylor et al. 1993), and other spatial features of the landscape.

Remotely sensed data on vegetation characteristics and land use can be incorporated into a habitat suitability model. This approach is valid if remotely sensed images are cross-checked with on-site surveys (ground-truthing). Movements within and between different habitat types can be assessed using radio or satellite telemetry for difficult-to-access individuals. Models describing movements of individuals within territories have been developed for insects and ungulates (Johnson, Milne, and Wiens 1992; Turchin 1991; Turner et al. 1993). These models are built upon actual movements of individuals as measured in the field with respect to resources. Movements in beetles and butterflies are nonrandom with respect to habitat characteristics (Johnson, Milne, and Wiens 1992; Turchin 1991), while movement between resource patches affected the mortality of ungulates (Turner et al. 1993). Species-specific movement studies could be used to measure habitat-use parameters for birds that might include residence time in resource patches or movements of individuals in relationship to resources such as territories or nest sites (Pulliam, Dunning, and Liu 1992). The relative area, spacing, and connectivity among important resource patches can then be incorporated into a GIS program and used to build descriptions of suitable habitat.

The Landscape Scale

The *landscape* scale is intermediate between the local and regional scales. The distinction between these arbitrary points along the "space continuum" can be clarified using an organism's perspective. The landscape can be defined as the scale over which a moving organism encounters a mosaic of habitat and "nonhabitat."

Nonhabitat may vary in its permeability to the movements of the species of interest, but it is always unsuitable for reproduction.

Conservation Issues and Previous Research

The Patch-Centered Approach

In spite of birds' mobility, avian ecologists have often assumed (implicitly or explicitly) that the populations and communities they studied belonged to closed systems. As a result, early conservation efforts often focused on habitat suitability for the target species, with little or no concern for the larger-scale dynamics of the target species or the disturbance regimes influencing habitat suitability itself. The field of landscape ecology reflects ecologists' recent recognition of the influence of spatial and temporal heterogeneity on ecological processes (Merriam 1988; Wiens 1995; Noss 1995).

Askins and Philbrick (1987) provided one of the clearest examples of the influence of spatial context on local dynamics in bird communities. Using a data set spanning thirty-two years, they documented the fluctuations in bird populations at the Connecticut Arboretum, an intensively studied, unmanaged forest. Although changes in abundance of some bird species could be explained by successional changes in the forest vegetation, the abundance of most forest long-distance migrant species was strongly linked to land-use changes within a two-km radius around the study site. Forest long-distance migrants declined as woodland was cleared for agriculture and urbanization in the surroundings, and increased when some of the agricultural land was abandoned and allowed to undergo natural regeneration. This study, among others, clearly showed that conservation cannot be conducted in isolation, especially for long-distance migratory birds. Population dynamics may be dependent on processes that occur at the landscape level or above in addition to local habitat. Stouffer and Bierregaard (1995a,b) documented similar patterns using a fragmentation experiment in the Amazon.

During the 1980s, many studies examined the effects of habitat fragmentation on forest birds (Walters, this volume). Most of these studies were conducted in agricultural landscapes and adopted a patch–centered approach inspired by the theory of island biogeography. The influence of island biogeography was so strong that the study landscapes were (and still are) often depicted as black-and-white mosaics of forests and nonhabitat. These studies were based on the selection of a series of patches of varying area and isolation, and the relative effects of these factors and vegetation characteristics on bird presence/absence and abundance were examined.

The patch-centered approach has been useful in pointing out the relative influence of local vs. context (landscape) effects on local population presence or abundance and in showing that context effects should not be neglected (Freemark et al. 1995). However, the stepwise multiple linear regressions used in many regional studies are sensitive to multicollinearity (Legendre and Legendre 1983), which is frequent among habitat variables. This problem can be minimized by replacing certain independent variables with the residuals of linear regressions between these variables and others strongly correlated to them, or by using principal components regression (Myers 1990).

The Landscape Approach

The transition from a patch-centered to a landscape approach was gradual and reflects not only the evolution of the new discipline of landscape ecology, but also the development and improvement of tools available to characterize landscape composition and structure. The shift to the landscape approach may have been accelerated by the "rediscovery" of the metapopulation concept (Merriam 1984; Hanski 1989). A metapopulation can be defined as a group of local populations connected by the dispersal movements of individuals (Hanski and Gilpin 1996). Local populations are groups of individuals who occupy distinct patches of habitat and whose probability of interaction with each other is "much higher" than their probability of encountering individuals of other local populations in their lifetime. The metapopulation concept is, therefore, intrinsically spatial since it implies that local populations interact to some degree through dispersal.

Since its rediscovery, the metapopulation concept has been applied to a wide variety of systems where habitat is naturally patchy or fragmented by human activities (Walters, this volume). However, the usefulness of the metapopulation approach may be limited in systems where: (1) habitat is fragmented at the scale of individuals (Haila, Hanski, and Raivio 1993); (2) dispersal among habitat patches is so frequent that local extinctions and recolonizations are virtually never observed (the "patchy populations" of Harrison 1991); or (3) the average adult tends to occupy several different patches in its lifetime (Harrison 1991). The subjectivity involved in the delimitation of local populations in many systems also tends to restrict the applicability of metapopulation models.

Perhaps the greatest contribution of the metapopulation approach to conservation biology is that it directed the attention of researchers toward the importance of considering spatiotemporal dynamics in the design of conservation strategies (Gutiérrez and Harrison 1996). The metapopulation approach requires that observers obtain information from the majority (or totality) of patches used by their target species in the focal landscape, instead of collecting information on a series of patches selected a priori, as was the case under the patch-centered approach. This has the benefit of emphasizing the potential importance of often overlooked landscape elements (e.g., small patches, or patches of marginal habitat) for the persistence of the metapopulation. Another important concept, that of *connectivity* (Merriam 1984, 1991), is also critical in metapopulation ecology. Connectivity can be defined as the degree to which organisms move among landscape elements. Thus, it stresses the importance of considering the entire landscape, including the matrix in which focal habitats are embedded. In other words, our mental image of the landscape must shift from a black-and-white picture to a complex mosaic with several different shades of gray (Wiens 1994).

The notion of connectivity led to the investigation of corridor or barrier effects. In spite of their physical ability to move over long distances, many species of birds seem to move preferentially along strips of hospitable habitat when they are available (Dmowski and Kozakiewicz 1990; Machtans, Villard, and Hannon 1996). Similarly, recent work by Desrochers and Hannon (in press) suggests that forest species may fly across open areas up to species-specific threshold widths, beyond which they will use alternate trajectories within forests, even when these trajectories correspond to substantially longer distances.

Modeling has been used to develop exploratory tools that can assist in the elaboration of conservation strategies. Useful models should integrate recent empirical findings on bird movements and demography. The studies that have been conducted on Bachman's Sparrow (*Aimophila aestivalis*) provide an excellent illustration of the procedure used in metapopulation modeling. This sparrow breeds preferentially in stands of longleaf and loblolly pines (*Pinus palustris* and *P. taeda*) with an open understory and a dense cover of herbaceous vegetation (Dunning and Watts 1990). These characteristics are found in mature pine stands that are regularly burned, as well as in recent clearcuts where some standing dead vegetation is retained. Patches of suitable habitat are illustrated in two shades of gray (mature stands and recent clearcuts) surrounded by white areas of unsuitable habitat (see figure 1 in Liu, Dunning, and Pulliam 1995).

Pulliam, Dunning, and Liu (1992) developed a spatially explicit, individual-based simulation model to explore the dynamics of Bachman's Sparrow in a hypothetical landscape inspired by those surveyed by Dunning and Watts (1990). A spatially explicit population model (SEPM) consists of a demographic model coupled with a landscape map. An individual-based model monitors the location and fate of each individual in the simulated metapopulation. In Pulliam et al.'s study, the hypothetical landscape was composed of equal-sized pine stands that were allowed to grow to commercial age (21 years old), at which time they were harvested. A variable number of mature stands (>80 years old) were also left uncut in the simulated landscapes. Since Bachman's Sparrows tend to preferentially occupy mature stands and recent clearcuts, the distribution of their habitat varied substantially over the 105-year simulation period. Dispersal of juveniles was modeled as a sequential search for suitable, unoccupied patches. Adults were assumed to remain in the same territory to breed year after year, until the local habitat became unsuitable. They then move to an adjacent unoccupied territory if one is available. Liu, Dunning, and Pulliam (1995) extended this model to investigate the effects of different timber harvest scenarios on sparrow metapopulations. Hanski and Gilpin (1996) consider this type of model "spatially realistic" since it can be linked to a GIS database illustrating a real landscape. Liu et al. modeled three harvest scenarios: (1) random harvesting of pine stands; (2) clustered harvesting; (3) harvesting of oldest pine stands first. They found that scenarios 2 and 3 allowed the sparrow metapopulation to reach the minimum number of pairs deemed necessary to maintain the species in the study area within the 50-year duration of the simulation.

SEPMs are still in their infancy. Their main limit resides in the ability of researchers to make reasonable assumptions about life-history parameters that have a major influence on the model's behavior. For example, the realism of models developed for the Bachman's Sparrow and Spotted Owl is limited by the incomplete knowledge of dispersal in these species (Liu, Dunning, and Pulliam 1995; Gutiérrez and Harrison 1996). Sensitivity analyses allow an assessment of the relative impact of our knowledge gaps, but they cannot compensate for them. Model validation can be used to assess the cost of simplifying assumptions used to cover knowledge gaps. To validate a model, researchers should compare model output to independent field observations of population parameters or ecological processes (Conroy et al. 1995). Unfortunately, this crucial step in the modeling procedure is often neglected.

The SEPM of Liu, Dunning, and Pulliam (1995) assumes that reproductive success is lower for pairs located in 3–5-year-old stands than it is for pairs nesting in 1–2-year-old or mature pine stands. Ultimately, some local populations may represent reproductive sinks whereby local reproductive success is insufficient to compensate local mortality, and they depend on immigration from reproductive sources to persist. Source-sink metapopulation dynamics have been proposed to approximate the dynamics observed in Neotropical migrant species in the U.S. Midwest (Faaborg et al., this volume).

SEPMs are very useful to support long-term inferences on the effects of different types of landscape alterations. Spatially realistic metapopulation models (Liu, Dunning, and Pulliam 1995; Holthausen et al. 1995) are particularly promising because they can be applied to a variety of real landscapes, or modified to examine the dynamics of several different species of management concern in a given landscape. However, predictions based on SEPMs have often proved unreliable because they are sensitive to both model structure and parameter estimation (Conroy 1993; Conroy et al. 1995). Therefore, SEPMs should not be viewed as a substitute for good empirical studies on population parameters and ecological processes. Although their predictions should be considered with caution, SEPMs are an extremely useful tool for conservation biologists since they provide crucial guidelines to conduct adaptive management when empirical information is incomplete.

To date, metapopulation models have largely been applied to resident species or populations (Verboom et al. 1991; Wooton and Bell 1992; Liu, Dunning, and Pulliam 1995). Villard, Freemark, and Merriam (1992); and Villard, Merriam, and Maurer (1995) examined the applicability of these models to Neotropical migratory birds. Some authors (Blake 1991; Haila and Hanski 1993) cite circumstantial evidence suggesting that habitat fragmentation does not significantly impede the dispersal of organisms as vagile as migratory birds. Therefore, in most fragmented habitats, the situation of these species would be analogous to a patchy population with extensive levels of dispersal among patches. Villard, Merriam, and Maurer compared observed numbers of population turnovers to those expected under different simulated scenarios of settlement into fragments by returning migrants in the spring. Models simulating random dispersal among fragments greatly overestimated the observed number of population turnovers, suggesting that fragments were not all equally accessible to migrants, or equally suitable. Fragment area, abundance of the target species in the fragments during the previous breeding season, and the distance from patches occupied the previous year explained a significant proportion of the variance in fragment occupancy, while vegetation characteristics were minor factors. These results suggest that even migratory bird species may be characterized by local populations in which most surviving adults return, and among which dispersal is influenced by the configuration of fragments of suitable habitat.

Research Needed to Further Conservation

A limitation of SEPMs is that they require the delineation of cells of homogeneous habitat and therefore are applicable only to relatively simple (or simplified) landscapes. In real-world landscapes, two types of difficulties may arise: (1) there may be a large number of habitat types for the species of interest, each one being characterized by different demographic parameters (e.g., for reproductive success and survival); (2) there may be few distinct habitat types, but these habitats may subtly intergrade into one another, making it difficult to delineate patches. The latter example is relatively common in some industrial forest landscapes under management. Different sylvicultural treatments (e.g., selective cutting at different intensities) may be applied over different areas and be superimposed over past interventions. The development of edge-detection techniques (Fortin 1994; Milne et al. 1996) suggests that SEPMs will soon be applicable to such complex landscapes. However, in the case of animal species, the critical information is the functional response of species to spatial and temporal changes in habitat characteristics. Thus, habitat edges apparent on aerial photographs or satellite imagery may be meaningless to some species, while the demography of other species will indicate a strong response to the same apparent edges. Studies examining habitat-specific reproductive success, survivorship, or movement patterns are thus vital to calibrate future SEPMs.

The Geographic Scale

Spatial modeling has the potential to make important contributions to our understanding of the biogeographical component of avian diversity (Maurer 1994). Biogeographical studies of avian diversity must be based on an understanding of why individual species are distributed in a characteristic manner across geographic space. Clearly, the geographical distribution of abundance of each species is determined by environmental factors that are unique to that species. However, recent advances in spatial modeling provide a general framework from which we can begin to develop a more complete understanding regarding how avian diversity is determined at large geographic scales. The key to developing such an understanding is to first develop a clearer picture about how geographic ranges of individual species are structured and how this structure varies from species to species (Maurer 1994). Geographic ranges encompass all of the processes we have discussed up to this point. The effects of these processes as they vary across large spatial and long temporal scales determine how the abundance of a species varies across geographic space (figure 4.1). Spatial modeling at the geographic scale seeks to obtain macroscopic descriptors of spatial patterns in geographic ranges that reflect these effects.

Conservation Issues and Previous Research
Properties of Geographic Distributions of Abundance
Recent development of the ideas regarding the distribution and abundance of birds originally discussed by Grinnell and others has stimulated interest in spatial modeling of geographic ranges (Bock 1997). Birds have provided ideal subjects for study because their biology is relatively well known and large databases describing their geographical distribution and abundance are available (Maurer 1994). The existence of such databases raises the possibility that conservation efforts for birds might be integrated across geographic scales. Indeed, recent experience with Neotropical migratory birds suggests that such an integration is critical (Maurer and Villard 1996).

Geographical ranges, although they come in many shapes and sizes, have several important properties that seem to be quite general. With the exception of localized endemics, these patterns are nearly universal among taxa for which we have adequate data. First, most species have relatively small geographic ranges, with relatively few having large ranges. Second, species are not uniformly distributed within their range. Abundance is generally concentrated within centrally located populations (Brown, Stevens, and Kaufman 1996). At range boundaries, populations are highly variable (Curnutt, Pimm, and Maurer 1996). Third, although there is a large degree of spatial variability in abundance within the range of a species, that variability is spatially autocorrelated. The pattern of this autocorrelation is not simple: sites relatively close to one another in the range are positively correlated, abundances at medium distances are negatively correlated, and abundances at distant sites are positively correlated (Maurer 1994; Maurer and Villard 1994; Brown, Mehlman and Stevens 1995). This complicated pattern of spatial autocorrelation across a geographic range has obvious implications for the validity of standard statistical techniques used to analyze spatial patterns of abundance. We address this further below.

Gap Analysis

In an effort to quantify, predict, and manage for areas of biodiversity, Gap Analysis was developed and has been implemented or is in the planning stages in forty-eight of the fifty states (see the Gap Analysis home page on the World Wide Web). In developing a coverage for a particular state under Gap Analysis, vegetation types, land-use types, and predicted animal distributions are mapped on a statewide grid (Scott et al. 1993). Vegetation maps are prepared from satellite imagery such as LANDSAT. Animal distributions are predicted from modeling of habitat affinities using range maps and other distributional data. Gaps in the protection of vegetation types and corresponding animal biodiversity are then indicated by the layering of these databases with land ownership maps.

The Gap Analysis in Utah has been completed and was used to test the predictability of the model (Edwards et al. 1996). Habitat relations were predicted using state-wide data excluding eight national parks. The predicted animal distributions were then compared with actual distributions based on species lists from each park. Errors of both omission, percentage of species not included on the gap-predicted list but present in the park-generated list, and commission, percentage of species incorrectly included in the gap-predicted list, were used as indicators of the strength of the model. Error rates were highest for amphibians and reptiles and lowest for birds and mammals and ranged from 0 to 25 percent for omission and 4 to 33 percent for commission. Error rates decreased as park area increased. Over all, the model performed well and provides a good tool for management of biological diversity.

Two concerns may apply: (1) Because Gap Analyses are limited by artificial boundaries, i.e. state borders, predictability may be lower for habitat areas that fall mostly outside the state. This is an argument for pooling information and coordinating efforts at a regional level. (2) Care should be taken in interpreting regions of high biodiversity as areas of conservation priority. It may be that these represent ecotones and the high diversity is a result of the proximity of several different habitats. Animal and plant populations in the area may represent edge populations that are non-self-sustaining. If so, such biodiversity hotspots may actually be biogeographic sinks. Conservation plans that preserved such sinks might be less successful than plans based on preserving important source areas.

Spatially Explicit Models of Geographic Ranges

In order to more fully understand what controls the distribution of individual species across geographic space, it is necessary to develop explicit models that consider how abundance can be modeled as a function of geographic space. Currently, no theoretical models exist that make explicit quantitative predictions of how abundance should vary geographically. Initially, it was thought that a bivariate normal distribution could be used, but since there are often several peaks of abundance within a geographic range, this distribution does not seem to be applicable. Hence, the prediction of abundance across geographic space is currently a statistical enterprise. The question that must be asked from such a perspective is not what specific mechanisms are causing abundance to vary across space, but rather, which statistical model best describes this variation.

Static models of the geographic pattern of abundance have centered around so-called "response surface" models (Hengeveld 1990). These models assume that abundance can be expressed directly as a function of latitude and longitude. Latitude and longitude cannot be used directly in these calculations because distances based on these variables (degrees) are not equivalent to true, linear distances (km). Hence, it is always necessary to transform from degrees of latitude and longitude to some distance- or area-preserving coordinate system, such as Alber's projection (Maurer 1994). The simplest model describing abundance would be a multiple regression using transformed latitude and longitude as independent variables. The model should include polynomial terms (e.g., x^2, xy, y^2, etc.) since abundance is unlikely to vary linearly with changes in latitude and longitude. Such a model is unsatisfactory because, as discussed in the previous section, a definite pattern of spatial autocorrelation exists among samples across geographic space. Since conventional statistical techniques do not account for this autocorrelation, their use in analyzing geographic scale abundance data is severely limited.

Modeling geographic variation in abundance using spatial autocorrelation models is the obvious alternative to multiple linear regression (Maurer 1994; Smith 1994). The simplest of these models are moving averages and similar techniques that use distance-weighting techniques to develop a description of the abundance distribution. The basic problem of these techniques is that the researcher must decide upon an objective criterion to use in weighting abundances at different distances. Ordinary kriging is the least arbitrary of these techniques and, although computationally intensive, can be performed with some GIS or statistical programs. Kriging uses estimates of spatial autocorrelation based on the sample semivariance (sometimes referred to as the variogram) to estimate weights used to predict abundance across a geographic range (Maurer 1994).

Mapping abundance across a geographic range may be enhanced if abundance can be correlated with environmental variation (Maurer 1994; Smith 1994). A number of techniques are available. The general statistical model for such techniques assumes a "multiscale" structure to the variation in abundance across space (Smith 1994). The modeling procedure consists of two steps:

1. Large-scale variation in abundance is modeled by relating abundance (or incidence, i.e., presence/absence) with environmental variables using multiple regression (for abundances) or logistic regression (for incidences). We have found that when using abundance, it is best to perform a logarithmic transformation on the data since abundances follow a hollow curve (Brown, Mehlman, and Stevens 1995; Brown, Stevens, and Kaufman 1996).

2. Once the relationship between abundance or incidence and environmental data is estimated, the residuals from these relationships are assumed to be spatially autocorrelated. This autocorrelation represents spatial variation in abundance due to processes that cause abundance or incidence within a geographic region to be correlated. Processes such as dispersal can result in such regional correlations. Modeling the autocorrelation structure of residuals can be done using a variety of techniques including kriging and spatially explicit regression (Cressie 1993). Significant spatial autocorrelation in residuals from a model that relates abundance or incidence to habitat variables would suggest that there are processes other than habitat-related ones that determine a species' abundance pattern.

Spatial Patterns of Population Change

In many applications, analysis of abundance patterns is not as important as estimating changes in abundance. This is particularly true in monitoring programs where the goal is to identify whether a species has declined for a significant amount of time to warrant special concern. Initial efforts to study patterns of decline in neotropical migratory birds used a technique called "route regression" (Geissler and Noon 1981; Geissler and Sauer 1990). The authors used linear regression on log-transformed count data to estimate whether a species had increased (positive slope), declined (negative slope), or remained constant over time at a route. To estimate population change within geographic regions, slopes from different routes were weighted within geographic regions by using different weighting procedures. The philosophy of these techniques was to find estimates of average trends while ignoring possible spatial variation in those trends.

Recently, the possibility that a knowledge of geographic variation in population changes is an important part of understanding conservation problems of Neotropical migratory birds has stimulated interest in techniques to model spatial variation in population change (Villard and Maurer 1996). James, McCulloch, and Wiedenfeld (1996) used a nonlinear smoothing technique (LOWESS) to estimate temporal trends in populations of wood warblers within physiographic regions (defined as regions within which vegetation important to birds is relatively uniform). They found that population trends of species varied geographically, with some regions showing substantial declines, and others substantial increases. Villard and Maurer showed similar results by manipulating trend surfaces estimated using ordinary kriging, irrespective of predefined physiographic units.

Relatively little data exist on geographic variation in population stability. In the single study available to date, it was found that populations had lower abundance and were relatively more variable near the range boundary than in the center (Curnutt, Pimm, and Maurer 1996). Curnutt, Pimm, and Maurer estimated the mean and standard deviation of abundance over years on BBS census routes with over ten years of data for several species of grassland nesting passerines. By plotting the standard deviation against the mean, it is possible to infer on which plots abundance is relatively more variable (Maurer 1994). Curnutt et al. showed that for nearly every species they examined, low-abundance BBS routes were relatively more variable than high-abundance routes. Since the low-abundance routes are, on the average, toward the periphery of the range, Curnutt et al. were able to conclude that peripheral populations were more variable. The conservation implications of these results are important. Although it has long been known that species are not equally abundant across geographic space, Curnutt et al.'s results show that all populations may not be equally valuable in terms of conservation. Clearly, peripheral populations that have low abundance and are subjected to wide fluctuations in abundance are less likely to persist over ecological time. Hence, conservation efforts to conserve particular species will be more effective if they focus on central, high-abundance regions of the geographic range. Recently, Lomolino and Channell (1995) showed that many species of rare mammals are concentrated in populations close to the boundaries of their former geographic range. This surprising result, in concert with the observation that peripheral populations are less stable than central populations, suggests that species may become endangered when demographically important central populations have declined or disappeared.

A few attempts have been made to develop analytical models of the dynamics of invasion or spread of an organism across a geographic region that it did not originally occupy. Skellam's (1951) pioneering paper approximated the spread of a species across a landscape as a simple diffusion process in a homogeneous medium. His model predicted that the square root of the area covered by the invading population should increase linearly with time. Although this prediction is approximately consistent with some data sets, it makes relatively simplistic assumptions regarding population processes underlying the spread. Recent workers have developed detailed life-history–based models that provide a more mechanistic approach to describing the spread of invading populations (Van den Bosch, Hengeveld, and Metz 1992; Veit and Lewis 1996). These models, although similar to Skellam's model in assuming a homogeneous diffusion medium, allow for a sufficient amount of demographic variability to provide closer descriptions of the spread of species like House Finches (*Carpodacus mexicanus*). The analysis of range expansion can yield interesting insight into the potential extent and frequency of demographic interactions among widely separated populations occurring within a species' range (e.g., long-term, large-scale movements indicated in figure 4.1).

Research Needs at the Geographic Scale

The question of where to focus conservation efforts within the geographic range of a particular species is a relatively novel one that has major implications. For a particular species, or group of species, it is necessary to identify important environmental variables that can be used to characterize those regions of the geographic range where population stability reaches a peak. Models that make such associations will have two important applications. First, they will allow prioritization of habitat needs for a species. For species of concern, identification of areas that have appropriate combinations of habitat characteristics may be critical in proactive responses to human impacts. Second, with reliable information on current associations of species with environmental variation, it may be possible to project, or predict, the future distribution of species in response to projected environmental changes (e.g., climate change induced by increasing levels of CO^2 in the earth's atmosphere). In order to be useful for either function, it is necessary to develop additional techniques that can be used to validate model predictions.

Life-history–based models of geographic range expansion offer a promising avenue for further exploration. To date, these models have focused on describing a continuous wave of population expansion. There is nothing in these models that explicitly halts the advancing wave of diffusion across geographic space. Eventually, however, the invading population front ceases to move, and a relatively stable geographic range boundary results. This phenomenon has not yet been incorporated into models of geographic range dynamics. Existing models assume that diffusion of individual birds occurs isotropically and therefore ignores the possibility that resistance to dispersal of individuals is greater the closer a population is to the geographic range edge. Furthermore, because geographic space is nonuniform (e.g., resources available to birds are affected by topography), diffusion is very likely anisotropic (heterogeneous in all directions). These factors might be profitably combined with life-history–based diffusion models to provide more accurate descriptions of geographic range dynamics.

Models developed to describe and predict population trends have focused in the past on examining the statistical reliability of the models in the face of known biases, such as variation among observers (Sauer et al. 1996). Although accounting for biases in existing data sets is critical to accurate description of past population trends, it is not clear that such efforts alone will improve our ability to predict population change into the future. As discussed in the previous paragraph, a new generation of models describing geographic variation in population dynamics might be brought to bear on the problem of population trends. Such models will be useful if they provide a bridge between lower-scale demographic processes and observed patterns of distribution and abundance of species.

CHAPTER 5

Using Geographic Information Systems and Spatially Explicit Population Models for Avian Conservation: A Case Study

Martin G. Raphael, Kevin S. McKelvey, Beth M. Galleher

Conservation Issue

Conservation planning often requires consideration of wide-ranging avian species occurring on large landscapes. In many cases, the geographic and temporal scale necessary to address critical issues such as population viability precludes field research and empirical study to answer conservation questions. Modeling, such as described by Villard, Schmidt, and Maurer (this volume), can be done at large scales and thus has an important role in helping inform management decisions.

Spatially explicit population models can be used to address a variety of questions in avian population viability assessment and the response of birds to perturbation of habitat (Dunning et al. 1995; Holthausen et al. 1995; Turner et al. 1995; Villard, Schmidt, and Maurer, this volume). In this chapter, we present a specific example of how such modeling has been used in conservation planning in the Pacific Northwest. The Northern Spotted Owl (*Strix occidentalis caurina*), a threatened species associated with late-successional and old-growth forests of the Pacific Northwest (Thomas et al. 1990; Forsman et al. 1996), has been a focus of controversy over the management of the species' remaining habitat (FEMAT 1993). As part of an analysis of management plans for forests of the Pacific Northwest, an individual-based and spatially explicit population model was used to evaluate the projected response of the Northern Spotted Owl population throughout its range on federal lands of Washington, Oregon, and California to a series of land management alternatives, each having different amounts and spatial patterns of forest reserved from timber cutting (Raphael et al. 1994). In this example of the use of spatially explicit population models, we illustrate how the population of owls might respond to changes in the amount and distribution of habitat as estimated from allocations of lands as either available for cutting or reserved under each of three land management scenarios compared to a control scenario where habitat was kept constant. We estimated both loss of habitat due to timber harvest and growth of habitat from maturation of currently immature forest.

Methods for Developing a Habitat Map

A geographic information system was used to prepare a habitat map and other spatial data for input into the owl population model. All GIS-based data for the model were drawn from a database created by the Forest Ecosystem Management Assessment Team (FEMAT 1993); other spatial data were generated or recoded from these data. The habitat map was originally developed from interpretations by USDA Forest Service and USDOI Bureau of Land Management biologists. These maps divided federal lands into areas of suitable and unsuitable owl habitat. Suitable habitat consisted of forest types used by spotted owls for nesting, roosting, and foraging (NRF). We modified these maps by adding an elevation cutoff to eliminate forest above the range of successful reproduction by spotted owls. We set the elevation cutoffs to 4,500 feet for the Olympic peninsula, 5,500 feet for the Washington Cascades, and 6,000 feet for Oregon and Northern California. The map was originally digitized in a vector (line) format but was generalized into a raster (grid) format for our analysis. The minimum resolution of the gridded maps (cell size) was 16 ha (400 m^2). Each cell on the resulting grid map was coded as suitable if half or more of the underlying map was suitable, and as unsuitable otherwise.

To project future habitat conditions, we developed a generalized harvest simulation in GIS using the macro programming capabilities of ARC/INFO (provided by Environmental Systems Research Institute, Redlands, California). A series of operations was carried out on maps of suitable habitat to simulate cutting of habitat at ten-year intervals. For each ten-year interval, we calculated the amount of habitat available for harvest and the amount of habitat remaining after a simulated harvest was performed. This was added to the amount of habitat in reserves, the total amount of which remained fixed over time. At successive time-steps, the amount of habitat available from the previous time step was used as input. The available habitat at the new time-step was then harvested and total remaining habitat determined.

To simulate harvest under each scenario, we first removed reserved lands (national parks, wilderness areas, late-successional reserves, riparian reserves) from consideration. We then calculated harvest on the remaining lands (matrix and adaptive management areas) under each of three land management scenarios. These scenarios were modeled after alternatives being considered in a *Final Supplemental Environmental Impact Statement* for management of federal lands in the Pacific Northwest (see USDA and USDOI 1994, for detailed descriptions of alternatives). Alternative 1 called for retention of all late-successional habitat, placing all such land in reserved status. In addition, this alternative called for retention of wide (generally 100 m) buffers along all permanent and intermittent streams. Under this alternative, 89 percent of federally administered land was in reserved status. Alternative 7 called for land allocations from the *Final Draft Recovery Plan for the Northern Spotted Owl* (USDOI 1992) and had a system of reserves similar to that recommended by the Interagency Scientific Committee (Thomas et al. 1990). Riparian buffers were smaller than those in other alternatives and were based on provisions of existing land management plans. Under this alternative, which allowed the greatest amount of harvest, 65 percent of federal land was reserved. Alternative 9 called for intermediate sizes of reserves and riparian buffers. This alternative was ultimately selected as the preferred alternative and has since been implemented as the Northwest Forest Plan. Under this alternative, 77 percent of the federal land base was reserved.

For the harvest simulation under each scenario, we calculated the amount of NRF habitat that would be cut per year in unreserved lands (matrix and adaptive management area) within each administrative unit (National Forest or Bureau of Land Management District) based on projections by Johnson, Crim, and Barber (1993). Because we did not have spatially explicit projections of harvest units by decade, we distributed harvest evenly across the unreserved lands. We did not have maps of all streams, so the riparian buffer system could not be explicitly mapped. Instead, we applied a riparian reserve factor to each land unit. This factor was developed by the FEMAT based on a set of sample watersheds. The factor was essentially an estimate of the percentage of land area that would be occupied by riparian buffers given the average density of streams in that area. We used the factor to calculate a limit to harvest in unreserved lands such that total acres of harvest would not exceed the total amount of land that would have been reserved within riparian buffers.

We implemented a habitat growth model by simulating the development of spotted owl habitat on previously harvested or burned forest. We assumed a constant rate of growth on all federal lands that are capable of regrowing trees. Because existing maps of habitat were coded as either suitable NRF or not (i.e., they did not contain information on existing seral stages, forest type, or productivity classes), we had to make some simplifying assumptions concerning the amount of habitat that would regrow within each 16-hectare cell (see Raphael et al. 1994 for a complete discussion of the development of the habitat map and other inputs used in this model). Our assumptions were guided by projections of habitat regrowth as provided in the spotted owl *Final Environmental Impact Statement* (USDA 1992,60) in which habitat was projected to increase by 83 percent on average after one hundred years within the owl's range. We created a growth constant of 0.1328 ha/year (or 0.083 percent of 16 ha/year) that would be applied to all federal lands not currently in congressionally withdrawn status (national park or wilderness area).

At the end of each ten-year time step, habitat remaining after growth and harvest was overlain with a map of hexagons simulating owl home ranges, and the amount of habitat within each hexagon was computed and output to the owl demography model (Raphael et al. 1994). We ran the GIS growth and harvest simulation for fifty years, thus producing maps of projected habitat at ten, twenty, thirty, forty, and fifty years from current conditions for each of the three alternative scenarios. For the control scenario, habitat was assumed to be constant over the entire period.

Methods to Simulate Owl Populations

To link the population of Northern Spotted Owls to the actual landscape as portrayed by our GIS maps of habitat, relationships were developed between the amount and distribution of habitat and the survival and reproductive performance of the owl. The model (OWL, Version 2.0a; McKelvey, Noon, and Lamberson 1993) allows for input of parameter estimates relating these vital rates to six classes of habitat. Our analysis was meant to compare the relative effects of the different harvest rates of these plans, not to predict the precise number of owls over time. Because there are such great uncertainties in such comparisons, it was advantageous to compare results under varying sets of assumptions. For this analysis, we developed three "rule sets," that is, three different sets of vital rates based on different assumptions about vital rates in relation to proportion of habitat within each hexagonal cell. The rule sets were derived from earlier work completed by McKelvey and others in cooperation with members of the Northern Spotted Owl Recovery Team (McKelvey, Noon, and Lamberson 1993). Rule Set 2 was designed to fit the demographic data summarized by Burnham, Anderson, and White (1996). Under this rule set, adult survival varied from 0.70 when NRF habitat occupied 20 percent or less in a cell to 0.95 when NRF habitat occupied greater than 40 percent of a cell. Juvenile survival was set as 0.29 for all habitat classes. Additional parameters set the likelihoods of settling on a territory or moving to a new territory in relation to percent habitat. McKelvey, Noon, and Lamberson, and Holthausen et al. (1995) describe and explain the full list of parameters used in running the model. Compared with Rule Set 2, Rule Set 1 had more pessimistic vital rates (required greater proportions of habitat for given rates of survival and fecundity); Rule Set 3 was more optimistic (required smaller proportions of habitat for those vital rates). See Raphael et al. (1994) for details on the parameters used to run this implementation of the simulation model.

Habitat Change, Simulated Owl Populations, and Occupancy Rate

Net change in amount of habitat varied under scenarios based on Alternatives 1, 7, and 9 (figure 5.1). Total increase in habitat was greatest under the Alternative 1 scenario and was nearly as great under the Alternative 9 scenario. The projected amount of habitat under the Alternative 7 scenario was only slightly greater than that of the control scenario and was nearly constant over the fifty-year simulation. Under this scenario, rates of harvest and growth of NRF habitat were nearly equal.

Mean simulated population size of Northern Spotted Owls showed different trends among the four scenarios and was highly dependent on the rule set used (figure 5.2). Under Rule Set 1, simulated populations declined over the entire one-hundred-year simulation run. Under Rule Set 2, simulated populations were essentially stable over the entire run. Mean population size was highest under the Alternative 1 scenario, nearly identical under the control and Alternative 9 scenarios, and smaller under the Alternative 7 scenario. Under Rule Set 2, the simulated owl population increased with the Alternative 1 scenario, declined with the Alternative 7 scenario, and was essentially stable with the Alternative 9 and control scenarios. The simulated population stabilized at a larger mean population under Alternative 9 than under the control (figure 5.2). Under Rule Set 3, simulated populations were stable (after initial increases) with Alternative 7 and the control, and increased with Alternative 1 and 9 scenarios.

Maps showing the spatial distribution of various levels of mean occupancy reflect the patterns described above (illustrated for Rule Set 2, figure 5.3). The Alternative 7 scenario resulted in few areas of high occupancy (>80 percent), and those areas that had cells with high occupancy were smaller in extent than those of other scenarios (figure 5.3). Compared to the other scenarios, the Alternative 1 scenario resulted in the largest areas of high occupancy. The Alternative 9 scenario resulted in large areas of high occupancy in the central Cascades of Oregon and Washington and in the Oregon Klamath and Olympic Peninsula. The control scenario resulted in a similar pattern, except that high-occupancy areas were somewhat smaller in the Washington provinces, similar in the Cascades of Oregon, and much smaller in Oregon Klamath and California provinces.

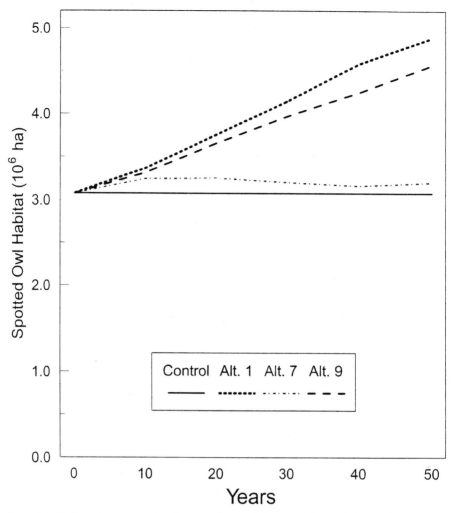

Figure 5.1. Projected amount of nesting, roosting, and foraging habitat of the Northern Spotted Owl over a fifty-year period under each of four land management scenarios. Control: no harvest or regrowth. Alternative 1: retain all current late-successional forest and regrow younger forest. Alternative 7: harvest in matrix following land allocations of the Final Draft Spotted Owl Recovery Plan (USDOI 1992) and regrow younger forest. Alternative 9: harvest in matrix following land allocations of the Northwest Forest Plan (FEMAT 1993) and regrow younger forest.

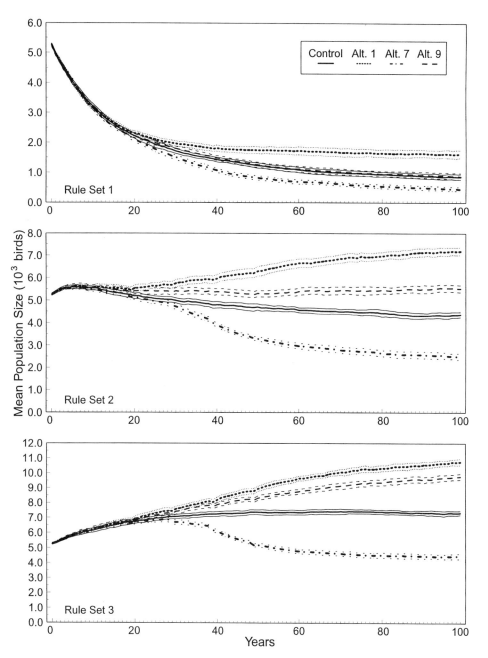

Figure 5.2. Simulated population trends (mean total population size +/− 95% Confidence Interval) of the Northern Spotted Owl with levels of harvest and growth of suitable owl habitat projected under alternative scenarios 1, 7, and 9 compared with a control scenario where habitat is held constant. Simulations were conducted using three sets of vital rates (Rule Sets, see text and Raphael et al. 1994 for details). Harvest and growth of habitat were modeled for the first fifty years for alternatives 1, 7, and 9.

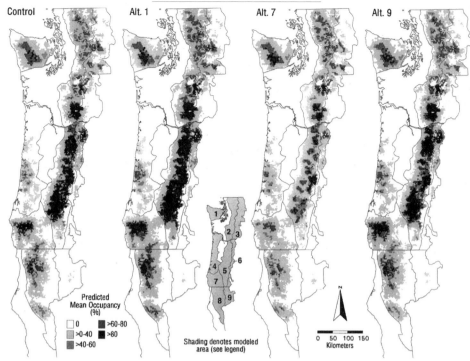

Figure 5.3. Map of mean occupancy by pairs of Northern Spotted Owls over ten replications of a 100-year simulation analysis of population dynamics under four scenarios of harvest and growth of suitable owl habitat. Control: no harvest or regrowth. Alternative 1: retain all current late-succes- sional forest and regrow younger forest. Alternative 7: harvest in matrix following land allocations of the Final Draft Spotted Owl Recovery Plan and regrow younger forest. Alternative 9: harvest in matrix following land allocations of the Northwest Forest Plan and regrow younger forest. Simulations using Rule Set 2 (see text) are illustrated; results using Rule Sets 1 and 3 are available from the senior author on request. Maps depict the range of the Northern Spotted Owl, which oc- curs along the northwest boundary of the United States: western Washington, western Oregon, and northwestern California. Numbers on the inset map denote physiographic provinces: 1 = Olympic Peninsula, 2 = Western Washington Cascades, 3 = Eastern Washington Cascades, 4 = Oregon Coast Range, 5 = Western Oregon Cascades, 6 = Eastern Oregon Cascades, 7 = Oregon Klamath, 8 = California Klamath, 9 = California Cascades.

None of the simulations or scenarios resulted in any areas of high occupancy in the Oregon Coast Ranges.

Features of Research That Improved Conservation

These simulations indicated that the amounts and distributions of habitat created under growth and harvest models will likely result in a more stable and well-distributed population of the Northern Spotted Owl under Alternative 1 and 9 scenarios than under the Alternative 7 and control scenarios. We do not contend that these results are sufficient as a viability analysis to predict the true likelihood that the owl will persist under the various alternatives. But the results of these simulations do indicate that under the conditions of the simulations there is little evidence to suggest the owl will go extinct under Alternative 1 and 9 scenarios using parameters fitted to known demographic rates (Rule Set 2). Under the Alternative 7 scenario the simulated population would decline to about half of its starting size, and a much smaller proportion of the range would be occupied at high rates, suggesting a much higher risk to population persistence. These results were conveyed to the decision makers charged with selecting a preferred alternative as part of the process leading to the selection of a final forest plan. The conservation of the Northern Spotted Owl was one of the primary objectives of the forest plan, and these results became part of the information that helped the decision makers with their selection of the final plan. The results also have been useful in displaying areas of population strength, such as the Cascade Range in Oregon (source habitat), as well as areas of weakness, such as the Coast Ranges of Oregon and Washington (sink areas). This knowledge, in turn, has helped in subsequent efforts to define the conservation value of additional habitat that might be retained on nonfederal land (Holthausen et al. 1995).

Research Needed to Further Conservation

Projecting both owl populations and forest conditions was valuable for visualizing the impacts of various management alternatives, but there were many weaknesses that need to be addressed. Better methods of projecting habitat change over time are needed. For example, in our growth and harvest model, we used a simplistic assumption that growth rate is constant over the entire range; we currently do not have spatially explicit stand-age data to better model the expected growth of habitat within each cell or province. This information is needed, and we recommend an effort to collect this data. Our habitat models were limited to federal lands. Development of vegetation maps that cover all ownerships would allow more realistic simulations. Dispersal of animals within and between patches of habitat is an important feature of spatial models, and more work to understand the dispersal behavior of individuals by sex and age class is needed. Further work is also needed to explore the relationship between vital rates and variation in habitat quality. Even though the Northern Spotted Owl has received more study than most other birds, our rule sets were based on limited empirical data. Spatial models are only as good as the data used to parameterize them; further refinements in parameters will improve model performance.

Making Research Effective for Conservation

The model demonstrated here has proved a very useful tool for evaluating the impact of future management plans. Although this application is the most extensive to date, the same model was used by the Bureau of Land Management to evaluate alternatives on its lands in Oregon, by the State of Oregon on the Elliot State Forest, by the U.S. Fish and Wildlife Service to evaluate potential contributions by private owners on the Olympic Peninsula, and by Weyerhaeuser Corporation in developing a habitat conservation plan for its lands in central Oregon. In all cases the model was useful because it allowed the landscape to be modeled in enough detail to separate various land management alternatives.

Because there are large uncertainties in model parameterization, the model was most useful for comparing between plans, rather than estimating probabilities of extinction associated with any given plan. For comparing plans, the model simply provides a neutral evaluation. All parties need to agree that the presentation of the landscape and the biological rules governing the organism are reasonable. The model then provides the manifestation of these understandings across time and space. If the model is used as a more formal population viability analysis, then it is important not only that the rules be reasonable, but also that they be right.

Given this approach, it is important that biologists understand the rules governing model output. In this regard, the individual-organism approach is valuable. The behavior of individual animals is governed by logical rules, rather than being abstracted into population-level equations. For instance, a rule might be that a dispersing female wanders randomly until she finds a territorial male and then pairs. A biologist will probably be able to offer valuable comments about the biological reality of this rule, comments that might be lost if we were to present these understandings as a system of diffusion-reaction equations. Because every rule needs general agreement, it is also important to be parsimonious concerning the number of rules.

Acknowledgments

We appreciate the support of the Interagency SEIS Team in providing technical information and partial funding, particularly Chris Hamilton, Cay Ogden, and Ken Denton for providing critical harvest estimates and other information. Additional funding was provided by the Pacific Northwest Research Station. Additional technical assistance was provided by Andy Wilson, Paul Newman, and Bob Varner. We also thank Richard Holthausen for his advice, counsel, and discussion of this work. Finally, we thank Dan Johnson and Janet Jones for administrative assistance.

CHAPTER 6

Using Landbirds As an Indicator Species Group

Richard L. Hutto

Conservation Issues and Previous Research

The broad goal of conservation biology is to maintain "biodiversity," which has been defined as the diversity of life at all levels of biological organization, from genes to landscapes (Office of Technology Assessment 1987). A variety of laws, beginning with the Organic Administration Act of 1897 through the Multiple-Use Sustained-Yield Act of 1960 to the National Forest Management Act of 1976 [16 USC 1604(g)(3)(B)], either encourage or mandate that the U.S. Forest Service provide for the diversity of plant and animal communities. The Code of Federal Regulations [36 CFR 219.19(a)(1 and 6)] specifies that this be accomplished by monitoring vertebrate indicator species as a means to ensure the maintenance of populations of all native vertebrate species. Numerous laws require essentially the same thing of the U.S. Fish and Wildlife Service, the Bureau of Land Management, and many state agencies.

Because of the operational difficulties of implementing the aforementioned legislation, agencies typically attempt to accomplish this task by assuring the maintenance of a small number of "indicator" species that, in turn, supposedly assure the maintenance of the complete range of vertebrate species (Severinghaus 1981). Agencies such as the U.S. Forest Service also typically work at a regional (or smaller) level to implement such legislation (Morrison and Marcot 1995).

Thus, I wish to focus attention here toward avian conservation efforts at the within-region level and away from avian conservation programs that are national in scope (e.g., BBS, CBC, BBIRD, MAPS; DeSante and Rosenberg, this volume), even though the latter may provide data that influence management decisions at the local level.

Which species should serve as indicators for implementation of the National Forest Management Act? The earliest suggestion (Graul, Torres, and Denney 1976; Graul and Miller 1984) was to use a series of the more stenotopic species. If we choose the most stenotopic species as indicators, however, it is unlikely that the maintenance of viable populations of all species can be assured throughout their historic ranges (as required by the National Forest Management Act of 1976, for example) unless we use enough of them to cover the entire range of ecological conditions.

Landres (1983); DeGraaf, Tilghman, and Anderson (1985); Roberts and O'Neil (1985); Fry et al. (1986); and Roberts (1987) subsequently suggested using representative species from different ecological guilds (guild indicators), but empirical data do not support the idea that population trends of species within a guild mirror one another at all closely (Mannan, Morrison, and Meslow 1984; Szaro 1986; Block, Brennan, and Gutiérrez 1987; Bayer and Porter 1988; Reader 1988). Consequently, the suggestion to use selected species from each of a variety of guilds has been met with criticism (Hutto, Reel, and Landres 1987; Landres, Verner, and Thomas 1988; Morrison, Marcot, and Mannan 1992).

Verner (1984) suggested using management-guild indicators, which he defined as groups of species that are suspected to respond in a similar way to changes in the environment. The latter, whole-guild approach avoids the problems inherent in the guild-indicator approach because it is designed to identify species that would be expected to share either the negative or positive effects of land management activity because they share a particular forest zone. Even here, however, the problem is that populations are affected by numerous factors that operate at different times of the year (Sherry and Holmes 1995). Thus, it is quite possible for the declining populations of one member of a management guild to be hidden by a general increase in the populations of others. In fact, using population trend data from the Breeding Bird Survey, Paige (1990) showed that there is really no group of species whose population trends mirror one another and that, therefore, would serve as a good group for a combined-species analysis.

Many are now coming to believe that we need some sort of ecosystem-level approach whereby we maintain and monitor the full range of "ecosystems." Some (Franklin 1993b, 1994) claim that this is the only way some species will be conserved because we cannot monitor all of them. Others argue that an ecosystem approach is destined to fail because we cannot even define an ecosystem (Orians 1993)—we have a hard enough time trying to define what a species is for conservation purposes (Rojas 1992). Still other discussions revolve around the recommended use of compositional indices (e.g., Anderson 1991; Götmark 1992; Karr 1987, 1991; Angermeier and Karr 1994; Kremen, Merenlender, and Murphy 1994).

Given the difficulties of working with indicators based on the ecosystem level of biological organization, sentiment seems to be converging toward something like that expressed by Noss (1990), who suggests using a hierarchical approach that includes monitoring compositional, structural, and functional elements at a variety of spatial scales. This plays on earlier ideas (expressed by Franklin 1988) that a preoccupation with compositional diversity has come at a cost in terms of awareness of structural and functional diversity. Neither Noss nor Franklin recommends creating a composite index of biological integrity; rather, they recommend monitoring a variety of parameters across combinations of elements and levels. Thus, at the species level of biological organization, we will need to know what is present (composition) and something about the demographics associated with those species (function), as advocated by DeSante and Rosenberg (this volume). While species are not likely to be the only indicators of ecosystem health, it is likely that we will continue to use them as indicators at that particular level of biological organization, which brings us right back to the question of which species to use.

The most recent recommendations include those of Kremen (1992), who suggests using ordination techniques to identify groups of species that might be best sets for monitoring purposes; Mills, Soulé, and Doak (1993), who suggest using species that come closest to being "keystone" elements; and Noss (1990), who suggests five types of species that should be included as indicators: (1) ecological indicators—species that speak for others, (2) keystones—pivotal species on which many others depend, (3) umbrellas—species with large area requirements, (4) flagships—species that serve as rallying points for conservation efforts, and (5) vulnerables—species most prone to extinction in human-dominated landscapes.

Features of Previous Research
That Improved Conservation

The indicator approach has clearly stimulated a lot of valuable discourse on alternative methods of indicator species selection, and while the methods of species selection were being debated, we have also learned a lot about the specific needs of selected indicator species. Specifically, we have used knowledge from field studies to assess the probable effects of alternative land-use practices on indicator species. These assessments have taken one of three approaches. One has been to build models that can predict the suitability of a patch of land for a particular species—e.g., Habitat Suitability Index (HSI) models (Fish and Wildlife Service 1981); Habitat Capability (HC) models (Hurley, Salwasser, and Shimamoto 1982); and Pattern Recognition (PATREC) models (Williams, Russell, and Seitz 1978). Another approach has been to build models that might predict effects on groups of species—e.g., Integrated Habitat Inventory and Classification System (Bureau of Land Management 1982); Life Form System (Thomas 1979); and various guild models (Severinghaus 1981; Verner 1984). The third approach has been to use habitat-analysis models—e.g., Wildlife and Fish Habitat Relationships program (Nelson and Salwasser 1982) or HSI-based Habitat Evaluation Procedures (Schamberger and Farmer 1978; Fish and Wildlife Service 1980). All three modeling approaches generally include information from a variety of spatial scales and have, at the very least, opened our eyes to the complexity of determining what constitutes suitable habitat for any given species.

Research Needed to Further Conservation

If we are going to retain the use of indicator species in conservation efforts, we must recognize that, from a purely theoretical standpoint, the indicator species approach to maintaining populations of all vertebrate species cannot be expected to work well (Hutto, Reel, and Landres 1987; Landres, Verner, and Thomas 1988). Specifically, because no two species occupy the same niche, the maintenance of several indicators cannot be expected to assure the maintenance of all other species, despite arguments to the contrary (e.g., Tracy and Brussard 1994). There is little reason to expect that a small group of species will serve as much more than a crude "coarse filter." Evidence from the Northern Spotted Owl (*Strix occidentalis*) scenario shows that current conservation plans do not come close to meeting needs of fish, the Marbled Murrelet (*Brachyramphus marmoratus*), and other species (Franklin 1994).

From a practical standpoint, the indicator species approach has not worked very well either. There are at least four reasons for the practical failure of this approach to conserve vertebrates in general, and birds in particular. First, because we cannot monitor all species, we spend excessive amounts of time trying to decide which species to monitor (Thibodeau 1983), and when all is said and done, the majority of indicator species are still relatively restricted to a combination of Threatened and Endangered species, and those taken for food, sport, or hides. The transition from an ecologically narrow "game production" mentality to truly broad-based conservation biology has been slow at best. In addition, we are destined to keep adding species to the list of difficult-to-monitor "indicators" because most agencies are required to include (the ever increasing number of)

Threatened and Endangered species. We cannot develop and maintain regional monitoring programs for an ever increasing number of rare species (Franklin 1993b).

Second, the only way we might expect a subset of species to represent the needs of all others is for the subset to subsume the ecological conditions of all others. Unfortunately, the indicator lists are almost certainly too short and too ecologically narrow to accomplish such a task. Some forests in the USFS (U.S. Forest Service) Northern Region, for example, have as few as five "Management Indicator Species," and no forest includes more than twenty-two on its list (table 6.1). Moreover, most indicators are traditionally managed game species and fur bearers, which, coupled with the small number, brings the efficacy of such indicator groups into question.

Third, the cost required to monitor traditional indicator species has been prohibitive because of the techniques needed to monitor rare species. Consequently, there is virtually no monitoring of either population trends or land-use effects on selected indicator species, even though such monitoring has been legally mandated for more than twenty years.

Lastly, despite what many view as an enormous success story associated with the few indicator species that *are* monitored (e.g., elk), numerous vertebrate species, including fish (Moyle and Williams 1990; Frissell 1993), amphibians (Baringa 1990; Blaustein and Wake 1990; Phillips 1990), and migratory songbirds (Terborgh 1989; Askins, Lynch, and Greenberg 1990; Robbins, Sauer, and Peterjohn 1993), are apparently falling through the cracks.

These limitations suggest that wildlife biologists in agencies such as the Forest Service, Fish and Wildlife Service, and Bureau of Land Management may need to change the approach they use to meet their legal mandates to maintain wildlife populations. While the agencies themselves are in the midst of changing their operational emphases away from maximizing the production of certain commodities toward both sustainable commodity production and the maintenance of ecological systems, this is an ideal time to either abandon (e.g., Morrison and Marcot 1995) or modify the current indicator species approach.

Proposed Modification of the Current Indicator Approach

What kind of change might serve to improve the existing indicator species approach? I argue here for inclusion of one or more indicator species "survey groups" as part of any comprehensive indicator scheme. I define an indicator species survey group as any group containing a large number of species that can be monitored simultaneously through a single survey method. Sparrow et al. (1994) provide arguments why butterflies might make a good indicator group in this sense. Specifically, I recommend broadening the list of desired indicator species to include most landbird species because most can be detected using a simple point-count survey methodology. In fact, I would suggest that there is no

Table 6.1. Numbers of big game, furbearer, endangered bird and mammal, non-endangered bird, amphibian, and fish species used as indicator species for management of U.S. National Forests (Northern Region)

Species Group	Forest												
	BE	BI	IP	CL	CU	DE	FL	GA	HE	KO	LC	LO	NE
Big Game Mammals	1	1	3	2	4	4	3	1	3	0	7	1	3
Furbearers	1	1	1	1	0	1	1	0	1	0	4	0	2
Endangered Mammals and Birds	4	4	3	3	5	2	4	2	4	4	3	4	4
Non-endangered Birds	3	1	2	2	11	3	2	1	3	3	4	2	2
Amphibians	0	0	0	0	0	0	0	0	0	0	0	1	0
Fish	2	1	0	0	2	1	2	1	1	0	3	1	2
Total Number of Indicator Species	11	8	9	8	22	11	12	5	12	7	21	9	13

Big Game Mammals are Mountain Goat (*Oreamnos americanus*), Bighorn Sheep (*Ovis canadensis*), Moose (*Alces alces*), Elk (*Cervus elaphus*), White-tailed Deer (*Odocoileus virginianus*), Mule Deer (*Odocoileus hemionus*), Black Bear (*Ursus americana*), and Mountain Lion (*Felis concolor*); Furbearers are Lynx (*Lynx canadensis*), Bobcat (*Lynx rufus*), Wolverine (*Gulo gulo*), Fisher (*Martes pennanti*), Pine Martin (*Martes americana*), and Beaver (*Castor canadensis*); Endangered Species are Caribou (*Rangifer tarandus*), Wolf (*Canis lupus*), Grizzly Bear (*Ursus arctos*), Black-footed Ferret (*Mustela nigripes*), Bald Eagle (*Haliaeetus leucocephalus*), Whooping Crane (*Grus americana*), and Peregrine Falcon (*Falco peregrinus*); Non-endangered Birds are Trumpeter Swan (*Cygnus buccinator*), Northern Goshawk (*Accipiter gentilis*), Golden Eagle (*Aquila chrysaetos*), Prairie Falcon (*Falco mexicanus*), Blue Grouse (*Dendragapus obscurus*), Ruffed Grouse (*Bonasa umbellus*), Sage Grouse (*Centrocercus urophasianus*), Greater Prairie Chicken (*Tympanuchus cupido*), Sharp-tailed Grouse (*Tympanuchus phasianellus*), Barred Owl (*Strix varia*), Hairy Woodpecker (*Picoides villosus*), Three-toed Woodpecker (*Picoides tridactylus*), Pileated Woodpecker (*Dryocopus pileatus*), Cassin's Kingbird (*Tyrannus vociferans*), Hermit Thrush (*Catharus guttatus*), Ovenbird (*Seiurus aurocapillus*), Yellow Warbler (*Dendroica petechia*), Spotted Towhee (*Pipilo erythrophthalmus*), Brewer's Sparrow (*Spizella breweri*), Lark Sparrow (*Chondestes grammacus*), and Northern Oriole (*Icterus galbula*); Amphibian is Tailed Frog (*Ascaphus truei*); Fish are Arctic Grayling (*Thymallus arcticus*), Brook Trout (*Salvelinus fontinalis*), Bull Trout (*Salvelinus confluentus*), Cutthroat Trout (*Oncorhychus clarki*), Rainbow Trout (*Oncorhychus mykiss*), Chinook Salmon (*Oncorhychus tshawytscha*), and Largemouth Bass (*Micropterus salmoides*). Forest Abbreviations are: BE=Beaverhead, BI=Bitterroot, IP=Idaho Panhandle, CL=Clearwater, CU=Custer, DE=Deerlodge, FL=Flathead, GA=Gallatin, HE=Helena, KO=Kootenai, LC=Lewis and Clark, LO=Lolo, and NE=Nez Perce.

better tool than a landbird monitoring program to enhance the effectiveness of wildlife conservation efforts. Why?

1. Landbirds are not only the most visible of vertebrate species, they also advertise their presence and identity through vocalizations. Thus, systematically collected field data are much easier and less expensive to gather for landbirds than for traditionally managed species that require trapping, radio tagging, locating, and so forth.

2. Because patterns of occurrence in the field are easily uncovered, the foundation of field data on which habitat suitability (HSI) models are built is potentially much stronger for landbirds than for most of the existing management indicator species.

3. Using a single survey method, one can collect data on nearly two hundred bird species simultaneously. Many species will not be monitored well, but having to manage for the maintenance of those that can be monitored will probably bring us much closer to maintaining populations of all vertebrates than would the still prevalent approach of managing entirely on the basis of a select few indicator (mostly game) species. This is especially true if we combine landbird monitoring with continued management for the traditional indicator species.

4. Having to manage for the maintenance of many landbird species will force movement toward management at broader spatial scales. This is because the indicator species list will now be large enough and ecologically broad enough to reveal some species that will benefit from, and others that will be harmed by, any proposed land-use activity. This would appear to lead managers into a no-win situation because any proposed land-use alternative will hurt *something*, but the way out of this apparent dilemma is to expand one's focus beyond the immediate project area. In fact, realizing that local populations of some species will invariably be harmed by any proposed land-use action forces us to expand our perspective toward broader landscapes. It is only at the landscape level that we can provide a plan or vision that will provide enough of each landscape element to maintain the populations of, and honestly claim "no effect" on, all vertebrate species. The local extinction of a species due to some land management activity is fine as long as the suitability for that same species is expected to increase at the same time in another part of the landscape (due to some other land-use activity or ecological succession, for example).

Because we will never fully understand the habitat requirements of all vertebrate species, I still believe the indicator approach is necessary; it merely needs to be applied in a way that avoids the pitfalls of managing entirely on the basis of the needs of just a few high-profile species. My goal in this chapter is to use a selection of preliminary results from a newly established landbird monitoring program (published in their entirety in Hutto [in press]) to illustrate why land-birds (primarily songbird species) are likely to be excellent conservation tools, and why they are likely to figure prominently in any change in the way wildlife biologists manage for the maintenance of all vertebrate species.

Methods

A complete description of the methods used to obtain the data summarized here is available in Hutto and Hoffland (1996) and Hutto (in press). Basically, a series of 646 10-point transects were geographically stratified by 7.5-minute topographic quad maps and were permanently marked in the field in 1994. Bird surveys were conducted at these points in the same year and at an additional 2,355 points distributed among 309 transects between 1989 and 1993 as part of an effort to acquire data from vegetation cover types that were likely to be undersampled on USFS lands. Thus, bird occurrence data were collected from 8,815 points between 1989 and 1994. A 10-minute point count was conducted at each of the sampling points along a transect. Points were visited once between mid-May and early July in a given year. On the return trip after all point counts had been conducted, observers stopped at each point again to record a variety of vegetation information within a prescribed area surrounding each point. For the purposes of this report, I present bird occurrence data in relation to a single vegetation variable—the vegetation cover type within which the point count was positioned (COVTYPE).

Each sample point, therefore, fell within one of a range of possible vegetation cover types defined according to a scheme based on a combination of the possible dominant plant species in the tallest vegetation layer and the possible vertical and horizontal vegetation structure. Thus, the basic cover type framework is one that included so-called "climax" vegetation types and, for the conifer forest types, which take on a very different structure from the climax type after disturbance, a series of successional stages (pre-shrub, low-shrub, tall-shrub, and pole-sapling stages). I then combined potential types to create a smaller series of eighteen vegetation types prior to model building. For conifer forest types, I defined eleven categories represented by six relatively mature and relatively undisturbed forest types (cedar-hemlock, spruce-fir, lodgepole pine, mixed conifer, Douglas fir, and ponderosa pine) and five early successional or post-treatment types (post-fire, clearcut, seed-tree cut, shelterwood cut, and group-selection cut). The remaining seven categories were the open and riparian types (sagebrush, grassland, agriculture, marsh, riparian shrubs, Cottonwood/Aspen, and residential).

Unfortunately, because sample points occurred across the landscape in clusters of 10 (per transect), multiple samples of a given cover type within a single transect are not statistically independent estimates of bird composition within that cover type. Nevertheless, I used individual points as sample units for calculating the probability of occurrence on a 10-minute point count in a given cover type because (1) transects themselves make meaningless habitat sample units when they cross multiple cover types, and (2) the danger of being misled because of pseudoreplication of points within a cover type has been largely eliminated by the tremendous geographic spread of points for each cover type (Hutto in press).

To determine habitat associations, I excluded points that were positioned within 100 m of the edge of another cover type to reduce the chance that birds would have been detected within a cover type that differed from that recorded at the census point. I had to include data from the entire set of points positioned within each of the three riparian types (marsh, riparian shrub, and riparian bottomland), however, because most of those cover type patches were so small or narrow, so that, by default, a point located within one of those types was also within 100 m of another cover type as well. To further reduce the chance of linking the occurrence of a bird species to a habitat within which it did not occur, I used only bird detections that were estimated to be within 100 m of the observer. Thus, the number of points used to calculate the probabilities of occurrence across cover types (4,097) was substantially less than the number actually conducted in the field (8,815), but there were still at least 50 points in each of the 18 cover types).

Results

We visited a total of 8,815 points, which were distributed among 955 transects on 13 national forests in the USFS Northern Region and on various BLM, Potlatch, Plum Creek, tribal, state, and private lands adjacent to these forests. A total of 186 bird species were detected, most of which (163) were those that the point count method was designed to detect—the smaller, diurnal, visually and vocally conspicuous landbirds. The total also included 15 waterfowl and 8 shorebird species. As an indication of the efficacy of this method at detecting species known to breed in a given area, in western Montana (where transect coverage coincides with coverage reported by the Montana Bird Distribution Committee in 1996) we detected 121 (91 percent) of the 133 known breeding landbird species (excluding hawks, grouse, and owls). A total of 91 species (68 percent of the potential breeding landbird species) were detected close enough (within 100 m) and frequently enough (on 30 or more points) to construct what we believe are meaningful habitat-relationship models (see Hutto et al. [1986] for a discussion of minimum number of point counts needed to generate reliable estimates of probabilities of occurrence in a given habitat type). I would emphasize here that, although the method does not provide information on all bird species, it allows one to monitor a much larger number of species than that typical of traditional indicator species approaches.

What generalizations emerged from these simple models of bird distribution among habitats? Some landbird species are very restricted to specific, naturally occurring ecological conditions that are themselves restricted in spatial extent, or at least less extensive than they were at the turn of the century. Obviously, the

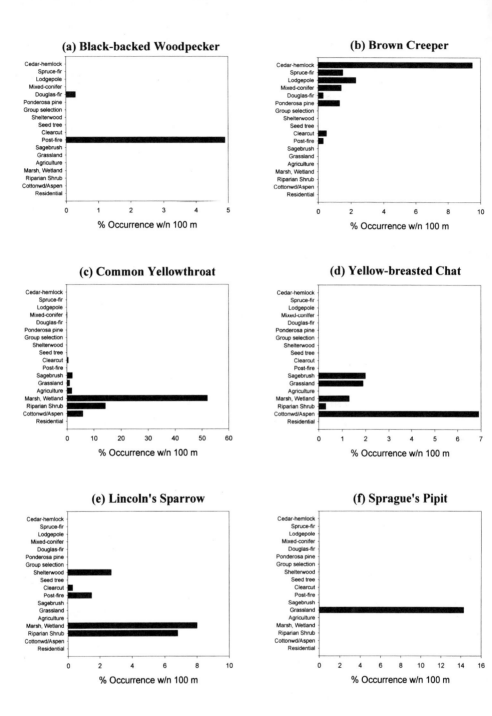

Figure 6.1. Examples of landbird species that are relatively restricted to each of several vegetation cover types. Only the latter two species are restricted to cover types that are products of human disturbance and, thereby, are more prevalent in the northern Rocky Mountain landscape now than in the pre-industrial past. Sample sizes (number of point counts) for each cover type are: cedar-hemlock (74), spruce-fir (134), lodgepole pine (215), mixed-conifer (1143), Douglas fir

(g) Brewer's Sparrow

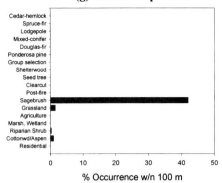

% Occurrence w/n 100 m

(h) Bobolink

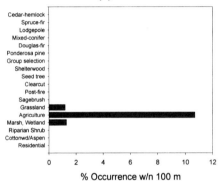

% Occurrence w/n 100 m

(i) Williamson's Sapsucker

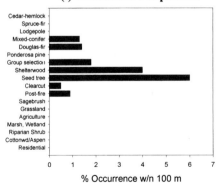

% Occurrence w/n 100 m

(292), ponderosa pine (77), group selection (112), shelterwood (75), seed tree (116), clearcut (365), post-fire (338), sagebrush (100), grassland (481), agriculture (56), marsh (75), riparian shrub (296), Cottonwood/Aspen (102), residential (46).

loss of any one of these cover types will mean the loss of those bird species that are relatively restricted to it. Thus, it should be clear that we need to maintain each of these cover types (defined at least as finely as defined here) on the broader landscape, although it is unclear how much of each needs to be retained to maintain viable populations of any given species. Even if we are not about to lose a given cover type from the broader landscape, land-use practices within and surrounding that type may have important implications, especially for species restricted to that cover type. Below, I provide examples of the more instructive distribution patterns that emerged from our survey work (a complete list of species that illustrate each pattern can be found in Hutto [in press]):

1. **Post-fire, standing-dead forests** (e.g., Black-backed Woodpecker [*Picoides arcticus*]; figure 6.1A)—The relatively restricted distribution patterns result from the fact that these bird species depend to a great extent on standing dead trees in burned forests for feeding and nesting. Not only do we have much less of this cover type than would naturally occur because of our fire prevention policies, but salvage logging what little does manage to burn will have a negative impact on species that are either restricted to, or relatively restricted to, early postfire conditions.

2. **Relatively uncut forests** (e.g., Brown Creeper [*Certhia americana*]; figure 6.1B)—Based on observed distribution patterns among cut and uncut forest types, the cutting (even light thinning) of dense, older forests (especially the cedar-hemlock type) will have negative effects on several species that are restricted to those conditions. If we break down the relatively uncut forest types into four age categories (young, selectively cut, mature, and old-growth), it becomes apparent that some of these species require not only relatively uncut, but relatively old forests as well (figure 6.2).

3. **Marshes** (e.g., Common Yellowthroat [*Geothlypis trichas*]; figure 6.1C)—The potential negative effects of wetland conversion on (mostly) private lands should be obvious.

4. **Riparian bottomlands** (e.g., Yellow-breasted Chat [*Icteria virens*]; figure 6.1D)—Numerous landbird species are relatively restricted to riparian bottomlands. This fact takes on special meaning when we consider that bottomland riparian cover types make up less than 0.5 percent of all land area in the Northern Region (Mosconi and Hutto 1982), and that they incur a disproportionate amount of human activity (i.e., home building, recreation, and livestock grazing) and cowbird activity. Much of this land base is private, making publicly owned land of this type much more important as potential refuges for wildlife that might be sensitive to the human activities listed above. We currently lack, but desperately need, information on cowbird parasitism rates in relation to the presence of livestock in riparian bottomlands, and we need information on the effects of vegetation alteration and livestock presence on nesting success of riparian bottomland birds.

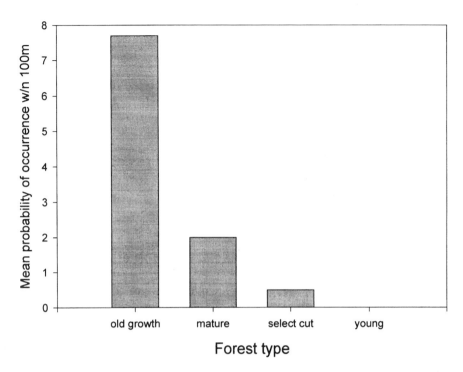

Figure 6.2. Within the relatively uncut conifer forest cover types, Brown Creepers are relatively restricted to the older, completely uncut category.

5. **Upland riparian stream environments** (e.g., Lincoln's Sparrow [*Melospiza lincolnii*]; figure 6.1E)—Species restricted to upland riparian streamside vegetation may be especially sensitive to so-called "best management practices," which have never been evaluated in terms of their effects on a wide variety of riparian-dependent terrestrial wildlife species.

6. **Grassland** (e.g., Sprague's Pipit [*Anthus spragueii*]; figure 6.1F)—If we couple the fact that many species are restricted to grassland with the fact that many of the same species are declining on a nationwide scale, the management of those lands becomes a pressing issue. Livestock grazing is a common land-use activity in grassland environments and may be incompatible with the needs of some of these bird species (Herkert and Knopf, this volume).

7. **Sagebrush** (e.g., Brewer's Sparrow [*Spizella breweri*]; figure 6.1G)—Once again, livestock grazing is a common land-use activity in sagebrush environments and may be incompatible with the needs of bird species restricted to such habitat (Rotenberry, this volume).

8. **Agricultural fields** (e.g., Bobolink [*Dolichonyx oryzivorus*]; figure 6.1H)—
This cover type is not "naturally occurring," but even though such environments are artificially created, the maintenance of viable populations of
species that are relatively restricted to such conditions may depend on the
management practices associated with such land. The main issue here is one
of whether mechanical disturbance from farm machinery interferes with
the reproductive biology of species that are relatively restricted to agricultural lands. If so, these environments may be acting as "ecological traps" that
attract individuals but do not allow them to be successful there (Rodenhouse and Best 1983).

9. **Harvested conifer forests** (e.g., Williamson's Sapsucker [*Sphyrapicus thy-
roideus*]; figure 6.1I)—No conifer-forest bird species appears to be restricted
to the harvested cover types, but several occur most commonly in the vari-
ously cut forests. The potential management issue is related to the fact that
harvested forests are "unnatural" in the sense that their structure consists of
combinations of elements (widely or evenly spaced live trees) that simply do
not exist in natural successional seres. A potential problem is that these
unnatural cover types may elicit settling responses by species that are "programmed" to respond to superficially similar, but fundamentally different,
early successional forest types. Thus, harvested forests could be acting as
"ecological traps" (e.g., Gates and Gysel 1978), where species are being
attracted to areas where suitability is poor because reproductive success
and/or adult survival is affected negatively by, say, inadequate food resource
levels, or abnormally high predation or parasitism rates. Because no forest
bird species is entirely restricted to harvested conditions, there will always
be "backup" bird populations in lightly harvested or unharvested forest as a
refuge from such a problem, should it exist.

Additional Research Needs

We need more data on occurrence of species among a broader range of existing
vegetation cover types, especially cover types that result from land-use activities.
As crude as such information might be, it is far better than the information provided by a typical field guide and should be the foundation beneath speculation
about the projected effects of any land management plan on a given wildlife
species. Armed with a solid understanding of which cover types are occupied by
a given species, we can proceed with comparisons of various measures of fitness
among the occupied cover types to ensure that presence is not a misleading
indicator of habitat suitability. Altered habitats need much more attention than
they currently receive because "unnatural" structural changes are likely to
uncouple habitat selection stimuli from factors that ultimately determine an individual's success, thereby creating "ecological traps."

Land management agencies should be actively engaged in the process of adaptive management, whereby effects on selected species are constantly monitored by agency and other researchers so that land-use practices can be modified on the basis of this continual appraisal of land-use effects. Finally, we should seek additional groups of species that can be monitored through single field methods so that habitat relationships can be built for, and land management decisions made on the basis of, as wide a range of species as possible.

Making Research Effective for Conservation

Many patterns of restricted habitat use have been common knowledge (e.g., Grasshopper Sparrow *(Ammodramus savannarum)* is restricted to grasslands; Brewer's Sparrow is restricted to sagebrush), but other patterns of relatively restricted distribution were probably not as evident prior to this work. As just one example, wildlife biologists never seriously considered standing dead forests created by stand-replacement fires as critically important wildlife habitat until data on landbird distribution patterns began rolling in. Black-backed Woodpeckers appear to depend on such habitats, at least in the northern Rocky Mountains (Hutto 1995). Attention to that particular nongame species exposes a clear conflict with post-fire salvage-cutting operations. It is with attention to nontraditional management species, not to traditional species of management concern, that biologists have begun to expose costs associated with this widespread and virtually unquestioned land-use practice. The relative restriction of Brown Creepers to comparatively uncut cedar forests is also impressive and serves to emphasize the value of this cover type to wildlife species. Thus, we can gain a new understanding of critical elements and processes through species-centered environmental analysis (James, Hess, and Kufrin 1997), and we stand to benefit by expanding such analyses beyond traditionally managed species.

Prior to this survey, it was also common knowlege that many bird species were widely distributed across cover types, but we had no knowledge of the relative abundance of these bird species among cover types, especially harvested forest types. It is now evident that Orange-crowned Warbler *(Vermivora celata)* and Solitary Vireo *(Vireo solitarius),* for example, occur not only broadly across forest types, but most commonly in harvested forest types (figure 6.3), and that Williamson's Sapsucker is even relatively restricted to such types (figure 6.1I). In short, the detail and region-specific nature of this information is noteworthy and should prove useful to individuals wishing to model probabilities of occurrence in planning areas that are projected to consist of alternative proportions of various cover types.

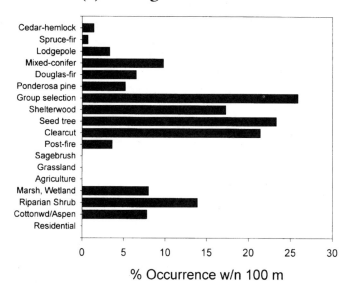

(a) Orange-crowned Warbler

% Occurrence w/n 100 m

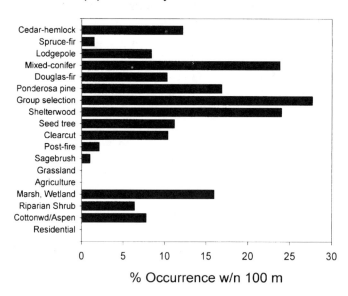

(b) Solitary Vireo

% Occurrence w/n 100 m

Figure 6.3. Two coniferous-forest species that illustrate monitoring data can expose subtle differences in probability of occurrence among the conifer cover types. Such resolution is essential for building accurate distribution maps in relation to cover types within a landscape.

I realize that measures of presence or probability of occurrence do not necessarily reflect suitability, but before we worry too much about whether two cover types are equally suitable, we first need to be able to predict where a species is likely to occur at all! The occurrence data described here have uncovered numerous (outlined in Hutto in press) potential "ecological trap" problems that need study. Thus, monitoring data like these can serve to focus future research efforts toward particular species or situations.

Even though populations of a species fluctuate from year to year, the relative abundances among broadly defined vegetation cover types do not (e.g., Sallabanks 1996). Thus, the value of a monitoring program such as this one should be readily apparent: objective results are possible from as little as a single year's field effort, and the results already demonstrate some clear limits within which any agency aspiring to maintain ecological integrity must work.

Programs like this one would benefit other USFS regions and (especially) other agencies, but I would in no way suggest that this sort of research should take priority over other kinds of research. In fact, the limiting factor in bird conservation seems to be a willingness to *use* information on land-use effects, not a shortage of such information. Wildlife conservation depends, ultimately, on attitudes of the public about the value of wildlife. And perhaps people's attitudes about wildlife conservation are less likely to be changed through results from largely descriptive research such as that which I've just described than from other kinds of research (e.g., behavioral studies). We should judge the conservation potential of proposed research on the basis of the stated research justification, period. Attempts to define research "needs" otherwise are largely misdirected; what we "need" is the widest possible variety of quality research.

Acknowledgments

I received financial support for developing the landbird monitoring program from the USFS Northern Region, the Bureau of Land Management, and Plum Creek Timber Company. This support would not have been possible without the efforts of Alan Christensen, Sally Sovey, and Lorin Hicks for the USFS, BLM, and Plum Creek, respectively. Indirect and in-kind support was provided by Montana Fish, Wildlife and Parks (thanks specifically to Dan Casey), and Potlatch Corporation (thanks specifically to Bill Wall). Sallie Hejl, of the USFS Intermountain Research Station, provided detailed consultation and guidance during the design phases of our field studies. Christine Paige hired the field crews and took the time to organize and clean the data that they submitted for analysis. The field crew included J. Adams, J. Balcomb, J. Barilla, D. Bergeron, A. Bosma, D. Casey, C. Couch, C. Cutler, J. Dodge, R. Fergus, J. Gallagher, T. Grunder,

D. Harvey, P. Heglund, S. Henderson, P. Hendricks, J. Hoffland, E. Loomis, G. Mazer, J. McBride, T. Mears, D. Miller, M. Mitchell, M. Munts, B. Pitman, S. Ritter, S. Reel, J. Ross, J. Slotterback, C. Snetsinger, B. Thompson, S. Winsor, and J. Young. Paul Hendricks, Andrew Bosma, Jock Young, and Wendy Williams helped with data management, analyses, and/or graphics production. Lastly, I wish to thank John Marzluff, Rex Sallabanks, Dick DeGraff, and Mike Morrison for taking the time to provide extensive comments on an earlier draft.

CHAPTER 7

What Do We Need to Monitor in Order to Manage Landbirds?

David F. DeSante and Daniel K. Rosenberg

Conservation Issues and Previous Research

A major issue in environmental biology concerns the worldwide decline of migratory landbirds (Robbins et al. 1989; Terborgh 1989; Kaiser and Berthold 1994; Martin and Finch 1995). Among northern-hemisphere species, long-distance migrants (i.e., those that breed in temperate and winter in tropical latitudes) appear to be declining most severely. For example, thirty years of data (1966–95) from the North American Breeding Bird Survey (BBS) indicate that 55 percent and 69 percent of Nearctic-Neotropical migrant species are declining in eastern and central North America, respectively, compared to 47 percent and 53 percent, respectively, for short-distance migrant species, and 45 percent and 56 percent, respectively, for permanent resident species (Peterjohn, Sauer, and Link 1996). Similarly, twenty-two years of migration-monitoring data (1972–93) from the Mettnau-Reit-Illmitz Program of the German Ornithological Institute "Vogelwarte Radolfzell" indicate that fifteen (71 percent) of twenty-one trans-Saharan migrant species decreased significantly while only two (4 percent) of the fourteen European-wintering species declined significantly (Kaiser and Berthold 1994). In addition, by 1990, data from the British Common Bird Census and Waterway Bird Survey showed that thirty of Britain's passerine species were experiencing long-term declines, as opposed to seventeen species whose populations appeared stable and eighteen species that showed population increases (Stroud and Glue 1991; Newton, this volume). In addition, seventeen (57 percent) of the thirty declining species were long-distance migrants, while only five (28 percent) of the eighteen increasing species were long-distance migrants (Stroud and Glue 1991).

The data that were used to describe the above temporal patterns were generated from broadscale, retrospective monitoring programs. Monitoring has been defined in many ways (e.g., Goldsmith 1991); here we refer to monitoring as the temporal assessment of demographic parameters of bird populations. A critical goal of any population monitoring program should be to identify the state of the population, that is, estimates of desired attributes, such as density, average productivity, or average survival rates for a given time period. Although detection of environmental influences on animal populations is difficult, especially considering the nature of time-series data, such as estimated population trajectories (Botsford and Brittnacher 1992), the detection process can provide information on changes in population parameters (Nichols, in press) and can be considered a preliminary search for patterns to be tested in detailed field studies (Holmes and Sherry 1988; Botsford and Brittnacher 1992). In this sense, monitoring facilitates applied research. Furthermore, monitoring, if done in an experimental or quasi-experimental manner (Nichols, in press), is necessary to determine the effectiveness of management actions designed to reverse population declines or bring about the recovery of small or threatened populations (Noon 1992).

Large-scale, long-term monitoring programs, such as those referenced above, are necessary to detect declining population trends over large geographic regions. Such programs require both large-scale coordination and cooperation. Generally, they rely on large numbers of trained volunteers and have only been implemented successfully in developed countries. Even there, many species are too rare or locally distributed to permit reliable identification of population trends. Moreover, interpretation of large-scale, long-term population data is not always straightforward; indeed, controversy still exists regarding interpretation of data from even the well-established BBS Program (Sauer and Droege 1990; James, McCulloch, and Wiedenfeld 1996; Peterjohn, Sauer, and Link 1996). This controversy exists largely because the probability of observing an individual bird, which potentially varies among observers, geographic areas, habitat, species, and time, is unknown. Clearly, the implementation of large-scale, long-term, population-trend monitoring is not a simple exercise.

Despite the general success of avian population-trend monitoring programs in identifying potentially declining species in certain well-studied countries, such monitoring programs provide little information as to factors responsible for population declines and even less direction as to appropriate management actions to reverse declines (Peterjohn, Sauer, and Robbins 1995). This is because they provide no information on the primary demographic parameters (productivity and survival) of the species monitored (DeSante 1995). Indeed, population-trend data by themselves provide no information at all as to the stage(s) in the life cycle that control(s) the population declines (Temple and Wiens 1989) and thus fail to distinguish problems caused by birth-rate effects from those caused by death-rate effects (DeSante 1992). As a result, the factors responsible for declining landbird populations have generally remained unclear (O'Connor 1992).

Features of Research That Improved Conservation

Broadscale, retrospective monitoring projects, such as the BBS, have heightened our awareness of possible population declines in landbirds. Although the magnitude of the declines is often not well estimated and even the direction of the changes is sometimes controversial (James, McCulloch, and Wiedenfeld 1996), these programs have provided the primary data used by decision makers to allocate additional effort toward further investigation of the potential declines or toward implementing conservation plans. The examples presented in this section highlight how monitoring has been used to aid conservation efforts.

Comparisons of BBS population-trend data among species having various life-history traits (i.e., migration strategy, habitat preferences on breeding and wintering grounds) provided indirect evidence that destruction and degradation of forested tropical wintering habitat could be a major cause of population declines in some species of eastern North American landbirds (Robbins et al. 1989). Other studies on some of these same species (i.e., Wood Thrush [*Hylocichla mustelina*] and Ovenbird [*Seiurus aurocapillus*]), however, suggested that reduced breeding success caused by high levels of brood parasitism by the Brown-headed Cowbird (*Molothrus ater*) caused, in turn, by fragmentation of temperate forest breeding habitat, could be a major cause of population declines in midwestern North America (Robinson et al. 1995; Faaborg et al., this volume). Obviously, both breeding-ground and wintering-ground processes could adversely affect population trends, and management actions could be suggested that would tend to mitigate against each of these processes (e.g., requiring overstory shading in tropical coffee plantations or implementing extensive cowbird-control programs on temperate breeding grounds). Because such management actions will likely be expensive and will involve major policy changes, public agencies and private organizations are reluctant to undertake them without considerable assurance that they will be successful. Knowledge of the rates of primary demographic factors throughout a species' range can facilitate the identification of key factors controlling its observed population trends. Monitoring demographic parameters helps achieve this latter knowledge.

A major advance in this direction was provided in a seminal paper by Baillie (1990) in which he advocated an "integrated population monitoring scheme" whereby various monitoring programs would address different aspects of the population dynamics of a suite of species over the same geographic area. In this scheme, which has been implemented by the British Trust for Ornithology (BTO), population trends are tracked by several programs, including the Common Bird Census and Waterways Bird Survey (Baillie 1990). Information on the potential proximal demographic causes of observed population trends in several habitat types is provided by the British Constant Efforts Sites (CES) Scheme (Peach, Buckland, and Baillie 1996), which monitors changes in productivity indices and survival-rate estimates through constant-effort mist netting. Finally, detailed, habitat-specific information on various aspects of reproductive success, including timing of clutch initiation, clutch size, brood size, and nesting success, are provided by the BTO's Nest Record Scheme (Baillie 1990).

Peach, Baillie, and Underhill (1991) and Baillie and Peach (1992) have used integrated population monitoring to better understand the potential causes of population declines in several trans-Saharan migratory species of European landbirds. They found, for example, using key-factor analysis (Varley and Gradwell 1960; Blank, Southwood, and Cross 1967; Krebs 1970; Southwood 1978), that variations in mortality of full-grown birds (individuals that have reached independence from their parents) explained most of the population fluctuations in all seven of the species investigated. Mortality of young birds during their first year of life was implicated as the key factor causing population declines in Sedge Warblers (*Acrocephalus schoenobaenus*) and Willow Warblers (*Phylloscopus trochilus*), while mortality of adult birds after their first year of life was implicated for Whitethroats (*Sylvia communis*). Moreover, for Sedge Warblers, Whitethroats, and Swallows (*Hirundo rustica*), fluctuations in mortality of full-grown birds were correlated with conditions on the wintering grounds. In the case of the first two species, both survival of full-grown birds and total population size were highly correlated with rainfall patterns on the species' sub-Saharan (Sahel), west African wintering ranges. Populations of these two species appear to be limited by competition for resources on the wintering grounds, and these resources are strongly dependent on rainfall during the preceding wet season. Thus, the population declines in British Sedge Warblers and Whitethroats appear to have been caused directly by the extensive Sahel drought. Conservation measures for these species, therefore, should be directed toward ameliorating the causes of drought in the Sahel or, at least, mitigating the effect of these droughts. These results suggest that conservation efforts that target the breeding ranges of these species may do little to reverse their population declines.

The concept of integrated population monitoring is beginning to be pursued in North America (see also Hejl and Granillo, this volume). Population trends are monitored by means of roadside point counts through the BBS (Robbins, Bystrak, and Geissler 1986; Peterjohn, Sauer, and Robbins 1995; Peterjohn, Sauer, and Link 1996). Productivity is monitored by means of constant-effort mist netting through the Monitoring Avian Productivity and Survivorship (MAPS) Program (DeSante et al. 1995; DeSante, Burton, and O'Grady 1996) and by means of direct nest monitoring through the Breeding Bird Research

Database (BBIRD) Program (Martin and Geupel 1993). Finally, adult survivorship is monitored by means of mark-recapture data from the MAPS Program.

The integration of BBS and MAPS data from the Sierra Nevada has shed light on the potential proximal causes of population decline in the Willow Flycatcher (*Empidonax traillii brewsteri*), a species that has shown drastic population declines in the Sierra over the past fifty years (Gaines 1988). In contrast, two similar species of Sierran flycatchers, Hammond's Flycatcher (*E. hammondii*) and Dusky Flycatcher (*E. oberholseri*), both show positive BBS population trends in the Sierra. MAPS data from ungrazed meadows in the Sierra, where all three species breed, show that both the productivity index and the annual adult survival rate estimate for Willow Flycatcher is as high or higher than those of the other two species (DeSante, unpublished data). It has been suggested from localized research efforts in the Sierra (Serena 1982; Gaines 1988) that the grazing of montane meadows, which results in defoliation of the lower portion of the willows, causes the habitat to become unsuitable for nesting Willow Flycatchers. MAPS data is consistent with this hypothesis by providing data to reject competing hypotheses such as low productivity in general (in ungrazed as well as grazed meadows) and low survivorship due perhaps to problems on the wintering grounds. MAPS data thus support research and management efforts for Willow Flycatchers aimed at reducing the impact of grazing Sierran montane meadows.

Research Needed to Further Conservation

An effective integrated monitoring program for landbirds should be able to accomplish three objectives: (1) identify species with declining population trends and describe these trends at multiple spatial scales; (2) identify reasonable hypotheses for the proximal demographic causes of population declines and suggest research activities to test these hypotheses; and (3) evaluate the effectiveness of local management actions and larger-scale conservation strategies implemented to reverse the declines. These objectives emphasize that the results of an integrated monitoring program will be most effective if reasonable hypotheses, backed by knowledge of demographic patterns derived from monitoring, can be formulated so that rigorous tests of the hypotheses can follow. In this sense, monitoring facilitates research.

In the past, monitoring, research, and management have generally been treated as independent activities. Only the first objective outlined above typically was considered to lie within the domain of monitoring. Objective 2 was more often considered to lie within the province of research, while objective 3 was relegated to the domain of management. Two major shortcomings of this traditional approach have been: (1) a general paucity of research efforts designed to investigate causes of population declines over large spatial scales; and (2) a lack of effective integration between monitoring, research, and management at both large and small spatial scales. Most research and management efforts directed at understanding and reversing declining populations of landbirds have focused on small or local scales (e.g., a particular refuge) or on particular habitats. This is not to disparage such research and management efforts; many such studies have added important information regarding factors influencing habitat choice and nesting success that has assisted in developing effective management actions. The real failure of such efforts, however, lies in the paucity of management actions following on the heels of successful research and the lack of follow-up monitoring of the effects of the management actions actually implemented. Too often, good research and well-intentioned management actions end up in a relative vacuum and are not integrated with each other or with continued monitoring. The interaction of monitoring, research, and management must be structured into an interactive loop for adaptive decision making.

Creation of an integrated monitoring effort must include the monitoring of productivity and survivorship as well as the population trends that result from the interaction of these primary demographic parameters (Hejl and Granillo, this volume). It is important to note that monitoring cannot, by itself, identify ultimate environmental causes of population change. Monitoring primary demographic parameters will, however, allow a temporal assessment of changes in these vital rates that, through correlation analyses, can identify hypotheses for further evaluation. Carefully controlled research efforts are then needed to test the hypotheses generated by monitoring in order to determine the environmental factors causing the observed changes (or differences) in the primary demographic parameter(s) responsible for the population changes (Nichols, in press).

Monitoring of primary demographic parameters also may be the most judicious way to determine whether or not management actions are working effectively (DeSante, in press). This is because management actions affect primary demographic parameters directly, and these effects can potentially be observed over a short time period (Temple and Wiens 1989). Because of buffering effects of floater individuals (Smith 1978) and density-dependent responses of populations, there may be substantial time lags between changes in primary parameters and resulting changes in population size or density (DeSante and George 1994). Moreover, because of the vagility of most bird species, local variations in population size may often be masked by recruitment from a wider region (George et al. 1992) or accentuated by lack of recruitment from a wider area (DeSante 1990). Thus, density of a species in a given area may not be indicative of population health, due to source-sink dynamics (Van Horne 1983; Pulliam 1988). Knowledge of primary demographic parameters is thus critical for understanding population dynamics and is directly applicable to population models that can be used to assess land-management practices (Noon and Sauer 1992), particularly when these parameters can be related to specific habitats or landscape features.

What Needs to Be Monitored and How Should it Be Done

Population Trends

An effective large-scale monitoring program must be able to provide reliable estimates of relative abundance and population trend over the entire ranges of many species. In general, the BBS currently has the capability of providing these estimates for a large number of North American species (Peterjohn, Sauer, and Robbins 1995; Peterjohn, Sauer, and Link 1996), although there is controversy as to how reliable the estimates are (Sauer and Droege 1990; James, McCulloch, and Wiedenfeld 1996).

A major shortcoming of the BBS program is that habitat-specific relative-abundance and population-trend data are not obtained. This could be rectified by a coordinated program of habitat-specific off-road point counts (Hutto, this volume) or area searches; or, perhaps, by incorporating remote-sensed habitat data associated with each BBS survey point. However, because of different detection rates in different habitats (Schieck 1997), relative abundance among habitat types will be difficult to estimate reliably, even for a single species.

Productivity

Productivity and survivorship are the major primary demographic parameters that provide critical information for understanding patterns of population change. Productivity has a number of components, including clutch size, egg and nestling survival, fledgling survival, and number of nesting attempts. Information on the component that most affects overall reproductive success will be very useful in assessing potential management actions designed to increase productivity.

Habitat- and site-specific estimates of several of these components (e.g., clutch size, egg and nestling survival) can be obtained from direct nest monitoring through the BBIRD Program (Martin and Geupel 1993), although these estimates may be difficult to obtain. Near the time of fledging, daily nest monitoring may be necessary to get the best estimate of nest success. If the birds are individually color marked and all nesting attempts are monitored, the number of nesting attempts per pair can be estimated as well. Direct nest monitoring, however, cannot provide an estimate of fledgling survival, which may or may not be correlated with survival of eggs or nestlings. Moreover, observer effects may bias the results of direct nest monitoring, especially if daily monitoring is necessary.

Indices of post-fledging productivity (the number of young per adult that reach independence from their parents) that integrate all of the individual components of reproductive success can be obtained from constant-effort mist netting through the MAPS Program (DeSante et al. 1995; DeSante, Burton, and O'Grady 1996). Data collected by mist netting are probably better suited for estimating productivity on a regional, rather than site- or habitat-specific, basis than are data from direct nest monitoring. Productivity indices obtained from constant-effort mist netting are likely to be biased, however, because of the lack of a well-defined sampling area and because of habitat- and species-specific biases in the capture probability of young compared to adults (DeSante et al. 1995). DeSante (1997) showed that habitat-specific biases indeed exist, but that consistent station operation and extensive sampling lessen their effect. DeSante (1997) also provided evidence that species-specific biases in productivity indices caused by differences in dispersal characteristics or foraging height between young and adults may be relatively small.

Direct nest monitoring and constant-effort mist netting thus provide information on different components of productivity at different spatial scales. As such, they provide complementary information; both methods, therefore, should be included in an effective integrated population monitoring scheme.

Survival Rates

Estimates of annual survival rates of adult birds can be obtained from mist netting (e.g., MAPS) using modified Cormack-Jolly-Seber (CJS) mark-recapture analyses (Clobert, Lebreton, and Allaine 1987; Pollock et al. 1990; Lebreton et al. 1992). Potential biases caused by including nonresident (transient) individuals in the sample of newly banded adults can be reduced using various transient models (Peach, Buckland, and Baillie 1990; Pradel et al. 1997). It is important to note that estimates of adult survival rates obtained in this way are actually estimates of apparent survival that include an unknown component of emigration (DeSante 1995).

Survival rates of young are difficult to obtain from capture-recapture studies because young birds typically have relatively large dispersal distances. Estimates of post-fledging, premigration survival rates and dispersal characteristics have been obtained by radiotelemetry for Wood Thrush (Anders et al. 1997), and such methods should be applied more broadly (Walters, this volume; Faaborg et al., this volume).

Considerations of Spatial Scale

An important aspect of monitoring is the ability to detect trends in selected parameters and investigate the scale at which they may be occurring. Regional trends in avian demographic patterns may occur due to large-scale weather changes or changes in the landscape that affect areas large enough to affect many local populations similarly. Local changes or trends, such as may occur in a specific national forest, may occur due to changes to habitat quality, for example, from tree harvest. If the pattern of local environmental change is pervasive, similar regional patterns may result. Understanding the scale of trends will thus be informative for determining future research needed to identify problems, and, once they are identified, to determine management solutions. Processes that affect patterns in demographic rates are likely to be scale dependent; management policies and activities often respond to relatively local issues, although concern over small-scale patterns may be motivated by documentation of larger-scale phenomena. Thus, monitoring must be effective at multiple scales ranging from "local" (e.g., national forest or park) through "regional" (e.g., physiographic strata) to "large" (e.g., eastern North America or the entire range of the species).

Avian populations vary spatially and temporally in regard to demographic parameters such as density (Brown 1995), productivity (Robinson et al. 1995; DeSante, Burton, and O'Grady 1996), and survivorship (Johnson, Nichols, and Schwartz 1992; Burnham, Anderson, and White 1996). Describing patterns of variation in primary demographic parameters is important for understanding dynamics of populations and for interpreting trends in population size that may reflect reduced viability (Wilcove and Terborgh 1984). Lack of knowledge on the spatial scale and the magnitude of temporal variation in demographic parameters often leads to incorrect conclusions regarding population health and makes it difficult to argue that specific population declines are noteworthy and deserve additional attention.

Although there is a tremendous effort to obtain data to monitor wildlife population trends at small spatial scales (e.g., an individual refuge), there are few programs that attempt to monitor trends at larger geographic scales (e.g., western North America). Understanding patterns at small scales is critical in guiding research and management actions to address local concerns. However, it is difficult to isolate reasonable hypotheses for the observed patterns when larger-scale patterns are unknown. The challenge is to design monitoring programs to provide estimates of parameters of bird populations at larger geographic scales, while maintaining adequate precision at smaller spatial scales to address local issues. Given a finite sampling effort, allocation of this effort can be extensive, intensive, or a combination of both approaches (figure 7.1). Unfortunately, there is a trade-off between precision of local estimates and potential bias of large-scale estimates when effort is fixed (figure 7.2): increasing the effort at a given site will increase precision of local site-specific estimates, but because the effort is apportioned into fewer sites, there is likely an increase in bias of average estimates across a large geographic area. The degree of bias is relative to the degree of geographic heterogeneity of the parameter of interest. The single most important consideration regarding allocation of sample effort in an integrated monitoring program should be that of geographic scale.

Considerations of Temporal Scale

Monitoring, by definition, is an assessment of specified parameters as a function of time. Any monitoring program must define the time frame of interest, as the success of the monitoring will be defined by the ability to detect changes. The probability of detecting change, if it is in fact occurring, is a statistical power issue (Steidl, Hayes, and Schauber 1997). Larger effect sizes (i.e., the magnitude of change from a null hypothesis) will require a shorter time frame for monitoring than would a smaller effect size given an equivalent statistical power. Statistical

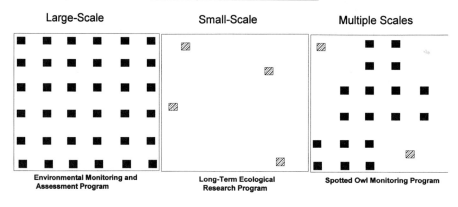

Figure 7.1. Hypothetical sampling strategies with finite effort. A sampling scheme can be envisioned in which samples are located systematically or with a probability-based scheme throughout a large landscape. In this large-scale approach (left panel), which is similar to the scheme initially used by the Environmental Monitoring and Assessment Program of the U.S. Environmental Protection Agency (NRC 1995a), many sites are sampled but with limited effort per site. In a small-scale approach (center panel), local sites are selected and sampled intensely. This scheme is similar to the National Science Foundation's Long-Term Ecological Research Program. In a multiple-scale approach (right panel), there are elements of both a large-scale approach in which many sites are sampled with limited effort, and a small-scale approach in which several sites are sampled intensively. This type of scheme was suggested for monitoring Spotted Owls (*Strix occidentalis*) in the Pacific Northwest (Bart and Robson 1992).

power also is a function of sample size. For example, the ability to detect a declining trend in survivorship is a function of the number of years of sampling (monitoring), the number of individuals sampled, and the effect size (percent annual change) given set recapture and survival probabilities (figure 7.3). Clearly, the spatial scale becomes important, as this dictates the ability to acquire large samples for a long period of time, the two critical factors for detecting trends in the demography of landbird populations.

Site Selection

A further issue relating to the effectiveness of extensive monitoring programs to provide unbiased estimates of large-scale trends is the representativeness of the samples, especially for large geographic areas. BBS attempts to sample randomly, although because all routes are along roads, there is a roadside bias. Most demographic monitoring programs, such as BBIRD and MAPS, use a nonrandom sampling framework. Sites selected are usually a function of local or habitat-specific interest or sites with large numbers (high density) of birds. Since most bird populations seem to be distributed with a high degree of heterogeneity, with most sites having few individuals of a given species and only a few sites with high densities (Brown 1995; Rosenberg 1997), it may not be feasible to have a prob-

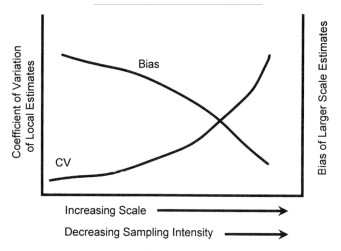

Figure 7.2. Trade-off in precision of local estimates and bias in large-scale estimates of parameters of bird populations. Given a finite sampling effort, there is unfortunately a trade-off in obtaining precise estimates for a given sampling area (e.g., a particular national forest) and obtaining unbiased large-scale (e.g., western North America) estimates. With a sampling scheme focused for estimating parameters of bird populations at small spatial scales (figure 7.1), precision for the local estimates will be much greater (and the coefficient of variation [CV] much lower) than under a large-scale approach in which the sampling intensity at a single site is minimal and thus produces estimates with larger CV. However, if average estimates over a large region are desired, sampling only a few sites may produce a region-wide estimate with large bias. The magnitude of the bias will be positively associated with the degree of spatial heterogeneity of the parameter being estimated.

ability-based sampling strategy over a large geographic area for demographic monitoring because of the large amount of effort required at each site. However, the nonrandom nature of the samples must be considered when making inferences on the larger-scale population parameters.

That few sites contribute most of the information for many species suggests that estimates from demographic monitoring programs that have a nonrandom type of sampling regime are unlikely to be representative for spatial scales larger than the actual study sites, and perhaps may be representative only of the study area of the few stations that contributed the majority of the data. Thus, if the average rate of a specified parameter for a given geographic area is the parameter of interest, then the estimate may be biased; the percent bias is unknown, although it is probably related to the level of geographic or habitat-specific variation in the parameter. If there is little or no variation, then estimates of the average rate may be unbiased. The bias in the average rate for a geographic area is primarily a concern because of the nonrandom sampling strategy. If a probability-based design were used, then it could be argued that the dominance of a few stations represents the true distribution of abundance; hence, there may be little bias in the average rates since they reflect the overall population. If there were common traits among study sites, then perhaps suggestive inferences

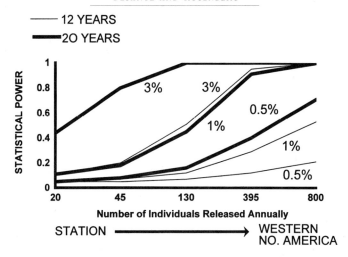

Figure 7.3. Simulation analysis of the statistical power to detect exponentially declining survivorship in relation to the number of individuals released annually, number of years of monitoring, and percent annual decline. The spatial scale reflects the average number of birds released annually in a geographic unit within a particular spatial scale (e.g., individual station). Data were simulated to reflect field data collected on Swainson's Thrush (*Catharus ustulatus*) from the MAPS Program (Rosenberg, De Sante, and Hines, in press). Initial survival rates used in the simulations were 0.45, and recapture probability was 0.54.

(hypotheses) could be made relevant to the common trait (e.g., a specific habitat type for BBIRD or landscape-level habitat information, such as extent of forest fragmentation, for MAPS). Such determinations potentially could be made with vegetation data from the study area (BBIRD) and from the landscape surrounding the station (MAPS).

Making Research Effective for Conservation

An effective integrated avian monitoring effort will be able to identify declining species, identify potential demographic traits responsible for the declines, and attempt to relate those traits to habitat and landscape characteristics. An integrated approach will also provide useful information to aid in identifying conservation strategies and management actions to reverse the declines, and will provide a means for evaluating the effectiveness of the strategies and actions implemented. Foremost will be the description of the temporal and geographic patterns of bird population parameters: relative abundance, productivity, and survivorship. Untangling the processes responsible for the observed patterns will require an experimental or quasi-experimental approach (Nichols, in press), which hopefully will be facilitated by knowledge of the temporal and geographic patterns found from monitoring.

Successful efforts to reverse population declines of landbirds will require an intimate integration of monitoring, research, and management. We believe that monitoring must play a central role in such an integrated process. Thus, it is the monitoring of population trends that defines the species of interest (i.e., declining species). It is the integrated monitoring of population trends and primary demographic parameters that provides information regarding the spatial and temporal patterns on which hypotheses of proximal demographic causes of population change are generated and intensive research at the local scale is based. Results from monitoring, coupled with the results of the implemented research, must then lead to suggested management actions at the local scale, and conservation strategies at larger scales, to reverse the population declines in species of concern. The integrated monitoring of population trends and primary demographic parameters, when done in an experimental framework, will then provide the means for evaluating the effectiveness of the implemented management actions and conservation strategies. We strongly recommend that each implemented management action be required to provide suggested monitoring efforts to evaluate its effectiveness at reversing the population declines in the species of concern. In this way, monitoring, research, and management truly can be integrated into an interactive sequence of activities.

An example of the operation of such an integrated monitoring, research, and management effort could be as follows. Assume BBS data show that a target species is declining in area A but has a stable population trend in area B. Assume that MAPS data show that survival rates for the species do not differ between areas A and B, but that productivity indices in area A are significantly less than those in area B and that the difference in productivity is sufficiently large to account for differences in population trends between areas A and B. Such a situation would suggest that low productivity is the proximal demographic cause of the population decline in area A. Habitat-specific information from BBIRD on nest success of the species in area A could provide information on the stage(s) in the nesting cycle that may be causing the low productivity (i.e., clutch size, number of nesting attempts, survival of eggs or young). This latter information would lead to research efforts to test hypotheses regarding the cause of the low productivity and to additional efforts to identify appropriate management actions to increase productivity at the local scale and to identify larger-scale conservation strategies to increase productivity over the entire range of the species where it is in decline. Continued integrated monitoring of population trends and primary demographic parameters would then determine the effectiveness of the management actions and conservation strategies implemented. This assessment would not only evaluate whether the declining population trend is lessening, but also whether productivity itself is being enhanced by the implemented actions and strategies.

Although the concept of integrated population monitoring is reasonably straightforward and analytical methods are well developed, implementation of integrated population monitoring is not a simple matter, especially over large geographic areas. Real-world issues of sample size (affecting precision of estimates), sampling strategy (affecting bias of estimates), and spatial scale are difficult and complex. The bottom line, inevitably, is that a large number of samples collected over a long time period will be necessary to provide precise and unbiased estimates of demographic rates and population trends. Coordination among the various programs will also be required for an integrated approach. Currently, there is little overlap of sampling sites among BBS, BBIRD, and MAPS.

A real commitment to continued, long-term monitoring on the part of agencies charged with the responsibility of managing bird populations is crucial. In addition, the large samples required can likely be obtained only by substantial volunteer effort. The recruitment, training, and maintenance of a network of volunteers is thus also crucial to the success of any large-scale, integrated, population-monitoring scheme. Ultimately, an interest in and appreciation of birds at the grassroots level must be cultivated if we are to succeed in identifying and describing avian population changes and in developing successful conservation strategies for reversing population declines.

Acknowledgments

We thank numerous colleagues, especially W. Link, T. Martin, J. Nichols, B. Noon, and J. Sauer, for many stimulating discussions of the concepts treated in this chapter. We thank R. Sallabanks and an anonymous reviewer for helpful suggestions on an earlier draft of this chapter. We thank the USDOI Fish and Wildlife Service, National Biological Service, Biological Resources Division of USGS, and National Park Service; the USDA Forest Service; the Department of Defense through its Legacy Resource Management Program; and the National Fish and Wildlife Foundation for financial support of the monitoring efforts on which this chapter is based. This is Contribution No. 70 of The Institute for Bird Populations.

PART III

Approaches for Conserving Endangered and Sensitive Species

Although many have argued that it is unrealistic to conserve avian biodiversity one species at a time, most programs to conserve birds are centered around one, or a few, conspicuous, highly endangered species. Even if such programs are not the best way to maximize our conservation of biodiversity, they are essential if we are to save a species that has progressed to the brink of extinction. The twenty years since Temple's (1978) seminal volume summarizing endangered bird recovery, *Endangered Birds: Management Techniques for Preserving Threatened Species*, have seen improvements in the status of nearly half of the species discussed and rapid evolution in captive husbandry and reintroduction techniques (Cade and Temple 1995).

In chapter 8, Derrickson, Beissinger, and Snyder discuss many of the approaches used and the decisions that must be made to recover endangered birds. They suggest that although single-species programs have demonstrated significant successes in many instances, multispecies or ecosystem approaches to the conservation of biological diversity can be expected to accelerate in the future. These approaches often take the form of spatially explicit models that incorporate habitat susceptibility, biophysical characteristics, and population viability. They note that the success of these management regimes will ultimately depend on improved knowledge, monitoring, and modeling of single-species populations. The development of spatially explicit population viability and ecosystem management tools will require a more rigorous understanding of the behavioral and ecological factors that influence dispersal, movements, habitat choice, and demographic parameters of individuals and populations through time.

Black (chapter 9) continues the dialogue on endangered birds by reviewing recovery efforts aimed at threatened and endangered waterfowl. He notes that less than one-third of the threatened species are receiving any kind of proactive management to forestall endangerment. Captive rearing and maintenance of captive stock is a common management strategy for threatened waterfowl, but it is one that will require long-term commitment by zoological institutions because few reintroduction programs are currently in place. The lack of reintroduction programs appears prudent at this time because limiting factors in the wild have not been corrected, and other, less costly, conservation measures have not been exhausted. The Hawaiian Goose has been the focus of extensive recovery actions for forty-five years. This effort has maintained the species in the wild but has not removed the threat of extinction. It has had the important benefit of postponing extinction and allowing managers to use behavioral research to guide future restoration efforts.

The application of intensive, active management to conserve endangered seabirds is discussed by Kress (chapter 10). Once superabundant puffins and terns have become increasingly rare, primarily because humans have introduced nonnative predators to their island nesting colonies. Nearly 10 percent of all seabird species and subspecies are endangered or threatened owing primarily to their colonial nesting habits, which make large numbers of breeders especially vulnerable to relatively localized disturbance. The plight of seabirds in general is discussed later by Boersma and Parrish (chapter 16).

Kress details the integrated use of predator control, chick translocation, and social attraction of breeders to reestablish breeding populations of seabirds off the northeastern U.S. coast. Extreme measures were needed to reduce predatory gull populations, but removal of this limiting factor was an essential and effective precursor to reintroduction. Detailed behavioral research was needed to understand how to move and hand-rear chicks and later attract them back to islands to breed. Interplay between research and management was effective; it enabled managers to learn from mistakes and modify methodology accordingly. Such "adaptive management" was necessary because most of the techniques to be used were untested under field conditions. Intensive management has increased the numbers of breeding seabirds on islands in this study. However, this success was possible only because of the long-term commitment of field crews and funding agencies. In some cases it took eight years before the first hint of success was apparent.

The challenges of endangered species management and restoration are immense, but an even greater challenge is to manage ecosystems so that species do not become endangered. This proactive approach to conserving avian diversity is discussed by Squires, Hayward, and Gore (chapter 11). Specifically, these authors discuss sensitive species management policies that were adopted by land management agencies in the United States to ensure that currently viable populations of animals and plants do not become threatened or endangered through management actions. This recognizes that recovery costs rise sharply and likelihood of success decreases as population abundance and species distribution contract. The authors note that changes in management philosophy have convinced some biologists and administrators that managing individual sensitive species is no longer necessary to ensure their conservation. Such managers contend that ecosystem management and other coarse filter approaches represent the best means of providing habitat and other life requisites for sensitive species. Squires et al. argue convincingly that this is unlikely to be the case. Rather, a combined approach that uses ecosystem management to prioritize conservation areas based on species richness and rarity and uses individual sensitive species' requirements to set the spatial scale of proposed areas appears better than a strictly ecosystem- or individual species-based approach.

The effectiveness of sensitive species management and its relationship with ecosystem management is clearly illustrated in chapter 11 with an example of Northern Goshawk management in the southwestern United States. There, management recommendations were formulated from the perspective of the goshawk, not from the general perspective of the ponderosa pine ecosystem. Specific recommendations were drafted based on detailed research on individual hawks. The habitat needs of the bird were provided; then the scope of the plan was expanded to the needs of the ecosystem. This resulted in real changes in ponderosa pine management that will improve the health of that ecosystem, such as thinning and restoring natural fire regimes. The goshawk served as a flagship with broad public appeal that galvanized the necessary resources to formulate a management plan. However, as the plan is implemented in daily forest management, it is viewed less as a plan for goshawks and more as an ecosystem management plan for ponderosa pine forests.

The chapters in this section as a whole make four general observations:

1. Successful conservation of rare species is dependent more on biopolitical concerns than on strictly biological concerns. Biologists need to be good politicians so that researchers and managers work together effectively, but we also need to enlist team members with diverse specialties (conflict resolution, sociology, psychology, policy formulation) to effectively conserve birds.

2. Conservation programs require long-term commitment by biologists and funding agencies.

3. The sheer number of sensitive, threatened, and endangered species mandates us to prioritize our conservation efforts. The lack of a mechanism to garner opinions on the relative importance of conserving various species from ornithologists in particular, and ecologists in general, is alarming. We cannot afford to continue to let priorities be set by independent lawsuits. Coordination is essential if we are to maximize the preservation of avian diversity. We discussed ideas for establishing objective prioritization in chapters 1 and 2. Specifics are given in chapter 11, including assigning priority based on rarity, genetic uniqueness, charisma, and ecological function; and an example of fine-scale prioritization within waterfowl is presented in chapter 9.

4. We must move beyond the ecosystem versus individual-species dichotomy to avian management. We need to manage to maintain functioning ecosystems, but we must make sure the needs of endangered, threatened, and sensitive species are met by the coarse filter approach of ecosystem management. In fact, as the Northern Goshawk example suggests, reliance on sensitive species management is likely to be essential to successful ecosystem management because individual species can garner public support and allow management recommendations to be based on defensible scientific assessment of demographics, distribution, and genetics.

CHAPTER 8

Directions in Endangered Species Research

Scott R. Derrickson, Steven R. Beissinger, and Noel F. R. Snyder

Conservation Issues and Previous Research

The present biodiversity crisis is a mass extinction event. This event differs from previous mass extinctions, however, in two important respects: (1) it involves diverse plant and animal taxa across a broad range of terrestrial and aquatic environments; and (2) it is human caused (McNeeley 1992; NRC 1995b). The primary anthropogenic factors underlying recent extinctions and ongoing species declines are habitat destruction and fragmentation, introduced non-native species (i.e., predators, competitors, diseases, and ungulates), and overharvesting (Diamond 1989; NRC 1995b; Steadman 1996). Extinction rates are now several orders of magnitude above background levels (Wilson 1992; May, Lawton, and Stork 1995; Nott, Rogers, and Pimm 1995; Pimm et al. 1995), and are likely to increase further given the large number of currently threatened taxa and accelerating environmental impacts associated with human population growth, resource use, and global commerce (Ludwig, Hilborn, and Walters 1993; Vitousek 1994; Noss, LaRoe, and Scott 1995; Myers 1996; Vitousek et al. 1996).

The current extinction crisis began with human colonization of formerly unoccupied areas (Milberg and Tyrberg 1993; Steadman 1995a). For birds, most recent extinctions have involved island endemics. In the tropical Pacific alone, more than two thousand species—or about 20 percent of the world's current avifauna—have been exterminated, and many others have suffered local or regional extirpations (Pimm, Moulton, and Justice 1994; Steadman 1995b, 1996). The impoverishment of island avifaunas has continued into the present, with island species accounting for about 90 percent of documented extinctions since A.D. 1600 (Johnson and Stattersfield 1990).

About 11 percent of the world's birds are now threatened with extinction. Habitat destruction and fragmentation and small range and/or population size remain the principal threats for about 75 percent of these species. Island species still comprise a large proportion of the total, but the number of threatened taxa in both temperate and tropical continental regions is growing (Collar, Crosby, and Stattersfield 1994). While endemic species with restricted ranges are clearly at the greatest risk of extinction (Bibby et al. 1992; Pimm and Askins 1995; Pimm et al. 1995; Simberloff 1995), populations of many widespread species are also declining (Howe, Geissler, and Harrington 1989; Sauer, Peterjohn, and Link 1994; Tucker and Heath 1994; James, McCulloch, and Wiedenfeld 1996).

The magnitude of the extinction crisis is moving conservation from the reactive management of threatened species toward the proactive management of ecosystems and landscapes (Scott et al. 1993; Grumbine 1994; Turner et al. 1995; Noss 1996). In the following sections, we review this trend but suggest that the preservation of avian biodiversity will require a variety of approaches, and that their successful application ultimately will depend on an increased understanding of single-species ecology and population biology.

Features of Research That Improved Conservation

Birds are well known and well studied in comparison with other animal groups. We know a great deal about their taxonomy, biology, distribution, and abundance, and they have proven useful as indicators of environmental change (Lynch 1980). Not surprisingly, avian studies have played a prominent role in our understanding of extinction processes and in the development of both the theoretical and applied aspects of conservation science (Soulé 1986; Simberloff 1988, 1995; Caughley and Gunn 1996).

Birds are also the only large animal group whose conservation status has been fully reviewed on four occasions (Vincent 1966–71; King 1978–79; Collar and Andrew 1988; Collar, Crosby, and Stattersfield 1994). These works, regional treatments for Africa (Collar and Stuart 1985), the Americas (Collar et al. 1992), and Europe (Tucker and Heath 1994), and national surveys and monitoring programs (cf. J. J. D. Greenwood et al. 1995; Robbins, Bystrak, and Geissler 1986) have drawn considerable attention to the magnitude of the extinction crisis. They have also stimulated a wide variety of conservation initiatives in both developed and developing nations.

The Single-Species Approach

Over the past three decades, avian conservation efforts have focused primarily on the recovery of single threatened and endangered species (hereafter termed simply threatened). These programs have attempted to answer the question of why a species is declining using a step-wise, "comparative" approach adopted from wildlife ecology and management: (1) study the natural history of the species to determine key aspects of its ecology and population biology; (2) list or diagram all potential limiting factors; (3) measure potential limiting factors across the present and former range; (4) identify the limiting factor(s) through comparison and correlation with population density, trends, and primary demographic parameters (i.e., productivity, fecundity, and survivorship; Temple and Wiens 1989); (5) conduct field experiments to confirm the limiting factor(s); (6) eliminate or ameliorate the cause(s) of decline; (7) monitor the population's response; and (8) if necessary, restock vacant areas within the former range (Green and Hirons 1991; Caughley 1994; Green 1995; James and McCulloch 1995).

The recovery programs for the Lord Howe Island Woodhen (*Tricholimnas sylvestris*) and Red-cockaded Woodpecker (*Picoides borealis*) are two of the best avian examples of successful application of this approach. In the case of the woodhen, comparative procedures were used to identify feral pigs as the cause of decline, and following their elimination, woodhens were restored to former habitats through captive breeding and release (Miller and Mullette 1985). In the case of the woodpecker, a variety of field studies and experiments were conducted to determine key aspects of the species biology (i.e., social structure, demography, dispersal, and genetic variation), ascertain the causes of decline (i.e., habitat loss and fragmentation, demographic isolation, and shortage of suitable nest sites), and develop suitable management techniques to increase population size, distribution, and viability (i.e., habitat protection and restoration, artificial nest cavity construction, and translocation) (Walters 1991; Allen, Franzreb, and Escaño 1993; Haig, Belthoff, and Allen 1993).

Because the biology of most threatened species is poorly known, gathering basic biological information and implementing effective countermeasures normally require many years (Green and Hirons 1991; Tear et al. 1995). While some recovery programs have clearly suffered from poor scientific method (Caughley 1994; Green 1995), a variety of significant problems have been encountered. Inadequate knowledge of a species' historic distribution, for example, has led sometimes to faulty assumptions about optimal habitat and thereby misdirected research and management (e.g., Nene [*Branta sandvicensis*] in Black, this volume; Takahe [*Porphyrio mantelli*] in Clout and Craig 1995). In other cases, the initial misdiagnosis of the causes of decline has resulted in the application of ineffective management actions (e.g., California Condor [*Gymnogyps californianus*]; Snyder and Snyder 1989). In other cases, it has proven impossible to eliminate or control the agent of decline (e.g., Brown tree snake [*Boiga irregularis*]; Rodda, Campbell, and Derrickson, this volume) or quickly restore critical habitat (e.g., Palila [*Loxioides bailleui*]; Lindsey et al. 1995). In other cases, inappropriate techniques were applied due to insufficient biological knowledge (e.g., cross-fostering of Whooping Cranes [*Grus americana*] to Greater Sandhill Cranes [*Grus canadensis tabida*]; Lewis 1990]. In still other cases, small population size alone has hampered effective application of comparative and/or experimental methods to identify limiting factors (e.g., Hawaiian Crow [*Corvus hawaiiensis*]; NRC 1992).

A variety of in situ and ex situ management techniques have been adopted to rescue threatened species when surviving populations are small and rapidly declining (table 8.1). Such measures are aimed at stabilizing or increasing population size by enhancing either fecundity or survivorship (Green 1995; Temple 1986). Some programs have used several procedures simultaneously with dramatic results. For example, the population of the Black Robin (*Petroica traversi*) was increased from 5 to 155 birds between 1980 and 1994 by combining translocation of the small remnant population with the provisioning of nest sites, supplemental feeding, and intensive clutch and brood manipulations (Butler and Merton 1992; Collar, Crosby, Stattersfield 1994). Similarly, the Mauritius Kestrel (*Falco punctatus*) population was increased from 6 to 222–286 birds between 1974 and 1994 through a combination of captive breeding, reintroduction, clutch and brood manipulations, predator control, and nest-box provisioning (Cade and Jones 1993; Jones et al. 1995).

Cade and Temple (1995) recently reviewed the effectiveness of thirty intensive management programs for threatened birds that were originally reported in Temple (1978) and involved a range of techniques and species. Thirteen (43 percent) programs were successful in increasing breeding population size, seven (23 percent) assisted in stabilizing population size or slowing the rate of decline, five (17 percent) were inconclusive, and the remaining 5 (17 percent) ended in failure. Translocation and captive breeding/reintroduction were the most successful techniques in this small sample.

Many programs for threatened species have attempted to establish, reestablish, or augment wild populations by translocating wild or captive-reared individuals. Griffith et al. (1989) recently conducted retrospective analyses of native bird and mammal translocations in Australia, New Zealand, Canada, and the United States between 1973 and 1986. They found that translocation success was closely tied to the numbers of animals released, habitat quality, location within the historic range, presence or absence of potential competitors and predators, animal origin (i.e., wild vs. captive-reared), and species status (i.e., threatened vs. nonthreatened). Among translocations of exclusively wild-caught animals, success was highest when source population density was high (77 percent success, $n = 109$) and the source population was increasing (83 percent success, $n = 93$)—conditions that rarely apply to threatened species. Overall translocation success for

Table 8.1. Intensive management techniques and examples of threatened birds that have benefited from their application.

Technique	Species	Reference
Supplemental feeding	Red-crowned Crane (*Grus japonensis*)	Masatomi (1991)
	Trumpeter Swan (*Cygnus buccinator*)	Archibald (1977)
Artificial nest sites	Osprey (*Pandion haeliatus*)	Poole (1989)
	Puerto Rican Parrot (*Amazona vittata*)	Viella and Arnizaut (1994)
Clutch/brood manipulation	Black Robin (*Petroica traversi*)	Butler and Merton (1992)
	Mauritius Kestrel (*Falco punctatus*)	Cade and Jones (1993)
Captive breeding/reintroduction	Peregrine Falcon (*Falco peregrinus*)	Cade (1990)
	Lord Howe Island Woodhen (*Tricholimnas sylvestris*)	Miller and Mullette (1985)
Translocation	Atlantic Puffin (*Fratercula arctica*)	Kress and Nettleship (1988)
	Kakapo (*Strigops habroptilus*)	Moorehouse and Powlesland (1991)
Predator control	Dark-rumped Petrel (*Pterodroma phaeopygia*)	Cruz and Cruz (1987)
	Aleutian Canada Goose (*Branta canadensis*)	Bender (1991)

threatened species was low (44 percent success, $n = 80$), even in high-quality habitat, and the translocation success for captive-reared birds was even lower (32 percent success, $n = 31$). Beck et al. (1994) also reported low success (11 percent success, $n = 145$) for programs involving the reintroduction of captive-bred animals, although their criteria of success were stringent and many programs were still in progress. The lower success of reintroductions involving captive-reared animals can be traced in many instances to behavioral deficiencies resulting from the captive environment (Snyder et al. 1994, 1996).

It is important to recognize that unless intensive management actions are tightly coupled with efforts to identify and ameliorate limiting factors, they can actually prolong rather than promote recovery (Frazer 1992; Meffe 1992; Scott et al. 1994). For example, the reproductive success of Kirkland's Warblers (*Dendroica kirklandii*) was improved by controlling Brown-headed Cowbird (*Molothrus ater*) nest parasitism, but significant population increases did not occur until large tracts of breeding habitat were restored through fire management and succession (Mayfield 1993). Many threatened species remain management dependent because recovery efforts have been initiated too late, resources have been insufficient, and population goals have been determined by political, social, and economic considerations rather than biological requirements and long-term population viability (Tear et al. 1993, 1995; Wilcove, McMillan, and Winston 1993; Murphy et al. 1994).

Ecosystem and Landscape Approaches

While the reactive approach of rescuing single, at-risk species has dominated the conservation arena, increasing attention is now being directed to proactive programs to protect and preserve habitats, ecosystem processes, and overall biodiversity (Noss 1987; Scott et al. 1987, 1993; Franklin 1993b; Grumbine 1994; Noss and Cooperrider 1994; Walker 1995). This change in focus has resulted from the recognition that: (1) the key to species conservation is the protection of natural habitats and species assemblages (Tear et al. 1993, 1995; Murphy et al. 1994); (2) the degradation of natural habitats, ecosystems, and evolutionary potential is accelerating (Noss, LaRoe, and Scott 1995); (3) there are "hot spots" of biodiversity and "megadiversity" nations (Myers 1988; Mittermeier 1988; Bibby et al. 1992; WCMC 1992); (4) protected areas are strongly influenced by their surrounding landscapes (Newmark 1995; Wiens 1995); and (5) existing reserve systems are inadequate for preserving overall biodiversity (Dinerstein and Wikramanayake 1993; Scott et al. 1993; McNeeley 1994). Although a variety of habitat-based approaches have been suggested and applied (e.g., preserving natural communities [Noss 1987]; multispecies habitat conservation plans [NRC 1995b]), the strategy of protecting, restoring, and sustainably managing ecosystems and landscapes has received the widest attention. This approach envisions a landscape mosaic that includes core reserves, buffer zones, and habitat corridors within a matrix of lands subject to more intensive human use (Grumbine 1994; Noss and Cooperrider 1994). While ecosystem management incorporates both resource use and biodiversity conservation, its core objective is to sustain ecosystem integrity and resilience (Grumbine 1994; Christensen et al. 1996). This approach has been endorsed by a number of conservation and scientific organizations and government agencies and is being implemented in a variety of forms at many sites (Mann 1995; Christensen et al. 1996; Yaffee 1996). Ecosystem management represents a major shift in traditional natural resource policy, yet we are still far from establishing a comprehensive biodiversity reserve system in the United States (Blockstein 1995) or other parts of the world (McNeeley 1994; Dinerstein and Wikramanayake 1993). And despite considerable headway in international agreements to preserve biological diversity (e.g., CITES, Rio de Janeiro Convention on Biological Diversity), many of the same social, political, and economic issues that have hampered single-species management efforts will still need to be resolved if this approach is to succeed (N. Williams 1995; Eisner et al. 1995; NRC 1995b).

Integrating Conservation Approaches

The single-species approach has been criticized as difficult, expensive, and unreliable (Csuti, Scott, and Estes 1987; Hutto, Reel, and Landres 1987; LaRoe 1993). These criticisms, however, ignore a number of important considerations. First, a number of single-species efforts have resulted in either increasing or stabilizing wild populations (Cade and Temple 1995; NRC 1995b). Second, single species can serve as an "umbrella" or "flagship" for the conservation of critical habitats for many other species (Butler 1992; Eisner et al. 1995). Third, many species will continue to require individual attention if they are to survive and their habitats are to be restored. Recovery and restoration will be expensive, but it cannot be simply dismissed (Blockstein 1995). Instead, we must use the insights gained from past experience to improve the processes of recovery planning and implementation (Tear et al. 1993, 1995; Clark 1995; NRC 1995b). Fourth, the preservation of biological diversity will continue to require a spectrum of conservation approaches, from protective legislation to sustainable ecosystem and landscape management, and each must be applied as appropriate (Cade and Temple 1995; NRC 1995b). Finally, and perhaps most important, single-species and ecosystem/landscape management should not be viewed as simple alternatives (Squires, Hayward, and Gore, this volume). Because populations are generally more sensitive indicators of environmental stress than ecosystems (Carpenter et al. 1995; Eisner et al. 1995), sustainably managing ecosystems and landscapes will require intensified knowledge and monitoring of single-species populations (Grumbine 1994; NRC 1995b). Furthermore, because populations of threatened species are already stressed by environmental change, carefully designed autecological studies of the factors affecting and regulating their populations can provide information relevant to both their recovery and the restoration and maintenance of community- and ecosystem-level processes (Woolfenden and Fitzpatrick 1991; Breininger et al. 1995; James, Hess, and Kufrin 1997).

Research Needed to Further Conservation

Below we discuss ten areas where further research is needed if we are to improve conservation of endangered species.

Documenting Distribution

Distributional information is essential for conservation planning whether it involves developing a recovery strategy for a single species, evaluating the adequacy of existing reserves, determining the impacts of different land-use or management regimes, or targeting areas for biodiversity protection. Species distributions vary through time in response to both natural and anthropogenic factors (Jehl and Johnson 1993; Keast 1995), and monitoring these changes can provide insight into both their underlying causes and species requirements. Avian distributions are relatively well known in comparison to other taxa, but many regions and species remain poorly sampled. For many threatened species, our knowledge of distribution is limited to the locales where specimens were collected decades or centuries ago. This can be corrected only through appropriately designed field surveys. The compilation of regional and national biodiversity inventories should also receive high priority (Scott et al. 1993, 1994).

Assessing Abundance

Although the technical literature on avian census techniques is large and well developed (DeSante and Rosenberg, this volume), abundance estimates are available for few threatened species. Green and Hirons (1991) found that estimates of abundance were available for only 202 (20 percent) of the 1,029 threatened species listed in Collar and Andrew (1988), and most of these species were not being monitored. This situation has improved (Collar, Crosby, and Stattersfield 1994), but estimates of population size and trend are still lacking for the vast majority of threatened birds, and surveys need to be undertaken. These surveys must employ methods appropriate to the species, be replicable, and be conducted at biologically meaningful intervals to provide trend information.

Determining Biological Characteristics

We know very little about the basic natural history, ecology, and demography of the vast majority of threatened species. Recovery efforts have been, and will continue to be, hampered by deficits in basic biological knowledge and the difficulty of obtaining this information when populations are small and declining. Unless we begin to gather basic biological data and document ecological requirements before species populations are severely reduced, most threatened species will likely remain conservation dependent for extended periods. Because it is easier to save species while they are still common (Scott et al. 1993), we must become more proactive in monitoring populations and gathering essential biological information. The establishment of regional and/or continent-wide monitoring programs that include habitat-specific surveys and also gather information on primary demographic parameters must receive a high conservation priority (DeSante and George 1994; J.J.D. Greenwood et al. 1995).

Determining Habitat Requirements

While the long-term maintenance of viable populations will ultimately depend on our ability to protect, restore, and create a biologically relevant mosaic of habitats within a human-dominated landscape, we lack the necessary information on habitat requirements to make even the simplest conservation decisions for most species, whether threatened or common. How do specific habitat features influence movements and dispersal (Wiens 1995)? Can losses of specific habitats be related to population declines (Rappole, Powell, and Sader 1994)? Can habitat selection be related to site-specific microhabitat requirements (McShea et al. 1995)? Can specific habitat characteristics be related to primary demographic parameters (Breininger et al. 1996)? Do certain habitats act as population sources and others as sinks, and are these patterns consistent through time (Brawn and Robinson 1996)? The rapid development of remote-sensing and geographic information system technologies has greatly improved our ability to effectively address these and other habitat-related questions at appropriate scales (Villard, Schmidt, and Maurer, this volume).

Assessing Genetic Variation

Molecular genetic techniques are being used both to explore questions of pragmatic conservation interest (e.g., taxonomic and systematic distinctiveness, hybridization and introgression, genetic effects of population fragmentation, etc.) and to develop specific management recommendations (e.g., genetic management of captive populations, identification of source populations for translocation, etc.) (Haig and Avise 1996; Fleischer, this volume). The rapid development of molecular techniques suggests that their applications in conservation will increase. The interface between the genetic and the demographic dynamics of populations remains an important topic for investigation.

Establishing Captive Populations

Ex situ populations have served an important role in the recovery of a number of critically endangered species and will undoubtedly have a role to play in the conservation of many others in the future. However, captive breeding as a recovery technique has a number of significant limitations, and its use should be restricted to situations where preferable ex situ conservation options are immediately unavailable or can be significantly enhanced (Snyder et al. 1996).

Improving Translocation Methods

Because populations of threatened species are often small, disjunct, and/or range-restricted, recovery plans frequently include proposals for translocation (Tear et al. 1993). Translocations can potentially serve a variety of conservation purposes, including augmenting population size; augmenting a population's genetic diversity when natural dispersal is prevented; reestablishing populations within their historic range; establishing satellite populations to minimize extinction risks; relocating individuals to suitable habitats following local extirpations; re-creating natural community assemblages; and identifying limiting factors in areas of former occupancy. While translocation will be an increasingly important conservation tool for managing species in reserves and fragmented landscapes, experience indicates that most translocations of threatened species will fail. Retrospective analyses provide some insight into the factors underlying translocation success (Griffith et al. 1989), but additional experimental studies are needed. Improving the translocation success of captive-raised birds will undoubtedly require detailed behavioral investigations and creative solutions (Költringer, Dodeikat, and Curio 1993; Maloney and McLean 1995).

Conducting Surrogate Studies

Many recovery programs have used biologically similar species to evaluate demographic parameters or develop rearing, propagation, marking, and reintroduction techniques (Scott and Carpenter 1987; Page, Quinn, and Warriner 1989; Snyder and Snyder 1989; Keuhler et al. 1994; Marzluff, Valutis, and Witmore 1995). Wiens (1995) has suggested that progress in understanding fragmentation effects could be hastened by targeting selected species for study that can serve as models for a larger suite of species that share ecological, life-history, or distributional features. Could surrogates be used effectively to address other important conservation and management issues?

Controlling Exotics

About 5.8 percent of the species presently listed by Collar, Crosby, and Stattersfield (1994) are threatened as a result of introduced predators, diseases, competitors, and grazing ungulates. To date, exotic introductions have primarily affected island birds, but their impact on continental species can be expected to grow as avian populations become increasingly restricted to habitat fragments in human-modified landscapes, and exotic organisms are redistributed across the globe at an increasing rate by human beings (Vitousek et al. 1996). Exotic species will likely become the most significant conservation problem to be faced in trying to preserve and restore native species and ecosystems. To minimize the risk of future invasions, national quarantine procedures and regulations will have to be improved, and effective detection, eradication, and control procedures will have to be developed and implemented.

Developing Population Models

Much work with threatened species has involved incorporating field data into several types of models:

1. *Matrix projection models* use a species life cycle and its deterministic demographic rates to estimate the population multiplication rate (l) and the factors most affecting it. They can be useful for evaluating the relative impacts of different limiting factors; however, inferences drawn from sensitivity analyses can be misleading and may not always indicate the most appropriate management strategy (Lande 1988; Green and Hirons 1991; Lebreton and Clobert 1991).

2. *Population viability analysis (PVA) models* project specific populations into the future to estimate the likelihood of extinction due to environmental variation and various management options. PVA models depend on detailed demographic and environmental information (Shaffer 1987, 1990; Soulé 1987; Burgman, Akçakaya, and Loew 1988; Boyce 1992; Heppell, Walters, and Crowder 1994). Some approaches try to establish direct links between extinction probabilities and ecosystem management practices, such as the response of the Snail Kite (*Rostrhamus sociabilis*) to hydroperiod characteristics and water management regimes (Beissinger 1995). If these models apply demographic measures to particular habitat patches, they are usually referred to as *metapopulation models* (Burgman, Ferson, and Akçakaya 1993).

3. Completely *spatially explicit models* follow individuals (rather than populations) across landscapes that are composed of many different kinds of habitats. Individuals are projected through time as they breed, disperse, or settle among territories depending on mating status and territory quality (Pulliam, Dunning, and Liu 1992; Dunning et al. 1995). Some forms of spatially explicit models that incorporate GIS-based data are already being used as land management and planning tools for particular species and landscapes (NRC 1995b; Pulliam et al. 1995; Noon and McKelvey 1996).

4. *Genetic models* estimate effective population size. These are used much less frequently than demographic models because their results cannot be related directly to extinction probabilities (Lande 1988).

Most modeling efforts suffer from a lack of quality information. This problem has been especially severe for PVA, which depends not only on accurate measures of survivorship and fecundity, but also on estimates of the variation in these rates. Accurate estimators of demographic rates and variances require that large data sets be collected over long periods. However, PVAs have been conducted for many species when best "estimates" have been used in place of accurate estimators. Predictions of population viability resulting from such analyses must be viewed with considerable skepticism (Caughley 1994). Furthermore, because available PVA models differ in their assumptions and internal structure and can provide markedly different predictions of population viability from the same set of demographic data (Lindenmayer et al. 1995; Mills et al. 1996), conservative interpretation of their results is also warranted.

Most models ignore behavioral variation instead of using it to structure the underlying life-cycle diagram or add complexity in order to improve model accuracy. Dispersal dynamics are very important to metapopulation and spatial models, but again, detailed data on dispersal rates, and mean dispersal distances (e.g., negative exponential, step function, uniform, etc.) are lacking for most species. Mathematical distributions of dispersal distances are often assumed on the basis of little information, and the effects of landscape elements and barriers on dispersal dynamics are usually ignored.

All of these factors can have important effects on the predictions obtained from these models. They can be magnified and compounded by PVA and spatially explicit models, which project errors into the future for decades and centuries. If too much is assumed, modeling becomes an exercise of "garbage in, garbage out." Nevertheless, PVA and spatially explicit models of populations and landscape dynamics appear to be the wave of the future and are going to be used more frequently in conservation planning (Dunning et al. 1995; NRC 1995b; Turner et al. 1995; Wiens 1995). If they are to achieve a high level of accuracy and precision, these models will require a much more detailed understanding of species behavior and ecological factors that influence demography, dispersal, movements, and habitat choice. It is a golden opportunity to meld behavioral ecology and theory into conservation biology, assuming that these data-intensive efforts are an improvement over less demanding population- and community-level approaches. To succeed, however, behavior must be translated across scales from individuals to populations to ecosystems and landscapes.

Making Research Effective for Conservation

With human populations now growing at a rate of about 1 million every four days, the destruction, fragmentation, and degradation of natural habitats can be expected to continue at an accelerating rate. The conservation of an increasing number of threatened species will require research of an unprecedented scale and effectiveness to diagnose limiting factors and develop effective countermeasures. High-priority research tasks for threatened species include: (1) documenting and monitoring distribution and abundance; (2) documenting basic biological characteristics and ecological requirements; (3) determining the factors that influence demography, dispersal, movements, and habitat selection; (4) developing effective translocation techniques; (5) developing effective eradication and control measures for exotic species; and (6) developing and refining PVA and spatially explicit population models for conservation planning and ecosystem management. When properly conceived and effectively implemented, research and recovery programs for threatened species can significantly assist broader efforts to preserve biodiversity and sustainably manage ecosystems.

Acknowledgments

We thank Chris Wemmer, John Rappole, and John Marzluff for reviewing and improving our manuscript.

CHAPTER 9

Threatened Waterfowl: Recovery Priorities and Reintroduction Potential with Special Reference to the Hawaiian Goose

Jeffrey M. Black

Conservation Issues and Previous Research

Waterfowl are an attractive and well-studied group of animals. A wealth of information about all aspects of waterfowl biology, ecology, and management has been generated by many dedicated workers that traipse through the varied wetland and inland habitats on which the birds depend (Weller 1988; Owen and Black 1990; Batt et al. 1992; Baldassare and Bolen 1994). The conservation and management of waterfowl and wetlands in North America is a multibillion dollar enterprise structured by intricate legislation, financed by taxes (Patterson 1995) and based on "political ecology" (Boyd 1991). Waterfowl and wetlands are the focus of large-scale international conventions, and the birds are conserved as a resource as well as for their own right (Moser, Prentice, and van Vessem 1993).

The majority of waterfowl taxa (≈ 180 of the 230) are considered to be safe in terms of their numbers, distribution, and threats to their conservation status (Green 1996). Healthy populations are generally influenced by changes in climate, food, water, and predators (Owen and Black 1990; Batt et al. 1992). For a range of reasons, 45 waterfowl taxa are currently globally threatened and at least 7 have gone extinct since 1860 (Green 1996; Callaghan, in press).

In this chapter I review the relative conservation effort that is being made on behalf of each of the globally threatened waterfowl, thus identifying shortfalls and priorities for the future. There are many steps in the process to restore small populations to viability. When populations fail to respond to initial protective measures, more intensive strategies may be required, culminating in captive propagation and reintroduction. I comment on the need and suitability of launching intensive captive breeding programs for each of the threatened waterfowl and describe the main requirements and guidelines for avian reintroductions. I will show that the Hawaiian Goose (*Branta sandvicensis*) has been the focus of an extensive recovery initiative, employing many proactive recovery strategies. Yet the species is still at risk of extinction. I outline some of the lessons that can be learned from this long-term program.

Features of Research That Improved Conservation

Broadscale approaches aimed at maintaining species diversity are likely to contribute more to nature conservation than are single-species reintroduction programs. Several worthy large-scale initiatives have begun (e.g., Global Biodiversity Assessment; Heywood 1995). Birdlife International's Biodiversity Project has identified 221 Endemic Bird Areas covering 5 percent of the earth's land surface, on which 75 percent of the world's 300 + threatened bird species occur (Bibby 1995). Bibby stresses that targeting conservation effort at these centers of endemism will be more effective than trying to save the hundreds of species one by one.

However, the ecosystem approach may not be effective in saving species or populations with precariously small numbers. If humans value these species or populations, direct and pragmatic management action has the potential of restoring them to viability. The process toward recovery of a threatened species begins with developing hypotheses about the limiting factors through counts and assessment of ecological and biopolitical threats. Then a stepwise series of recovery action initiatives are required, beginning with natural recovery strategies that treat the root cause for the species' decline, through protective legislation and refuge establishment. Examples where natural recovery action was successful in enabling natural recovery include European populations of Peregrine Falcons (*Falco peregrinus*) and Sparrowhawks (*Accipiter nisus*) after passage of protective legislation and a ban on DDT (Newton 1986; Ratcliffe 1993).

Natural recovery strategies are the most influential and should obviously be the first actions to be initiated. Should the population fail to respond, increasingly intensive actions may be required. Subsequent actions include a phase of detailed research programs, the development of an Action Plan (and/or a Population and Habitat Viability Assessment), and a series of proactive recovery strategies. Proactive strategies include managing habitat or supplying food, reducing predators and competitors, providing nest sites, and translocating eggs or birds (Cade and Temple 1995). Should all else fail, the final proactive recovery strategy, reintroduction of captive-bred birds and subsequent intensive monitoring, may be appropriate.

Temple's (1978) seminal book, *Endangered Birds: Management Techniques for Preserving Threatened Species*, reviewed and encouraged the use of proactive recovery strategies on behalf of seriously endangered birds. Cade and Temple's (1995) recent review of the thirty proactive projects that were outlined in the book revealed that 43 percent of the initiatives were largely successful. Captive propagation and reintroduction programs were the most successful methods; three of four captive-breeding and five of seven reintroduction programs met their criteria for success. Supplemental feeding, provision of nests, and fostering and cross-fostering of eggs and young were the least successful methods.

Massive supplemental feeding programs in the form of growing food on North American and European wildlife refuges has been extremely successful in enhancing goose numbers (Ankney 1996; Cracknell, Madsen, and Fox 1997). Threatened goose populations that had been overexploited have been restored to viable levels through a combination of natural and proactive recovery strategies. For example, the initiation of hunting regulations, breeding sanctuaries, and habitat enhancement was highly effective in increasing numbers of the Svalbard Barnacle Goose (*Branta leucopsis*) population from three hundred in the 1940s to over twenty thousand in 1995 (Owen and Norderhaug 1977; Black 1995a). A parallel North American story of overexploitation and subsequent recovery concerns the Trumpeter Swan (*Cygnus buccinator*), whose numbers recovered from fewer than one hundred in the 1930s to nearly ten thousand in 1985, through stopping the trade in swan pelts, comprehensive protection from illegal hunting, winter feeding, and an extensive translocation program (Banko 1980).

Captive propagation and reintroduction come into their own when a species is otherwise beyond recovery; when natural recolonization is unlikely and other more effective management and conservation measures have been exhausted or are overdue (Derrickson, Beissinger, and Snyder, this volume). Reintroduction programs that are initiated under appropriate conditions and in a timely fashion are potentially effective conservation tools. Unfortunately, this potential is apparently rare, as reintroduction programs may require more energy, finance, and long-term commitment than program participants are able to provide. Reviewing programs for all types of animals, Griffith et al. (1989) reported a 32 percent success rate from 31 captive-rearing, reintroduction projects. Using more stringent achievement criteria (populations increased to five hundred individuals), Beck et al. (1994) calculated only an 11 percent success rate for 145 reintroduction projects using captive-bred animals.

The success rate for birds may well be greater than the values reported above, as there are several additional avian reintroduction programs that have resulted in solid recoveries. For example, reintroductions of Peregrine Falcons in the United States and Europe (Cade et al. 1988; Cade and Temple 1995); Goshawks (*Accipiter gentilis*) in Britain (Marquiss and Newton 1982); and Mauritius Kestrels (*Falco punctatus*) in Mauritius (Jones et al. 1995) have all resulted in viable populations. The White-headed Duck (*Oxyura leucocephala*) in Spain has recovered from a low of 22 in 1977 to over 900 in 1996, after a range of recovery initiatives, including the release of 270 captive-reared birds between 1987 and 1993 (A.J. Green, personal communication). The Giant Canada Goose (*Branta canadensis maxima*) was reestablished to healthy status through extensive translocation and reintroduction programs in numerous states in America (Hanson 1965; Szymczak 1975; Lee et al. 1984). Similarly, the Aleutian Canada Goose (*Branta canadensis leucopareia*) in Alaska has had good fortune through the removal of introduced Arctic Foxes (*Alopex lagopus*), revised hunting regulations, translocation of wild-caught birds, and release of more than 1,000 captive-reared birds. Numbers have increased from less than 800 in 1975 to 20,000 in 1995 (Byrd, in press). Success is also likely to be achieved in the reintroduction programs for Atlantic Puffins (*Fratercula arctica*) in Maine (Kress, this volume), White-tailed Sea Eagles (*Haliaeetus albicilla*) in Scotland, Eagle Owls (*Bubo bubo*) in Germany, and Griffon Vultures (*Gyps fulva*) in France (Cade and Temple 1995).

Conservation Priorities for Threatened Waterfowl

Ranking species and subspecies according to conservation criteria is the first step in identifying priorities, causes of decline, and actions required for recovery. The process was initiated for waterfowl taxa by Kear and Williams (1978) and Kear (1979). This work was revised, resulting in several publications with increasingly accurate assessments as further data became available (Ellis-Joseph, Green, and Hewston 1992; Green 1992b; Callaghan and Green 1993; Green 1996; Callaghan, in press).

Green (1996) lists several key features that may explain the cause of original decline and current limiting factors affecting threatened waterfowl ($n = 48$ in his list). Habitat loss affects 73 percent of the threatened taxa, hunting affects 48 percent, and introduction of exotic species affects 33 percent (see table 9.1). Threatened waterfowl are more likely to be nonmigratory and breed south of 55° latitude. Island species are also more likely to be threatened than continental species. Russia ($n = 14$) and China ($n = 11$) claim more threatened waterfowl than other countries, whereas New Zealand has the largest number of highly threatened taxa (excluding vulnerable, $n = 4$). An exceptional concentration of seven threatened, migratory taxa are found in the east Asian flyway (Green 1996).

Research Needed to Further Conservation

Based on the most up-to-date method for classifying threatened species, there are currently 45 threatened waterfowl in the world (Callaghan, in press). Figure 9.1 ranks them in order of the recovery effort that has been initiated on their behalf.

Threatened waterfowl	Number of action strategies

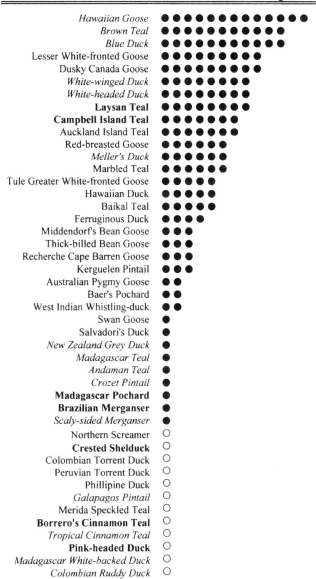

Hawaiian Goose	● ● ● ● ● ● ● ● ● ● ● ● ●
Brown Teal	● ● ● ● ● ● ● ● ● ● ● ●
Blue Duck	● ● ● ● ● ● ● ● ● ● ● ●
Lesser White-fronted Goose	● ● ● ● ● ● ● ● ●
Dusky Canada Goose	● ● ● ● ● ● ● ● ●
White-winged Duck	● ● ● ● ● ● ● ●
White-headed Duck	● ● ● ● ● ● ● ●
Laysan Teal	● ● ● ● ● ● ● ●
Campbell Island Teal	● ● ● ● ● ● ●
Auckland Island Teal	● ● ● ● ● ● ●
Red-breasted Goose	● ● ● ● ● ●
Meller's Duck	● ● ● ● ● ●
Marbled Teal	● ● ● ● ● ●
Tule Greater White-fronted Goose	● ● ● ● ●
Hawaiian Duck	● ● ● ● ●
Baikal Teal	● ● ● ● ●
Ferruginous Duck	● ● ● ●
Middendorf's Bean Goose	● ● ●
Thick-billed Bean Goose	● ● ●
Recherche Cape Barren Goose	● ● ●
Kerguelen Pintail	● ● ●
Australian Pygmy Goose	● ●
Baer's Pochard	● ●
West Indian Whistling-duck	● ●
Swan Goose	●
Salvadori's Duck	●
New Zealand Grey Duck	●
Madagascar Teal	●
Andaman Teal	●
Crozet Pintail	●
Madagascar Pochard	●
Brazilian Merganser	●
Scaly-sided Merganser	●
Northern Screamer	○
Crested Shelduck	○
Colombian Torrent Duck	○
Peruvian Torrent Duck	○
Phillipine Duck	○
Galapagos Pintail	○
Merida Speckled Teal	○
Borrero's Cinnamon Teal	○
Tropical Cinnamon Teal	○
Pink-headed Duck	○
Madagascar White-backed Duck	○
Colombian Ruddy Duck	○

Figure 9.1. Threatened waterfowl ranked according to implementation of recovery actions. Rankings are between 0 (little or no recovery action) and 13 (all forms of recovery action in place, culminating in the active monitoring of individuals and the environmental situation after reintroductions). Critically endangered species are in bold, endangered in italics, and vulnerable in a normal font.

Table 9.1. Globally threatened Anseriforme species and subspecies, their threats, recovery effort, and captive propagation and potential.

Common name[1]	Scientific name	Distribution	Threats[2]	Recovery effort & strategies[3]	Captive stocks (ease of keeping)[4]
Critically Endangered					
Crested Shelduck	*Tadorna cristata*	E Asia	US	0	None (1)
Laysan Teal	*Anas laysanensis*	Hawaii	LIS	1–6, 8, 11	Yes (1)
Campbell Island Teal	*Anas nesiotis*	S New Zealand	I	1–6, 11	Yes (2)
Pink-headed Duck	*Rhodonessa caryophyllacea*	Myanmar	UL	0	None (2)
Borrero's Cinnamon Teal	*Anas cyanoptera borreroi*	Colombia	LH	0	None (1)
Madagascar Pochard	*Aythya innotata*	Madagascar	LH	0–1	None (2)
Brazilian Merganser	*Mergus octosetaceus*	E South America	L	1	None (3)
Endangered					
Madagascar White-backed Duck	*Thalassornis leuconotus insularis*	Madagascar	HLI	0	None (3)
Hawaiian Goose	*Branta sandvicensis*	Hawaii	IL	1–13	Yes (1)
Blue Duck	*Hymenolaimus malacorhynchos*	New Zealand	LI	1–6, 8, 10–13	Yes (3)
White-winged Duck	*Cairina scutulata*	S Asia	L	1–6, 11–12	Yes (1)
Meller's Duck	*Anas meller*	Madagascar	LH	1–5, 11	Yes (2)
New Zealand Grey Duck	*Anas superciliosa superciliosa*	New Zealand	IZ	1	Yes (2)
Madagascar Teal	*Anas bernieri*	Madagascar	L	11, 4, 11	None (2)
Andaman Teal	*Anas albogularis*	Andaman Islands, India	LH	1	None (2)
Galapagos Pintail	*Anas bahamensis galapagensis*	Galapagos Islands	LS	0	None (2)
Brown Teal	*Anas chlorotis*	New Zealand	IL	1–8, 11–13	Yes (2)
Tropical Cinnamon Teal	*Anas cyanoptera tropica*	Colombia	LH	1	None (1)
Crozet Pintail	*Anas eatoni drygalskii*	S Indian Ocean	I	0	Yes (2)
Scaly-sided Merganser	*Mergus squamatus*	SE Asia	ILH	1	None (3)
White-headed Duck	*Oxyura leucocephala*	Mediterranean to W. Asia	LHZ	1–6, 11–13	Yes (1)
Colombian Ruddy Duck	*Oxyura jamaicensis andina*	Colombia, South America	L	0	None (1)
Vulnerable					
Northern Screamer	*Chauna chavaria*	NW South America	L	0	Yes? (2)

Common name	Scientific name	Distribution	Current threats	Recovery action required	Ease of captive breeding
Recherche Cape Barren Goose	*Cereopsis novaehollandiae grisea*	SW Australia	C	1, 3	Yes (1)
Swan Goose	*Anser cygnoides*	E Asia	LH	1-3	Yes (1)
Middendorf's Bean Goose	*Anser fabalis middendorffi*	E Asia	LH	1-3	Yes? (1)
Thick-billed Bean Goose	*Anser fabalis serrirostris*	E Asia	LH	1-3	Yes (1)
Tule Greater White-fronted Goose	*Anser albifrons gambeli*	W North America	S	1-5	Yes (1)
Lesser White-fronted Goose	*Anser erythropus*	E Europe & Asia	ULH	1-6, 11-13	Yes (1)
Red-breasted Goose	*Branta ruficollis*	E Europe & W Asia	LS	1-6	Yes (1)
Dusky Canada Goose	*Branta canadensis occidentalis*	W North America	LH	1-4, 5, 9-11	Yes (1)
Peruvian Torrent Duck	*Merganetta armata leucogenis*	C South America	LPI	0	None (3)
Colombian Torrent Duck	*Merganetta armata colombiana*	N South America	LPI	0	None (3)
Australian Pygmy Goose	*Nettapus coromandelianus albipennis*	NE Australia	LI	1-2	Yes (2)
Salvadori's Duck	*Salvadorina waigiuensis*	New Guinea	LH	0, 2	None (3)
Hawaiian Duck	*Anas wyvilliana*	Hawaii	LHI	1-2, 6, 11-12	Yes (1)
Philippine Duck	*Anas luzonica*	Philippine Islands	LH	0	Yes (1)
Auckland Island Teal	*Anas aucklandica*	S New Zealand	I	1-7	Yes (2)
Baikal Teal	*Anas formosa*	E Asia	HP	1-2, 4-6	Yes (2)
Merida Teal	*Anas andium altipetens*	W South America	LH	0	Yes (1)
Kerguelen Pintail	*Anas eatoni eatoni*	S Indian Ocean	I	1, 4, 10	None (2)
Marbled Teal	*Marmaronetta angustirostris*	Mediterranean to W. Asia	LH	1-6	Yes (1)
Ferruginous Duck	*Aythya nyroca*	S Europe & Asia & Africa	LH	1, 4-6	Yes (1)
Baer's Pochard	*Aythya baeri*	E Asia	LH	1, 3	Yes (1)

[1] Listed according to latest methods according to Callaghan (in press).

[2] Current threats: U = unknown; L = loss of habitat; H = hunting; I = introduced species; Z = hybridization; S = small range, highly concentrated; P = pollution (from Green 1996).

[3] Numbers refer to—**Recovery action required**: 0) Little or no current action; **Natural recovery stategies**: 1) Monitoring distribution and population assessments, 2) Legislation and protection from limiting factors, 3) Refuge establishment; **Detailed research programs**: 4) Ecological needs and constraints, 5) Mating system and social structure, 6) Action plan documented and/or population viability assessment; **Proactive recovery strategies**: 7) Habitat management, restoration, and/or supplemental feeding, 8) Alleviating competition, predation, etc., 9) Providing and enhancing nest sites (boxes, platforms, ledges, etc.), 10) Translocation, fostering or cross-fostering eggs and/or birds, 11) Intensive and comprehensive captive breeding, 12) Reintroduction, and 13) Monitoring individuals and situation after release.

[4] Numbers in parentheses refer to the relative ease of keeping and breeding birds in captivity: 1 = straightforward, 2 = intermediate, 3 = difficult.

The shading on the figure indicates three levels of effort: programs that are well established ($n = 21$ taxa), those in which initial counts and some legislation or protection have occurred ($n = 12$), and those without any form of recovery effort ($n = 12$). The comparison shows that only 2 of the 7 critically threatened and 6 of the 15 endangered waterfowl have a sizeable recovery program in place. The lower 24 taxa (53.3 percent) are in need of the most basic recovery initiatives.

Eighteen taxa (40.0 percent) have a full complement of the natural recovery strategies (table 9.1). Detailed research into the ecology and behavior of threatened taxa has been initiated for 17 taxa (37.7 percent), the Blue Duck (*Hymenolaimus malacorhynchos*), Hawaiian Goose, White-headed Duck, and Tule White-fronted Goose (*Anser albifrons gambeli*) being foremost in this area of concern. Recovery action plans have been made for 16 taxa (35.6 percent); suggestions for preparing such plans are provided by Green, Black, and Ellis-Joseph (1993). Thirteen taxa (28.9 percent) have some form of proactive recovery action, 12 (26.7 percent) have captive-rearing programs, and six (13.3 percent) have had active reintroduction programs.

Feasibility of Captive Propagation

It has been argued that captive propagation of threatened species can buy time to initiate or correct management error, restore habitat, and stop human overexploitation. Establishing such programs should be initiated before wild stocks dwindle to critical levels, thus reducing controversy involved in the capture of a species' last survivors (Flesness and Foose 1990; Seal 1991). In this section I review the suitability of each of the threatened waterfowl as candidates for captive propagation prior to reintroduction.

Rearing waterfowl in captivity is a well-rehearsed business employed by many zoos and aquaria. Species-specific methods are adjusted according to several phenotypic and genetic traits (Kear 1975, 1977). Through the auspices of the International Union for the Conservation of Natural Species Survival Commission (IUCN/SSC) Captive Breeding Specialist Group (CBSG), institutions that keep waterfowl have been provided with a preliminary set of guidelines to help determine which taxa are best to keep in terms of contributing to conservation aims (Ellis-Joseph, Green, and Hewston 1992; Green and Ellis 1994). The guidelines recommend that some level of captive propagation should be instigated for 150 of the 234 taxa and that 84 taxa should be managed to extinction in captive collections to free up space for higher-priority species. They identify 39 taxa for which intensively managed captive programs should be stepped up or instigated within ten years. The level of the programs should ensure the preservation of 90 percent of the average heterozygosity of the wild gene pool for one hundred years. The actual number of individuals required will vary, but this will probably be in the region of fifty to two hundred birds that actively breed; actual flock sizes may need to be much larger, as it has been shown that only a small proportion of birds in a population actually succeed in breeding (Newton 1989).

Captive stocks are being held for seventeen of the forty-five threatened waterfowl, for which there are twelve active propagation programs for reintroductions (table 9.1). Only two of the seven critically threatened and eight of the fifteen endangered waterfowl are held in captive collections.

Based on experiences at the Wildfowl & Wetlands Trust, I ranked each threatened taxa according to the ease with which they have been (or probably could be) kept and bred in captivity. Rank 1 means that given appropriate conditions the birds will readily breed and survive in captivity. Rank 2 is intermediate, often indicating some problems with susceptibility to disease, predation, or hybridization. Rank 3 means they are very difficult to keep alive and breed in captivity, requiring special and often expensive conditions.

The Madagascar White-backed Duck (*Thalassornis leuconotus insularis*), the mergansers, and all four river-dwelling ducks were ranked difficult; fifteen were ranked intermediate; and twenty-three were more straightforward to keep and breed in captivity. Our successful experiences with the river-dwelling Blue Duck and other more difficult species, like Freckled Duck (*Stictonetta naevosa*), indicate that most problems can be overcome. Therefore, captive propagation would be feasible for most waterfowl once an appropriate number and quality of birds are in captivity and adequate techniques are practiced.

There is much to accomplish in relation to initiating and managing adequate captive stocks of threatened waterfowl. Since the circulation of the CBSG's recommendations for waterfowl (Ellis-Joseph, Green, and Hewston 1992), few expeditions have been mounted to collect birds for captive programs for the forty-five threatened waterfowl; there are no birds in captivity for twenty-eight taxa (62.2 percent). The Jersey Wildlife Preservation Trust and the New Zealand Department of Conservation are among the few organizations that have recently attempted to secure threatened waterfowl for captive programs (Young and Smith 1989; Goudswaard 1991; Wilme 1994; Williams and Robertson 1997).

Reintroduction As a Conservation Tool

Reintroduction programs are complex affairs demanding input from several disciplines all at once, e.g., education, public relations, fund-raising, captive breeding and care, genetics, biology, ecology, behavior, population dynamics, modeling, and conservation politics (Stanley Price 1989; Black 1991; Gipps 1991; Olney, Mace, and Feistner 1994). By considering the assessment criteria that are now available (Kleiman, Stanley Price, and Beck 1994), it is possible to determine what aspects of a potential program are liable to succeed and fail before releasing any birds. Such assessment exercises may help to evaluate whether reintroduction projects are feasible and may save valuable resources.

When the species or population in question is fortunate enough to be held in high esteem by adjacent human populations, the usefulness of reintroduction programs to ecosystem conservation escalates. It is likely that support for biodiversity conservation will be boosted when "flagship" reintroduction programs have succeeded in restoring healthy populations. However, there is a danger that the publicity from such programs can mislead decision makers into thinking the problem is solved before results have been actually assessed (Ounsted 1991).

Guidelines, Precautions, and Risks

Reintroductions of birds are taking place throughout the world, often without ensuring that the appropriate groundwork has been done, or worse still, that the original cause for the demise of a population is still active. Releasing captive-bred animals into the wild is risky for the sole reason that disease, parasites, and/or lethal genes may be introduced into remnant wild populations (CBSG 1991a,b). Releasing captive-bred birds should, therefore, be undertaken with the utmost care and consideration (Woodford and Rossiter 1994).

In recent years several sets of guidelines have been drafted to enable managers to assess and plan reintroduction programs (Cade 1986; Stanley Price 1989; Black 1991; Chivers 1991; Wilson and Stanley Price 1994). The IUCN/SSC's Reintroduction Specialist Group was formed in 1988 in an attempt to facilitate collaboration and guidance for prospective projects (IUCN/SSC 1997). The guidelines stress that reintroduction should be the last conservation measure that is initiated—that all other appropriate measures should have already been instigated. A feasibility study on the bird should be undertaken, looking at its ecology and status. The bird's environment must be assessed in terms of abiotic, biotic, and socioeconomic factors that have been active in the past and are likely to be active in the foreseeable future. Provided suitable release stock is available, reintroductions should be undertaken only when the original limiting factors have been removed and an adequate quality and size of original or restored habitat exists. The next section covers the feasibility study process.

Reintroduction Criteria

Kleiman, Stanley Price, and Beck (1994) recommend that reintroduction/translocation of endangered species is most appropriate when all feasibility criteria are met. There are four general areas of concern: condition of the species, environmental conditions, biopolitical conditions, and biological and other resources. A considerable amount of information about the species (e.g., its environment and biopolitical situation) is required to make accurate assessments. Therefore, some research will be required during the initial feasibility study.

Once the reintroduction method is deemed appropriate and adequate funding and permission have been secured, the program should be approached like a scientific experiment, with hypotheses and predictions based on previous work and theory, well-rehearsed methodology, assessment, and scrutiny. A phase of planning and preparation should be undertaken to ensure that: (a) an appropriate health and genetic screening is undertaken; (b) a release strategy is prepared (e.g., pre-release training; post-release acclimatization techniques; released group composition, number, and timing); and (c) pre- and post-release monitoring programs are instigated. Monitoring the birds and the environment throughout the project, and especially after release, allows participants to revise and plan for a continuum of phases during the long-term project (Derrickson, Beissinger, and Snyder, this volume).

As indicated in the next section by the Hawaiian Goose example, managers in the 1960s may have thought enough was known to initiate an intensive reintroduction program, whereas further ecological and behavioral research in the 1990s revealed that the crucial environmental conditions were still not met.

Hawaiian Goose Recovery Program

Elsewhere I reviewed the results of a research program aimed at examining each of seven potential limiting factors that influence the low productivity and survival of released Hawaiian Goose (or Nene) populations and inhibit success of the reintroduction program (Black 1995b). Six limiting factors are likely to be still affecting recovery of the species (table 9.2). These problems may be addressed by integrating the research findings into future recovery planning. There is scope, therefore, for maximizing genetic diversity in wild flocks through strategic translocations of eggs and adults, training captive-reared goslings in predator avoidance skills prior to release, intensive habitat management to enable the acquisition of adequate fat and nutrient reserves, intensive predator control, community education, and further financial support (Banko 1992; Black and Banko 1994; Black et al. 1994, 1997; Black 1995b). Only one of the suspected limiting factors may not be currently inhibiting recovery; parasite burdens in the birds are probably low (Bailey and Black 1995). However, the effect of prevalent avian pox–like disease is still unknown (Kear and Berger 1980).

Apparently, with the available information in the 1960s most of the thirteen assessment criteria were met and reintroduction was appropriate for the Hawaiian Goose (table 9.3). The one criterion that was not met, reintroduction technology, was refined during the project. Various styles of making the birds flightless were tried so they could get accustomed to the habitat (Kear and Berger 1980). Higher mortality rates in the first releases were linked with the initial methods (Black et al. 1997). The major criticism of the original program was that insufficient priority and funding were directed toward detailed monitoring of the behavior and ecology of the birds after release and analyses of existing data (see Banko and Elder 1990). A retrospective analysis revealed that the majority of releases in upland sanctuaries perished because the habitat apparently could not support the birds, especially during drought years (Black et al. 1997).

With current information in the 1990s the criteria for reintroduction are clearly not met (table 9.3), failing on crucial environmental conditions criteria. On the islands of Hawaii and Maui, introduced predators are still having an impact in the nesting and brood-rearing periods (Banko 1992; Black and Banko 1994), and current unmanaged habitats do not contain sufficient quality or quantity of food to provide adequate fat and nutrients for the prebreeding,

Table 9.2. Limiting factors, proposed by Stone et al. (1983), that were assessed in the Nene Recovery Initiative research program, 1990–96.

Potential limiting factor	Limiting?	Explanation	Recommendation	S
Inbreeding	Yes	low fertility and survival	optimization genetic diversity in flocks with few founders and emigration	h
				i
				l
Disease/parasites	No	low infestation levels	survey/research on avian pox	g
Loss of adaptive skills				
foraging	No	captive diets not a problem		f
social	Yes	parent-reared birds are best	research on methods for training prior to release	d
predator detection	Yes			j
Diet/nutrition deficiency	Yes	exotic plants are a bonus but not enough high-quality food available	habitat management required	f
				k
Predation	Yes	40% of nests destroyed by mongoose, lowlands worse than highlands	intensive predator control	a
				c
				e
Poaching/road kills	Yes	isolated events	further education	m
Inadequate funding	Yes	shoestring budget	further fundraising and collaboration	b

S = Source: a = Hoshide, Price, and Katahira (1990), b = Banko and Elder (1990), c = Banko (1992), d = Marshall and Black (1992), e = Black and Banko (1994), f = Black et al. (1994), g = Bailey and Black (1995), h = Rave et al. (1994), i = Rave (1995), j = Rojek and Conant (1996), k = Black et al. (1997), l = Rave et al. (1997), l = Rave et al. (submitted), m = unpublished anecdotes.

Table 9.3. An assessment of the criteria for reintroduction/translocation of Nene for past and current perspectives (scale = 5 is best).

	1960s perspective	1990s perspective	Comments/comparison
Condition of species			
1 Need to augment wild population	Yes	Yes	Still declining
2 Available stock	Yes	Yes	Improved/best available
3 No jeopardy to wild population	Yes	?	
Environmental conditions			
4 Causes of decline removed	Yes?	No	New evidence
5 Sufficient protected habitat	Yes?	No	New evidence
6 Unsaturated habitat	Yes	No	New evidence
Biopolitical conditions			
7 No negative impact for locals	No	No	Could benefit
8 Community support exists	1	4	Education needed
9 GOs/NGOs supportive/involved	Yes?	Yes	Improving
10 Conformity with all laws/regulations	Yes	Yes	
Biological and other resources			
11 Reintroduction technology known or in development	Yes	Yes	Still refining
12 Knowledge of species' biology/ecology	2	4	
13 Sufficient resources exist for program	No	No	
Recommended reintroduction/ translocation?	Yes	No	Habitat enhancement, predator control, and maximization of

nesting, and brood-rearing periods (Banko 1992; Black et al. 1994). There is evidence that some local populations on these islands may die out without continued releases (Black and Banko 1994; Black et al. 1997). The one environment where Hawaiian Geese are flourishing is on the island of Kauai, where Mongooses (*Herpestes auropunctatus*, the main predator) are absent and pasture grasses (a rich food resource) are plentiful; this flock has increased from an initial 12 geese in 1982 to more than 150 in 1996. Based on these findings, therefore, the translocation method would not be appropriate on Hawaii and Maui until suitable habitat is established and predators are removed.

However, a powerful argument in favor of the initial reintroductions as a conservation tool is that after forty-five years of Hawaiian Goose recovery efforts, a wild population still exists and has increased from thirty to seven or eight hundred individuals. Many of these are surviving in the remote highlands of Hawaii and Maui that most resemble the varied, pristine habitats that once predominated. In effect, extinction has been postponed, giving managers a chance to address the problems that have been highlighted by the ecology and behavior research (Black 1995b).

Making Research Effective for Conservation

This review has shown that there is still much to accomplish on behalf of the world's threatened waterfowl. Very few or no recovery actions have been undertaken for twenty-four of the forty-five taxa. Whereas all five threatened waterfowl in North America and the five in Europe have established programs, those from South America ($n = 8$), Africa, Northeast Asia, Philippines, and India ($n = 1$ each) lack any recovery action. Yet, as the Hawaiian Goose example indicates, even in the well-established recovery initiatives, further work is required to ensure recovery of viable populations.

Whereas this chapter outlines the achievements and shortcomings in recovery efforts, Callaghan (in press) has assembled a catalog of recovery priorities that await implementation for each taxa. He lists over one hundred actions for all taxa in table 9.1, outlining details, costs, and contacts for the required research. The project briefs include a broad range of research, from basic surveys to detailed ecological studies and community education projects. The aim of the publication is to foster global collaboration on behalf of threatened waterfowl.

The community of institutions that keep captive waterfowl can make a valuable contribution by initiating and working together on comprehensive captive programs that ensure healthy and genetically viable stocks (Seal, Foose, Ellis 1994; de Boer 1994). A Population Habitat Viability Analysis workshop (Seal, Foose, Ellis 1994) should eventually be held for each species to establish a network of collaborators and a plan of action. Zoos and aquaria should facilitate and increase support for sister organizations within the threatened birds' country of origin. Further expeditions will be needed to establish adequate captive stocks. Should reintroduction programs become necessary and feasible in terms of fulfilling reintroduction criteria, the global network can expand its efforts toward supporting this much larger and consuming recovery strategy (Seal 1991; Seal, Foose, Ellis 1994).

Although keeping and breeding waterfowl in captivity is usually achievable, reintroduction is currently not appropriate for most threatened waterfowl. This is because few recovery actions have been undertaken for the majority of taxa. The fact that habitat loss is the predominant cause for initial declines reduces the usefulness of reintroduction in the recovery process. That the remaining habitats are often inadequate in quality and size is why so few endangered animals are recommended for reintroduction (Stuart 1991).

The need for an international secretariat for captive breeding programs and international legislation that allows the transport of surplus birds or eggs becomes paramount when reintroduction into the wild is not feasible (de Boer 1994). For example, there has long been a captive breeding program for the White-winged Duck (*Cairina scutulata*), which readily breeds in captivity (Tomlenson et al. 1991), yet reintroduction is not an option because many of the preconditions are not met (Green 1992a): the original causes for its decline have not been removed;

there is insufficient protected habitat; the health and genetic makeup of the captive stock are in doubt; little is known about the birds' ecological requirements; there is insufficient involvement from local human populations; and financial resources are not secure. While it is important to maintain and perfect the global captive program for this and other threatened waterfowl, immediate attention should focus on implementing the natural recovery strategies and active research programs.

The most comprehensive recovery program has been for the Hawaiian Goose. Many recovery strategies have been employed since its initiation in 1949 (Kear and Berger 1980; Banko and Elder 1990; Black 1995b). The New Zealand programs on Blue Duck and Brown Teal (*Anas chlorotis*) have also been intensive (Hayes and Dumbell 1989; Williams 1991). The Aleutian Canada Goose program, which recently resulted in the race being removed from the threatened species list, was equally comprehensive (Byrd, in press).

With regard to the Hawaiian Goose recovery program, in order to achieve a self-sustaining set of populations throughout the Hawaiian islands without the necessity for further releases, it seems that intensive management will have to be implemented and sustained. The remaining genetic stock will have to be managed in order to maximize genetic diversity and reduce inbreeding. On Hawaii and Maui, emphasis should be given to creating ample high-quality habitat, preferably adjacent to scrubland nesting habitat that could be made predator free (Banko 1992; Black et al. 1994; Black 1995b). Once these areas are established, a second intensive release/translocation program may be needed to allow sufficient numbers of animals on which natural mutation rates can act, thus enhancing genetic diversity. A culturally transmitted set of seasonal movements between upland and lowland refuges and between islands should be taught, thus reducing the risk of starvation during drought years and enabling gene flow between areas.

Our simulation models suggest that it is likely that with these managed sites, flocks will swell (Black and Banko 1994), so the birds and humans will probably meet more frequently. Hawaiian Geese are perhaps the tamest of all waterfowl, an attribute that could be useful to the conservation effort if refuges could serve as focal points for community education programs. In order to save the Hawaiian Goose, I suspect managers will need to have the support of the Hawaiian people at all levels. When the state bird of Hawaii—the Hawaiian Goose—begins to recover, broad support for the conservation of all of Hawaii's threatened species and habitats is more likely to be achieved.

Clark, Reading, and Clarke (1994) perceptively verbalize why some recovery programs flounder. They explain that, "Because extinction is viewed largely as a biological phenomenon, the dominant professional and organizational response has been to focus on biology, obscuring nonbiological dimensions. Thus most endangered species programs are staffed by professional biologists. . . ." They quote G.B. Schaller as noting, "Conservation problems are social and economic, not scientific, yet biologists have traditionally been expected to solve them."

Clark et al. (1994) call for an integrated, interdisciplinary approach to threatened species recovery—an approach incorporating biologists, policy experts, sociologists, psychologists, organizational consultants, and, last but certainly not least, conflict managers. Beck et al. (1994) apparently concur by suggesting that biopolitical conditions and long-term funding are perhaps more important to the success of a reintroduction program than knowledge of the species' biology. I suspect progress will be achieved by those programs that are aware of this perspective and take heed.

Acknowledgments

I thank Glynn Young, Murray Williams, Dave Price, Andy Green, Des Callaghan, and Paul Banko for their help and the British Ornithological Union for permission to update tables 9.2 and 9.3.

CHAPTER 10

Applying Research for Effective Management: Case Studies in Seabird Restoration

Stephen W. Kress

Conservation Issues and Previous Research

Nearly all seabirds nest on islands where, prior to human presence, their populations were regulated mainly by available food and nesting habitat. In remote, largely predator-free islands, enormous populations can result. For example, Wilson's Storm-Petrel (*Oceanites oceanicus*) and Antarctic Fulmar (*Fulmarus glacialoides*) lay just one egg per year yet are among the most abundant birds on earth, with world populations of 5–10 million (Croxall, Evans, and Schreiber 1984).

While island nesting habitats often permit seabird populations to grow to great size, the confines of island life soon work against survival when humans and human-introduced predators such as cats, rats, and dogs arrive. No fewer than 200 of the 217 species or races of birds known to have become extinct in the last four hundred years were island nesting birds (Rodda, Campbell, and Derrickson, this volume). The rate of loss for island nesting birds continues today, with two-thirds of all threatened species occurring on islands (Collar and Andrew 1992). Island nesting increases vulnerability to predation because most species have had no opportunity to evolve defenses against predators. Large seabirds on remote islands (e.g., albatross, boobies, frigates, and penguins) are usually naive to the threat of humans, who can easily approach incubating birds and kill them at their nests. Even burrow-nesting seabirds are vulnerable to mammals, which can readily locate active burrows by smell and then proceed to dig out eggs and chicks. Narrow-bodied nest predators such as mongoose and rats are even more lethal, since they can enter most burrows, where they can kill adults and chicks.

When mammals are introduced onto islands, large seabird populations may dwindle to relics and local extinctions usually result. For example, in 1949 five house cats brought to the meteorological station on Marion Island (subantarctic Indian Ocean) to control house mice founded a population of about 2,200 feral cats by 1975. These kill about 450,000 seabirds each year, resulting in the disappearance of Common Diving Petrels (*Pelecanoides urinatrix*) throughout the island (van Aarde 1979). Similar examples are also documented of the devastating effect of cats on populations of Blue Petrels (*Halobaena caerulea)* at Macquarie Island (Jones 1977); of Polynesian rats (*Rattus exulans*) on Bonin Petrel (*Pterodroma hypoleuca*) at Kurre Island in the Leward Island archipelago of Hawaii (Woodward 1972); of Black Rats (*Rattus rattus)* on Dark-rumped Petrels (*Pterodroma phaeopygia*) in the Galapagos Islands (Harris 1970); and of the arctic fox *(Alopex lagopus*) on a variety of seabirds in Alaska (Bailey 1993). Mammal predators introduced to remote oceanic nesting islands have been the single greatest threat to seabirds (Atkinson 1978; Cooper, Hockey, and Brooke 1983; Moors and Atkinson 1984).

Excessive hunting for meat and feathers (Fisher and Lockley 1954; Cramp, Bourne, and Saunders 1974; Nelson 1979) has led to the depletion of many species with severe range contractions and occasional extinctions, the most recent of which was the Great Auk (*Pinguinus impennis*), which was hunted to extinction in 1844 (Allen 1876). Although protective laws and sanctuaries have helped to reduce seabird losses in some developed countries, exploitation for food and sport (Elliot 1991) and unregulated seabird egg collection continue throughout much of the world. In addition to direct hunting by humans and predation from introduced mammals, other anthropogenic forces also deplete seabird populations (Boersma and Parrish, this volume). These include both major oil spills (Piatt et al. 1990) and chronic oil contamination (Boersma 1986), entanglement in fishing gear (King 1974), depletion of forage fish (Furness 1982), and competition and predation by gulls, whose numbers have increased due to abundant food supplies at garbage dumps (Hunt 1972) and of fisheries offal dumped at sea (Hudson and Furness 1988).

The techniques for seabird management fall into two broad categories: (1) protection at the ecosystem level; and (2) active management at the species or colony level. Examples of ecosystem management include establishment of marine sanctuaries and regulations that reduce adult mortality—for example, losses due to pollution, entanglement, and hunting. Effective fisheries management that favors conservation of forage fish also benefits seabird management at the ecosystem level.

Active seabird management usually consists of protecting existing seabird colonies through acquisition and posting to reduce human disturbance during nesting seasons. Occasionally, seasonal wardens and public education interpreters are also employed to further the cause of seabird conservation. Such limited measures help to hold the line against the degradation of island habitats from direct human exploitation, introduction of additional predators, and assaults to the marine environment. Too often, however, funding limits adequate protection of existing colonies and long-term management is compromised by short-term management efforts.

Protection of large, productive seabird colonies through effective ecosystem management is necessary to reduce anthropogenic mortality, but management at this level does not necessarily lead to expansion of constricted ranges and reestablishment of seabird colonies extirpated at specific historic sites. This chapter describes several long-term case studies that show how existing knowledge of seabird behavior and life histories can be used to: (1) reestablish extirpated seabird colonies to islands; (2) expand species ranges to former distributions; and (3) increase diversity of nesting seabird communities.

Features of Research That Improved Conservation

In the case studies that follow, I have relied on data collected from other researchers to develop seabird restoration plans and reestablish historic Atlantic Puffin (*Fratercula arctica*) and Tern (*Sterna SPP.*) colonies. Key areas of research are: (1) studies of historic distribution and status; (2) studies of gull predation on island-nesting seabirds in the northeastern United States; (3) life-history studies that focus on nesting ecology; (4) studies of philopatry in seabirds; and (5) social facilitation studies.

Historic distributions were considered to determine which sites would be most appropriate for restoration. The presence of known colonies was seen as evidence that such sites once had suitable nesting habitat and that the marine environment was recently suitable for supporting nesting seabirds. Early accounts by Norton (1924a,b) provided much detail about occurrences of seabirds at specific Maine Islands. Likewise, Palmer (1949) provided an overview of fluctuating tern populations through the 1940s, while Drury (1973, 1974) summarized changes in Maine seabirds through the early 1970s.

Studies of gull predation on seabirds demonstrated the need to incorporate a gull control program into efforts to restore puffins and terns. Papers by Drury (1965), Nisbet (1973, 1975), and Hatch (1970) described how Herring Gulls (*Larus argentatus*) and Great Black-backed Gulls (*Larus marinus*) can displace terns by nesting earlier and by preying on tern eggs and chicks. Drury (1973, 1974) discussed the importance of gull predation as the principal factor causing the recent disappearance of many tern colonies and the principal cause of a forty-year decline in New England tern populations. Studies in Newfoundland by Nettleship (1972) described predation and cleptoparasitism on puffin chicks. These studies identified the need for a gull control program prior to active attraction of nesting puffins and terns. From these studies it was clear that any effort to establish puffins and terns without gull control would fail, since the gulls compete for nesting habitat and prey on puffin and tern adults, young, and eggs.

Life-history studies proved essential for designing restoration plans. For example, knowledge of the puffin's life history from Bent (1919) and Lockley (1953) demonstrated that young puffins show a high degree of natal philopatry, returning to their natal island home as prebreeders and usually nesting near the burrow where they hatched. Lockley found that parent puffins carry loads of fish for their young and drop them on the burrow floor. This information guided our development of chick feeding methods in which assistants placed meals on the burrow floor. Field studies such as Nettleship (1972) described the frequency of feedings and weight of meals in Newfoundland, and aviculturists at the Bronx (New York) Zoo provided recommendations on vitamin dietary supplements (J. Bell, personal communication). Our management methodology, therefore, arose from combining field studies with practical experience gleaned from aviculturists.

Previous studies of homing in seabirds have shown that adult seabirds have remarkable abilities to return to their nesting places when transported far from home. For example, a Manx Shearwater (*Puffinus puffinus*) was transported by airplane from its breeding island off Wales and released in Boston, over 5,100 km from home. Twelve and a half days later the bird was back in its burrow (Mazzeo 1953). Clearly, moving adult birds to new homes would prove futile since seabirds are well known for their abilities to travel between hemispheres and to locate their homes, even in habitats dominated by fog. Although few experimental studies have attempted to manipulate natal philopatry, several studies have suggested that young seabirds learn the location of home by using proximate navigation clues acquired during the early days of chick development.

Fisher (1971) relocated 3,124 fledgling Laysan Albatross (*Diomedea immutablis*) from Midway Atoll to other Laysan Albatross colonies up to 400 km away and found that most returned to Midway Island. Serventy (1967) experienced similar results with fifty fledgling Short-tailed Shearwaters (*Puffinus tenuirostris*) that were moved to another island. Both Fisher and Serventy reported, however, that when young were translocated well before they reached fledging age, some of the birds would return to the release site, rather than their natal home.

We applied these studies of philopatry to management by reasoning that if seabirds learn the location of their natal home sometime during early chick rearing, we might be able to move the chicks before they learned the location of "home," release them at a historic site, and have them return to the site when they approached breeding age.

Darling (1938) was one of the first to propose that the social facilitation provided by established breeding of conspecific seabirds would attract first-time breeders. Gochfeld (1980) found that copulations and courtship displays in Common Terns (*Sterna hirundo*) and Caspian Terns (*Hydroprogne caspia*) spread from one pair to the next through the colony as "contagious courtship displays." Likewise, Southern (1974) found that when one pair of Ring-billed Gulls (*Larus delawarensis*) was engaged in copulation, the behavior of neighboring pairs would change from body maintenance behavior to more courtship.

To adapt social facilitation for management purposes, we used decoys, mirrors, and audio playbacks of colony sounds to encourage prospecting-age birds to visit and linger at historic nesting sites. Our hope was that a few live birds landing with and resting among our artificial attractants would help to attract additional birds and that eventually colonies would result. Decoys, audio recordings, and calls are well known for their use to lure game birds into shooting range, but this was the first use of decoys and calls for encouraging birds to colonize historic nesting habitat. Likewise, mirrors are commonly used for amusing and relieving boredom in caged psittacines, but this was the first use of mirrors in the field for conservation and management purposes (figure 10.1). The adaptation of social facilitation for management purposes is referred to as social attraction.

Figure 10.1. Two-year-old translocated puffins with mirror box at Eastern Egg Rock, Maine.

Case Studies

Eastern Egg Rock (located at 43° 52' N, 69° 23' W) is a 2.9-hectare island located about 10 km east of New Harbor, Maine, USA. It is owned by the Maine Department of Inland Fisheries and Wildlife and managed by the National Audubon Society. The island is treeless and oval in shape with a central meadow of grass and shrubs, surrounded by large granite boulders. Mammals do not occur on the island. Two to four student warden-biologists occupy a small field camp on the island from June to August. Atlantic Puffins once nested under the granite boulders but disappeared due to hunting about 1885 (Norton 1923). About a thousand pairs of Common and Arctic Terns nested there in 1880, and "hundreds of pairs" still nested there in 1914 (Norton 1924a,b). Competition with Herring Gulls reduced the colony to just a few pairs in 1936 (J. Cadbury, National Audubon Society, personal communication).

Seal Island is a National Wildlife Refuge owned by the U.S. Fish and Wildlife Service and managed by the National Audubon Society. The forty-hectare island is located at 43° 52' N, 69° 23' W, 28 km southeast of Rockland, Maine. Two to six student warden-biologists are placed on the island from mid-May to mid-August each year. The island is treeless with low, alpine vegetation, pasture grasses, and boggy patches. Puffins once nested under boulders until 1887, when the last were trapped by entangling the birds in fishing nets placed over burrows (Norton 1923). Arctic Terns nested in large numbers on the island (Allen and Norton 1931) but were eliminated in 1887 by millinery collectors (Palmer 1949). Terns reestablished in the early 1900s following the end of millinery hunting, but they declined again about 1940 due to increased competition with Herring and Great Black-backed Gulls (Drury 1973, 1974) and were excluded from the island by the mid-1950s (W.H. Drury, personal communication).

Gull Control

Control of nesting Herring and Great Black-backed Gull populations was necessary as the first step in puffin and tern restoration. Control was necessary because New England Herring Gull populations increased from about 11,000 pairs on 17 islands in 1901 to about 89,500 pairs nesting on 305 islands in 1972 (Drury 1973, 1974). During this period, Great Black-backed Gulls showed similar growth, increasing from only 30 pairs on 12 islands in 1930 to about 12,400 pairs on 177 islands in 1972 (Drury 1973, 1974). Mechanized fishing and the growth of nearshore lobstering provide abundant food during the summers near nesting islands for the opportunistic gulls (Hudson and Furness 1988).

Increasing gull populations reduce tern populations because the gulls nest earlier than terns, setting up nesting territories before the more migratory terns return from the southern hemisphere. In addition to competition for nest sites, gulls also prey on tern chicks and eggs (Drury 1965). This conflict between gulls and terns led to a fifty-year decline in nesting populations of Common Terns, Arctic Terns (*S. paradisaea*) and Roseate Terns (*S. dougallii*) (Korschgen 1979).

Arctic Terns declined from about 4,500 pairs on Maine islands in 1940 (Drury 1973, 1974) to 1,300 pairs in 1982 (Kress, Weinstein, and Nisbet 1983). At the same time, Common Terns declined from 8,000 pairs in 1940 to only 2,095 pairs in 1977 (Korschgen 1979). Terns displaced by gulls often attempted to nest at sites near or on the mainland, where they were rarely successful in rearing young (Nisbet 1973).

In 1974, prior to gull control, about 200 pairs of Great Black-backed Gulls nested on Eastern Egg Rock (Kress 1983). Prior to the beginning of island-wide gull control in 1986, 1,106 pairs of Herring Gulls and Great Black-backed Gulls nested on Seal Island. A total of 127 adult Herring and Great Black-backed Gulls were killed at Eastern Egg Rock using avicide and shooting between 1974 and 1996. One hundred were killed in the first four years; none were killed between 1978 and 1983; and only an average of two per year were killed between 1984 and 1996.

On Seal Island, a total of 3,683 gulls (1,431 Herring and 2,252 Great Black-backed Gulls) were killed between 1984 and 1996, with 84 percent destroyed between 1986 and 1988. Only 3.5 gulls were killed per year between 1991 and 1996. Gulls that loaf in the intertidal zones or in areas remote from tern nesting habitat were not disturbed on either island. Since 1991, only gulls that specialize as puffin and tern predators have been killed. Avicides have not been used at Eastern Egg Rock since 1975.

Each year Herring and Great Black-backed Gulls attempt to nest at Eastern Egg Rock and Seal Island, and their nests are destroyed before eggs hatch. This disruption, combined with the presence of resident warden-biologists who shoot predatory gulls, has been enough to prevent the colonies from increasing and becoming reestablished.

Translocation of Puffin Nestlings

In cooperation with the Canadian Wildlife Service and the province of Newfoundland, a total of 1,904 puffin chicks were translocated from Great Island, Newfoundland, to the Maine coast. Of these, 954 were brought to Eastern Egg Rock in the years 1973–86 and 950 to Seal Island between 1984 and 1989. The chicks averaged seventeen days old and ranged from two to forty days old. They were removed in early morning from their natal burrows on Great Island, Newfoundland (47° 11′ N, 52° 49′ W) and placed in artificial sod burrows at Eastern Egg Rock and Seal Island in early evening. On average it took about seventeen hours to relocate the chicks from Great Island to the study sites in Maine.

After several pilot burrow designs failed due to problems with drainage and accumulation of excrement, surface sod burrows were devised and found effective for housing the chicks. These were L-shaped with an entrance tunnel leading to a nest chamber about 0.8 meters deep (figure 10.2). The burrows were constructed of 15-cm-thick blocks of sod and connected by a common wall, about twenty to a row. The walls were then covered by larger sheets of sod supported by wood lath braces and vinyl-coated wire mesh. Hardware-cloth door covers prevented chicks from leaving their burrows for the first seven days following translocation. After this time, the gates were open, and the birds were permitted to come and go.

Most of the chicks were fed one to three meals each day, consisting of about 100 grams of 3-inch-long smelt (*Osmerus mordax*) or silverside (*Menidia menidia*). Based on recommendations from aviculturists at the Bronx Zoo (J. Bell, personal communication), we supplemented the chick diet weekly with vitamins B1 and E and a multiple vitamin capsule with B12 (multiple maintenance mixture by Richlyn Laboratories, Philadelphia, Pennsylvania). The vitamin supplements were placed inside the mouth or body cavity of fish that were then placed with the meal. Between 1973 and 1981, most chicks at Eastern Egg Rock were not handled, weighed, or measured, in order to reduce disturbance. Between 1984 and 1986, puffin chicks at both Eastern Egg Rock and Seal Island received varied meal sizes and food types and were weighed and measured daily as part of a study of diet. Beginning in 1977, wooden decoys and a four-sided mirror box were set on conspicuous outcrops to encourage prospecting puffins to stay at the island (figure 10.1).

Of the 954 puffins translocated to Eastern Egg Rock, 914 (96 percent) successfully fledged. Of the birds that died prior to fledging, 32 died within ten days of arrival at Egg Rock. Translocated puffins from the 1975 cohort began

Figure 10.2. Artificial sod burrows used at Eastern Egg Rock and Seal Island National Wildlife Refuge for rearing translocated puffin chicks. The photo shows first level of construction prior to covering with wooden lath, vinyl-coated wire, and sod roofs.

returning to Eastern Egg Rock when they were two years old. A total of 149 translocated puffin chicks were subsequently seen at least once at Eastern Egg Rock or another puffin colony in the Gulf of Maine. The return rate varied between cohorts from 0 percent to 56 percent.

When the translocated puffins were two to four years old, they moved frequently among puffin nesting islands in the Gulf of Maine, but most were sighted at Eastern Egg Rock (Kress and Nettleship 1988). The return of 56 percent of the 1977 cohort marked a great increase in prospecting individuals during 1979 and 1980, which culminated in breeding in 1981. In that year four pairs bred, including one pair of unbanded birds. These were the first puffins to breed at the island in nearly a century. The colony quickly increased to nineteen pairs in 1985 as the majority of translocated birds reached breeding age. The colony stabilized at fifteen or sixteen pairs for the next ten years and slowly increased to nineteen pairs by 1996 (figure 10.3). During most of this period, translocated puffins from Newfoundland have made up the majority of nesting pairs, but the number of nontranslocated breeding birds has steadily increased, and by 1996, the number of nontranslocated puffins had exceeded the number of breeding translocated birds (figure 10.4).

Most new recruits are likely native birds produced on the island, but the nesting in 1996 of a Machias Seal Island (44° 30' N, 67° 06' W) native puffin is evidence that some young may be joining the colony from as far as 176 km to the east.

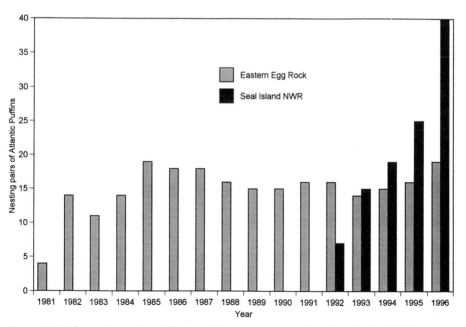

Figure 10.3. Changes in nesting puffin populations at two restored puffin colonies in Maine.

A total of 155 (17 percent) of the 912 chicks translocated and fledged from Seal Island were subsequently sighted either at Seal Island or at other puffin colonies in the Gulf of Maine. None of these birds were encountered away from the Maine coast. Return rates of the different cohorts varied greatly from year to year, ranging from 0 percent to 47 percent. The strong return of the 1988 and 1989 cohorts (47 percent and 27 percent, respectively) resulted in many pre-breeding-age puffins prospecting for nest sites on Seal Island in the years 1990–1991.

In 1992, seven pairs of puffins nested near each other in boulder habitat on the north end of Seal Island. These were the first puffins to nest at Seal Island in 105 years. The colony increased to fifteen pairs in 1993, nineteen pairs in 1994, twenty-five pairs in 1995, and forty pairs in 1996 (figure 10.3). In 1992, ten of the first colonists were translocated birds, and four (29 percent) were non-translocated immigrants from other colonies. The proportion of nontranslocated puffins has continued to increase, and by 1996, 68 percent of the breeding puffins were nontranslocated birds (figure 10.4).

The return of translocated puffins to Eastern Egg Rock and Seal Island demonstrates that some of the birds moved as nestlings learned to recognize the release islands and vicinity as their natal home and eventually returned to recolonize these historic nesting islands. The colony at Eastern Egg Rock increased rapidly for the first few years until most of the surviving translocated puffins were incorporated into the breeding pool, but stabilized to about fifteen or sixteen pairs for the next ten years while most of the populations remained composed of translocated birds. By 1996, 45 percent of the breeders were still translocated Newfoundland puffins, but now the majority of puffins are nontranslocated birds that were either produced as young at Egg Rock or were recruited from other Maine colonies.

This study demonstrates that it is possible to restore puffin colonies using translocation, but a long-term effort, moving hundreds of young over at least six years was necessary since survival of translocated cohorts varied greatly across years. Likewise, a long-term monitoring effort that extends beyond the translocation period is necessary to assess the outcome of the project.

Among seabirds, puffins are ideally suited for translocation because: (1) large thriving colonies existed from which young could be removed and transported to the release site within a day; (2) nestling puffins are fed whole food (which they can pick up off the burrow floor) rather than predigested food; (3) fledglings are not typically fed by their parents after they leave their nesting islands. Translocation might not work as well with species that do not share the above characteristics.

Social Attraction of Terns

Common, Arctic, and Roseate Terns are surface-nesting colonial seabirds that feed their young for several weeks after fledging. Translocation would not be fea-

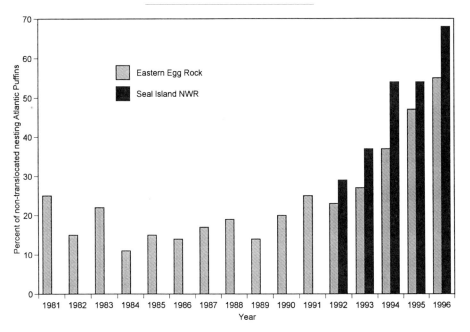

Figure. 10.4. Changes in the percent of nesting nontranslocated puffins at Eastern Egg Rock and Seal Island NWR.

sible for this life history because parent terns would not be available to feed the fledglings. As an alternate attraction technique, social attraction (use of decoys and audio) was used to attract adult birds without moving chicks (Kress 1983).

Following the use of gull control, a total of thirty-three Arctic Tern decoys were placed in suitable tern nesting habitat on Eastern Egg Rock during the months of June and July, 1978–80. During these years a sound system broadcast endless recordings of Arctic Tern colony sounds. In 1978 and 1979, the sound system was played only during daylight hours, but in 1980, it played twenty-four hours a day.

Both Arctic and Common Terns began landing among the decoys soon after they were positioned at Eastern Egg Rock. During 1978 and 1979, terns courted, copulated, built nest scrapes, and offered fish to prospective mates, but no breeding occurred. Similar behavior continued in June and early July of 1980, with about thirty pairs of Arctic Terns and fifty pairs of Common Terns nesting by July 20. A pair of Roseate Terns associated with the new colony in 1980, but the first breeding of Roseate Terns did not occur until the following year.

The tern colony on Egg Rock increased between 1980 and 1983, reaching 1,002 nests in 1983. In that year, the colony included mostly Common Terns. The colony declined from 1984 to 1986, following an outbreak of avian cholera in 1984 and Black-crowned Night-heron (*Nycticorax nycticorax*) predation in 1985 and 1986. Restoration of nesting Common, Arctic, and Roseate Terns at Petit

Manan Island National Wildlife Refuge in 1984 may have also influenced the new Egg Rock colony (Drennan 1987), since tern numbers at Petit Manan increased just as the Egg Rock colony declined. From a low of just 57 Common and Arctic Terns in 1986, the colony grew in ten years to 1,374 pairs of Common Terns, 79 pairs of Arctic Terns, and 126 pairs of Roseate Terns. By 1996, Eastern Egg Rock was the largest Common and Roseate Tern colony in Maine, containing 24 percent and 79 percent, respectively, of the state's breeding Common and Roseate Terns. Following gull control and recolonization by terns, Laughing Gulls *(Larus atricilla)*, also recolonized. Laughing Gulls nested on the island prior to 1895 but were first displaced by sheep and later by Herring and Great Black-backed Gulls (Norton 1924a,b). Three pairs of Laughing Gulls recolonized the island among the growing tern population in 1983, and the colony steadily increased to 460 pairs by 1996.

The restoration of terns at Eastern Egg Rock using gull control and social attraction was the first use of these combined techniques for encouraging tern recolonization at a former nesting site. The removal of the nesting gull population alone did not lead immediately to tern restoration. Gulls abandoned the island by 1976, but terns did not land and prospect in 1976 or 1977. When decoys and recordings were used in 1978, terns immediately began to land and court but did not nest until three years later, in 1980. Although the entire island was free of nestsite-competing gulls, the terns chose to nest among the decoys, often positioning their nests within inches of the decoys.

The rapid decline of the colony between 1984 and 1986 (from 1,002 to 56 pairs) illustrates how factors beyond the control of the manager can often affect the outcome. This decline, however, and the subsequent rapid growth to 1,579 pairs by 1996 illustrates that longterm management is necessary to document the outcome of restoration projects such as the puffin and tern recolonization at Eastern Egg Rock. These fluctuations also point to the need to have many large, regionally dispersed and productive tern colonies. Such dispersed colonies benefit the population by spreading the risk of catastrophic events (e.g., disease and predation) among islands.

Making Research Effective for Conservation

The seabird restoration case studies described above used basic research concerning historic distributions, population trends, and life history and specific behavior studies such as research into natal philopatry and social facilitation. Practical knowledge obtained from aviculturists was also useful in the early chick-rearing stages of the puffin translocation projects. This mixed background was essential to the development of specific management techniques. These were tested and improved in the field (sometimes midstream) as necessity required.

Midstream changes in management techniques were necessary because most of the methods described in these case studies were new and required testing under field conditions. For example, burrow design for translocated puffin chicks went through three major types before an ideal design was achieved. The original design, consisting of ceramic chimney tiles, was rejected when we found that the burrows overheated and that poor drainage for excrement resulted in severe feather fouling, a condition that likely would affect waterproofing. In the second year of the program, we replaced the ceramic burrows with hand-dug earthen burrows modeled carefully after natural burrows, only to find that these filled with water following severe rains. In that year several chicks died from aspergillosis, a fungus infection that thrives in wet soil. Midstream in the second year, we developed our first sod burrows built on the soil surface. These solved previous problems and were subsequently used for all translocated puffins reared at Eastern Egg Rock and Seal Island National Wildlife Refuge.

Several lessons emerge from these studies that are applicable to the transfer of these case studies to wildlife managers:

1. Seabirds are long-lived species with delayed breeding, and this requires restoration projects that are defined by the decade rather than by the year. Projects concerning translocation and social attraction of seabirds need to be long term to allow the development of management techniques. Also, variations in the marine environment such as weather and available food can affect survival of young birds. The monitoring phases of the program should be long enough to effectively evaluate the project and make mid-course changes in procedures when necessary.

2. Because of the long-term nature of the management, the work should be carried out by individuals with extraordinary commitment and institutional support for administration and funding that is likely to last the course of the project.

3. Selection of an appropriate restoration technique or combination of techniques should take into consideration the specific local environment, including such factors as: vulnerability to predators, disturbance from humans, logistics associated with staffing the project, and public relations. For example, sites near the mainland and large human populations would likely require more resident wildlife wardens during the nesting to reduce human disturbance. Such sites would also require a more concerted effort to explain the need for removal of predators such as gulls and owls.

4. Selection of appropriate restoration sites should include: long-term history as a productive nesting island; knowledge of food base, knowledge of vulnerability to predators and human disturbance, and practicality of staffing for research and follow-up monitoring. For example, islands with introduced predators would usually be unsuitable for restoration until the predators were removed. Likewise, islands that were once important seabird

communities, but are now occupied by large human habitations or by people who are not in support of restoration would be unlikely places to begin a restoration program. Suitable sites for restoration should be in conservation ownership with an institutional commitment for long-term funding for management.

Acknowledgments

The case studies described in this study were funded largely by contributors to the National Audubon Society's Seabird Restoration Program. The fieldwork was carried out by interns and volunteers too numerous to list, but whose efforts and dedication were essential to these projects. I also thank David N. Nettleship, the Canadian Wildlife Service, and the Province of Newfoundland for permission to translocate puffin chicks to Maine. I thank the staff of the Petit Manan National Wildlife Refuge for their assistance and permission to work on Seal Island National Wildlife Refuge and thank the Maine Department of Inland Fisheries and Wildlife for permission to work on Eastern Egg Rock. I also thank Donna Ramil for assistance compiling data and preparing the figures and acknowledge James Lowell for creating the map. Dave Manuwal, John Marzluff, and an anonymous reviewer made helpful suggestions on the manuscript.

CHAPTER 11

The Role of Sensitive Species in Avian Conservation

John R. Squires, Gregory D. Hayward, and James F. Gore

Conservation Issues and Previous Research

In the logging community of Saratoga, Wyoming, a bumper sticker on a pickup reads, "Biologists are Predators." This message illustrates the frustration that some people have with sensitive-species management. It suggests that the providers of biological knowledge prey on members of the community who depend on extracting natural resources for their livelihood. Species such as the Northern Spotted Owl (*Strix occidentalis caurina*) epitomize the political, social, and biological challenges that face society when managing species of concern. Their dependence on old-growth forests (Forsman, Meslow, and Wight 1984; Carey, Reid, and Horton 1990; Thomas et al. 1990) makes conflict inevitable between powerful economic interests and local communities that face job losses, and preservationists who wish to conserve ancient forests. Land management agencies were caught between economic and conservation interests, at first denying the Spotted Owl problem, only to play catch-up when public pressure demanded conservation action (Yaffee 1994). These events demonstrate that wildlife is best managed before becoming endangered. Management options that balance development with conservation become increasingly limited as species become rare. Remedial actions under the Endangered Species Act (ESA) of 1973 are not triggered until populations become low (Orians 1993).

To be proactive, the U.S. Forest Service (USFS) initiated a sensitive-species program with the explicit goal of developing and implementing management practices to ensure that species do not become threatened or endangered through the agency's actions (USDA 1995b). Other goals of the program included maintaining viable populations of all native and desired non-native wildlife, fish, and plants in habitats distributed throughout their geographic range, and developing and implementing management objectives for populations and/or habitats of sensitive species.

To prevent species from becoming endangered biological evaluations (BE) are written to examine how proposed management activities will impact sensitive species. These evaluations support planning under the National Environmental Policy Act (83 Stat. 852 as amended: 42 U.S.C. 4321, 4331–4335, 4341–4347). Biologists writing BEs face the difficult task of assessing whether management activities will adversely impact the population or habitat within the area of concern and the persistence of the species as a whole (see Ruggiero, Hayward, and Squires 1994 discussing problems with scale). If projects will have significant impacts on a sensitive species, management objectives may be developed in cooperation with the state wildlife agency. An important outcome of USFS policy regarding sensitive species is the consideration these species receive in impact analyses.

Our goal in this chapter is to discuss how sensitive-species management and research contribute to avian conservation. Because both operational and biological challenges influence the efficacy of these programs, we will discuss both but will stress biological issues. We will review characteristics associated with successful sensitive-species programs, including potential problems. Finally, we will explore the role of research in the conservation of sensitive species.

To gain a broader appreciation of the issues that confront biologists when managing sensitive species, we sent an informal questionnaire to district biologists of the USFS during September 1996. Our questionnaire consisted of fourteen short-answer questions that addressed the following topics: magnitude of their sensitive-species program; how funds were distributed among species; how sensitive species impact management actions in their districts; their opinion of how the program should be improved; and their opinion of how research can best help management efforts. We encouraged biologists to participate, but we stressed that their participation was optional. Questionnaires were sent by e-mail on region-wide mailing lists, so we do not know how many biologists actually received them. We received thirty-seven responses from five regions (R2 Rocky Mountain, 12 responses; R3 Southwest, 2; R4 Intermountain, 7; R5 Pacific southwest, 6; R6 Pacific northwest, 10). This was not a random sample of opinion (our sample may include systematic bias because participation was voluntary), but we believe it was sufficient to identify important issues that concern biologists who are responsible for sensitive-species management. However, we cannot estimate the prevalence of issues.

Designating and Defining Sensitive Species

Federal agencies have differing criteria for designating and defining sensitive-species. In the USFS, regional foresters designate sensitive species within their region (nine regions within the United States). Each region develops its own criteria to evaluate proposed species. In most regions, a species is considered sensitive if a decline in either population abundance or habitat conditions suggest it is trending toward endangerment. However, determining population trends is difficult, especially for species that are rare. The individual regions of the USFS gather data from many sources that include state and federal biologists, research scientists, academics, conservation organizations, and others. Both qualitative and quantitative information are synthesized to determine if a species' population is declining and the "sensitive" status is warranted. For many sensitive species, trend data for populations are either lacking, limited, or in many cases are flawed. Determining habitat trends is also difficult but is becoming easier as agencies adopt geographic information systems technology.

In 1996, the nationwide list of sensitive species for the USFS included 2,339 species: 5.6 percent birds, 4.4 percent mammals, 5.0 percent reptiles/amphibians/snails, 4.8 percent fish, 2.9 percent clams/crustaceans, 3.7 percent insects, and 73.6 percent plants (L. Fenwood, USFS, personal communication, 1996). Our questionnaire of district biologists indicated their districts averaged 12 species of terrestrial vertebrates (range 2–49 species, SD 12, $n = 36$) listed as sensitive; approximately 8 species were birds (range 0–29 species, SD 7, $n = 36$). These biologists spent approximately 52 percent (range 1–100 percent, SD 34 percent, $n = 33$) of their total threatened/endangered/sensitive-species budget managing sensitive species.

According to district biologists responding to our survey, 35 percent (range 0–90, SD 31, $n = 36$) of management actions (e.g., timber harvest, road construction, recreation facilities) were altered because of sensitive-species considerations. Although sensitive-species management almost never resulted in projects being denied ($\bar{x} = 0.6$ percent, range 0–10, SD 2, $n = 37$), considerations for sensitive species resulted in approximately 25 percent (range 0–90 percent, SD 28, $n = 37$) of projects having timing restrictions and approximately 25 percent (range 0–80 percent, SD 25, $n = 37$) undergoing design modifications.

The U.S. Fish and Wildlife Service (USFWS) is responsible for declaring species as threatened or endangered under the Endangered Species Act of 1973 (16 U.S.C. 1531 et seq.). All federal departments and agencies must ensure that their activities are consistent with conserving endangered and threatened species, including their critical habitats. In 1996, in the United States there are 215 endangered vertebrates, including 74 species of birds and 89 species of threatened vertebrates, of which 16 are birds (USDOI, Fish and Wildlife Service 1996). The USFWS also lists 182 species (including plants) as "candidate species"; this designation includes species that existing data suggest should be listed in the near future.

Prior to 1997, the USFWS maintained a "Category 2" list that included species whose status was unknown but of concern due to declines in population trend or habitat. This list is somewhat analogous to the sensitive-species lists of the USFS, because species were identified as possibly needing management consideration before becoming endangered. The decision to drop Category 2 listings may significantly impact sensitive-species management. For example, the Bureau of Land Management (BLM) identifies "Special Status Species" of plants and animals that may need additional management consideration (USDOI 1988). These include species that either are listed under the Endangered Species Act or were listed as Category 2 species by the USFWS. Currently, many BLM biologists still give special management attention to species on the old Category 2 list, but this may decrease over time. In our opinion, the FSFWS's decision to drop Category 2 designations may decrease the political impetus to manage sensitive species before they become endangered.

Management Context of Sensitive-Species Programs

Sensitive-species management has evolved from intra-agency programs to an integral part of ecosystem management. The sensitive-species program of the USFS is an extension of both past and present conservation efforts that stress biodiversity. The Northern Spotted Owl and old-growth forest issues during the 1980s demonstrated that land-use agencies needed a new way of viewing land management and forestry. In response, the USFS adopted a philosophy called "New Perspectives" that promised to foster a more sustainable and resilient forest management (Salwasser 1991; see Frissell, Nawa, and Noss 1992 and Lawrence and Murphy 1992 for dissenting views). One theme of New Perspectives was the concept of adaptive management that encourages a fluid and dynamic land management philosophy that evolves when new information warrants change (Walters 1986). New Perspectives provided the philosophical shift that has recently evolved into a new paradigm called ecosystem management. In 1992, the USFS officially embraced ecosystem management as a new management philosophy that impacts 77 million hectares of National Forest System land (Robertson 1992, as cited by Salwasser 1992). Ecosystem management seeks to achieve an ecological approach to achieve multiple-use management of National Forests and Grasslands as sustainable ecosystems. Ideally, the application of ecosystem management accounts for the needs of sensitive species while managing sustainable systems through an adaptive approach.

Research plays a central role in the adaptive management of sensitive-species. As management plans become more rigorous and complex, there is an increasing need for sound empirical data. Land managers, physical scientists, and biologists who must determine how management actions impact sensitive species rarely have complete answers to central questions concerning ecosystem function and restoration. Yet informed decisions are made as needed; this process involves risk. Adaptive management facilitates adjustments as new information is derived through the management process (Walters 1986). Managers should not be judged right or wrong unless they fail to make any decision or fail to amend land management plans in light of new data. Thus, research plays a critical role in evalu-

ating how management actions impact sensitive species so that adaptive management is possible.

Challenges of Managing Sensitive Species

Managers face significant operational challenges when managing sensitive-
species. Some of the most formidable challenges include managing with limited
funding, fostering an environment that embraces change, enhancing intra-agency
cooperation and line-officer support, and developing interagency cooperation.
Determining the true "cost" of managing sensitive species is difficult. Many costs,
such as the time agency personnel dedicate to surveying, monitoring, evaluating,
and writing management plans for sensitive species, are not delineated as separate
operational costs. The USFS spends approximately $12.5 million (1992–95
average) on sensitive-species management (L. Fenwood, personal communication 1996). In addition, it receives approximately $2.8 million from partnerships
with private organizations and state agencies to help fund conservation efforts.
Although the USFS commitment to fund sensitive-species management is
substantial, efforts to fully develop and implement conservation measures for
sensitive species still lack adequate funding and staffing. For example, the districts
responding to our questionnaire indicate that approximately one management
plan ($\bar{x}$ = 0.8 plans, range 0–5, SD 1, n = 37) has been written per district for
sensitive-species. Thus, most sensitive species in the districts responding to our
questionnaire lacked management plans because of limited personnel and
funding.

Sensitive-species programs also suffer from inconsistent coordination among
agencies and organizations. Within agencies, the quality of interaction between
organizational regions is highly variable. Much information regarding the distribution of sensitive species is organized according to state boundaries, even for
species with multistate ranges. This may be due to state and federal agencies frequently relying on the Natural Heritage databases as maintained by The Nature
Conservancy on a per-state basis or to the state government's jurisdiction of the
wildlife resource. Some states develop panels of species "experts" from various
state and federal agencies that use a delphi-process to assess the biological status
of sensitive species.

Efforts to revise the Tongass Forest Plan in Alaska provide a striking example of
effective intra-agency coordination when managing sensitive-species. In 1993,
the regional forester in Alaska, in cooperation with USFS Research evaluated potential impacts of management actions on several sensitive-species (e.g., Marbled
Murrelet [*Brachyramphus marmoratus*], Northern Goshawk [*Accipiter gentilis*],
American Marten [*Martes americana*]). Conservation assessments were then
written for each species by a team of both researchers and managers (Iverson et al.

1996). This close cooperation provided a unique perspective throughout the forest planning process and helped ensure that the resulting plans were scientifically sound. This example also illustrates that line supervisors need to be fully committed to sensitive-species management before meaningful intra-agency cooperation can be achieved.

As the number of sensitive species increase, more administrative boundaries between agencies are crossed, which hampers coordination and communication. Therefore, close cooperation between agencies is critical when conserving sensitive-species. Conservation agreements are valuable tools for sensitive-species management because they encourage interagency communication throughout the planning process. A conservation agreement is a pledge from a land management agency to a regulatory agency (i.e., U.S. Fish and Wildlife Service, National Marine Fishery Service) to faithfully apply mutually agreed upon conservation practices. To insure accountability, the management agency reports regularly to the regulatory agency regarding conservation activities. As long as the conservation agreement is implemented and surveys show the species habitat and/or population is stable, the regulatory agency will not list the species. For example, in Idaho, the management of four sensitive species representing diverse taxa—Idaho ground squirrel (*Spermophilus brunneus brunneus*), Christ's Indian paintbrush (*Castilleja christii*), Stanley Whitlow-grass (*Draba trichocorpa*), and guardian buckwheat (*Eriogonum meledonum*)—are protected with approved conservation agreements. Both state and federal agencies have vested interests in these species, including the Idaho State Governor's Office. All parties hold frequent meetings to review species status and to develop conservation strategies. Idaho also has formalized steering/oversight committees of key managers and people knowledgeable of the species to help with the effort.

Although operational challenges make sensitive-species management difficult, land managers who are committed to conserving sensitive-species do initiate meaningful actions. When expertise or funding is lacking, partnerships are formed with other agencies, conservation organizations, and industry to help forward sensitive-species management.

In addition to substantial operational challenges, managers of sensitive species must consider significant biological challenges. Several management approaches exist that enhance or hamper conservation efforts depending on implementation.

Single versus Multiple Species

The sheer number of sensitive species requires prioritization from both research and management perspectives. Tight monetary constraints make it impossible for intensive autecological studies of all sensitive species. Biologists responsible for evaluating potential impacts and for prescribing management actions are overwhelmed by the enormity of their task. The responsibility of managing so many sensitive species on dwindling resources heightens the dichotomy between wildlife managers who advocate single-species management (fine filter) and those who promote approaches that stress ecosystem patterns and processes for conserving biodiversity (coarse filter). Some land managers believe that managing "sensitive or impacted ecosystems" will be easier than managing individual sensitive-species. They argue that less empirical data will be needed or that management actions will be less contentious if we manage at the ecosystem level. Assessing strengths and weaknesses associated with these different philosophies is important because they impact future funding, listing, and the very existence of sensitive-species programs.

As currently practiced, sensitive-species management by federal agencies is a fine-filter approach that seeks to provide requisite habitat needs on an individual-species basis. Franklin (1993b) argues that efforts to preserve biodiversity on a species-by-species basis will fail because we lack the time, money, societal patience, and scientific knowledge (Wilcove 1994a).

However, single-species management is not necessarily a one-species-at-a-time approach (Tracy and Brussard 1994). Managing for charismatic species such as Northern Spotted Owls, Northern Goshawks, California Condors (*Gymnogyps californianus*), and grizzly bears (*Ursus arctos*) may also provide habitat for hundreds of other species. The challenge is to identify appropriate "umbrella" species that have life histories (long generation times, low rates of population increase) that are sensitive to anthropogenic habitat alterations (Tracy and Brussard 1994). However, this hypothesis needs careful consideration. Franklin (1994) points out that despite the large size of conservation areas for Northern Spotted Owls (USDA 1992), these areas do not provide habitat for anadromous fish or Marbled Murrelets; they also excluded the most significant remnants of old-growth forests.

Ecosystem Management

Lesser-known species as well as "endangered ecosystems" may be better protected by focusing on the rarity, distribution, and species diversity of ecosystems rather than identifying conservation areas based on the presence of charismatic species (Franklin 1993b). Invertebrates represent over 90 percent of total biodiversity; these species will be ignored using single-species approaches. Franklin believes we should establish habitat reserves and give special attention to the seminatural landscape matrix that surrounds these refugia. Habitat refugia that are isolated in a hostile matrix will fail to preserve biodiversity for the long term. We often find ourselves in the ironic position of using a species as an indicator of an ecosystem, for example, Spotted Owls for old growth, rather than managing the extent and distribution of the forest itself (Orians 1993). Despite the appeal of ecosystem management, there are significant challenges with this approach.

Selecting conservation areas that represent ecosystems will require prioritization and is scientifically challenging (Pressey 1994). Williams et al. (1996) reviewed how well the biodiversity of British birds was conserved using areas selected based on richness hot spots (Prendergast et al. 1993; Sisk et al. 1994), rarity hot spots (Terborgh and Winter 1983; N. Myers 1990; Sisk et al. 1994), and complementary areas (areas high in both richness and rarity; Ryti 1992; Kershaw, Mace, and Williams 1995; Humphries et al. 1996). Although species richness hot spots were easily identified in Britain, correlations between diversity and rarity were low. Rare species were not nested within distributions of more widespread species; thus, these two measures were not surrogates of one another and left some species without protection. Richness hot spots included 89 percent of British breeding bird species, while rarity hot spots included 98 percent. Complementary areas included all species but still would not fully serve conservation until factors affecting viability, threat, and cost were incorporated. Similarly, in South Africa, identifying hot spots of species richness was not an efficient method to identify ideal reserves that protect vertebrates because areas of high species richness were not coincident with areas of endemism (Lombard 1995). These studies illustrate the lack of well-established methods for identifying and ranking the importance of conservation areas. The degree to which selection models are appropriate will vary depending on the distribution of biodiversity within given ecosystems. Combined models that incorporate both richness and diversity are promising, but they will need to be evaluated on an ecosystem-by-ecosystem basis. In addition, individual species act as templates that spatially set the scale of proposed conservation areas (Wilcove 1994a; Kochert and Collopy, this volume). The spatial arrangements of conservation areas must be consistent with the demographic and habitat needs of the sensitive-species most threatened.

Beyond technical considerations, aesthetic preferences of the public may differ from the opinions of scientists interested in conserving biodiversity (Shrader-Frechette and McCoy 1994). Laypersons are often most interested in protecting charismatic birds and mammals (Franklin 1987). Ecologists may argue the scientific merits of a given program but often fail to address whether the action is politically feasible (Shrader-Frechette and McCoy 1994). Ecosystem-level protections that fail to conserve high-profile species like spotted owls and grizzly bears will be regarded a failure by the public regardless of the total biodiversity preserved. Although, in some cases, conserving ecosystems may protect the most total biodiversity, it may not be the best choice in terms of policy as it relates to the chance of implementation (Shrader-Frechette and McCoy 1994).

Multispecies Management

Between the single-species and ecosystem extremes are multispecies management philosophies such as those of indicator species (Szaro and Balda 1982; Landres, Verner, and Thomas 1988; Williams and Marcot 1991), guilds (Severinghaus 1981; Verner 1984; Block, Brennan, and Gutiérrez 1986), and keystone species (Paine 1969; Estes and Palmisano 1974; Power et al. 1996), which vary in the resolution of management. These management philosophies are appealing because they potentially reduce cost by considering a subset of species that may be easier to manage and monitor. Hutto (this volume) discusses the merits and short-comings of these approaches.

Rectifying Single- versus Multispecies Management

We agree with Wilcove (1994a) that contrasting "single-species management" with "ecosystem management" creates a false dichotomy; both are necessary components for protecting biodiversity. Ecosystem management approaches that maintain large conservation areas could be important for conserving the many species with ecologies that are completely unknown, especially plants and invertebrates. But, no matter how rigorously ecosystem management is applied, the needs of some sensitive species will remain neglected. Researching their auto-ecology as it pertains to population persistence will still be an important part of conserving biodiversity. Multiple-species management may have a role in con-serving sensitive species, but these management philosophies must be cautiously applied. No grouping of species will serve as the perfect surrogate for an individual species. Finally, we recognize that highly charismatic species have an important role in conserving all sensitive species because they capture public support and galvanize conservation action.

Features of Previous Research That Improved Conservation

Boreal Owl and Northern Goshawk management illustrate some of the strengths and weaknesses of sensitive-species programs. Boreal Owls were first recognized as breeding residents in the United States south of the Canadian border about the time sensitive-species programs were initiated by the USFS (Palmer 1986; Hayward, Hayward, and Garton 1993). Just four years after breeding populations were discovered, managers and research biologists outlined management directions in a proactive context. Consequently, Boreal Owl management has avoided some of the conflicts associated with other sensitive-species programs. The Northern Goshawk is a sensitive species that epitomizes many of the contentious issues confronting land-use agencies. Northern Goshawks often nest in old-growth or mature forests (Hayward and Escaño 1989; Speiser and Bosakowski 1987; Squires and Ruggiero 1996) that are impacted by forest management (Reynolds 1983; Crocker-Bedford 1990). The conflict between producing forest products and managing goshawks was most acute in the ponderosa pine forests of the southwestern United States. Goshawk management in the Southwest pro-vides an interesting model of sensitive-species management because research had a central role in helping resolve this conflict by offering scientifically based management recommendations. In this section, we identify the characteristics of effective sensitive-species programs based on our experience with these two forest raptors.

Involvement of Researchers throughout the Planning Process

Research scientists should be involved in formulating management strategies of sensitive species at all stages of the planning process. When the Northern Goshawk was listed as a sensitive species in the southwestern United States, the USFS regional forester established the Northern Goshawk Scientific Committee in 1990. This nine-person committee was composed of foresters and silviculturalists, program coordinators, staff officers, district biologists and research scientists, and academics. Together they published *Management Recommendations for the Northern Goshawk in the Southwestern United States* by Reynolds et al. (1992), hereafter referred to as the *Recommendations*. Research scientists had a central role in developing the goshawk *Recommendations*, which greatly enhanced their credibility (D. Garcia, USFS, personal communication 1996).

Formulating scientifically based management recommendation of sensitive species requires diverse professional expertise. Only a varied array of scientists and managers can provide the broad knowledge necessary to develop ecologically sound management recommendations. In the Southwest, wildlife biologists and researchers helped ensure that habitat requirements of Northern Goshawks were provided in a manner that was spatially consistent with their biology, whereas foresters and silviculturalists helped ensure that management prescriptions were consistent with forest ecology and disturbance patterns of ponderosa pine forests.

Managers and researchers also developed a close working relationship during the early stages of Boreal Owl management. A coordinated program was formalized at high levels in the USFS soon after Boreal Owls were located in the Rockies. This coordination and high-level administrative involvement greatly contributed to the success of the Boreal Owl program. However, recent changes caused by reductions in federal budgets have stagnated efforts to develop conservation strategies based on national conservation assessments written for the species (see Hayward and Verner 1994). Budget reductions also hindered progress with other national sensitive-species programs that were initiated concurrently with the Boreal Owl assessment (see examples in Ruggerio, Hayward, and Squires 1994; Young 1995).

Recent developments in Boreal Owl management demonstrate the problems that can develop when management is not coordinated across regions for widespread species. Interregional coordination of owl management has been lost. Given the metapopulation structure exhibited by Boreal Owls (Hayward, Hayward, and Garton 1993), local management actions may have broadscale impacts; these impacts cannot be addressed without communication among regions. Furthermore, biologists in individual national forests are once again faced with the task of developing individual management plans; they must reinvent the wheel for each administrative unit without linkage to neighboring units and the broader metapopulation. The absence of a coordination team leading efforts toward a broadscale conservation strategy may also hinder long-term monitoring efforts. Because the demographic patterns in Boreal Owls spans broad geographic scales, monitoring patterns of abundance requires information from dispersed sample points. Thus, monitoring is difficult, if not impossible, without close coordination among administrative units.

Recommendations Consistent with Ecosystem Structure and Function

Effective sensitive-species programs are firmly grounded in ecological knowledge that supports management recommendations. Understanding the ecological characteristics associated with a given ecosystem, such as food webs, predatory relationships, disturbance patterns, vegetative structure, and landscape characteristics is essential for providing the specific habitat needs of sensitive species within the constraints of ecosystem function. Autecological information at varying spatial scales is also valuable because management actions may encompass several spatial scales. For example, the Northern Goshawk recommendations discuss nesting habitat at three spatial scales—nest area (approximately 30 ha), post-fledging area (approximately 170 ha), and foraging area (approximately 2,200 ha) (Kennedy et al. 1994; Reynolds et al. 1992). The most effective efforts to manage sensitive species combine autecology with community ecology from a broad information base.

Given that Boreal Owls were only recently discovered south of the Canadian border, our understanding of their ecology is limited. To date, only four major published investigations from North America provide the ecological basis for management planning (Bondrup-Nielsen 1978; Palmer 1986; Hayward, Steinhorst, and Hayward 1992; Hayward, Hayward, and Garton 1993). None of these investigations represent experimental approaches to ecological questions, none of these were designed to directly address forest management issues, and all extended for four years or less, a temporal scale insufficient to address important issues in forest management or the ecology of a long-lived vertebrate. However, existing information suggests that populations of Boreal Owls may be threatened by habitat change resulting from certain forest management practices. Therefore, four National Forest Regions list the owl as sensitive, and the bird was the focus of a national conservation assessment (Hayward and Verner 1994).

An important aspect of early Boreal Owl research was its focus on the ecological system that supports owl populations. Initial studies examined the dynamics of small mammal populations as well as habitat associations of the owl. Current research extends from the fungi and lichen species important to prey populations to landscape-scale examination of owl habitat associations (Hayward, in press). Research also examined the geographic variation in Boreal Owl ecology. This helps avoid problems with applying management recommendations to very different ecological settings from the one for which they were developed. (Iverson et al. 1996). Boreal Owl management will suffer until a coordinated research program is implemented that relates owl ecology to ecosystem processes.

Our limited understanding of the ecology of many, if not most, sensitive-species represents the single most important barrier to developing ecologically based management recommendations. For most species, we lack a basic understanding of distribution and abundance, limiting factors, response to changes in landscape pattern, and natural disturbance patterns. This information is needed when developing defensible conservation strategies. For example, we do not know if prescribed changes in forest structure will favor Great Horned Owls (*Bubo virginianus*), a potential predator of Northern Goshawks, or how habitat alteration affects goshawk foraging and wintering biology. Understanding the dynamics of long-lived sensitive species, like Northern Goshawks and Boreal Owls, requires long-term research (Valiela, Parsons, and Johnson 1989). Likewise, understanding habitat associations and population dynamics of mobile vertebrates requires that we address these questions at appropriate geographic scales (Wiens 1989). Species with large home ranges and high vagility require researchers to consider very broad geographic scales, which is difficult and expensive. Field experiments that account for broad ecological scales are difficult, which in turn hinders our attempts to implement adaptive management. Until further field research is completed, managers are forced to extrapolate from limited information collected at few study sites, an uncomfortable situation at best.

In order to cope with our uncertainty, assumptions must be clearly identified so they can be tested and changed if warranted. For example, the scientific committee that developed Northern Goshawk recommendations made three assumptions (Reynolds et al. 1992): (1) they assumed that forests are dynamic and that goshawk nesting habitat should be a mosaic interspersed with many different structural classes from young to old forests; (2) they acknowledged the poor understanding of the extent to which southwestern forests were modified by Native Americans prior to settlement by Europeans but assumed that land-use practices such as timber harvest, grazing, and fire suppression by Europeans changed forest structure; and (3) they assumed that large trees, snags, and large downed logs provide habitat for goshawk prey and that every hectare of goshawk habitat needs to include clumps of large trees that are allowed to mature, die, and become snags that fall and decompose. These assumptions were clearly articulated in the management recommendations and could be tested if necessary.

Proactive versus Reactive Management

The Boreal Owl represents an example in which potential threats to persistence were identified soon after the species was discovered breeding in the western

United States. The species enjoyed "sensitive-species" status before a majority of its breeding range south of Canada was documented. Within two years of the discovery, managers began a proactive program that included researchers from Colorado and Idaho. In 1984, just four years after breeding populations were first documented, nearly sixty biologists in four Rocky Mountain states participated in surveys designed to document the distribution of Boreal Owls in the Rockies (Hayward et al. 1987). Thus, land managers in coordination with researchers began active owl management prior to public pressure.

In 1992, the proactive approach to Boreal Owl management was formalized when funding was provided by the USFS to develop a national conservation assessment (Hayward and Verner 1994). This assessment, initiated jointly by research and management staffs in Washington, D.C., represented the first critical step toward defining a coordinated conservation strategy for the species. By evaluating current understanding of the species' ecology, conservation status, and information gaps, the assessment provided a foundation from which to build an adaptive approach to conservation based on efforts from both research ecologists and managers.

Adequate Funding and Line-Officer Support
Effective programs have funding available for research, habitat manipulations, computer modeling, and other conservation work. The amount of funding required depends on the complexity of the issues and on our understanding of the species' biology. Surprisingly little funding is sufficient for some species that require limited management programs; other species with life histories that are more difficult to manage require substantial funding.

Successful sensitive-species programs also depend on a strong commitment by line officers at all levels (district rangers, resource-area managers, forest supervisors, state directors). To foster that commitment, researchers must communicate with line officers throughout the planning process; participation builds ownership. For Boreal Owls, a close cooperative relationship developed between research and management during these initial surveys and continued for the next decade as studies expanded to examine habitat use and test monitoring methods and evaluate demography. Close cooperation between research scientists and managers may represent the most important factor contributing to managing this sensitive-species. As line officers watch their staffs become involved in coordinated efforts to manage species like Boreal Owls, they are more likely to provide the necessary funds for sensitive-species management.

Charismatic Species Galvanizing Conservation Action
If success is measured in terms of actual conservation actions implemented on the ground, then the *Recommendations* by the Goshawk Scientific Committee were successful. Forest plans have been amended throughout the Southwest to

accommodate habitat prescriptions for Northern Goshawks and Mexican Spotted Owl (*Strix occidentalis lucida*). Forest management on the N. Kaibab National Forest has changed from a 150-year rotation, even-aged, evenly spaced forest before the *Recommendations*, to a 250-year rotation, mature forest with uneven age structures and a clumped tree distribution that is more consistent with the forest ecology pre-European settlement (K. Menasco, USFS, personal communication 1996). Prior to the *Recommendations*, it was difficult for biologists to convince line officers of the need to retain snags. After the *Recommendations*, forest planners now consider both micro-habitat elements and landscape characteristics in an ecosystem context (K. Menasco, personal communication 1996). The *Recommendations* have encouraged forest managers to view the ecosystem as an assemblage of interacting species. Goshawk management in the Southwest has helped to change an agency's culture and to begin achieving very real steps toward ecosystem management.

The *Recommendations* were formulated from the perspective of the goshawk rather than a general ecological-based plan for ponderosa pine ecosystems (R. T. Reynolds personal communication 1995). It was a top-down rather than a bottom-up approach. The habitat needs of goshawks were provided first in the landscape, before the scope of effort was expanded to ecosystem-level considerations. We believe that management would have changed little had the Scientific Committee simply authored a general ecosystem plan for ponderosa pine forests. The goshawk served as a flagship with broad public appeal. It provided the catalyst that galvanized the necessary resources to formulate a management plan. However, as the plan is implemented in daily forest management, it is viewed less as a plan for goshawks and more as an ecosystem plan for ponderosa pine forests (K. Menasco, personal communication 1996).

Partnerships

Successful sensitive-species programs have active, willing, and in some cases adversarial partners. Potential partners include state government, industry, and private organizations. Partners with national standing are most effective because they have the political and financial clout to lobby Congress and generate funding. For example, the Pacific River Council was interested in stream management, which led to the formation of the Columbia River Basin Initiative. This effort, initiated to manage aquatic species, could stimulate sensitive species management for forest carnivores, raptors, and big game. Partners in Flight (a nongame landbird conservation program based upon cooperative partnerships among an array of federal and state agencies, NGOs, private industry, and academic institutions that fosters coordinated avian conservation efforts through its framework of working groups at the international, national, regional, and state levels) has the backing of several conservation groups that can directly benefit sensitive Neotropical migrants. The proposed federal Teaming with Wildlife tax on bird seed, binoculars, and field guides may provide partners in industry that will help fund sensitive-species conservation. Often adversarial partners "encourage" the management of a sensitive-species. Adversarial partners may encourage agencies to alter their operations in ways more consistent with sensitive-species management and ecosystem function. For example, goshawk management is a hotly contested issue (Hitt 1992; Henson 1993). The management recommendations for goshawks in the Southwest describe ways that timber harvest can be used as a tool for restoring a forest structure that was present pre-European settlement; this will help prevent catastrophic crown fires while maintaining goshawk habitat. In this case, adversarial partners helped encourage agencies to actively manage this sensitive-species.

Multiple Benefits
Finally, successful sensitive-species programs usually benefit other species of concern. Often sensitive, Neotropical migrants are impacted by the degradation of riparian areas. If management recommendations for riparian birds also benefit species such as bull trout, the program has a greater chance of implementation. Multiple benefits may include commodity production such as timber or grazing that is extracted in an ecologically sensitive manner. For example, thinning trees of small diameter may help Northern Goshawks by restoring ponderosa pine forests while producing a valuable commodity (Reynolds et al. 1992).

Future Research Needed to Further Conservation

The availability and quality of knowledge accessible to biologists and policy makers plays a significant role in determining the success of sensitive-species management. As mentioned earlier, close cooperation between management and research is a common characteristic of strong sensitive-species programs. Thus, research must play a critical role in future sensitive-species management. But what focus should research adopt to most effectively meet the information needs of managers of sensitive species?

We summarized the responses of thirty-five wildlife biologists who replied to our questionnaire request to list two general research needs that would help them manage sensitive species. The fifty-five ideas voiced by the biologists resulted in eighteen research needs (table 11.1). Information regarding the range of natural variation in population characteristics along with research on autecological habitat relationships were listed as the top research priorities by biologists responding to our questionnaire. Many biologists also asked for help monitoring sensitive species and expressed a need for information on habitat relationships at a landscape scale.

Table 11.1. Responses of district biologists employed by the U.S. Forest Service when asked to list two general research needs that will help them manage sensitive species. Thirty-five biologists provided 55 responses; some only offered one response.

Research need	Percent of responses
Define the range of natural variability for species and autecological habitat requirements	27
Assist with monitoring and surveys	13
Determine necessary habitat conditions at landscape scale	9
Develop cumulative effect models at landscape scale	7
Determine effects of disturbances on sensitive-species	7
Conservation guidelines of management strategies	5
Determine effective management techniques	5
Dependable methods of determining population viability with limited data	4
Determine effects of fire suppression	4
Study laws and regulations regarding protection and management	2
Add spatial features to computer programs used by agency	2
Review sensitive-species lists	2
Collect prey base data	2
Research peripheral populations	2
Develop database of ecological information of sensitive species	2
Study impacts of grazing	2
Study the importance of management indicators	2
Ecosystem management as it relates to human-use factors	2

Obtaining information on responses to disturbance, population trend, and dispersal is important for sensitive-species management, but these topics are discussed in detail elsewhere (chapters 1, 5, 6, and 11). Below, we discuss additional research needs.

Distribution

Knowledge of how a species' geographic range changes over time provides managers with an initial screen to evaluate potential impacts of proposed management. Management actions scheduled outside the current and historic distribution are of low concern unless the distribution of the species is expanding and will reach the impacted site during the temporal life of the project. Without knowledge of current and past distribution, management cannot determine potential impacts.

Limiting Factors

Wildlife research often focuses on identifying important habitat characteristics, but few studies are designed to determine what factors limit populations. Our poor understanding of limiting factors may result from the difficulties associated with conducting field experiments with large mobile vertebrates. Therefore, limiting factors must be inferred from an understanding of the natural history and ecology of species. Efforts directed toward identifying factors that likely limit sensitive-species are important despite the difficulties with designing and implementing these studies.

Employing a combination of envirograms (Andrewartha and Birch 1984) and sensitivity analysis of life history represents one approach for identifying potential limiting factors. An envirogram is a graphical description of environmental characteristics that directly and indirectly influence population persistence. Sketching an envirogram for a sensitive species provides a graphical hypothesis of potential links between the species and other parts of the ecosystem that "limit" population growth. Demographic sensitivity analysis can further refine hypotheses concerning limiting factors by identifying life-history characteristics that have the strongest influence on population growth. Based on the insights gained from these exercises, research priorities can be identified to test hypotheses regarding potential limiting factors.

Ranking Habitat Quality

When managing habitat for sensitive species, we assume that individual organisms select particular habitats because doing so enhances their fitness (Rosenzweig 1985). Habitat loss has been associated with declines in many species as a consequence of a variety of mechanisms (Hunter 1996). Most studies of wildlife habitat associations focus on identification of habitats "selected" from some set of "available" habitats. However, management of habitat for sensitive species requires more than just identifying good versus bad habitat. Knowledge of the relative ranking of habitat is necessary to evaluate the short- and long-term impacts of proposed management actions. Therefore, future habitat studies for sensitive species should strive to rank habitats in terms of quality for particular functions. This is a tall order, because habitat quality must be examined at a number of spatial and temporal scales. Furthermore, identifying quality habitat depends on measuring demographic rates (survival and/or reproduction) of individuals occurring in particular habitats and comparing those with rates observed in alternative habitats.

Demographic Analysis

"Demography is a tool for understanding population-level dynamics in terms of events . . . at the level of the individual" (McDonald and Caswell 1993, 139). Demographic analysis, in particular sensitivity analysis, can be a powerful tool in allocating research effort and directing management. Sensitivity analysis pinpoints the most ecologically important life-history stage for a population. Therefore, the analysis of a life-cycle graph or projection matrix can identify those life-history stages that must be understood most completely in order to understand threats to population persistence. After identifying critical life-history characteristics, researchers and managers can coordinate programs that examine the environmental factors most important for that life-history stage.

Knowledge of Environmental Interactions at Varying Spatial Scales

Ecologists and managers increasingly recognize the importance of geographic and temporal scales in determining the processes that affect species abundance (Wiens 1989). Likewise, biologists recognize that diverse biological interactions (mutualism, competition, predation, parasitism) affect species abundance (examples in Ehrlich 1994; Estes 1995). Despite this knowledge, management planning often ignores scaling issues and complex interactions. For instance, in 1991 we reviewed management recommendations for Great Gray Owls, (*strix nebulosa*) which are considered sensitive in two national forest regions. Recommendations for eight of nine forest-level plans provided direction to protect nest sites or to protect raptor nests in general. These plans ignored processes occurring at geographic scales broader than the nest site or those operating during the nonbreeding season. Great Gray Owls interact with an array of small mammals at differing spatial and temporal scales, each of which must be understood for effective management.

Biological invasions threaten the long-term persistence of some sensitive species and need to receive much greater research attention. The range expansion by Barred Owls (*Strix varia*) into the range of the Northern Spotted Owl (Taylor and Forsman 1976) and the consequence of exotics to island avifauna (Rodda, Campbell, and Derrickson, this volume) provide examples of how invading species can disrupt ecological communities. Managers will increasingly need to predict how habitat alterations impact the ecological relationships that determine the assemblages of species.

Evaluation of Threats to Persistence

Ideally, a population viability analysis (PVA) should be conducted for each sensitive species. PVAs consider all factors that threaten a species with extinction (Gilpin and Soulé 1986). A formal (one that is analytically comprehensive) PVA integrates factors such as demographics, genetics, and environmental stochasticity, with life-history and habitat-use information. In addition, ecological processes like dispersal, competition, and predation are evaluated. This comprehensive process is expensive in terms of time and money. Thus, our ability to conduct formal PVAs for most sensitive species is highly unlikely.

Given the usual dearth of empirical information that is necessary for PVAs, resource managers face considerable difficulties when evaluating threats to species persistence. We need to learn more about how best to evaluate threats to the persistence of sensitive species in a manner that is economically achievable. Several tools are available that may help biologists approach the issue of assessing persistence in the absence of local demographic and ecological information. We suggest an approach that integrates demographic sensitivity analysis (McDonald and Caswell 1993) and evaluation of an envirogram in the context of ecological theory. Ruggerio, Hayward, and Squires (1994) discuss several guidelines that can aid in this evaluation. Ecological understanding can be incorporated into the evaluation most rigorously by applying the hypothetico-deductive approach to management (Murphy and Noon 1991). Through a rigorous assessment of the assumptions that form the basis of management, they reduce the uncertainty clouding an evaluation of the efficacy of various management options. Caughley and Gunn (1996, 223–270) provide an extended discussion of a similar approach to evaluate population declines. They demonstrate the application of hypothetico-deductive logic at several stages of management planning that may be applicable to sensitive species.

Abundance, Persistence, and Distribution

Fundamental to sensitive-species programs is the notion that wildlife should be actively managed before reaching low population densities. Thus, determining how abundance impacts the persistence and distribution of wildlife is relevant when evaluating sensitive-species management. The abundance of species and the extent of their geographic distribution appear correlated (Hanski 1982; Brown 1984). Within taxa, species with large geographic distributions tend to have greater local abundance at sites where they do occur compared to geographically restricted species (Lawton 1993). Brown (1984) hypothesized that species with wide-niche breadth (i.e., that can exploit many resources) become both widely distributed and locally abundant; empirical support of the hypothesis is equivocal (Lawton 1993). If widespread environmental changes such as pollution or the introduction of a predator or competitor cause populations to decline, then overall ranges are expected to contract, even without habitat destruction. If the species' original distribution has a well defined center, we expect that range contraction is expected to compress distribution toward the core; if the original distribution had multiple modes, ranges are expected to fragment and to contract into former hot spots (Lawton 1993).

The broadscale relationships between abundance and distribution are poorly understood and need additional research. We have only a short time to learn about these effects, as anthropogenic factors increasingly impact most vertebrate populations. However, our limited understanding of these relationships does reaffirm the importance of active intervention through management before population abundance is reduced to the point of range collapse.

Prioritizing Research Dollars

Land management agencies are confronted with an increasing number of sensitive species to manage with declining budgets. Funding conservation actions for all sensitive species is impossible. Some land managers argue that we should abolish lists because we have progressed beyond a "listing mentality" through ecosystem management. Although ecosystem management may render listing individual species unnecessary in the future, we are not there yet. We fear that abolishing sensitive-species lists will reduce management's focus as more pressing issues capture its attention. We believe the most cost-effective way to protect biodiversity is to actively manage wildlife of concern before they become endangered. Thus, programs that manage sensitive species are cost effective as well as ecologically defensible.

As we have stressed throughout this chapter, efforts to conserve biodiversity need to include both single-species and ecosystem management. Each approach has strengths and weaknesses. Ecosystem management needs to provide the necessary habitat elements and ecosystem structure for many, if not most, sensitive-species. Species that are obscure to the public, such as some plants and invertebrates, will not garner public support for large-scale management actions. In addition, often the ecologies of these species are poorly understood and little empirical data are available for developing management plans. Ecosystem management may represent the best chance for conserving this element of biodiversity. Managing entire ecosystems may include establishing conservation areas imbedded within an ecologically favorable matrix. Possibly, conservation areas that include hot spots of both rarity and species richness will prove most valuable for conserving maximum biodiversity (Williams et al. 1996). However, conservation efforts that embrace ecosystem or other multispecies management philosophies, regardless of how comprehensive, will fail to protect all species. Individual species act as templates, spatially setting the scale of landscape conservation efforts (Wilcove 1994a). Determining the autecological habitat needs of rare and specialized species will continue to be an important part of conservation planning.

We previously described some important autecological research needs that are important to conservation. These included determining changes in species distribution, population trend, limiting factors, habitat quality, demography, dispersal, and responses to disturbances. Clearly, meeting these needs is beyond the financial means of many organizations. This is more than a question of allocation; funding detailed research on a few species or general research on many. There are other considerations that may prove helpful when prioritizing how we spend our limited research dollars for sensitive species.

Highest research priority should be assigned to those sensitive species that are most threatened in terms of rarity or sharply declining trends in habitat quality. Establishing population trends for sensitive-species that occur at low population densities is difficult (Verner 1984) and requires efforts from both management and research. For some low-density or secretive species, a delphi-process involving researchers familiar with the species' life history and habitat associations may begin to address population trends. Although species can decline without habitat alteration, research priority should be given to species that inhabit ecosystems that have undergone significant changes in size, structure, or distribution.

Keystone species have a larger effect on communities and ecosystems than is expected based on their abundance (Power et al. 1996). Keystone modifiers can impact community structure in ways that are critical to maintaining biodiversity. We reiterate the concern of Mills, Soulé, and Doak (1993) that species are not inherently keystones, that their keystone status depends on ecological context. An important role of research is to determine the ecological context in which species serve as keystones. Once keystone species are determined for a given ecosystem, we believe that allocating funds to research their ecologies is highly justified and ecologically important.

Although we recognize that charismatic species are not perfect surrogates for other species or "endangered" ecosystems (Franklin 1994), we believe they have an important role in sensitive-species management. Our credibility with the public demands that we maintain charismatic species throughout their range. Management plans that fail to provide for these species will be viewed as a failure. In addition, ecosystem plans that focus on a high-profile species have a better chance for effecting real change (see Reynolds et al. 1992). Charismatic species, especially those whose habitat requisites will serve as an umbrella for other species, should be given priority for management and research funding.

Preserving both within and between species genetic variation is important in conservation planning (Lesica and Allendorf 1995). Should land management agencies spend dwindling conservation dollars on species considered sensitive at the periphery of their distribution but secure at the center? The answer to this question impacts sensitive-species programs across the country. Some land managers believe that peripheral populations, which often occur at low population densities and in ecologically marginal habitats, are less valuable to conservation compared to those at the center of a species' distribution. However, sensitive species at the periphery of their ranges may be disproportionately important for conserving genetic diversity relative to their size and frequency (Ehrlich 1988; Lesica and Allendorf 1995). Peripheral populations that differ genetically from parent populations are most valuable. Priority for funding autecological studies should target the most genetically divergent sensitive-species populations.

Close coordination between research and management is required at all stages of conservation planning. Conservation strategies that involve both research and management throughout all stages of the planning process are cost effective compared to efforts that lack coordination (Young and Varland, this volume; Kochert and Collopy, this volume). Wildlife biologists, often frustrated with the basic lack of biological knowledge of sensitive-species frequently initiate small, poorly funded investigations that have insufficient power to answer the important questions they must address. These well-intended efforts often lack coordination and may result in research that is conducted at inappropriate spatial and temporal scales. A joint implementation team should set research priorities, play a strong role in formulating research hypotheses, and coordinate information sharing. Close coordination between research and management is also a necessary component of adaptive management, so that both small- and large-scale experiments are tested under alternative management schemes using rigorous scientific methods.

Land management agencies have a legal and an ethical obligation to be actively involved in conserving biodiversity. The philosophy behind sensitive-species management recognizes that wildlife populations need management before becoming rare and endangered. This basis has considerable social and ecological merit. Sensitive-species programs should continue to be a cornerstone of agencies' efforts to conserve biodiversity. We are concerned that agencies are under increasing political pressure to reduce their commitment to sensitive-species management. Agency leadership is critically important to avian conservation.

Acknowledgments

We wish to thank the district biologists of the U.S. Forest Service who responded to our questionnaire. We realize it is difficult for biologists to make time for such requests, and we deeply appreciate their input. We thank L. Fenwood from the Washington Office of the U.S. Forest Service for information of national scope. We also thank Rex Sallabanks, J. Marzluff, T. Hoekstra, and an anonymous reviewer for providing helpful suggestions and comments.

PART IV

Conservation in Forested Landscapes

The bulk of avian conservation research has been done on terrestrial systems, especially those covered by commercially valuable forest. Being primarily terrestrial organisms ourselves, humans have likely had their most significant impacts on birds inhabiting terrestrial systems. Here, authors discuss issues that relate to the conservation and management of avifauna in terrestrial, forested ecosystems of temperate North America.

Walters (chapter 12) and Faaborg et al. (chapter 13) address one of the most widespread threats to bird populations: habitat fragmentation. Whereas Walters offers a general theoretical discussion of the need to better understand why some bird species are more sensitive than others, Faaborg et al. reflect on their own research experiences in fragmented midwestern landscapes to review previous work and set future goals. Both chapters identify dispersal and demography as critical areas for future research if we are to fully understand the effects of habitat fragmentation. Dispersal behavior is a prime candidate as a general factor involved in differential sensitivity of birds to fragmentation (Walters) and is central to understanding (and to some extent verifying) regional source-sink models (Faaborg et al.).

Although landscape-level processes such as fragmentation have been heavily implicated in the regulation of bird populations, Irwin (chapter 14) provides evidence that we still lack a general understanding of the factors that influence habitat selection by birds (a point echoed by Walters). Irwin takes a different viewpoint from that of Walters and Faaborg et al. and considers bird-habitat models at a finer scale. New avenues of research, that dovetail those suggested in the two preceding chapters, are recommended. Specifically, the role of abiotic factors is emphasized by Irwin, and evidence is presented that suggests that greater predictive precision with respect to avian habitat selection may be achieved by the inclusion of such traditionally neglected variables.

On a more practical note, Haufler (chapter 15) describes a strategy for monitoring avian distribution, abundance, and population viability in forested ecosystems of the western United States. Much like Faaborg et al., Haufler's chapter is presented as a case study from a specific region of the United States, but with clear applicability to other habitats and geographic locations. The strategy described by Haufler takes a more management-oriented approach than the other chapters in part IV but also notes the need to identify mechanisms underlying regional patterns at the landscape scale.

Taken together, these important contributions provide a mix of theoretical, conceptual, and practical discussions of factors that influence the distribution, abundance, and health of bird populations. They review existing knowledge of bird-habitat relationships and landscape-level effects and provide suggestions for future research. The concepts presented in these chapters are applicable not only to forested ecosystems within temperate North America; many of the issues raised by these authors are repeated elsewhere throughout this volume by authors working in other continents, with other genera, and in entirely different habitats. The need for long-term investigations of avian demographic responses to human disturbance is emphasized by all authors. The benefit and utility of such work is exemplified by Faaborg et al.

Four important messages are emphasized in this section. First, habitat fragmentation is potentially the most important factor threatening the preservation of viable bird populations. The influence of fragmentation is well studied and appears to depend on the type of edge (e.g., forest-urban; forest-agricultural; forest-forest) created, but the ecological basis of differences between species in sensitivity to fragmentation has not been a focus of research. Although we have documented many effects of fragmentation, we still have a poor understanding of the mechanisms that cause these effects. The actual role of fragmentation per se, independent of associated habitat removal, is also poorly understood. Second, gaining a better understanding of the dispersal behavior and metapopulation dynamics of birds is perhaps the most important contribution that research biologists and land managers can make to the field of avian conservation. Such knowledge would enable us to better understand why habitat fragmentation appears to reduce avian population viability. However, it requires a major shift in our thinking about avian population dynamics. The predominant theory of competition-driven, density-dependent, closed populations needs to give way to a perhaps more general theory of metapopulation dynamics where source and sink populations are linked by dispersal. Third, site-specific physical characteristics of the environment should not be overlooked when describing bird-habitat relationships, regardless of species and geographic location. Lastly, a two-phased research strategy that first documents distribution and then identifies viability problems and causal mechanisms is an effective approach, but especially so if performed in a context compatible with existing management operating and classification systems. Researchers need to understand and speak the manager's language.

CHAPTER 12

The Ecological Basis of Avian Sensitivity to Habitat Fragmentation

Jeffrey R. Walters

Conservation Issues and Previous Research

Habitat fragmentation has been called "perhaps the most significant challenge to the development of models applicable to wildlife management, if not ultimately to the survival of wildlife altogether" (Temple and Wilcox 1986). Fragmentation is a fact of life in the modern world and will only become a more predominant feature of the world's landscapes in the forseeable future. The challenge to conservation is not only to limit fragmentation, but also to understand how animals respond to it so that adverse effects can be avoided or reduced.

A species is considered sensitive to fragmentation if density or fitness of individuals within remaining patches of its habitat changes as fragmentation of the surrounding landscape changes. Defined in this way, sensitivity to fragmentation is part of the more general phenomenon of sensitivity to landscape structure, which encompasses relationships among within-patch density and patch sizes, shapes, and configurations within the landscape. Fragmentation effects must be distinguished from effects of habitat degradation, which often accompanies fragmentation and may produce adverse effects that are independent of landscape structure (Lovejoy et al. 1984; Lynch and Whigham 1984; Saunders, Hobbs, and Margules 1991).

The ecological basis of differences between species in sensitivity to fragmentation has not been a focus of research. Instead, the emphasis has been on determining whether adverse effects of fragmentation occur, and on identifying factors that might cause observed effects. Experimental demonstrations of sensitivity of birds to fragmentation specifically or landscape structure generally (Bierregaard and Lovejoy 1989; Bierregaard et al. 1992; Darveau et al. 1995) have been limited compared to work on plants and invertebrates (Kareiva 1990; McGarigal and McComb 1995). What the literature on birds provides is an abundance of studies that demonstrate correlations between patch area and isolation and species number, and between within-patch abundance or fitness of particular species and level of fragmentation (for recent reviews see Robinson and Wilcove 1994; Faaborg et al. 1995; Freemark et al. 1995). There is little question that landscape structure affects within-patch abundance and fitness of birds, or that fragmentation reduces within-patch abundance or fitness in some species.

The mechanisms producing adverse effects of fragmentation have been extensively studied in birds as well. Initially, there was especially intense focus on fragmented forests in the northeastern and midwestern United States because of alarm about declines of Neotropical migrants in these landscapes (reviewed by Askins, Lynch, and Greenberg 1990; Paton 1994; Robinson and Wilcove 1994; Faaborg et al. 1995; Freemark et al. 1995). These studies identified elevated levels of nest predation and nest parasitism in small fragments due to proximity to edges between forest and agricultural or residential lands as major factors producing reduced fitness of Neotropical migrants on fragmented landscapes (Faaborg et al., this volume). Reduced pairing success in small fragments and near edges is a factor in some species as well (Gibbs and Faaborg 1990; Villard, Martin, and Drummond 1993; Porneluzi et al. 1993; Van Horn, Gentry, and Faaborg 1995; Hagan, Vander Haegen, and McKinley 1996).

Studies in Europe (reviewed by Opdam 1991) reinforced the notion of isolated woodlots in agricultural or urban landscapes as the classic image of fragmentation. In recent years, however, studies of a wide variety of landscapes have provided a much richer context for identifying pattern and process related to fragmentation effects. Reductions in density in remaining habitat of at least some species in fragmented or otherwise disturbed landscapes is widespread, being found in systems as disparate as eucalypt woodlands in Australia (Ford, Barrett, and Howe 1995), shrub-steppe communities in the central United States (Knick and Rotenberry 1995), old-growth forests in the Pacific Northwest (McGarigal and McComb 1995), and cloud forests in Colombia (Kattan, Alvarez-Lopez, and Giraldo 1994). However, the consequences of patchiness vary widely among landscapes. One general pattern is that the magnitude of edge effects generally, and changes in nest predation and nest parasitism specifically, depend on the type of edge (here defined as the boundary between any two habitat types), the contrast between habitat types within the landscape, and perhaps the existence of particular habitats (specifically anthropogenic ones) in the landscape. For example, in forested landscapes in which the alternative (to old-growth forest) habitats are forest stands of different ages created by timber cutting, effects of fragmentation and edge typically are less apparent than in midwestern and northeastern forests fragmented by agricultural development (Haila 1986; McGarigal and McComb 1995; Schieck et al. 1995; Hagan, Vander Haegen, and McKinley 1996; reviewed by Freemark et al. 1995). In these forests, contrast between patch types is much less, and edges generally are gradual rather than abrupt. (Indeed, some would not qualify as edges where high contrast is part of the definition of edge.) Saurez, Pfenning, and Robinson (in press) found that in a midwestern forest nest predation rates on Indigo Buntings (*Passerina cyanea*) were higher along abrupt edges (i.e., agricultural land, wildlife openings, roadsides) compared to gradual edges (i.e., treefall gaps, selective cuts, riparian habitat) and along exterior (i.e., agricultural land) compared to interior edges (i.e., treefall gaps, wildlife openings, selective cuts, riparian habitat). Variation among edge types in their effect is evident in other habitat types as well (Small and Hunter 1988; Gibbs 1991; Darveau et al. 1995; Latta et al. 1995).

Recent studies of avian communities in bottomland forest in the southeastern United States provide an additional example. In this system there are two primary habitat types, cypress-gum swamp forest and levee forest (a mixed hardwood type) (Schafale and Weakley 1990), inhabited by a rich avifauna, many species of which strongly prefer one habitat type over the other (Pashley and Barrow 1992). Scott K. Robinson (personal communication) studied this system in southern Illinois and documented sensitivity to patch size and adverse edge effects along abrupt exterior edges between forest and agricultural lands. He found no edge effects along abrupt interior edges between forest and river, however. In Georgia, Hodges and Krementz (1996) showed that densities of several species are reduced in bottomland patches less than 350 m wide between the river and pine plantations. Our research group studied this system along the Roanoke River in North Carolina, where the forests are much wider and most edges are gradual interior ones between levee and swamp, rather than abrupt, exterior ones between bottomland habitat and agricultural land (Illinois) or pine plantations (Georgia). Here patch-size effects were absent, even in the species shown to be sensitive in Illinois and Georgia, and edge effects were lacking along levee-swamp edge, as well as forest-river edge. There were, however, edge effects along exterior boundaries with agricultural land (Sarraco and Collazo, unpublished data).

These studies illustrate the fairly obvious, yet critical, point that small size and isolation of habitat patches are not invariably linked with reduced abundance or fitness. Some habitat types by nature occur in small, isolated patches, yet there are species that thrive in them. Thus, adverse effects of fragmentation are attributable not to patch size and patch isolation alone, but also to other features that characterize fragmented landscapes but not naturally patchy ones, such as increased contrast between patch types, creation of novel edge types, and presence of habitat types associated with human activity. To foreshadow what is discussed below, I will mention two other possibilities. First, species may be adapted to the historic distribution of their habitats, and those adapted to more continuous habitat types may suffer especially from fragmentation (Temple and Cary

1988). Second, comparisons among landscapes suggest that dispersal behavior may be critical to fragmentation effects, because patch size appears to matter less where intervening habitat facilitates rather than inhibits movement between patches.

The patterns just described apply primarily to static landscapes in which fragmentation is well established. Some studies have focused instead on responses of birds during the process of fragmentation. These studies show that abundance may increase in fragments initially, presumably due to an influx of displaced individuals from habitat that is destroyed (Lovejoy et al. 1984; Bierregaard and Lovejoy 1989; Saunders, Hobbs, and Margules 1991; Darveau et al. 1995). Where fragmentation is dynamic, such as in industrial forests, continuing displacement may maintain high densities in fragments, perhaps even high enough to disrupt reproductive behavior (Hagan, Vander Haegen, and McKinley 1996). Again, movement is a critical factor.

Features of Research
That Improved Conservation

Previous research has been successful in elucidating some of the mechanisms that reduce abundance and fitness in certain fragmented landscapes, and thus in indicating to managers desirable landscape configurations and potentially effective remedial measures in those systems (i.e., control of cowbirds and human-dependent nest predators). Initially, it was critical to document adverse effects of fragmentation in order to justify efforts by conservationists to institute minimization of fragmentation as an objective in land-use planning, and this has been accomplished. What managers are offered based on initial work is a uniform, general strategy, which could roughly be described as minimize patchiness, connect patches with corridors, and control nest predators and parasites. This strategy may work well in some cases, but it will be inefficient or ineffective in others. For example, nest predation may have nothing to do with adverse effects of fragmentation in some systems, and contrast between adjacent patches may matter more than patch size in others. Our current understanding of fragmentation effects is too poor to be the basis of consistently effective management. The contrast between application and theory is perhaps best illustrated by the widespread efforts to design and create corridors linking habitat patches despite the lack of direct evidence of their effectiveness (Simberloff and Cox 1987; Hobbs 1992), and the possibility that they are of little consequence to birds (Faaborg et al. 1995).

The ultimate level of precision, currently lacking, is to be able to predict the severity of fragmentation effects in a particular system and the particular species that will suffer most. Such a theory would enable managers to obviate adverse impacts, as well as repair them, and to do so in systems not yet studied. This would be invaluable, given that the pace of development over the next fifty years surely will exceed our ability to study individual systems, particularly in the third world. To progress toward this goal requires increased attention to differential sensitivity to fragmentation.

Evidence of differential sensitivity can be found in virtually every study of fragmentation or landscape structure, beginning with the pioneering work of Whitcomb et al. (1981). The Australian landscape of eucalypt woodland mixed with cleared or partially cleared pastoral land studied by Ford and colleagues is typical. There, some species of birds are found only in the largest remaining woodland patches, whereas others, often from the same genera, are as common or more common in small patches as large ones (Barrett, Ford, and Recher 1994; Ford, Barrett, and Recher 1995). In the classic landscapes of the Northeast and Midwest, that Neotropical migrants were much more sensitive to landscape structure than residents was immediately recognized (Whitcomb et al. 1981; see also Askins, Lynch, and Greenberg 1990 for a recent review). Even in landscapes where fragmentation effects are reported as lacking generally, there are some sensitive species. For example, McGarigal and McComb (1995) reported that the within-patch density of most species associated with old-growth forest in Oregon was unaffected by landscape structure, but densities of Winter Wren (*Troglodytes troglodytes*) and Brown Creeper (*Certhia americana*) were dependent on the amount of old growth in the surrounding landscape.

In all avian communities, it seems, there are large differences among species in their sensitivity to fragmentation effects. Although variation in sensitivity is well documented, the cause of this variation is not. A variety of patterns have been noted and hypotheses proposed about differential sensitivity to fragmentation, but none have been tested. In South American forests, foraging guilds differ greatly in their sensitivity (Lovejoy et al. 1984; Kattan, Alvarez-Lopez, and Giraldo 1994; Stouffer and Bierregaard 1995a,b), whereas in Chile, endemics are most sensitive (Willson et al. 1994). The greater sensitivity of Neotropical migrants compared to residents in the northeastern and midwestern United States has been related to differences in nesting behavior and movement patterns (Whitcomb et al. 1981; Lynch and Whigham 1984; Askins, Lynch and Greenberg 1990; Holmes and Sherry 1992). Presumably, differential sensitivity is based on the fact that the mechanisms responsible for fragmentation effects—elevated levels of nest predation, for example—apply to some species but not others (e.g., cup nesters but not cavity nesters), or translate into larger changes in population dynamics of some impacted species compared to others (e.g., Blake, Niemi, and Hanowski 1992). Differential sensitivity may be heightened in some systems compared to others, and sensitivity of a particular species might change from place to place. Possibly each system is unique; that is, each pattern is an unpredictable result of the interaction between changes in particular landscapes and the ecological requirements of particular species. More likely, there are at least some generalities across systems. Generalities may emerge from examination of patterns in ecological correlates of sensitivity as more systems are studied. Increased examination of all manner of ecological correlates of sensitivity (e.g., nesting habits, foraging habits, size, rarity) is needed. Here I focus on one particularly important variable, dispersal behavior.

Research Needed to Further Conservation

To further conservation, what is required is a theory that enables one to explain effects of fragmentation in terms of changes in landscape structure and their interaction with the varying population dynamics of bird species. Three kinds of research are required to build such a theory.

Research in Landscape Ecology

The first area of research that is needed to further avian conservation is the one that has dominated studies of fragmentation to date: investigations of effects of landscape structure on avian populations and the mechanisms involved in producing those effects. The model developed by Robinson et al. (1995; Faaborg et al., this volume) for midwestern forests illustrates the kind of approach required. It relates adverse effects of fragmentation to landscape structure and population dynamics in a way that could explain differential sensitivity, and that results in management recommendations specific to the system (Thompson et al. 1996). Equally sophisticated models of additional systems, varying in degree of fragmentation, are required to assess general patterns.

The ideal product of this research would be a comprehensive framework into which the particular phenomena that characterize habitat fragmentation could be placed. Such a framework would need to include not only basic effects such as those of patch size, patch isolation, and edge, but also more complex effects such as how basic effects vary as a function of contrast between patch types, and how these landscape variables interact with population dynamics. Again, Robinson et al. (1995) provide an excellent precedent: in their model, severity of edge effects depends on landscape-level habitat coverage patterns. Progress in this area will come through the development of general theory within landscape ecology, as well as specific studies such as that of Robinson et al. (1995).

Research on Dispersal Behavior

Understanding how landscape structure interacts with population dynamics is necessary to explain differential sensitivity to fragmentation, but it is not sufficient. An improved understanding of variation in population dynamics generally and dispersal behavior specifically is also necessary.

Table 12.1. Differences between contiguous and
fragmented woodland in population structure of
Brown Treecreepers.

	Contiguous	Fragmented
Groups	15	11
Mean Group Size	3.0	2.1
% Groups without Female	7%	45%
% Groups Nesting	87%	36%
% Nests Successful	85%	100%
Fledglings/Nesting Group	1.8	2.8

The critical importance of dispersal in understanding population dynamics in complex landscapes has long been recognized (Kareiva 1990). Basic to the idea that landscape structure around a habitat patch matters is the implicit assumption that structure affects movement into and out of the patch, and that this has important consequences. Variation among species in dispersal behavior is a prime candidate as a general factor involved in differential sensitivity to fragmentation. Disruption of movement has emerged as a likely basis of sensitivity to fragmentation in a variety of systems (Saunders, Hobbs, and Margules 1991; Matthysen, Adrianensen, and Dhondt 1995; Stouffer and Bierregaard 1995b). It was proposed as a factor in the earliest studies documenting sensitivity of Neotropical migrants in the United States (Whitcomb et al. 1981; Lynch and Whigham 1984), and continues to be an important hypothesis in this case (Askins, Lynch, and Greenberg 1990; Saunders, Hobbs, and Margules 1991; Holmes and Sherry 1992; Faaborg et al. 1995; Hagan, Vander Haegen, and McKinley 1996). Migrants and residents obviously differ systematically in their movement patterns, but the critical differences lie not only in their seasonal movements, but also in how young birds select their initial breeding site and the level of site fidelity subsequently (Holmes and Sherry 1992), details about which we know much less.

The Australian landscape studied by Ford and colleagues (see above) may provide another example of variation in sensitivity related to gross differences in dispersal behavior. One of the genera in which variation among species in sensitivity to fragmentation has been observed is Climacteris, the treecreepers. The Red-browed Treecreeper (C. erythrops) remains only in the largest woodland patches, whereas the White-throated Treecreeper (C. leucophaea) continues to thrive in the same fragmented landscape, and the Brown Treecreeper (C. picumnus) is widespread but declining in small patches (Barrett, Ford, and Recher 1994). An initial study indicated that reproductive success of Brown Treecreepers in isolated woodland patches in fragmented landscapes is equal to that of those in habitat patches within wooded landscapes. However, many males are unpaired in fragmented landscapes but not in wooded ones, suggesting that dispersal of females is disrupted by fragmentation (Walters, Ford, and Cooper, unpublished data) (table 12.1). Brown Treecreepers and Red-browed Treecreepers are cooperative breeders, which differ systematically from other birds in their dispersal behavior, specifically in their greater reliance on short-distance forms of movement (Emlen 1991). An obvious hypothesis is that cooperative breeders are more sensitive to habitat fragmentation because they are relatively poor at long-distance movement.

Cooperative breeders represent an extreme in dispersal behavior. Whether smaller differences in dispersal among the myriad of noncooperative species relate to differential sensitivity to fragmentation is the issue. Generalizations such as "migrants are more sensitive" and "cooperative breeders are more sensitive" are of limited use but suggest that the benefits of acquiring detailed information about dispersal to understanding sensitivity to fragmentation may be great. Indeed, in those few cases where such information has been acquired, the effect on understanding of the interaction between population dynamics and landscape structure is dramatic (Stith et al. 1996; Martin, Stacey, and Braun, in press). What is needed is detailed information about movement, at least at the level of probability of dispersal as a function of distance and intervening habitat type, such as has been obtained for the Florida Scrub Jay (*Aphelocoma coerulescens*) (Stith et al. 1996).

Of course, the primary reason such information is scant is not because it previously was viewed as unimportant, but because it is notoriously difficult to obtain. The logistics of studying dispersal are improving, however (Faaborg et al., this volume). It is imperative that field biologists increase their efforts to obtain information about dispersal. This is a critical contribution that field biologists can make to solving the habitat fragmentation problem currently.

Research on Avian Population Dynamics

To obtain the information about dispersal required to understand differential sensitivity to habitat fragmentation is difficult logistically but simple conceptually. In contrast, progress in the third research area, avian population dynamics, requires a fundamental change in conceptual approach. The recent alarm about population trends of Neotropical migrants has caused ornithologists to attempt to interpret population behavior at an unprecedented level of precision, with little success. Our poor understanding of avian population dynamics can be attributed, in part, to lack of long-term population data, which are hard to obtain. But one can reasonably conclude that it results more from inadequacies of the competition-driven, density-dependent, closed-population thinking that underlies our approach to population dynamics.

The traditional paradigm is well grounded in elegant theory and in the thinking of some of our most influential avian ecologists, particularly David Lack and Robert MacArthur. There is convincing evidence of density-dependent population regulation (e.g., McCleery and Perrins 1985; Tinbergen, van Balen, and van Eck 1985; Newton and Marquiss 1986), and the closed-population paradigm works quite well for some populations, notably Song Sparrows (*Melospiza melodia*) on Mandarte Island in British Columbia (Arcese et al. 1992). However, it appears that such cases are the exception rather than the rule. The issue is not the relative importance of density dependent and density independent factors, both of which clearly operate (Arcese et al. 1992 provide a recent review), but the importance of movement. Immigration into the Song Sparrow population on Mandarte Island is rare (Arcese et al. 1992), whereas evidence is accumulating (see below) that immigration into most bird populations is sufficiently frequent to match or exceed the importance of regulating factors within the population, whether density dependent or density independent, such that one can question whether regulation at the local level even exists.

There have been indications of serious problems with traditional thinking about bird populations for some time. Perhaps the best-known example is the work on shrub-steppe birds by Wiens and Rotenberry (1980), whose observations of population dynamics were difficult to reconcile with the traditional paradigm (Wiens 1983). Recently there have been even more dramatic examples. Stacey and Taper (1992) found that the persistence of a small, isolated population of Acorn Woodpeckers (*Melanerpes formicivorus*) in New Mexico was completely dependent on regular immigration. Martin, Stacey, and Braun (in press) found that 94 percent of the females recruited into five neighboring populations of White-tailed Ptarmigan (*Lagopus leucurus*) in central Colorado were immigrants from beyond that neighborhood. In contrast to the Mandarte Island population (see above), in other Song Sparrow populations on nearby islands closer to the mainland and on the mainland itself, immigration is frequent and critical to population dynamics (Smith et al. 1996). Territoriality results in important density-dependent effects in Sparrowhawk (*Accipiter nisus*) populations, but population trend is driven by external recruitment (Newton and Marquiss 1986; Wyllie and Newton 1991). Persistence of many migrant species in small forest patches in the Midwest may be highly dependent on immigration from distant populations (Faaborg et al., this volume), and this may be true in Europe as well (Hinsley et al. 1996). In none of these cases could one predict numbers accurately based on processes within the population alone (see Sherry and Holmes 1992 for another example). Indeed, a traditional density-dependent, closed-population model predicts extinction of populations that are in reality quite persistent (Stacey and Taper 1992; Donovan, Thompson et al. 1995).

A paradigm better suited to such cases than the traditional one is the meta-population paradigm (Stacey and Taper 1992; Donovan, Lamberson et al. 1995; Smith et al. 1996; Stacey, Johnson, and Taper 1997). This paradigm is well suited for describing interaction between subpopulations in the remaining patches of a formerly more continuous habitat in fragmented landscapes (Villard, Freemark, and Merriam 1992; Stacey, Johnson, and Taper 1997). Rather than considering the population dynamics in such landscapes to be an abnormality caused by alteration of the environment, they may be viewed as a special case of the norm that involves changes of parameters but no fundamental change in process.

Classic metapopulation structure (Levins 1969, 1970), which involves a set of small patches that are individually prone to extinction but close enough together that recolonization balances extinction, is probably rare in birds. However, the broader model of interacting subpopulations of varying sizes and degrees of isolation described by Harrison (1991) and expanded by Stith et al. (1996) includes several types of population structure that appear to be common in birds. These kinds of metapopulations are characterized by rescue rather than extinction of subpopulations due to regular movement between subpopulations (Stith et al. 1996; Martin, Stacey, and Braun, in press). Both source-sink dynamics (e.g., Donovan, Lamberson et al. 1995) and mutually dependent subpopulations (e.g., Stacey and Taper 1992) may occur in birds. In some species, such as the ptarmigan studied by Martin, Stacey, and Braun and perhaps many migratory species (Donovan, Lamberson et al. 1995), the metapopulation may include all the birds in an entire region. In other species, such as the Florida Scrub Jay studied by Stith et al. (1996), the metapopulation is restricted to subpopulations within fairly short distances of one another (e.g., 12 km in this case) (see Marzluff and Balda 1989 for another example).

If we are to understand how population dynamics interact with landscape structure in fragmented landscapes, we must first increase our understanding of population behavior, and this requires empirical and theoretical studies of the metapopulation dynamics of birds. For empirical studies this means changing priorities with respect to measuring demographic variables. In a metapopulation model, numbers in a particular patch of habitat depend not only on mortality and reproductive rates within the patch (the variables traditionally emphasized in empirical studies), but also on mortality and reproduction in other patches within the metapopulation, the degree of synchrony in demographic variability among patches, and movement between patches. Again, data on dispersal are critical, particularly information about how movement depends on density within a patch, distance between patches, and the nature of intervening habitat. Also, it may be important to measure mortality and reproduction in different habitat types and understand habitat selection among these types (Holmes, Marra, and Sherry 1996). The few data available indicate a surprising degree of asynchrony in demographic variability among neighboring patches, which supports the idea that regular rescue is a feature of avian metapopulations (Smith et al. 1996; Stacey, Johnson, and Taper 1997; but see Sherry and Holmes 1992 for an exception).

For theoretical studies, adopting the metapopulation paradigm means discarding a well-developed theory for one in its infancy. Population models associated with the traditional paradigm are primarily analytical ones based on the logistic equation and life table analysis. There is a rich literature on this topic. Analytical models of metapopulation dynamics are developing rapidly (see reviews by Hanski and Gilpin 1996; Hanski 1994), but it will take some time for this theory to acquire the complexity, detail, and sophistication of the traditional approach. Such models are not yet realistic enough to be usefully applied to management of actual populations in most cases.

In contrast to the analytical models are the complex simulation models suited to metapopulation modeling that are available. Of most interest are spatially explicit, individual-based models (Villard, Schmidt, and Maurer, this volume). These models track individuals on the landscape, and their parameters are probabilities of mortality, reproduction, and movement of individuals that can vary with their age, sex, location, or other attributes. Population behavior is a product of what happens to individuals. These models are especially well suited to metapopulation dynamics generally and dynamics in fragmented landscapes specifically because they can incorporate effects of landscape structure on movement and habitat quality on demography. They can also incorporate another element critical to population behavior in many birds—territoriality. Territorial behavior lends an element of density dependence to population behavior, and this may account for much of whatever success the density dependent, closed-population paradigm has enjoyed in its application to birds (Newton and Marquiss 1986; Smith et al. 1996). Unfortunately, the usefulness of the simulation models is limited to an even greater extent than is the usefulness of the analytical models by availability of data to estimate parameters. In fact, it may prove impossible to collect sufficient data to accurately estimate all the parameters of these models, and therefore to ever place much confidence in them (Murdoch 1993; Conroy et al. 1995). However, recent models developed for especially well-studied species provide cause for optimism (Stith et al. 1996), although these are limited to cooperative breeders with restricted dispersal.

Making Research Effective for Conservation

Research on metapopulation dynamics in birds, dispersal behavior, and landscape ecology is necessary to develop an understanding of how different species will respond to fragmentation and other changes in the landscape. Such a theory is required if conservation is to be proactive, and developing it is the primary contribution of researchers to conservation efforts related to the fragmentation problem.

However, if the theory envisioned is to be useful to managers, researchers will need to become involved in its application to real landscapes as well. The objective is to be able to inform managers of the likely consequences to the avian community of management alternatives that involve differing landscape structures (Raphael, McKelvey, and Galleher, this volume). It seems inevitable that a specific analysis of the particular bird community present and landscape changes considered will be necessary to evaluate potential effects of fragmentation on a specific management unit. As is so often the case, a combination of basic and applied research is necessary for success, and researchers interested and able to carry their basic work through its applications will make especially important contributions to conservation.

Acknowledgments

The Nature Conservancy, the National Biological Service (now the Biological Resources Division of the U.S. Geological Survey) and the U.S. Fish and Wildlife Service funded our research on avian communities along the lower Roanoke River, and North Carolina State University and the Bailey Fund of Virginia Polytechnic Institute and State University the work on Australian treecreepers. I thank the editors, John Faaborg, and Scott Robinson for comments on a previous draft of the manuscript.

CHAPTER 13

Understanding Fragmented Midwestern Landscapes: The Future

John Faaborg, Frank R. Thompson III, Scott K. Robinson,
Therese M. Donovan, Donald R. Whitehead,
and Jeffrey D. Brawn

Conservation Issues and Previous Research

Much of the focus on breeding Neotropical migrant birds has centered on the negative effects of habitat fragmentation on breeding success (Walters, this volume). Review papers (most recently Faaborg et al. 1995) have shown overwhelming evidence that habitat fragmentation results in reduced nest success rates and higher parasitism rates, particularly in the northeastern and midwestern United States. Comparisons of nest success rates between fragmented and less-fragmented habitats have shown that fragmented habitats are often likely population sinks, where reproductive success is not high enough to replace adult mortality. In contrast, less fragmented habitats may serve as sources, where breeding success exceeds mortality and excess birds are produced that can help maintain sink populations through regional dispersal (Donovan, Thompson et al. 1995; Donovan, Lamberson et al. 1995).

Features of Previous Research That Improved Conservation

Many studies of fragmentation have shown that high rates of nest predation, parasitism, or both, are related to increasing amounts of edge associated with the fragmentation process. Exceptions (Paton 1994; Hanski, Fenske, and Niemi 1996) suggest that the severity of edge effects may stem from landscape-level patterns of habitat coverage. In our first regional study, Robinson et al. (1995) found that levels of nest predation and parasitism were highly correlated with several habitat features at the landscape scale (measured in 10-k-radius circles from the study sites). These habitat measures include the percent forest cover, mean forest patch size, and percent forest interior. In general, fragmented regions with low overall forest cover, patch size, and percent forest interior had high nest predation and parasitism levels, whereas unfragmented regions had higher nest success and little parasitism.

The initial conservation implications of these findings are fairly clear: bird populations in highly fragmented landscapes have low enough reproductive success that they are probably population sinks and are possibly maintained or recolonized only by regular dispersal from population sources in landscapes with greater forest cover. Protection of source populations may be critical to maintenance of the regional population of a species, whereas relatively little may be done to protect some nesting species in forest fragments of even relatively large size if they occur in highly disturbed environments.

Research Needed to Further Conservation

Because geographic and regional-landscape patterns may constrain populations, local management decisions should acknowledge the landscape and regional context in which they occur (Thompson, Probst, and Raphael 1993; Freemark et al. 1995; Probst and Thompson 1996). Thompson et al. (1996) developed management recommendations for migrant birds in central hardwood forests that acknowledge the importance of the regional patterns in forest fragmentation and reproductive success reported by Robinson et al. (1995). The first priority for these species is to protect or restore the large, heavily forested landscapes that likely support important regional source populations. Second, it is important to provide the appropriate mix of forest habitats within landcapes. Robinson (1996) termed these large forested areas that are likely population sources "macrosites," and medium and small sites "mesosites" and "microsites," respectively. We are currently working on conservation guidelines that acknowledge the different potentials or constraints of these different types of sites or landscapes (Thompson 1996).

Although designing management guidelines for midwestern forests from these studies is valuable, there are several deficiencies in our knowledge that must be addressed to identify the most effective management scheme in this region. Here we discuss some of these deficiencies and what is needed to address them so that we can provide guidelines that allow managers to maximize the value of their managed areas within the constraints determined by the regional landscape.

Dispersal and the Regional Source-Sink Model

Recent fragmentation studies (Donovan, Thompson et al. 1995; Robinson et al. 1995; Brawn and Robinson 1996) hypothesized that populations are structured as sources and sinks because some populations have much lower reproductive success than that needed to replace losses due to mortality of the breeders. Yet these populations have shown no tendency to decrease over time. In contrast, source populations produce a large surplus that could easily account for the population stability of sinks, given adequate dispersal. In many cases, particularly for Neotropical migrant species, the isolated sink populations are several hundred miles from source populations. Given that migrants cover thousands of miles between breeding and wintering grounds, however, such hypothesized dispersal distances do not seem unreasonable. Two questions need to be addressed to assess these dispersal issues: (1) Does dispersal occur at a regional scale (and what is the appropriate scale)? and (2) Is dispersal from sources to sinks evident?

While it is easy to envision that the population stability of sinks is the result of colonization from source populations, there is no direct evidence supporting dispersal of migrants from sources to sinks for any midwestern species, and there is limited information about avian dispersal in general (Walters, this volume). Banding has offered little evidence about the natal dispersal of Neotropical migrants, i.e., the distance from the birth location to the first breeding location, although we know that most migrants show low philopatry (Weatherhead and Forbes 1994). For example, several hundred nestlings of the Black-throated Blue Warbler (*Dendroica caerulescens*) and American Redstart (*Setophaga ruticilla*) have been banded at Hubbard Brook Experimental Forest in New Hampshire, but less than 1 percent have returned to breed on the study area there (R.T. Holmes and T. W. Sherry, personal communication). None of these banded nestlings have been caught anywhere else. Because natal dispersal is thought to be the most significant geographical movement of individuals (Greenwood and Harvey 1982), the inability to track natal dispersal limits our ability to test source-sink structure in our region.

A quantitative understanding of dispersal in any given species through classical banding would require marking thousands of nestlings or fledglings within a source population (to compensate for high juvenile mortality rates) coupled with an incredibly broad netting or resighting effort the next breeding season throughout the potential dispersal range (Brawn and Robinson 1996). Given time and money constraints, we doubt banding is the solution for understanding dispersal in small forest migrants.

Are there other direct means of measuring dispersal of birds from their natal areas? Radio transmitters are barely small enough to follow the larger of the Neotropical migrants for a few months; it seems unlikely that there will ever be a radio transmitter small enough to put on a 5–20-gram migrant that also has a battery life of nine or ten months and enough power to be picked up over a large area. Technological advances may help us here.

Two biochemical techniques might provide a breakthrough in direct measurement of dispersal in migrants. Highly polymorphic molecular markers such as DNA microsatellite markers have been used to measure gene flow between populations (Fleischer, this volume); a modified approach might illustrate general patterns of genetic structure across a region. This could demonstrate gene flow across regions, which is consistent with, but does not prove, regional source-sink models. Preliminary work on Ovenbird (*Seiurus aurocapillus*) populations in fragments of mid-Missouri showed high levels of band-sharing among populations using DNA fingerprinting, which suggests high levels of dispersal among populations (Arguedes 1992). We hope that recent advances in the use of DNA sequencing techniques will result in significant new information in the near future.

Alternatively, recent work has suggested that isotope tracers might be used to identify regional characteristics of bird populations and some measures of dispersal (Hanson and Jones 1976; Hobson and Clark 1992; Mizutani, Fukuda, and Kabaya 1990; Hobson and Wassenaar 1997). As birds develop new feathers, either in the nest or after breeding, they ingest isotopes from their food and water. Levels of the isotopes of hydrogen, carbon, and strontium vary systematically along both latitudinal and longitudinal gradients across North America (Chamberlain et al., 1997). It may be possible to develop a base map for levels of these isotopes in feathers for a species; one could then determine the natal location of second-year birds on the breeding grounds, as these birds would have feathers showing the isotope concentrations of their natal area. This approach has been used to link breeding and wintering populations of migratory birds (Chamberlain et al., 1997; Hobson and Wassenaar 1997) and has the potential to give us general information about breeding ground dispersal patterns.

Other field measurements could help test the hypothesis of source-sink population structure without measuring dispersal. For example, several years of detailed demographic measures across a variety of landscapes might show correlations between productivity in source populations and population change in sinks. Even without measures across landscapes, detailed observations on sink populations in relation to local production of young would help test the model, although Brawn and Robinson (1996) have shown how difficult understanding population variation may be in a source-sink scenario. There is a particular need for long-term monitoring of the interactions among population variation, demography, and nesting success on various scales.

Without some sort of effective individual marking program, the best for which we can probably hope in the future to test the regional source-sink hypothesis is the combination of more general biochemical markers (DNA or isotopes) combined with detailed local demographic measures, including population variation, survival rates of resident birds, and reproductive rates.

Assessing Source-Sink Status

The five thousand nests monitored by Robinson et al. (1995) provided clear general patterns with regard to predation and parasitism rates in relation to amount of regional forest cover. Yet, even statistically significant regressions and correlations have variation, and we should not ignore factors that may affect local variation even if our options for success are constrained by landscape level processes. For example, the Robinson et al. (1995) study combined data from many sites and several years, which may mask characteristics of breeding success of importance to managers in developing specific management goals for a site. Additionally, estimating source-sink status requires more than just nesting success data. Ideally, we need data on the number of young produced per female per year (annual fecundity), the probability that those young will survive to reproduce in the future, and the probability that an adult will survive to breed in future years. These parameters are difficult to measure, and better methods of estimating these parameters are needed.

Estimating Annual Fecundity

Most studies of reproductive success, including Robinson et al. (1995), find as many nests as possible, monitor them, then compute an overall nesting success rate using the Mayfield technique. In most cases, the researchers are content to have an adequate sample size to say something about average nesting success for several species for a year or multiple-year study. In comparative studies, these nesting success measures most likely provide a reasonable estimate of reproductive output relative to other study locations. However, it is important to note that estimates of fecundity based solely on measures of nesting success are not adequate to estimate the overall reproductive rate of a population.

There are two reasons for this. First, partial nest loss due to nest parasitism by the Brown-headed Cowbird (*Molothrus ater*) is prevalent in our study region, and this partial nest loss is not incorporated into typical nest success measures (Donovan, Thompson et al. 1995; Porneluzi 1996). In Mayfield estimates of nest success, if at least one host survives the nesting cycle, the nest is scored as "successful." Thus, successful nests that fledge four hosts are scored the same way as those nests that fledge only one host. Yet successful nests may vary in the numbers of young fledged due to partial nest loss because of parasitism, and this partial nest loss significantly affects fecundity per female per year (Donovan,Thompson et al. 1995).

Second, many birds renest after a failed nesting attempt, and these renests must be accounted for when estimating fecundity. For example, a recent study (Porneluzi 1996) compared marked birds in fragmented and unfragmented habitats and followed them throughout the breeding season. Porneluzi found that although average nest success was low on the fragments, 50 percent of males were eventually successful; nesting success was higher in the unfragmented plots, and 70 percent of the males there were eventually successful.

A third factor affecting estimated fecundity is the timing of nesting; we have some evidence that late-season nests are more successful than early-season nests. Given that late-season renests may be harder to find due to increased vegetation, lower densities of nests, and, perhaps, reduced enthusiasm by field workers, we suggest that more detailed studies of marked birds be done to determine the difference between nest success averages and individual season-long success rates (Morse 1996; Trine 1996). This may be particularly true of those species that are multiple brooded (such as the Indigo Bunting [*Passerina cyanea*], Wood Thrush [*Hylocichla mustelina*], and Acadian Flycatcher [*Empidonax virescens*]), which may show high nest loss and parasitism rates early in the breeding season but produce young more successfully later in the summer, when these abnormally high predation and parasitism rates seem to decline.

Direct measurements of fecundity are needed for species to establish source-sink status, but this requires following marked birds throughout the breeding season and making detailed measures of all nesting attempts they make. Because this is not an easy chore, little information exists on the season-long fecundity of most migrants. In the absence of direct measures, indirect measures can be useful but must be estimated with caution. We developed a model to indirectly estimate fecundity for species in our study region (Donovan, Thompson et al. 1995). This model, however, requires that assumptions be made regarding the number of possible broods per year, the average number of renesting attempts, and survival estimates of juveniles and adults. It then uses direct field observations based on nesting success and the average number of young produced per successful nest.

A second method for determining source-sink status involves estimating the number of renests needed for a population to be stable (Brawn and Robinson 1996). In some fragmented habitats, over twenty attempts are needed to ensure success; these populations are undoubtedly sinks. Because most researchers do not have the luxury or time to directly measure fecundity, we suggest that these and other models (e.g., Pease and Grzybowski 1995) need to be validated in terms of how successfully they estimate fecundity.

Estimating Survival

As noted above, the equations used to determine source-sink status (Pulliam 1988) involve three parameters: fecundity per year, annual adult survival, and juvenile survival. Although we are making headway in our understanding of reproductive rates with regard to different breeding situations, we still have remarkably poor data on survival rates of migrant birds and virtually no data on juvenile survival rates.

Population models for Neotropical migrant songbirds frequently assume juvenile survival is 0.31 (Greenwood 1980; Temple and Cary 1988; Howe, Davis, and Mosca 1991; Thompson 1993; Donovan, Thompson et al. 1995; Donovan, Lamberson et al. 1995). This estimate was calculated indirectly from known adult survival and fecundity data, assuming the population is at equilibrium. A recent radiotelemetry study on juvenile survival in Wood Thrush in the Missouri Ozarks found a 0.42 survival rate for just the first eight weeks post fledging (Anders et al., 1997); post-fledging mortality may play an important role in our understanding of source-sink dynamics and habitat requirements (see below) and requires more study.

Our knowledge of survival rates of adult Neotropical migrants is not much better, although some reliable estimates using modern Jolly-Seber techniques have appeared in recent years (Nichols et al. 1981; Faaborg and Arendt 1995). We need much better data on survival rates, including how survival rates may differ by the sexes, in sources or sinks, and just regionally. In this case, banding studies are valuable, although survival rate estimates based solely on return rates should be viewed with caution. For example, the study of Ovenbirds mentioned above (Porneluzi 1996) found that paired male Ovenbirds that did not nest successfully during a breeding season tended not to return to a territory the next year (only two of twenty-two unsuccessful males returned, in contrast to forty-five of ninety-four successful or unpaired males). Thus, dispersal away from a territory can confound survival estimates based on return data. We must recognize these sorts of factors as we compute our survival rates, as counting these birds as mortalities (which would be the case in most capture-recapture models) undoubtedly overestimates mortality rates.

Estimating Annual Variation in Source-Sink Status

Annual variation in numbers seems to be a fact of life for many migrant bird populations. Such variation may reflect the previous year's reproductive success (Sherry and Holmes 1992), exceptional conditions during either fall or spring migration or both, unusually good wintering habitat conditions, good conditions on the study site that attract an inordinate number of breeders that year, or chance. A lone researcher is lucky to gain insight into even two of these possible limiting factors.

Even if population size does not vary annually, fitness traits can. We find significant annual variation among mid-Missouri forest fragments in nesting success and parasitism rate, and even have found striking variation among our presumed source populations in the Missouri Ozarks. For example, during 1991 through 1993, fifty-six Wood Thrush nests had an average success rate of 0.414 (Donovan, Thompson et al. 1995), but eighty-six nests monitored in 1994 and 1995 averaged only a 0.266 success rate (Anders et al. 1997). The decline in success was due to higher predation rates, but the important implication is that even within the large Ozark forests, some migrants may have years when they do not replace estimated annual mortality. Thus, source locations may be so only as an average over years or, perhaps, as an average over many sites within a year.

Uncovering the Mechanisms Underlying Regional Patterns: Landscape versus Edge Effects

Although many studies have shown the deleterious effects of edges on nesting success and parasitism rates, not all studies show clear edge effects for both these parameters (see Paton 1994 for a review). The patterns of Robinson et al. (1995) suggest the overriding importance of landscape and regional patterns on such factors as parasitism rate and nest success, but they do not necessarily identify the mechanisms involved. Because the habitat characteristics evaluated by Robinson et al. are highly correlated (e.g., percent forest cover, average patch size, interior habitat, and edge density), a major challenge is to tease apart the factors to uncover the mechanisms actually at work.

Cowbirds serve as an example of the dilemma of whether landscape effects or edge effects are affecting abundance patterns in fragmented areas. Are cowbirds drawn to edges per se, resulting in increased parasitism levels in edge-dominated fragmented landscapes? Or, does fragmentation maximize the juxtaposition of cowbird feeding and breeding habitats and result in high numbers in fragmented landscapes? Similarly, if parasitism is high near a clearcut, does this mean the clearcut draws cowbirds there, or are cowbirds drawn to the area because there is a nearby pasture with suitable cowbird breeding resources? Our current research is directed toward disentangling these mechanisms.

Landscape Effects and Edge Effects

We have been examining the variation in edge effects and landscape effects in two different situations in the Midwest. Recent intensive studies in south-central Indiana have examined the role of clearcuts, forest openings, regeneration openings, and external edge within a fairly large tract of forest with over 80 percent forest cover in the region (D. Whitehead, University of Indiana, unpublished data). A broader-scale study looked at the effects of edge (mature forest-field) on predation and cowbird numbers. This study attempted to tease apart landscape and edge effects by examining one type of edge in multiple landscapes (Donovan et al., 1997). The researchers evaluated if edge effects in predation and cowbird numbers varied with landscape composition, predator community, nest-site vegetation, or local host abundance.

The Indiana study showed positive edge effects for cowbird abundance and also increasing abundance of cowbirds as openings increased in size. Parasitism rates were also related to the presence of both external and internal edge when compared to interior forest sites, although many of the differences were not statistically significant. The data on nest depredation rates showed even fewer edge effects and more variation among sites; some situations were characterized by higher nest losses near edges, others were not. Overall, there was evidence that edges sometimes were associated with lower nesting success rates through parasitism and depredation, but the pattern was not always strong or clear, which is perhaps what one should expect from a heavily forested landscape.

The regional study (Donovan et al. 1997) of thirty-six randomly selected sites found predation levels varied with landscape. Highly fragmented landscapes had the highest rates of nest predation; moderately fragmented landscapes had moderate levels of nest predation; and unfragmented landscapes had low levels of nest predation. These patterns essentially mimic those documented by Robinson et al. (1995). Edge effects, however, were documented only in the highly and moderately fragmented landscapes, suggesting that predation pressures differ among landscapes and also that predators in fragmented landscapes focus along edges. A similar study in a moderately fragmented landscape, however, failed to find significant edge effects (Marini, Robinson, and Heske 1995). A current study of nest predators in the Midwest (Dijak 1996) supports these results; there were increased raccoon and opossum populations in fragmented landscapes compared to contiguous landscapes, and raccoons appeared to forage along agricultural and riparian edges. Raccoons, therefore, could contribute to the "edge effect" found in highly fragmented and moderately fragmented landscapes.

In contrast to predators, the regional study showed a significant effect of landscape on cowbird abundance but did not show evidence of edge effects in cowbird distribution (Donovan et al. 1997). Cowbirds were most abundant in highly fragmented landscapes and least abundant in the unfragmented landscapes. The absence of edge effects in any landscape suggests that the matrix of habitats at the landscape scale may be a strong predictor of cowbird abundance, not edge per se. This is a logical conclusion because cowbirds use different breeding and feeding habitats within a landscape, and both types of habitat are needed for cowbird occurrence (Thompson et al. in press).

Edge Effects and Populations

Although we are starting to understand the mechanisms of edge effects, we still do not know what impact edges have on a regional population of birds. To begin answering this question, we have been modeling a generic migrant population under a variety of scenarios (F. Thompson, unpublished data). We are modeling the potential sensitivity of net reproductive rate to variation in demographic factors and habitat factors. Demographic factors included in the model include nest success, number of renesting attempts, number of broods, juvenile and adult survival, and carrying capacity. These values can be habitat specific so we can determine net reproductive rate in a habitat matrix. We are using this model to investigate the possible impacts of edge effects by varying the amount of edge in a habitat matrix, as well as species preference or avoidance of edge habitat, and the difference in nest success between edge and core habitat. Early results suggest that: (1) net reproductive rate is particularly sensitive to adult survival and number of nesting attempts; (2) populations with nest success of 30–40 percent and adult and juvenile survival of 0.6 and 0.3, respectively, have a net reproductive rate of 1; and (3) landscapes with more than 35 percent edge have net reproductive rates of less than 1.

Obviously, we need to understand better the interplay of regional landscape patterns and local habitat juxtaposition to manage for the highest nesting success possible, particularly in fragmented habitats. For example, knowledge about the detrimental effects of breaks in the forest such as the classic wildlife openings in various landscape contexts is critical for proper management. In some cases, such clearings may have little effect on saturated populations of predators and parasites, but in other cases such clearings may extend the detrimental effects deeper into the forest, perhaps turning a potential source population into a sink. Once again, the manager must evaluate the habitat traits of the area under his guidance with regard to the occurrence of internal and external edges, but also within the characteristics of the regional landscape in which the area occurs.

Habitat Considerations in Different Landscapes

The focus of Robinson et al. (1995) and most fragmentation studies has been on so-called forest interior species, as those are the ones showing the most negative responses to fragmentation and/or its associated edge effects. The general conservation suggestion for these species is to maximize forest cover and to minimize the amount of edge so that the negative effects of edge are reduced. The data of Robinson et al. (1995) do not change this conservation implication, but they suggest that regional landscape traits constrain the success managers can expect with their local manipulations. First of all, forest tracts vary in the microhabitats they can support as mature forest; a manager must recognize this variation and its effects on avian species composition before any management activities are done. Second, disturbance to these forests through natural or human-caused factors affects the distribution of forest and edge in varying ways, depending on the landscape in which this disturbance occurs. Here we discuss the role of habitat quality in affecting species composition, then we consider the trade-offs associated with nonforested habitats in differing situations.

Habitat Quality and Forest Structure

Although studies such as Robinson et al. (1995) try to control for habitat quality by using only mature oak-hickory forests, we must recognize that there is still much variation within these sites in terms of habitats and microhabitats offered. Such variation may affect the densities of forest-dwelling species found in a specific location and the possible goals of a manager on that site. For example, we have shown that central Missouri forest fragments have generally similar patterns in vegetational structure (Wenny et al. 1993), but there is still variation in the proportions of different microhabitats between sites, and this variation will affect species densities. Additionally, a species such as the Worm-eating Warbler

(*Helmitheros vermivorus*), which requires steep slopes for nesting, will not occur in fragments without such slopes, even when they seem to be sufficiently large and provide otherwise appropriate oak-hickory forest habitat.

Obviously, the first task of a manager of a site is to determine the bird species and microhabitats found in this site and how management actions could affect these habitats. The manager must recognize that a site full of riparian forests ideal for Kentucky Warblers (*Oporornis formosus*) cannot be turned into Ovenbird (*Seiurus aurocapillus*) habitat. Of equal importance is a regional plan that takes into account the potential for local management to affect or be affected by regional patterns of abundance and distribution and regional management goals. Then, taking into account both regional constraints and local habitat considerations, the manager can make a long-range plan. For migratory species that use mature habitats, such a plan may call for protection of the forests, but for those requiring earlier successional habitats or higher stem densities, some manipulation may be in order (see below). A manager must become aware of how regional population dynamics interact with local conditions to develop the proper management plan for the site over time.

Successional Stages of Forests

In most cases, the production of edge within managed sites is related to either the creation of so-called "wildlife openings" or the result of timber harvest techniques, especially clearcuts but also including small openings from selection cuts. Such openings have three effects: (1) to the extent they remove mature trees and replace them with early successional stages, they reduce the amount of habitat available to species requiring mature forests; (2) they create edge along or within the mature forest, which we have already discussed; and (3) they provide habitat for species that require edge or early successional habitats. A variety of bird species of second-growth are also Neotropical migrants with declining populations due, in part, to high nest predation and parasitism rates. Maximizing habitat quality for forest interior birds results in trade-offs of habitat for birds of early successional or second-growth habitats. Extreme minimization of edges could effectively reduce the populations of these species locally, if not regionally.

We also need to understand better the influence of regional forest cover on the reproductive success of second-growth species so that we can better manage them. By their nature, bird species of second-growth habitats live near edges more often than forest interior species, so they are naturally exposed to the high predation/parasitism rates associated with these habitats. We also assume that they have evolved mechanisms to deal with them as best they can, which may be in contrast to forest-interior species that have evolved mostly with lower predation/parasitism pressures. To date, we have little data on these species in different landscapes to guide our management decisions, but what we have suggests that second-growth species occurring in highly forested regions such as the Missouri Ozarks show much higher nesting success than their counterparts living in highly fragmented regions of central Missouri. The rates of parasitism and nest loss in forested regions may be higher in second-growth habitats than in the interior forests (due to edge effects as noted above), but they are still low enough that these species are doing much better reproductively in the forested regions than they are in fragmented regions. This presents the intriguing possibility that source-sink dynamics for second-growth birds also operate on a regional basis, such that these species receive protection from predation/parasitism by living in forest-dominated landscapes. Obviously, to the extent that they live in restricted stages of forest regeneration, these forested regions can support only limited populations of second-growth species. Yet these areas may serve as refugia from the almost pathological conditions of highly fragmented regions, where high densities of these species may occur but reproduction is exceedingly difficult.

While remaining aware of the possible costs of such clearings to nearby forest-dwelling birds, we must recognize the high value of such second-growth areas as clearcuts for some species. Additionally, because some species are area sensitive, they are not attracted to the very small clearings associated with single-tree or group-selection cuts (Annand and Thompson 1997). Clearcuts of at least five hectares within forested landscapes may be necessary to maintain populations in these potential source areas. The value of clearcutting, however, depends on the still uncertain costs of this silvicultural practice for forest birds. Finally, we have recent evidence that many forest breeding birds disperse to second-growth habitats after breeding; Anders (1996) radiotracked Wood Thrush fledglings and found that juveniles dispersed from the parental territories at about three weeks of age and settled in second-growth areas such as clearcuts for the next few weeks of their lives. We also found that netting such second-growth areas in July and early August catches many purportedly forest-dwelling species, including both adults and young of the year. Successional mosaics may be a natural condition, one that we need to reproduce.

The data show that birds of second-growth in fragmented regions are hard hit by predation and parasitism, at least early in the breeding season. Managers are likely to be very limited in their options when attempting to promote reproductive success of birds in second-growth habitat, but some options may exist. A study of Indigo Buntings in central Missouri showed the highest parasitism rates in nests within or very close to forest edge, whereas those away from the forest in a second-growth field had much lower parasitism rates (Burhans 1996). Perhaps, if they are necessary, wildlife openings or clearcuts should be fewer and larger in fragmented regions, effectively minimizing the edge effects around these habitats. This might produce both the largest amount of contiguous mature forest and the best second-growth nesting conditions within this location, even if both are much less than ideal because of uncontrollable regional pressures. More detailed behavioral research such as that of Burhans' is needed.

The other issue we need to address with second-growth bird species is the seasonal fecundity effect; these species differ from their forest-interior Neotropical migrant brethren by generally being multibrooded, with breeding seasons lasting well into the summer. We have preliminary data from central Missouri suggesting that, despite very high parasitism and predation rates early in the breeding season, these species may be successfully producing young with nests initiated after the middle of July, a time when cowbirds have generally stopped breeding and nest predation rates seem to decline. Obviously, a single successful brood at this time matches the maximal success of their single-brooded forest counterparts.

One must wonder about the relative value of young produced so late in the season, which may depend on aspects of wintering ecology that differ between forest-dwelling and second-growth species of birds. One cannot help but wonder why a Worm-eating Warbler that fledges its young in late May does not appear to nest again, whereas an Indigo Bunting might fledge young as late as mid-September. We encourage researchers to keep monitoring birds through the breeding season to record these late breeding events.

Fire and Midwestern Habitats

Evidence is increasing for hardwood forests that fire was historically a major source of periodic disturbance (Fralish et al. 1994). Without fire, oak forests in the Midwest may not be self-sustaining owing to encroachment by mesophytic, shade-tolerant species such as maples (*Acer* spp.). The problem is especially severe for open-canopy forests, and, recently, restoration or conservation of oak savannas/woodlands has generated considerable interest.

Historical and paleoecological evidence indicates that these habitat types were once pervasive in the Midwest and may have covered up to 13 million hectares at the time of European settlement (Nuzzo 1986). Whereas definitions for these habitats have proved elusive (White 1994), it is clear that these open-canopy forests—be they called savannas, woodlands, glades, or openings—are now far less extensive. One estimate indicates less than 20 percent of the presettlement acreage remaining, and most of that is in degraded condition owing to over-grazing, fire suppression, and agriculture; less than 1 percent of the former acreage remains in a high-quality state (Nuzzo 1986).

Appreciation of this problem has led to widespread use of prescribed fire in oak forest ecosystems (Botts et al. 1994). This prescription has the potential to change fundamentally habitats and landscapes by using a natural process as a management tool. Depending on the scale of application, fire represents a unique opportunity for control and management of a terrestrial ecosystem. Clearly, an effort as potentially influential as savanna restoration will have strong effects on the conservation status of certain bird species. Several studies are in progress to assess these effects on birds; we anticipate that there will be trade-offs because some species will likely benefit from fire while others are adversely affected.

Associations between fragmentation, habitat restoration via fire, and the viability of bird populations are unknown. Savanna or woodland habitat is obviously more fragmented than before. Yet, historically, savannas and their constituent bird communities may have been represented by islands of habitat and pervasive expo-sure to edges. Studies of simple area effects on the reproductive biology and com-munity structure of savanna birds are needed. Well-documented adverse edge effects on forest birds suggest that savanna restoration in comparatively con-tiguous landscapes such as those found in parts of Wisconsin and Missouri may yield the most positive results for breeding birds.

Another use of fire is the management of edges. Historically, savannas were often juxtaposed between prairies and wooded habitats. Thus, there may have been a gradual transition between nonwooded and wooded, closed-canopy habitat. In the fragmented regions of the Midwest, sharp transitions between agriculture and forest create edges that are poor breeding habitat for Neotropical migrants. Fire may serve to soften the transition and create edges that are more productive for birds.

Making Research Effective for Conservation

While we believe conservation recommendations based on a regional source-sink model of avian demography for the Midwest are appropriate, we are also aware that this is still just a hypothesized model. Evidence supporting dispersal on the regional scale is needed to prove that a regional approach is valid and that the protection of large, contiguous habitats will likely support populations in sink habitats.

Certainly, areas with limited habitat fragmentation are critical for the preser-vation of Neotropical migrants; even if they do not serve as source areas for vast regions of fragmented habitat around them, managers would be foolish not to focus on protecting those areas where we know that birds are reproducing at replacement level or above, on average. Future research should show us the rel-ative costs and benefits of internal openings to both forest interior and second-growth bird species.

The Robinson et al. (1995) model suggests that we accept that the ability of managers to improve breeding conditions for some species in highly fragmented regions is limited. Yet we also must be careful not to forsake smaller blocks of habitat until we have a better understanding of avian demography. Although some of these populations may appear to be "sinks" due to high average nest pre-dation or parasitism rates, in only the very extreme cases are the reproductive rates near zero for most species. Rather, because of the persistence of many breeders on habitat fragments, they exhibit reproductive rates that are below those of replacement level but perhaps not by as much as we have initially suggested. In these cases, local management practices might make enough of a difference to allow populations to cross the source-sink threshold. Such success could improve the local balance of sources and sinks and reduce the size of the drain in population sinks, and is the most prudent approach given that the regional source-sink model has not yet been validated. Sink populations may also serve important functions in terms of maintaining overall population size and genetic diversity levels (Howe, Davis, and Mosca 1991); these values are undoubtedly maximized with maximal populations. Even if these populations are nonviable, these birds may play important roles in ecosystem processes. Finally, the role of small habitat patches as migratory stopover sites needs to be studied.

The studies described above will help us understand some of the mechanisms associated with forest fragmentation and edge effects and how they vary with landscape characteristics. As we develop an understanding of these patterns, it is our responsibility to translate these results into practical management guidelines. Before any management can be undertaken, however, a manager must first determine what is living on the area being managed, how the habitats that exist might change, and how the populations of concern fit within the regional framework of habitat distributions. The manager also must remember that virtually any manipulation of a forest for management purposes involves trade-offs of a variety of traits. A wildlife clearing intended to provide habitat for quail or deer or a clearcut for timber harvest might also support early successional forest species, but possibly at the cost of reduced breeding success for birds in the surrounding woodlands. There are costs and benefits involved on all sides, and the manager must make the most enlightened decision possible when managing an area for the various concerns involved. The role of the researcher is to provide better information on these costs and benefits and to help managers identify the highest-priority species and habitats within their sites and the constraints imposed by the landscape in which the sites are located.

Acknowledgments

The authors would like to thank all of the funding agencies who have supported our research; we hope they understand that they are too numerous to list individually here. We also would like to thank the many students whose dedicated work in the field has provided new insights into migrant bird ecology, many of which are mentioned here.

CHAPTER 14

Abiotic Influences on Bird-Habitat Relationships

Larry L. Irwin

Conservation Issues and Previous Research

Various conservation strategies have been developed to avoid or arrest loss of biological diversity that may result from reduction and fragmentation of forests. The dimensions of such strategies frequently are predicated on wildlife habitat association models, while accounting also for historic disturbance regimes (Hansen et al. 1993; Bunnell 1995) and providing connectivity among a network of reserve areas (Ruggiero et al. 1994). Most wildlife habitat models that support conservation strategies emphasize vegetation because selection of vegetative structures is believed to link bird populations with their environments (Cody 1985).

However, agencies that developed conservation strategies often mapped vegetation cover types as surrogates for vegetation structure (Thomas et al. 1990; FEMAT 1993). For example, Ruggiero et al. (1994) recommended using cover types as surrogates for evaluating the likelihood for multispecies persistence, or viability. Also, the U.S. Department of Interior's GAP program employs vegetation cover maps to identify areas of high biological diversity not contained within current conservation networks (Scott et al. 1993). Wildlife habitat models that use vegetation cover types may well be useful for predicting a species' presence or absence, or possibly predicting species distributions, because cover types tend to combine many habitat variables, including factors that limit populations (Short and Hestbeck 1995).

Wildlife habitat association models based on vegetation cover types provided a useful place to begin. However, few vertebrate species are so closely associated with individual cover types to allow accurate predictions of population responses to proposed conservation strategies (Short and Hestbeck 1995). As a result, wildlife habitat models generally have performed poorly in predicting animal population changes from vegetation changes (Noon 1993; Conroy and Noon 1996). One consequence of predicating conservation strategies on wildlife habitat association models could involve allocating reserve areas in the wrong places (Irwin and Wigley 1993). New research is needed that links bird population dynamics with habitat conditions in forested environments. Here, I provide a perspective for scientific inquiry that might improve reliability of models that predict bird population changes relative to habitat modifications and thereby increase confidence in associated conservation strategies.

Features of Research That Improved Conservation

Significant improvement in predictive capability of bird-habitat models may be gained by adding indices of landscape structure (Saunders, Hobbs, and Margules 1991; Block and Brennan 1993; Donovan, Lamberson, et al., 1995; With and Crist 1995; Faaborg et al., this volume). Numerous studies have linked bird population declines to forest fragmentation (Walters, this volume). Such declines are widely believed to be caused by edge effects that result in high rates of nest predation, brood parasitism, and interspecific competition and lower rates of nesting success and pairing success (Walters, this volume; Faaborg et al., this volume). Certainly, recent conservation strategies attempt to account for possible effects of forest fragmentation on biological diversity (FEMAT 1993).

Although forest fragmentation frequently results in increased negative influences from species that usually live outside old-forest patches, the jury is still out regarding fragmentation effects across North America (Walters, this volume). This may be particularly true in managed forest landscapes, where the forest edges are not permanent, as they are in agricultural landscapes with forest fragments (Paton 1994). For example, Hanski, Fenske, and Niemi (1996) found no edge effects on nesting success of breeding birds in managed forest landscapes in northern Minnesota. Schieck et al. (1995) found no relationship between forest patch size and richness or abundance of birds associated with old-growth forests in British Columbia. Flather and Sauer (1996) found few associations between population trends of Neotropical birds and landscape structure in different portions of the eastern United States. In fact, Flather and Sauer found that estimates of trends in Neotropical bird abundance were lower in landscapes with larger forest patches and higher in those landscapes with more edge environment. Thus, the effects of fragmentation probably will vary depending on regional variation and species life-history traits (Hansen et al. 1995). These findings preclude confident predictions of the consequences to bird populations that may occur following implementation of major forest conservation strategies in some areas.

Because landscape structure and vegetation attributes, by themselves, may not accurately portray the multiple and interacting influences that regulate bird populations, some investigators recommend accounting for the effects of natural disturbance. In so doing, one might assume that native faunas and floras are adapted to natural patterns of disturbance (although few tests have been made). Indeed, Bunnell (1995) found that proportions of fauna breeding in only the early or only the late seral forest stages were significantly correlated with both fire size and burn rate (the latter defined as average fire size divided by average fire return interval). Those breeding in early seral stages tended to increase with increasing fire size and burn rate; those breeding in late seral stages tended to decrease. However, there is no compelling evidence that supports an assumption that vegetative habitat measures, plus disturbance regimes, can account for a large proportion of the variation in a species' demography. As Short and Hestbeck (1995) noted, we are still in need of a coherent theory that links the generic term "habitat" with a species' demography.

Evidence for Abiotic Influences

There are numerous reasons to suggest incorporating attributes of the physical environment in bird-habitat models. Brown, Mehlman, and Stevens (1995) examined national Breeding Bird Survey data for spatial variation in abundance of common breeding birds. This revealed a pattern in which individual bird species commonly were represented by only a few individuals in most samples where each species occurred, but were orders of magnitude more abundant in a few "hot spots." Further, the hot spots tended to occur in the same locality for each species. Such spatial patterns of abundance possess several deterministic features that beg a mechanistic ecological explanation. This led Brown et al. to develop a niche-based explanation: that this pattern largely reflects the extent to which local sites satisfy the niche requirements of multiple species assemblages. They hypothesized that multiplicative combinations of several independent environmental variables, including abiotic factors, caused the hot spots. If so, bird niches are probably concentrated at high biotic resource levels along resource gradients and optimal levels of regulator gradients such as energy or food supply (Currie 1991; Huston 1994).

Ecologists have provided good reason to marry components of the physical environment with vegetative conditions. Scientists have known for at least three decades that species diversity is related to gradients of ecosystem productivity (Abrams 1995). Bird diversity increases with increasing potential evapotranspiration, or PET (Currie 1991), or net primary productivity (Huston 1994). In fact, terrestrial vertebrate richness is less closely related to tree species richness and vegetation productivity (for which cover types are surrogates) than it is to PET.

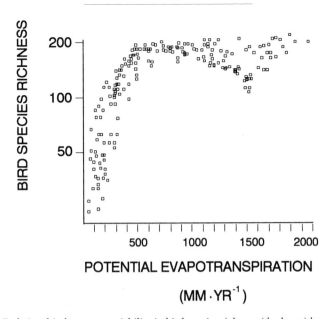

Figure 14.1. Relationship between variability in bird species richness (the logarithm of the number of species plus one) and potential evapotranspiration in North America. Adapted from figure 6 in Currie (1991).

Considering only the physical environment, with no explicit information on history, biotic interactions, or disturbance, Currie was able to account statistically for over 91 percent of regional species richness, and he was able to account for 80 percent of bird species richness using PET alone (figure 14.1).

Precipitation, vegetation, topography, lithology, and soils interact in multiple ways to cause differences in rates of nutrient cycling and energy flow through ecosystems, as well as the relative number of functional redundancies within them. These interactions determine the productive capacity of a physiographic region. As productivity rises in a region, bird and mammal diversity rises and then falls (Rosenzweig 1992). Increased productivity can raise biological diversity by increasing the local abundance of rare species or by improving the combinations of resources and conditions required by specialist species, allowing them to compete successfully (Abrams 1995).

Including attributes of the physical environment in models that might allow predictions of bird population trends is only recently established in the ecological literature. For example, in describing the habitat concept as it is used in ornithology, Block and Brennan (1993) devoted only one paragraph (p. 47) to abiotic factors, that involving bird distributions as constrained by physiological tolerances to the physical environment (e.g., Root 1988a). However, extant literature indicates vertebrate bird distribution, diversity, and abundance are strongly influenced by abiotic factors (Haufler and Irwin 1994).

Climatic factors are the most frequently cited abiotic variables influencing distributions and diversity of bird communities. For example, Root (1988b) documented temperature limits on distributions of 60 percent of 113 bird species wintering in North America, and Emlen et al. (1986) related latitudinal distributions in abundance of some forest birds to climatic factors. O'Connor et al. (1996) noted that climatic variables accounted for more variation in bird diversity than land cover and landscape pattern metrics. Wagner et al. (1996) found that annual variation in mean fecundity rates of Northern Spotted Owls (*Strix occidentalis caurina*) were negatively correlated with precipitation during September-April. On the other hand, LaHaye, Gutiérrez, and Ackakaya (1994) found a positive correlation between fecundity of California Spotted Owls (*S. o. occidentalis*) and precipitation in southern California. Similarly, annual variation in abundance of long-distance migrant bird species was correlated with drought in northern Wisconsin and Upper Peninsula, Michigan (Blake, Niemi, and Hanowski 1992). There, long-distance migrants, most of which nested in June when drought effects were most severe, were more affected by drought than resident or short-distance migrants. Failure to account for such abiotic factors as covariates can result in misleading conclusions and can produce opposing management recommendations (Schueck and Marzluff 1995).

A few investigators measured influences of soil types and topography on bird-habitat relationships. Rice et al. (1993) predicted Bobwhite Quail (*Colinus virginianus*) population responses to land management based upon precipitation and soil factors, as well as vegetation. Dipper (*Cinclus cinclus*) clutch size and body mass are inversely related to stream acidity, a function of soil composition (Ormerod et al. 1991). Moreover, Martin (1990) wrote that Red Grouse (*Lagopus lagopus*) densities were highest where the underlying rock types produced base-rich, fertile soils. These soils were associated with increased nutritional value of heather, the primary food source for Red Grouse. Similarly, Owen and Galbraith (1989) determined that soil conditions influenced densities of American Woodcock (*Philohela minor*). Light-textured, moderately drained soils in forested areas that had been previously farmed supported the greatest biomass of earthworms, the Woodcock's primary food source. Also, James, McCulloch, and Wiedenfeld (1996) suggested that correlates of elevation should be considered as factors that may regulate populations of warblers and other landbirds in eastern and central North America.

Research Needed to Further Conservation

Despite the fact that ecologists have long known that bird species distributions, habitat selection, and diversity are influenced by physical environmental attributes such as ecosystem productivity or PET, less is known about whether vital demographic rates of individual populations can be predicted by attributes of the physical environment. Research that evaluates the influences of the physical environment on animal-vegetation relationships might spawn niche-based models with increased reliability for predicting demographic responses (Irwin 1994). For example, differences between source and sink populations might result from spatial variation in essential nutrients in soils. Therefore, I agree with Brown, Mehlman, and Stevens (1995) and O'Connor et al. (1996), who recommended developing more realistically complex, niche-based models. Such models, however, must confront the problem that bird populations and their habitats are embedded in a hierarchy of processes operating at increasing scales of time and space (Conroy and Noon 1996). Although the processes are inherently complex, ecologists should be able to produce spatially explicit models that simulate patterns of trends in bird abundance and population dynamics across heterogeneous landscapes.

Such hierarchical, niche-based models, whether constructed by scaling up from population processes at the habitat-patch scale, or scaling down from indices of diversity or metapopulation structure at the landscape scale, can be developed by investigating the multiple interactions among environmental correlates of the dynamics and trends in bird populations. Doing this would involve using census and demographic data to characterize distribution of abundance and geographic variation in population trends. Variation in demography and population trends would be analyzed relative to spatial variation in the biotic and abiotic environment. Appropriate data on climate, geology, soils, vegetation structure, and pattern are now increasingly available from direct measures and remote sensing. The appropriate analyses could involve GIS, geostatistics, and multivariate statistical techniques such as those described below. Combining these measured variables should identify important environmental correlates of the niche dimensions and reveal scale-dependent influences of environmental factors and processes that limit bird distribution, abundance, and demography. By extension, the same approach can be applied to temporal variation and to questions about influences of other species (i.e., prey, competitors, and predators).

Building bird-habitat models for predicting population responses to conservation strategies must optimize among data-hungry population models (Anderson and Mahoto 1995; Wennergren, Ruckelshaus, and Kareiva 1995) and complex ecological relationships that must be simplified for broad acceptance and application. The key thus involves collapsing the number of parameters that must be estimated, yet adequately representing spatial patterns of vegetation, geomorphology, and climate as well as estimates of natality, mortality, and density. Scaling up from individuals to populations and metapopulations may involve quite different combinations of biophysical variables. For example, processes that operate at larger scales, such as demographic rescue effects associated with metapopulations (Gilpin 1996), are likely to be driven by factors other than those that determine fecundity of a local population.

Individual-based models might be scaled up via links with ecologically based, hierarchical land classification systems. Such classifications provide an opportunity to predict future populations by emphasizing contingencies among vegetation patterns and biophysical processes. The tricky part is identifying the multiple combinations of vegetation and physical environmental factors that set the conditions under which populations are regulated and through which metapopulations are maintained. Topography, geomorphology, and PET should be involved because they control the distribution of water and soil nutrients, solar energy, and precipitation, thereby influencing the distribution and continuity of areas of high and low bird population productivity (i.e., sources and sinks). Hansen and Rotella (in press) recommend incorporating spatial variation in net primary production (NPP), because NPP indexes interactions among climatic, topographic, and edaphic conditions.

Of course, interpretations of any approach to hierarchical, niche-based modeling must involve interactions between demography and biophysical habitat variables on a scale of resolution that matches that of the animal species and process being modeled (Rosenzweig 1991; Schulz and Joyce 1992). Large grain size, such as an annual home range, may not adequately capture the spatial juxtaposition of important habitat characteristics. For example, Bingham and Noon (1997) recommended linking demography with habitat and environmental conditions within core areas, which for Northern Spotted Owls were less than 25 percent of breeding season home ranges. Indeed, Laymon and Reid (1986) found that a sample frame of 25 percent of home ranges of Northern Spotted Owls was sufficient to predict presence, although a smaller sampling unit, 1 percent of home range size, was necessary to predict habitat selection. Such individual-animal processes must be linked with processes that operate at larger scales, such as dispersal.

Promising Developments in Wildlife–Habitat Modeling

Several developments in statistical applications promise to clarify bird-habitat relationships, because they have an ability to account for the fact that environmental correlates of bird demography may be manifest only in a contingent manner. For example, resource selection probability functions, or RSPFs (Manly, McDonald, and Thomas 1993), employ logistic regression to calculate coefficients for log-linear models. For example, Boyce, Meyer, and Irwin (1994) developed an RSPF to support a habitat-based population viability analysis (PVA) model for Northern Spotted Owls in western Oregon. By linking the model with multiple regression analyses of likelihood of occupancy, forest-growth models, and various landscape management scenarios in a GIS environment, the procedure can predict future distributions and population sizes and identify locations where populations might exist as reproduction sinks. Spatially explicit models would be required to include larger-scale processes, such as dispersal.

Another recent statistical tool concentrates on the statistical modeling of spatial dependence—geostatistics, which focuses on the detection, modeling. and estimation of spatial patterns (Rossi et al. 1992; Villard, Schmidt, and Maurer, this volume). Geostatistics can model the joint spatial dependence between a species' distribution patterns and components of its biotic and abiotic environment. Villard and Maurer (1996) used geostatistics (universal kriging) to reveal that considerable variability existed in the direction of changes in bird abundance, both geographically and temporally, even for species where significant long-term declines have been noted.

Yet another statistical approach that can accommodate multiple ecological pathways to predicting bird population trends includes classification and regression trees, or CART (Hollander, Davis, and Stoms 1994). CART recursively groups sets of independent variables according to the strength of relation between each independent and the dependent variable (Breiman et al. 1984). This grouping could lead, for example, to a representation of the likelihoods for reproduction or mortality relative to alternative forest landscape configurations. Hollander et al. used CART, wildlife-habitat modeling, and a multitiered database to illustrate the concept of spatial hierarchy as it applied to distributions of a teiid lizard in southern California and Baja California. CART has inherent ability to detect contingencies in effects, such as accounting for effects of climate, while allowing for an examination of other factors. For example, O'Connor et al. (1996) used a CART model to relate the variation in bird species richness as determined by eleven groups of unique sequences of hierarchically constrained independent variables. Then, within the hierarchy, climatic data accounted for more variability in the bird data, followed by land cover percentages and then landscape pattern metrics.

Making Research Effective for Conservation

Linking biophysical habitat conditions with bird species' demography requires incorporating more sources of variation in a species' spatial patterns of abundance and dynamics than vegetation cover type. Such influences include attributes of soils, weather and climate, geomorphology, vegetation structure, vertical height diversity, and landscape structure. This suggests that the goals for conservation purposes are unlikely to be satisfied from investigations in a single study area that cannot incorporate sufficient environmental variation (Root 1988a; Marzluff and Sallabanks, this volume). Short-term, local fluctuations in bird demography may or may not reflect those of regional or long-term trends (Gilpin 1996), so meta-analyses of repeated studies might be helpful (Fernandez-Duque and Valeggia 1995). Gathering the required data with sufficient accuracy and precision forces an enhanced emphasis on manipulative experiments that link population ecology with ecosystem ecology. Such manipulations properly should include scientists and managers from private, state, and federal forest management organizations.

Acknowledgments

This paper benefited from discussions with several colleagues, most notably J.B. Haufler, B.G. Marcot, F.B. Samson, W.A. Wall, and T.B. Wigley. The editors of this volume and A.J. Hansen, and R.J. O'Connor provided helpful criticisms that significantly improved upon earlier drafts.

CHAPTER 15

A Strategy for Bird Research in Forested Ecosystems of the Western United States

Jonathan B. Haufler

Conservation Issues and Previous Research

Considerable attention has been directed toward bird populations in recent years, sparked by concerns over reported declines in abundances of some species (Robbins et al. 1989; Terborgh 1989) and the establishment of the Partners in Flight program. Much of this attention has focused on bird habitat relationships and human interactions with bird populations. Despite this increased activity, two basic questions remain unanswered: (1) How can we quantitatively describe and predict the occurrence and abundance of birds across a planning landscape? and (2) Are there problems with bird productivity and/or survival rates within a landscape, and if so, what factors are influencing those population parameters? Understanding these two questions is essential to effectively prioritizing avian conservation activities. This chapter describes a strategy for addressing these two key questions of significance to avian ecologists and land managers.

The rationale behind the questions outlined above is that managers first need to be able to identify where and in what abundance we can expect to find an avian species within a landscape. Reproductive performance or other population parameters are of little relevance unless it is first known where a species will occur and in what breeding abundance. Therefore, I describe a need for classifying and mapping a landscape such that bird occurrences and abundances can be related to and predicted from the classification. Once the distributions and abundances of species can be quantified, it is then appropriate to determine if problems exist in population parameters, such as low rates of reproductive success or survival. If problems are found, the causes for these problems need to be identified and, if human related, corrective actions developed.

Vegetation and ecological community classification systems have been developed and described for a considerable number of years. Two of the more prominent systems were Kuchler's (1964) and Bailey's (1980). These efforts were designed to describe vegetation distributions and were not focused on wildlife associations. Several other classification systems have been developed with specific application to wildlife populations and are described below.

Thomas (1979) developed a landscape planning approach for the Blue Mountains of Oregon and Washington that incorporated groupings of wildlife species he termed "life forms," which were based on feeding and reproductive similarities. He then associated each life form with plant communities that he defined as a vegetation complex driven by site conditions. Fifteen plant communities (e.g., dry meadow, sagebrush-bitterbrush, mixed conifer, lodgepole pine, alpine meadow) were included. He further associated wildlife species with six successional stages from grass-forb to old growth. He used these associations to identify a versatility index for each wildlife species depending on the number of communities and successional stages judged to be used by the species.

DeGraaf and Rudis (1983) described bird and other vertebrate population associations with a landscape classification system for New England. Their classification system covered terrestrial vegetation types including fields, alpine, and forests, as well as wetland and special features such as cliffs and caves. The forest classification was based on descriptions of existing stands of vegetation characterized by the dominant tree species and the size of the trees in five size classes.

Hoover and Wills (1984) associated wildlife species with seventeen forested plant series for Colorado, which they condensed into nine ecosystems by combining plant series that "generally occur as a complex." They further described five structural stages (grass-forb to old growth), with the sapling-pole and mature stages further divided into three canopy classes. They then rated the habitat quality for each structural stage of each ecosystem for selected species based on feeding or cover requirements.

Hamel (1992) described bird relationships to vegetation types for the southeastern United States. He identified twenty-two vegetation types characterized by dominant tree species and further divided each of these into four successional stages. He associated bird species use with each of these vegetation type/successional stage combinations.

O'Neil et al. (1996) used wildlife species to group 130 vegetation types into 30 wildlife habitat types that they delineated for Oregon using satellite imagery. These habitat types were general descriptions of existing vegetation based on the dominant species (e.g., Big Sagebrush [*Artemisia tridentata*], Douglas Fir [*Pseudotsuga menziesii*] and mixed conifer). O'Neil et al. did not attempt to incorporate a temporal component, such as successional change, into their classification.

Finally, gap analysis (Scott et al. 1993) has attempted to identify vegetation communities within many states and to link wildlife species to these associations through habitat models. The vegetation communities are based on a classification of existing vegetation (Jennings 1993) and do not incorporate a successional component.

Features of Research That Improved Conservation

All of these previous efforts at associating the occurrence or preference of wildlife species with habitat descriptors increased the abilities of resource managers to consider the effects, to varying degrees, of management actions on avian populations. O'Neil et al. (1996) and Scott et al. (1993) both addressed habitat classification for large areas (states) but did not attempt to include a temporal or successional aspect. Their efforts were directed more toward assessing status in terms of acreage or ownership of existing stands of vegetation that were mapped from satellite imagery at fairly coarse scales. The other classification efforts (Thomas 1979; DeGraaf and Rudis 1983; Hoover and Wills 1984; Hamel 1992) incorporated a temporal, structural component into their habitat descriptors, so some dynamics of the vegetation communities could be considered.

Thomas (1979) and Hoover and Wills (1984) both attempted to use an ecological description of site potential as one component of their habitat descriptors, with generalized growth or successional stages as a further modifier. DeGraaf and Rudis (1983) and Hamel (1992) used existing vegetation descriptors rather than a site potential classification to associate species habitat, further modified by the age or size of the dominant overstory vegetation. Thomas identified the Blue Mountains of Oregon and Washington as the landscape of focus, while Hoover and Wills considered all of Colorado. DeGraaf and Rudis included all of New England, while Hamel included all of the southeastern United States. Thus, the scale of area being considered varied considerably among these approaches, as did the classification system. All used fairly general classifications of habitat, and all but Thomas attempted to include relatively large areas in their classifications. Some attempted to include a chronosequence in their classifications, but none attempted to identify a full range of successional trajectories through which plant communities in their classification system could pass.

All of these efforts provided a description of a landscape and linked avian and other wildlife species to the classification. Thomas's classification included the framework for incorporating an ecological classification of site potential, and clearly distinguished between disturbance-created differences and differences caused by ecological site characteristics. This was a major contribution for describing species relationships to a landscape classification.

Research Needed to Further Conservation

Breeding bird populations will be distributed across landscapes primarily in response to available habitat conditions. How we display and quantify these habitat conditions is dependent on how we classify or characterize the landscape.

For best application to management, a classification system should incorporate three characteristics. First, the classification system should allow for current distributions and abundances of birds to be described in terms of a landscape classification that can be used for management planning. The classification should allow for current distributions of bird populations to be quantitatively described in adequate detail such that linkages to required habitat are not overlooked. This is a function of both the classification system used and the resolution used in mapping. Second, classifications should allow for predictions or projections of future habitat conditions to be identified and modeled, so that future distributions and potential abundances of bird populations can be incorporated into management plans. Thus, the classification system should allow for stands with different successional trajectories and stand hazards to be distinguished from each other. Third, the classification system should allow for an understanding of historical disturbance regimes (Morgan et al. 1994) and the relationship of bird species distributions and abundances to these historical conditions. This allows for bird occurrences and abundances to be related to specific ecological processes. It also allows for an evaluation of the amounts of specific ecological communities in the classification system that may have occurred under historical disturbance regimes, and can provide insights of the importance of each ecological community to historical populations of birds. This historical perspective can provide a better understanding of present bird distributions and abundances and their relationships to potential management activities.

As mentioned previously, a two-tiered approach to avian research is proposed. The first phase is to relate bird distributions and abundances to stands in the landscape using a classification system that provides the capabilities outlined above. The second phase addresses bird population parameter responses to landscape conditions produced by combinations of stands.

An Example Classification System for Landscape Planning

An approach to land classification for use in ecosystem management has been proposed by Haufler, Mehl, Roloff (1996). This approach is based on an ecosystem diversity matrix (Haufler 1994) that is developed for a specified planning landscape, such as the Idaho Southern Batholith Landscape (Haufler, Mehl, Roloff 1996). This delineation conforms to broader ecological boundaries described by the national hierarchy of ecological units (Ecomap 1993), and by McNab and Avers (1994).

The ecosystem diversity matrix uses a classification of site potential across one axis and a description of successional change, as expressed by vegetation structure and alternative disturbance regimes, down the other axis. Haufler, Mehl, Roloff (1996) described the use of habitat typing (Daubenmire 1970; Steele et al. 1981) as the basis for site potential categorization, although other classification systems of site potential also could be used. For example, delineating a landscape in terms of areas of similar slope, aspect, and elevation could provide a similar stratification of the ecological complexity of communities contained within the landscape. In any such efforts, a balance between providing a classification with sufficient resolution of stand conditions to adequately distinguish bird distributions must be identified without having such a complex classification system that it becomes impossible to map and assign attributes to stands over any sizeable landscape.

Habitat typing (Daubenmire and Daubenmire 1968) is a land classification system that integrates the effects of many physical site factors resulting in a classification of areas that have the capability of supporting similar successional trajectories of vegetation, as well as having similar historical disturbance regimes. Habitat typing characterizes site conditions but does not describe existing vegetation. Existing vegetation is influenced by the habitat-type of each delineated site but can be in a variety of successional stages dependent on past disturbances and management activities. Therefore, within each habitat type, existing conditions of stands can be described in terms of a description of successional trajectories. For example, successional trajectories of all Grand Fir (*Abies grandis*) habitat types might be described according to vegetation-growth stages (Haufler, Mehl, Roloff 1996) created by two different possible fire regimes (see left column of figure 15.1).

Habitat types are a fairly fine ecological classification. For example, Steele et al. (1981) identified sixty habitat types for central Idaho. These specific habitat types can be grouped into classes having similar successional pathways and historical disturbance regimes (Haufler, Mehl, Roloff 1996). These classes, combined with a classification of existing stand conditions (figure 15.1), allow for existing vegetation conditions to be mapped and described as well as for interpretations of past disturbance regimes and future successional change (Haufler, Mehl, Roloff 1996). All of these are essential functions of management planning.

This landscape classification can be used as a basis for investigating and describing the distributions and abundances of bird species across a planning landscape. It should be noted that the important features of this classification are the description of the ecological complexity of the landscape provided by one axis, and the temporal dynamics and responses to disturbance of the other axis. While habitat-type classes and vegetation-growth stages were selected for the ecosystem diversity matrix displayed in figure 15.1, as mentioned, other classifications could also function in a similar manner. In addition, it should be noted that the matrix displayed in figure 15.1 is for forested ecosystems in the planning landscape. An additional matrix for riparian/wetland ecosystems allows for quantification and delineation of those areas, while a matrix of shrub/grassland ecosystems would allow for description and quantification of drier sites.

ECOSYSTEM DIVERSITY MATRIX—
IDAHO SOUTHERN BATHOLITH LANDSCAPE
FORESTED SYSTEMS

Habitat Type Class

VEGETATION GROWTH STAGES	Warm, Dry Douglas-Fir Moist Ponderosa Pine — Potential Cover Types	Acres	Dry Grand Fir — Potential Cover Types	Acres	Cool, Moist Grand Fir — Potential Cover Types	Acres	Warm, Dry Subalpine Fir — Potential Cover Types	Acres	High Elevation Subalpine Fir — Potential Cover Types	Acres
Grass/Forb/Seedling										
Shrub/Seedling	Pinus ponderosa / Populus tremuloides		Pinus ponderosa / Pseudotsuga menziesii		Pseudotsuga menziesii / Pinus ponderosa		Pseudotsuga menziesii / Pinus contorta		Pinus albicaulis	
Sapling; shrub/seedling	Pinus ponderosa / Populus tremuloides		Pinus ponderosa / Pseudotsuga menziesii		Pseudotsuga menziesii / Pinus ponderosa		Pseudotsuga menziesii / Pinus contorta		Pinus albicaulis	
WITH HISTORICAL UNDERSTORY FIRE REGIME	Understory Burn 10-22 Years		Understory Burn 10-30 Years		Understory Burn 10-30 Years		Fire Mosaic 50-90 Years		Understory Burn 25-70 Years	
Small Trees Multi-story	Pinus ponderosa / Pinus contorta		Pinus ponderosa		Pinus ponderosa / Pseudotsuga menziesii		Pseudotsuga menziesii / Pinus contorta		Pinus albicaulis	
Medium Trees Multi-story	Pinus ponderosa / Pinus contorta		Pseudotsuga menziesii / Pinus ponderosa		Pinus ponderosa / Pseudotsuga menziesii		Pseudotsuga menziesii / Pinus contorta		Pinus albicaulis	
Large Trees Multi-story	Pinus ponderosa		Pinus ponderosa		Pinus ponderosa / Pseudotsuga menziesii		Pinus contorta		Pinus albicaulis	
WITH HISTORICAL STAND-DESTROYING FIRE REGIME	Historical Stand-Destroying Wildfire Unlikely		Historical Stand-Destroying Wildfire Unlikely		Historical Stand-Destroying Wildfire Unlikely		Fire Mosaic 50-90 Years		Stand Destroying Wildfire	
Small Trees Multi-story	Pinus ponderosa / Pinus contorta		Pinus ponderosa / Pseudotsuga menziesii		Abies grandis / Pseudotsuga menziesii		Pseudotsuga menziesii / Pinus contorta		Pinus albicaulis	
Medium Trees Multi-story	Pinus ponderosa / Pseudotsuga menziesii		Pinus ponderosa / Pseudotsuga menziesii		Abies grandis / Pseudotsuga menziesii		Pseudotsuga menziesii / Picea engelmannii		Abies lasiocarpa / Pinus albicaulis	
Large Trees Multi-story	Pinus ponderosa / Pseudotsuga menziesii		Pinus ponderosa / Pseudotsuga menziesii		Abies grandis / Pseudotsuga menziesii		Picea engelmannii / Abies lasiocarpa		Abies lasiocarpa / Pinus albicaulis	
Old Growth	Pinus ponderosa / Pseudotsuga menziesii		Abies grandis		Abies grandis		Abies lasiocarpa / Picea engelmannii		Abies lasiocarpa / Pinus albicaulis	

Figure 15.1. A partial example of the ecosystem diversity matrix developed for the Idaho Southern Batholith planning landscape, after Haufler, Wehl, and Roloff (1996). Dominant tree species typically occurring in each cell or ecological land unit are listed. Species in columns where stand-destroying wildfire is unlikely were predominantly influenced by understory burns during historical fire regimes. Species in columns indicating stand-destroying wildfire is likely were predominantly influenced by this disturbance, with far less influence of understory burns.

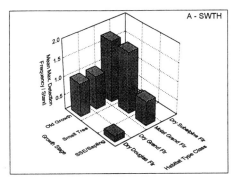

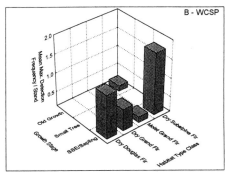

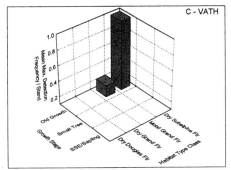

Figure 15.2. Relative abundance of (A) Swainson's Thrush (*Catharus ustulatus*), (B) White-crowned Sparrow (*Zonotrichia leucophrys*), and (C) Varied Thrush (*Ixoreus naevius*) expressed as mean maximum detection frequencies for bird sample plots in ecological land units investigated in the Idaho Southern Batholith planning landscape. From Sallabanks (1996).

Examples of Bird Distributions and Abundances

Habitat selection of bird species can be described in terms of occurrence and/or abundance in relation to the ecosystem diversity matrix. For example, bird surveys conducted in Idaho (Sallabanks 1996) on selected replicated cells (termed ecological land units) of the ecosystem diversity matrix (figure 15.1), found that the matrix could be used to differentiate habitat associations and relative abundances of many bird species. The matrix was able to account for significant differences in bird abundances attributable to both vegetation-growth stages and habitat-type classes. Some of these differences occurred among habitat-type classes that would have been considered the same using a landscape classification based only on vegetation types of dominant overstory species composition.

Several examples of bird abundances are displayed in (figure 15.2). The Swainson's Thrush (*Catharus ustulatus*) (figure 15.2A) was found to be significantly more abundant in one particular habitat-type class, the cool-moist Grand Fir class. Other species, such as the White-crowned Sparrow (*Zonotrichia leucophrys*) (figure 15.2B), were significantly more abundant in a particular vegetation-growth stage, in this case the seedling-sapling stage. Many species occurred across many ecological land units in the matrix, but a few species, such as the Varied Thrush *(Ixoreus naevius)* (figure 15.2C), were restricted to a limited range of the ecological land units investigated.

Ecological land units can be mapped with a GIS by overlaying a map of polygons of habitat-type classes with a map of polygons of existing vegetation categorized into vegetation-growth stages (Haufler 1994). The resulting map of ecological land units displays their size and location across the planning landscape. The resolution of these maps can have a significant influence on the accuracy of associated bird distributions and abundances with the mapped ecological land units. Using a resolution that is too coarse (such as a 1-km pixel size) will result in poor accuracy of describing species distributions or abundances within the planning landscape, due to the wide range of habitat conditions that will occur within an area of this size. Finer-scale delineations (such as the approximately 2-ha resolution used in the data collection displayed in figure 15.2) will result in much more accurate information on bird locations within the planning landscape.

By knowing the association of species with each ecological land unit, the map of ecological land units can be used to assess current distributions of most breeding birds. Species that require more than one ecological land unit in an appropriate proximity or juxtaposition to meet habitat requirements can have their occurrences or abundances described through habitat modeling programmed into the GIS and linked to habitat characteristics of each ecological land unit (Haufler, Mehl, and Roloff 1996). However, the distribution of the majority of avian species can be adequately described through association with characteristics of each cell in the matrix.

Investigations that evaluate species locations across planning landscapes described and delineated in a matrix fashion are needed in order to: (1) quantify existing species locations and numbers across landscapes in terms of the ecological complexity of the landscape and successional stage distributions; (2) predict future species abundances and distributions based on vegetation change; and (3) relate species occurrences to historically occurring populations. Investigations should strive to collect sufficient data about study sites to allow for the sites to be classified and characterized in a compatible manner. A key component to collaborative ecosystem approaches is for classification schemes used in a landscape to be compatible (Haufler 1995), thus allowing the capability to consider information across ownerships. For example, if data on slope, aspect, elevation, rainfall, and soils were recorded at a site, but the habitat-type class was not, most sites could be evaluated and equated to a habitat-type class description, providing useful comparisons with studies that utilized habitat typing characterizations.

All ecological land units occurring within a planning landscape should eventually be described in terms of bird distributions and abundances. However, a priority rating system of habitat-type classes and/or vegetation-growth stages of initial or primary management concern can be developed. This would allow for a planned research prioritization as research funds become available. Habitat-type classes that are felt to be the most different in the current distribution of vegetation-growth stages compared with the vegetation-growth stages produced by historical disturbance regimes would be of high priority. For example, low-elevation conifer forests in the Dry Douglas Fir and Dry Grand Fir habitat-type classes historically supported Ponderosa Pine (*Pinus ponderosa*) maintained by frequent understory fires (Agee 1993). Suppression of these understory fires has changed the species composition of these stands by allowing a larger component of Douglas Fir and Grand Fir. Thus, these habitat-type classes have exhibited a substantial change in the distribution of the vegetation-growth stages occurring historically compared to the present. An additional priority ranking could be given to habitat-type classes that are relatively rare in the landscape and may support many species that also are relatively rare in the landscape.

Determining Bird Population Parameter Relationships

With a knowledge of where species occur across a landscape, an additional concern is whether or not they are successfully maintaining their populations. In other words, is the population of a species in any area a recruitment source for the population, or a population sink (Faaborg et al., this volume)? This will depend on the reproductive success of the population in an area and the survival of individuals. Factors that will influence these will be the habitat quality of the particular area in terms of such attributes as cover and food availability, and the potential occurrence of decimating factors such as pollutants, diseases, predators, competitors, or brood parasites such as the Brown-headed Cowbird (*Melothrus ater*). A first step in this aspect of the research is the determination of whether or not any avian productivity problems are present. If they are found to occur, then factors influencing this population parameter can be investigated more fully.

Of primary importance from a land management standpoint would be factors that can be influenced by management actions. Of key significance are influences such as natural disturbances (fire frequencies and intensities) and human activities (agriculture, ranching, timber harvesting, development, and recreation). To understand the implications of these, research is needed that includes a larger-area perspective (Freemark et al. 1995). Each of these variables may independently influence decimating factors and must therefore be factored into a research design in a larger-area analysis. Stands to be monitored for population parameters such as reproductive success must be carefully selected and replicated on the basis of the surrounding area characteristics. Surrounding area descriptions may need to include radii of several km or greater to adequately reflect the influencing variables. For example, cowbirds may be closely associated with agriculture, ranching, towns, and parks, and this association may extend for at least 3 km (Stribley 1993). The potential influence of cowbirds on nesting success of other species would need to be assessed (if productivity problems were found to occur) relative to appropriate scales and attributes. Thus, any appropriate study design must factor in these types of attributes at a scale of analysis appropriate for the questions being addressed, with replication of similar analysis area configurations. For example, a study may select a design that targets a specific stand condition within replicated study areas that include agricultural lands and ranches within several kilometers of the study stand and compare these to replicated study areas that do not have agricultural lands or ranches within several kilometers of the study stands.

The first question is to determine if and where any avian productivity problems occur within the planning landscape. If they are found to occur, a second step is to determine what specific decimating factors are responsible. For example, if nesting success is found to be low for a species in certain areas of a planning landscape, then the decimating factor should be identified. If the decimating factor is found to be cowbird parasitism, this may point to one set of solutions, while if Red Squirrels *(Tamius hudsonicus)* are the problem, then an entirely different set of solutions may be appropriate. The third step is then to determine what landscape factors are influencing the occurrence or effect of the decimating factor. With this information, the particular factors in any areas where avian productivity problems are found to occur can be managed to reduce the problems.

Larger-area investigations (influences of areas encompassing at least several kilometers in distance from a particular stand) that have been experimentally designed to address questions relating to bird population parameters have generally not been undertaken, to date, in the intermountain west. Freemark et al. (1995) reviewed avian studies conducted at this scale that have been conducted in the Pacific Northwest, and reported on a number of studies conducted west of the Cascades. However, they stated that research at this scale is needed in the intermountain area before any conclusions could be made concerning bird populations.

Making Research Effective for Conservation

I have emphasized the need for avian research to be linked directly to compatible classification systems that allow for distributions and abundances of bird populations to be quantified at a planning landscape scale. To be most useful, the classification system selected should allow not only for quantification of existing vegetation, but also for an understanding of future projections of existing vegetation and an interpretation of the relationship of existing conditions to those produced by historical disturbance regimes. Research directed in this manner will allow for managers to ascertain the present distribution and abundances of species in a quantifiable manner relative to past species distributions.

Based on the distributional and abundance information generated from research utilizing appropriate landscape classifications, avian population parameters can be researched and evaluated and, if needed, corrective management actions identified. This research should first focus on whether or not problems with avian productivity, survival, or other population parameters for a given species even occur in the landscape of interest. This would involve evaluating reproductive rates and/or survival rates using a population analysis to identify source versus sink parameters. If reproductive or survival rates are found to be a problem, then the specific decimating factors causing the problem can be identified. The habitat attributes contributing to the decimating factors (i.e., feeding areas for cowbirds) can then be evaluated and appropriate management actions identified and implemented.

Research on avian populations needs to be focused on a region, and even more specifically to the planning landscape of interest (Freemark et al. 1995). While findings generated in other regions or landscapes may help direct the design of research, researchers and managers must resist overgeneralizations and misapplication of findings from extrapolations of research findings in one region to other regions where conditions or relationships may be significantly different. This can result in management actions that may create or exacerbate problems rather than result in solutions. Rather, we should organize our research efforts in a manner that allows us to combine research results from various research studies in a planning landscape or region and integrate results into landscape planning and adaptive management strategies. Compatible landscape classifications that allow us to quantify present and future distributions and abundances of avian species are a critical step for this to occur. Based on this distributional information, research addressing population parameters of concern can be appropriately focused and addressed.

Avian research, particularly in those regions of the country or world where less research has been conducted, needs to be approached in a planned, scientific manner. If we are to obtain the information we need to assure bird conservation in these areas, we need to work better at collaborative, effective research efforts directed at key questions and problems with management solutions as a planned product, not a chance by-product of our research efforts.

Acknowledgments

Rique Campa, Bill McComb, Bill Wall, Gary Roloff, John Erickson, Carolyn Mehl, and Rex Sallabanks provided helpful reviews of this manuscript. Carolyn Mehl assisted with preparation of the graphics.

PART V

Conservation in Nonforested and Urban Landscapes

Aquatic, riparian, shrub, and grassland habitats are easily and frequently altered by human activities. Many of these habitats are highly prized by humans for recreation, housing, and commodity production, and alteration of these areas often receives less attention than extremely visible alteration of forested landscapes. In part V, authors present case studies of conservation efforts in some of these habitats to emphasize the special needs of species in such ecosystems. They discuss the uniqueness of the habitat and birds while emphasizing their importance and critical information needs. The section concludes with a general discussion of urbanization and the effects of this extreme habitat modification on birds.

Seabirds often occur in spectacularly high densities and nest in remote, isolated colonies. However, these birds are now often threatened by our increasing population and per capita consumption of resources. Previously, Kress (chapter 10) discussed intensive management to restore endangered seabirds. Here, Boersma and Parrish (chapter 16) review the major sources of anthropogenic threats to seabirds. Discharge of oil, plastic, and heavy metals along with other persistent chemicals has resulted in pollution problems for a number of seabird species. Fisheries impacts, as evidenced by the removal of 80 million metric tons of fish annually by the late 1980s, may have altered the population dynamics of some seabird species through direct competition with fishers, ecosystem change, and seabird by-catch. Human exploitation coupled with species introductions has resulted in increasing disturbance to nesting colonies, habitat degradation, and depredation. Most seabirds are exposed to multiple threats as well as the direct and indirect effects of human activities. Although many studies have linked persistent declines or increases in seabird populations to human modifications on a local scale, regional and global effects are becoming more apparent. Long-term studies of seabirds have identified problems and provided a database to assess the health of seabird populations. These data may be of use in developing an understanding of species interactions and community structure. Research results provide some general predictions about how biological parameters will change with modifications in the physical environment. Research is needed in four basic areas: (1) understanding marine system function; (2) seabird response to environmental change; (3) the impact of human activities on the marine environment; and (4) seabird management strategies. Until interactions between the physical and biological systems are better understood and predictions tested, seabird management cannot become proactive. More important, Boersma and Parrish note that management alone is unlikely to solve seabird conservation problems. Human behavior must change if seabird conservation is to be successful. They suggest four avenues along which societal activities should be modified: education, energy use, ecological function, and ethics. These 4Es need to be as familiar as the 3Rs for conservation to be truly effective.

Semiarid, montane regions of the world are typically covered with structurally simple, but compositionally diverse shrublands. Rotenberry (chapter 17) discusses the plight of one such shrubland, the western North America shrub steppe.

Here the most pressing conservation issue relating to birds is the loss and concomitant fragmentation of native shrubland habitats. This loss has been precipitated by a complex interaction among agriculture, livestock grazing, and the invasion of exotic annual plants, especially cheatgrass (*Bromus tectorum*). Cheatgrass, which now occupies millions of hectares of western rangelands, both in monoculture and as a major understory component, has greatly increased fire frequency and has substantially, and perhaps permanently, altered post-fire successional pathways. Although research on the ecological mechanisms by which shrubland-obligate birds are adversely impacted by fragmentation is essential to understand how to maintain these populations through the short term, it is critical that we learn how to protect native shrublands and to restore them to western landscapes if we are to conserve avian diversity in the long term. Rotenberry notes that successful long-term preservation of shrublands will require removal of the exotic annuals that have, in essence, become self-perpetuating.

The physical factors that allow for the development of rich natural grasslands are perfectly suited for the development of productive agricultural lands. Throughout the world such areas are intensively cultivated for cereal grains and managed for livestock production. Later, Newton (chapter 20) discusses the effects of agricultural intensification on birds in Europe. Here, Herkert and Knopf (chapter 18) use the plains of North America to illustrate the effects of the conversion from natural to agricultural grasslands on birds. They note that many species of grassland birds have declined recently. While there is compelling evidence that declines in breeding habitat have influenced grassland birds, there are also indications that reproductive failure (due primarily to high rates of nest predation and occasionally nest parasitism) and problems on the North and South American wintering grounds are also potentially influencing population declines. Efforts to identify which factors are most important in limiting grassland bird populations are hindered by limited data on several aspects of their ecology. Primary information needs for improved grassland bird conservation are: (1) research on grassland bird demographics (i.e., nest success, renesting frequency, dispersal, survivorship); (2) research on landscape structure and its effects on grassland bird populations; (3) research on the winter ecology of grassland birds; (4) research on the effects of grazing and other common grassland management activities on breeding birds (especially for shortgrass prairie); and (5) research on the effects of grassland restoration on birds. Herkert and Knopf suggest that until more information identifying specific problem areas for particular species becomes available, grassland bird conservation efforts should address a variety of potential limiting factors including breeding season habitat availability, reproductive failure, and winter ecology.

A consequence of the exploding human population and increasingly industrialized society is urbanization. Despite the obvious impact of urbanization on birds, this phenomenon has attracted relatively little research interest. Marzluff, Gehlbach, and Manuwal (chapter 19) review some of the impacts of urbanization on birds and suggest future research needed to improve our understanding of urbanization. Urbanization affects birds directly by: (1) interrupting ecosystem processes (fire suppression alters nutrient cycling; water demands shorten water cycles); (2) removing and modifying habitat (forest are cleared and fragmented for houses and roads; climate is ameliorated in cities); and (3) subsidizing some birds with extra food, water, and the removal or suppression of their natural predators. Urbanization affects birds indirectly by: (1) changing predator communities (supplemental food may allow nest predators to increase in abundance); (2) concentrating birds, thereby making them susceptible to disease outbreaks; (3) unbalancing competitive interactions (mesopredator release; increasing competition for nest cavities); (4) altering community composition by favoring more generalist and less specialist species and providing a haven for non-native species; and (5) selecting for new behavioral responses (nest placement; aggressiveness or tolerance of humans). Exemplary past research used some combination of: (1) long-term investigation; (2) rigorous experimental design; (3) identification of mechanisms that produced effects of urbanization; and (4) quantification of demography. Future research needs to continue these qualities and do simultaneous comparisons of avifaunas along several points in the continuum from urban to rural areas. Important future questions are: (1) How do we best characterize urbanization? (2) How do bird communities change with increasing urbanization? (3) How does the type of urban development affect birds? (4) What types of urban developments are most compatible with native birds? (5) How do nest predators influence nesting success in urban areas? (6) How do birds respond to the urban/rural interface? (7) How does the interspersion of urban, suburban, exurban, rural, and wildland affect avian diversity and population viability? (8) How will projected human growth in an area affect birds? Successful implementation of research into management and policy will require conservation biologists to work with geographers, social scientists, and land-use planners so that important suggestions for minimizing urbanization effects can be incorporated into long-term, spatially extensive land-use plans.

Several common themes come from the chapters in this section. They echo the need for long-term, demographic studies called for in previous chapters. As suggested earlier, testing various management practices is best done in an experimental or "adaptive" framework. Researchers in nonforested, terrestrial ecosystems are concerned with the effects of fragmentation, but its impacts are much less understood here than in forested systems. Shrublands, grasslands, and riparian areas are badly degraded in the United States, and restoration of these habitats is required. Unfortunately, development of techniques to reestablish plants in these arid lands is in its infancy. The importance of understanding how biotic and abiotic factors interact to affect bird abundance, productivity, and diversity is emphasized in the reviews of seabirds, shrubland birds, and urban birds. It is especially important to test the interactive effects of the many different human-caused factors. Understanding the effects of a single factor helps, but in reality birds are simultaneously influenced by a myriad of factors and we must un-tangle this web to fully quantify and rectify our influences on avian diversity.

The importance of politics and general public education to avian conservation is again emphasized, and four general areas, outside of research, that need our attention are explained by Boersma and Parrish. Throughout this book it is apparent that we must often step outside of our role as researchers to educate the public and policy makers if our research results are going to be used to conserve birds.

CHAPTER 16

Threats to Seabirds: Research, Education, and Societal Approaches to Conservation

P. Dee Boersma and Julia K. Parrish

Conservation Issues and Previous Research

Seabird colonies are found in areas where oceans are regionally productive but locally patchy: coastal margins, zones of upwelling, and high latitudes. Small clutch size, slow growth of chicks, delayed maturity, and high adult survival are all conservative life-history traits allowing survival in these dynamic environments. Many seabirds are adept at taking advantage of whatever prey is available. Although confined to a relatively small portion of the ocean during breeding, seabirds are highly mobile and may disperse widely during the nonbreeding season. For centuries, seabirds were protected by their life-history traits, relatively remote breeding sites, and broad diet; however, this is no longer the case. Local, regional, and global pollution; fishery interactions; indiscriminant hunting and egging; species introductions; and habitat degradation are the legacy of increasing human population size and increasing per capita consumption. Each of these environmental problems drives change in seabird populations. We review the conservation problems faced by seabirds in the past and present, provide an overview of how we have attempted to preserve and protect seabirds, and suggest what we need to do to be successful in achieving that goal.

Pollution

Pollution in the marine environment is increasing as the human population increases, because we produce and discard more material per capita, and because durable materials such as plastics persist in the environment long beyond the date of discard. Many persistent pollutants can spread widely from the pollution source, through atmospheric and oceanic processes, as well as via migration of contaminated biota, making source identification difficult. In coastal areas marine pollution levels are correlated with human population density and land-use patterns (i.e., industrial, municipal, agricultural) with occasional pollution "hot spots" such as well-settled estuaries (Elliott and Noble 1993). Although offshore islands are less contaminated, this may change as persistent pollutants migrate throughout the world's oceans. Adult seabirds may be able to ameliorate the effects of pollutants through behavioral and physiological mechanisms; however, chicks are often extremely vulnerable, as they may receive high levels of pollutants from the yolk sac as well as subsequent doses in their food. Bourne (1976), Nisbet (1994), and Furness (1993) provide overviews of seabird and pollution problems. Several types of pollutants may cause conservation problems for seabirds, notably oil, plastics, persistent chemicals, and heavy metals.

Oil affects seabird thermoregulation, digestion, osmoregulation, blood chemistry, immune function, and reproductive success (Burger and Fry 1993). Burger (1993) assembled data on large oil spills from 1937 to 1991 and found that for the thirty-four spills with carcass counts 122,000 seabirds were recovered; for the twenty-six spills with total mortality estimates 1,200,000 seabirds may have died. He compares these data to six small spills from unexplained sources (i.e., no leaking vessel found) in which approximately 120,000 carcasses were found or estimated to have died (Burger 1993). Although oil spills occasionally kill a substantial number of seabirds (e.g., the *Exxon Valdez* carcass toll was more than 30,000; Piatt and Lensink 1989), the majority do not. Most authors have concluded that there is no evidence that a single oiling event has had any lasting effect on seabird populations (Burger 1993; Furness 1993).

By contrast, chronic oil pollution sometimes considered "small spills" from such diverse sources as leaky tanks and valves, accidents during onloading and offloading, bilge dumping, and urban runoff can have far greater cumulative effects on seabird populations than isolated large spills (Clark 1984; Camphuysen 1989; Furness 1993). Oil pollution accumulates along shipping lanes and in high-traffic coastal areas worldwide, where breeding, wintering, and migrating seabirds may encounter it. Gandini et al. (1994) estimated that thousands of Magellanic Penguin (*Spheniscus magellanicus*), juveniles and adults, are oiled annually off the coast of Chubut, Argentina, by chronic sources, with potentially serious demographic consequences to the Argentine colonies. Boersma (1986) suggested that the sublethal ingestion of petroleum compounds in Procellariids be used as a monitor of environmental quality.

Plastic particles in the marine environment, including industrial pellets and random broken pieces, are an increasing problem throughout the world's oceans (Ryan 1987a,b). Ship-based trash surveys indicate that plastics are the major component of human-made debris (65 percent Morris 1980; 86 percent Dahlberg and Day 1985). Most plastic is found around important fishing grounds and well-traveled shipping corridors and in the vicinity of ocean dumping sites (Laist 1987). Seabirds may become entangled in strapping, plastic line, or netting (see "By-catch" later in this chapter); they may ingest plastic, or they may use it as nesting material (Podolsky and Kress 1989), increasing the risk of later entanglement or ingestion by either parents or chicks (Montevecchi 1991). Furness (1993) suggested that plastic availability in the marine environment has increased based on increasing use by Northern Gannets (*Sula bassana*) of plastic as nesting material (from 20 percent cited by Bourne 1976, to 100 percent cited by Montevecchi 1991).

Ingestion of plastics happens when a seabird mistakes a fragment for food, or when the plastic is attached to, or ingested by, a bird's prey item (Ryan 1987a,b). Rates of ingestion are related to foraging habits: dippers and patterers are most likely to ingest plastics, followed by surface seizers and pirates, and finally by pursuit and plunge divers (Ryan 1987a,b). Ocean currents concentrate plastic debris, making seabirds that feed along frontal zones more susceptible (Furness 1993). Plastic ingestion is also higher in omnivores (Ryan 1987b). Once ingested, threats are mainly mechanical: blockage of digestive tract, damage to internal organs, or lessening of apparent hunger (Laist 1987). Although adults may be able to eliminate plastic by regurgitating, this is often not possible for chicks (Fry, Fefer, and Sileo 1987).

Many types of synthetic chemicals initially manufactured for a variety of terrestrial uses have accumulated in the marine environment and have bioaccumulated and biomagnified in seabirds. A range of studies shows pollutant accumulation to sublethal-effect levels is widespread in the world's seabirds (Walker 1990). The common denominators of these toxicants seem to be that these chemicals persist in the environment for decades, and that they are lipophilic. Many persistent pollutants have been implicated in reproductive, teratogenic, and immunological dysfunction (Livingstone, Donkin, and Walker 1992; but see Furness 1993). Seabirds are most susceptible as chicks and young-of-the-year (Walker 1992). Sublethal effects include depressed reproduction via eggshell thinning (Parslow, Jefferies, and Hanson 1973; Cooke 1979; Elliott, Norstrum, and Keith 1988) and aberrant parental behavior (Cooke, Bell, and Prestt 1976; Mineau et al. 1984; Kubiak et al. 1989).

Routes of transport of persistent chemicals into the marine environment are varied. All compounds are subject to aerial transport on a global scale (Livingstone, Donkin, and Walker 1992) at the location where they enter the ocean by rainfall or dry deposition (Walker 1990). Seabirds such as storm petrels, which forage in the surface microlayer, may be particularly susceptible to contamination, as the air-water interface is the atmospheric entry point and accumulates volatile compounds (Seba and Corcoran 1969). In coastal areas, agricultural and industrial runoff (Elliott and Noble 1993) and incineration of municipal and industrial wastes (Hites 1990) are likely sources. Ocean dumping and leakage of ship hydraulics and other fluids are also problematic (Walker 1990). Because many seabirds migrate great distances, accumulated toxicants can be dispersed far from the initial entry point (e.g., into polar regions; Nettleship and Peakall 1987).

Because these chemicals persist in fatty tissue, levels of accumulation can be monitored (Furness 1993) and may indicate local to regional levels of pollution. Levels of DDE and PCBs in seabird eggs follow local production curves (in Common Murres; Olsson and Reutergårdh 1986), and declining toxicant levels have been linked to increasing restrictions on chemical use and production (in Northern Gannet; Chapdelaine, Laporte and Nettleship 1987). Uptake and metabolism of persistent pollutants appear to be species and individual specific. Toxicant accumulation in Atlantic Puffins (*Fratercula arctica*) is lower than other co-occurring species perhaps due to the former's ability to induce HMOs (hepatic microsomal monoxygenases). Within populations chronically exposed to persistent pollutants, HMO levels are high in some individuals, suggesting that selection for increasing elimination capability may be operating (Walker 1992).

As awareness of the deleterious effects of persistent pollutants has risen, governments have begun to restrict use, production, and methods of disposal. However, not all compounds have been restricted at once, restrictions have not usually been total but have been phased in, and not all countries within the range of a seabird species enacted restrictions simultaneously (Elliott and Noble 1993). Given the additional facts that these chemicals persist for decades, that many tons of now banned chemicals are improperly stored for long-term safety, and that the dispersal routes of these chemicals make persistent pollutants a global issue, seabird populations worldwide are at risk of exposure and at potential risk of adverse effects.

Heavy metals, both essential (e.g., copper, selenium) and nonessential (e.g., cadmium, lead, mercury) trace elements are found in seabird tissue, occasionally at levels that threaten the health of the bird (Ohlendorf 1993). As with persistent chemicals, heavy metals bioaccumulate and biomagnify, making upper-trophic-level species particularly susceptible. Levels of cadmium (Lee et al. 1989), mercury (Ohlendorf and Fleming 1988), and selenium (White et al. 1989) have been found in seabirds at concentrations causing sublethal effects in other species of aquatic birds. Physiological effects of heavy metal poisoning include loss of appetite and muscle control, anemia, decreased egg size, fertility, and hatchability and retarded chick growth (Ohlendorf 1993). Heavy metal poisoning may also affect parental (Barr 1986) and chick (Heinz 1979) behavior, although these effects have not been specifically investigated in seabirds.

In areas surrounded by heavy industrial use (e.g., Baltic Sea, Mediterranean), heavy metal loads are higher than in less populated areas (Appelquist, Drabaek, and Asbirk 1985; Thompson, Hamer, and Furness 1991; Thompson, Furness,

and Walsh 1992). Although these studies suggest that human activities have increased heavy metal loading in seabirds, Furness (1993) points out that this conclusion may be misleading. Based on analysis of museum skins, upper-trophic-
level seabirds such as Wandering Albatross (*Diomedea exulans*) have had
consistently high levels of mercury over the past 150 years, suggesting that bio-
magnification is "normal."

Fishery Interactions

By the late 1980s, humans were removing almost 80 million metric tons of fish
from the world's oceans on an annual basis. As upper-trophic-level components
of coastal ecosystems, seabirds also remove a substantial fraction of secondary production; estimates range from 5–10 percent in the North Sea (Bailey 1986; Tasker
et al. 1989; Bailey et al. 1991) to almost 30 percent off South Africa (Furness and
Cooper 1982). As humans continue to exploit marine ecosystems, conflicts with
seabird populations will become more pronounced. Nettleship, Sanger, and
Springer (1984), Montevecchi (1993), and Duffy and Schneider (1994) are useful
reviews of seabird-fishery interactions.

Fisheries interact with seabird populations in a myriad of ways, most of them
negative on an individual and occasionally population level. We will consider
three main categories of human-seabird interaction: (1) direct competition of
fishery and seabird activities (2) ecosystem change as a result of human fisheries,
in which the consequences to seabirds can be positive or negative, and (3) by-
catch of seabirds by the fisheries.

Direct Competition

Many of the world's fisheries may compete directly with seabirds and other
upper-trophic-level predators for access to schooling fish, squid, or krill, although
the evidence is correlative. There is often a relationship between the percent of a
prey species in the seabird diet and the commercial fisheries yield (e.g., Cape
Gannets [*Sula capensis*] and pilchard, [*Sardinops ocellata*] Berruti and Colclough
1987; mackerel [*Scomber scombrus*] and Northern Gannets, Montevecchi and
Myers 1992). In some cases, the overlap in fish size classes caught by humans and
seabirds is high (e.g., Northern anchovy [*Engraulis mordax*] and Elegant Tern
[*Sterna elegans*] Schaffner 1986). These types of data suggest that if the fishery becomes overexploited, both fish and seabird biomass can drop substantially. Examples include the capelin (*Mallotus villosus*) fishery and murre (*Uria* spp.) populations in Norway (Vader et al. 1990) and the pilchard fishery and Cape Gannet and
Cape Cormorant (*Phalacrocorax capensis*) populations in the Benguela Current region (Crawford, Shelton, Batchelor, and Clinning 1980; Crawford, Shelton,
Cooper, and Borrke 1983; Burger and Cooper 1984). In some cases, competition
with the fishery was inferred because depressed seabird populations did not

rebound after other stressors had been removed; e.g., Double-crested Cormorant (*Phalacrocorax auritus*) and Tufted Puffin (*Fratercula cirrhata*) populations in the California Current system remained low after hunting and egging stopped, presumably due to overfishing of Pacific sardines (*Sardinops caerulea*) (Ainley and Lewis 1974). Exploitative competition may not always be the case, however; Rice (1992) has suggested that fishing amplifies the large natural variation in schooling stock abundance, driving prey to exceptional lows to the detriment of both humans and seabirds.

Is overfishing unequivocally responsible for seabird population declines? The fact that schooling stocks such as anchovies and sardines are prone to dramatic population fluctuations in the absence of humans (Soutar and Issacs 1974; Baumgartner, Soutar, and Ferreira-Bartrina 1992), coupled with the observation that not all fisheries collapses necessarily lead to declining seabird abundance (Barrett and Furness 1990) suggest the answer is no. Climate events and oceanographic processes can also produce stock collapse (Radovich 1981). In the California Current system, a 90 percent decline in Sooty Shearwater (*Puffinus griseus*) abundance (over 4 million birds) has been attributed to a long-term shift in physical oceanographic factors (Veit et al. 1997). These data suggest caution be used in interpreting correlation as causative.

Ecosystem Change

A large portion of the world's marine fisheries do not take species of seabird prey, or take the larger-size classes of seabird prey. Although these fisheries are not in direct competition with seabirds and other top marine predators, a heavily exploited fishery can dramatically alter ecosystem structure and diversity (e.g., change in Georges Bank benthic community from predominantly gadoids to predominantly elasmobranchs; Murawski 1991) with potentially destructive consequences to seabird populations (Rice 1992; Springer 1992). However, the converse is also true. Larger-size-class fisheries may enhance seabird prey stocks either because the smaller-size classes are released from cannabalism (e.g., adult Walleye pollock [*Theragra chalcogramma*] are major predators on the juvenile-size classes also taken by seabirds; Hatch and Sanger 1992) or because other potential seabird competitors are also directly affected by the larger-size-class fishery (e.g., declining marine mammal populations release co-occurring seabirds from competition; Swartzman and Haar 1983; Alverson 1991). Seabird populations can also be enhanced by competitive release if the competitors are fished directly, as was the case with baleen whales in the Southern Ocean (May et al. 1979). Finally, because of the large amount of discarded fish and fish waste in mixed stock and/or high by-catch fisheries, large-scale operations with at-sea processing can boost seabird populations by creating a new food source (Burger and Cooper 1984; Hudson and Furness 1988; Furness, Ensor, and Hudson 1992). Increases in gull species' populations may have deleterious effects on small alcids and storm petrels (Boersma and Groom 1993).

Until recently, many fisheries stocks were managed as single species with only fishing mortality taken into consideration. More recently, natural mortality including predation has been incorporated in fishery models (Anderson and Ursin 1977), and multispecies models that incorporate the effects of humans, predators, and trophic cascades have been developed to more accurately determine sustainable yields (Pope 1991; Sissenwine and Daan 1991). In at least one case, stocks are explicitly managed to provide adequate prey for seabirds (e.g., northern anchovy stock quotas off California must be set relative to the needs of Brown Pelicans [*Pelecanus occidentalis*]; Duffy 1994a).

By-catch

In all fisheries nontarget catch occurs, including fish, seabirds, and marine mammals (Alaska Sea Grant 1996). The proportion of target to nontarget species is a function of behavior and ecology of the target species, gear type, fishing practices, and economics. Although fisheries managers have long been aware of fish by-catch, mainly because nontarget species of one fishery are often a target species for another (Murawski 1996), by-catch of marine mammals and seabirds is a relatively recent political issue. Seabirds become entangled in fishing gear by taking bait or by failing to avoid nets. With the development of monofilament gear, by-catch became a more serious issue because lost gear continued to catch fish for weeks to months. Estimates of annual seabird deaths from ghost fishing nets in the North Pacific ranged from 200,000 to 750,000 (U.S. Department of Commerce 1981), prompting widespread calls for limitation of the high-seas driftnet fishery (Jones and DeGange 1988). Seabird by-catch has been regarded as an ethical problem (i.e., it is wrong to capture and kill seabirds), a logistical problem (i.e., capturing nontarget species fouls gear and wastes time and effort), and a conservation problem (i.e., seabird by-catch threatens and endangers populations). We will concern ourselves with the last issue.

By-catch of seabirds is ubiquitous and has been reported in the scientific literature for many fisheries (e.g., North Pacific [Jones and DeGange 1988; Ogi et al. 1993]; North Atlantic [Piatt and Nettleship 1987]; Mediterranean [Guyot 1988]; Southern Ocean [Brothers 1991]). Unlike many other anthropogenic threats, by-catch is almost always fatal. Conservation attention has been paid to gear types particularly prone to seabird capture (e.g., monofilament gillnets, longlines) as well as to fisheries in locations in which endemic, threatened, or endangered species are found (e.g., Short-tailed Albatross [*Diomedea albatrus*], Marbled Murrelets [*Brachyramphus marmoratus*]). In some cases, by-catch has been cited as a significant factor in local population depression or extirpation. Common Murre populations in central California declined 50–100 percent within a four-to-six-year period in which the most severe declines occurred at colonies located nearest to an intensive nearshore gillnet fishery. Estimates of total gillnet mortality from 1979 to 1987 were 70,000 to 75,000 murres (Takekawa, Carter, and Harvey 1990).

Human Exploitation

Humans have killed seabirds for a variety of purposes: food and oil (e.g., alcids [Salomonsen 1970; Nettleship and Evans 1985]; penguins [Cox 1990; Crawford et al. 1995]; eggs [Cott 1953–55]), bait (Feare 1984), feathers for the millinery trade (e.g., terns [Doughty 1975]), removal of a fisheries competitor (e.g., Double-crested Cormorants, [Carter, Sowls et al. 1995]), culling for disease prevention (Coulson 1991), and to clear habitat for other uses (e.g., removal of seabirds on Midway by the military [Harrison, Naughton, and Fefer 1984]; removal of gulls to provide habitat for other species [Coulson 1991; Kress, this volume]). Overviews of human predatory pressure on seabirds can be found in Cott (1953–54); Cline, Wentworth, and Barry (1979); Feare (1984); and Burger and Gochfeld (1994).

Feare (1984) credits industrialization and increasing urban development with the relaxation of human predatory pressure on seabirds, noting that hunting and egging continue today in remote locations. Human exploitation has caused the actual or near extinction of several species or populations including the Great Auk (*Pinguinus impennis*) in the North Atlantic, the Cahow (*Pterodroma cahow*) in Bermuda, Audouin's Gull (*Larus audouinii*) in the Mediterrannean, the Short-tailed Albatross in the Bonin Islands, Sooty Terns (*Sterna fuscata*) and Common Noddies (*Anous stolidus*) in the Abrothos Islands of western Australia, and the Spectacled Cormorant (*Phalacrocorax perspicillatus*) in the northern Pacific (Cline, Wentworth, and Barry 1979; Moors and Atkinson 1984).

Egging also seems to have occurred everywhere humans have come into contact with colonial seabirds: the South American coast (Murphy 1936), western North America (Vermeer 1963), the Carribbean (van Halewyn and Norton 1984; Haynes 1987), South Africa (Frost, Siegfried, and Cooper 1976), and the Indian Ocean (Feare 1976) are a few specific examples. Cott (1953–54) lists eighty-two seabird species exploited for egging, including Sooty Terns, at over 1 million eggs annually, and several species for which hundreds of thousands of eggs were (are) collected on an annual basis: Jackass Penguins (*Spheniscus demersus*), Herring Gull (*Larus argentatus*), Arctic Tern (*Sterna paradisaea*), Brown Noddy (*Anous stolidus*), Thick-billed Murres (*Uria lomvia*), and Common Murres (*Uria aalge*).

There are several interacting reasons that seabirds are vulnerable to human hunters. Combined with their relative terrestrial immobility and tendency toward synchronicity, coloniality, and nesting in unprotected locations, seabirds are attractive targets. Once humans and predators introduced by humans discovered seabird colonies, mortality increased and the indirect effect of concentrated predatory visitation during the breeding season further depressed reproductive success as birds were chronically flushed from nests. Unlike other predators, humans had (and have) the ability to preserve meat and eggs, allowing harvests far in excess of the short-term consumption needs of the hunters. Thus, intense predatory pressure on long-lived, low-fecundity species drove lambda below 1.0 and populations quickly decreased (e.g., colonies on St. Helena and Ascension islands; Olson 1977). Overexploitation through hunting and egging continues today. In northern West Greenland, human hunting is implicated in an 80–90 percent decline of murre populations (Kampp, Nettleship, and Evans 1994).

Despite case studies to the contrary, subsistence and commercial hunting and egging can be conducted sustainably. Harvests of Atlantic Puffins and Common Murres on the Faeroe Islands are regulated by a cooperative and limited entry system in which islanders compete to maintain their section of cliff-face at maximum production (Nørrevang 1986). Approximately half a million fledgling muttonbirds *(Puffinus tenuirostris)* were sustainably harvested each year from the Flinders Island group (Cline, Wentworth, and Barry 1979). Serventy, Serventy, and Warham (1971) suggested that this harvest actually protected the species, as habitat would otherwise have been converted to sheep ranching.

Species Introductions

With humans have come a variety of other species, which fall into three basic categories: (1) escaped pets such as cats *(Felis catus)* and dogs *(Canis familiaris)*; (2) accidental releases, mainly rats *(Rattus norvegicus* and *R. rattus)* and mice *(Mus musculus)* from ships, shipwrecks, and, more recently, aircraft; and (3) intentional releases, including food and sport animals, fur bearers, and biocontrol agents. Overviews of interactions between introduced species and native seabirds can be found in Jones and Byrd (1979), Moors and Atkinson (1984), and Burger and Gochfeld (1994). Introduced species may become a conservation problem if they are habitat changers or predators. We define the former category as animals whose actions affect seabird habitat to such extremes that nesting is no longer possible.

Ungulate introductions, including cattle *(Bos taurus)*, caribou, or reindeer *(Rangifer tarandus)*, deer *(Odocoileus hemionus)*, elk *(Cervus canadensis)*, musk oxen *(Ovibos moschatus)*, sheep *(Ovis aries)*, and goats *(Capra hircus)*, have all caused habitat destruction via overgrazing and resultant erosion, as well as burrow trampling (Jones and Byrd 1979). Ungulate introductions were most often the result of ships in the 1500s–1800s dropping off goats, sheep, and pigs *(Sus scrofa)* to serve as food sources for future human colonization or in case existing supply lines broke down (Jones and Byrd 1979). Rabbits *(Oryctolagus* spp.) and hare *(Lepus europaeus)* can also cause extensive habitat degradation via overgrazing and erosion (e.g., Laysan Island; Warner 1963); however, these species are not always destructive (e.g., Triangle Island [Rodway, Lemon, and Summers 1990]; Manana Island, Hawaii [Tomich, Wilson, and Lamoureaux 1968]). Occasionally, ungulate introductions have been credited with contributing to local seabird extinctions, as is the case with goats introduced to Guadalupe Island, Baja Mexico, which together with cats, caused the extinction of the Guadalupe Storm Petrel

(*Oceanodroma macrodactyla*) (Howell and Cade 1954; Jehl 1972). Pigs, in addition to altering habitat, are also known to root up and eat petrels (Moors and Atkinson 1984).

Introduced predators are not always devastating to a seabird population; the strength of the interaction appears to be influenced by several factors:

1. *Seabird behavior and evolutionary experience.* Docile species are more at risk than active defenders (e.g., mobbing by gulls and skuas). Seabird populations that evolved in the absence of terrestrial predators are also more apt to decline after introductions (Burger and Gochfeld 1994; Kress, this volume).

2. *Predator size and behavior.* Introduced terrestrial predators can access previously "safe" habitats such as burrows. Larger predators obviously have a greater effect, eating more biomass of a wider species range than small predators.

3. *Topography.* Island topography also influences the degree of seabird vulnerability, mainly through limiting access of introduced predators.

4. *Alternative food.* If alternative food sources are available during the seabird nonbreeding season, predator populations are buoyed and predatory pressure during the breeding season can be intense. Low levels of food during the winter can eventually cause introduced predators to die out. Conversely, if more easily obtainable food sources are available during the seabird breeding season, interaction between otherwise serious introduced predators and seabirds can be low (e.g., cats on Dassen Island, South Africa, ate rabbits rather than seabirds; Cooper 1977). Finally, if otherwise available alternative foods are suddenly rare, low-level interactions can become intense (e.g., egg predation by Black and Polynesian rats [*Rattus exulans*] increased following storm events; Woodward 1972; Moors and Atkinson 1984).

Burger and Gochfeld (1994) estimated that indigenous predators can remove between 1 and 12 percent of their prey populations annually. However, a single cat can kill two hundred seabirds per year (Marion Island [van Aarde 1980; Kress, this volume]). Cats are widely seen as the most universally damaging introduction to seabirds worldwide (Moors and Atkinson 1984; Burger and Gochfeld 1994), although rat introductions may be slightly more common (Jones and Byrd 1979).

Other species of particular conservation concern include foxes released on seabird islands in the Northern Hemisphere to augment the fur trade, and on islands off Australia for hunting (Moors and Atkinson 1984). In Alaska, foxes were introduced to more than 450 islands, mostly between 1900 and 1930 (Bailey 1993). By 1936, the Aleutian archipelago had produced over 25,000 fox pelts valued at over 1 million dollars. Roughly half of the fox food, as ascertained by scat samples, was seabirds (Murie 1959), and seabirds were seriously reduced if not extirpated on many islands (Bailey 1993). Mustelids, including American mink (*Mustela vison*) and stoats (*M. erminea*), have also been introduced as part of the fur trade, with similar consequences to local seabird populations (Moors and Atkinson 1984).

Many species of predator brought to islands to control pest species introduced previously (e.g., rats and rabbits) preyed on native birds, including seabirds, instead. Mongooses were introduced to control rats in crops like sugar cane; mustelids, including ferrets (*Mustela furo*) and weasels (*M. nivalis*), were introduced to control rabbits, and owls, including Barn Owls (*Tyto alba affinis*), Tasmanian Masked Owls (*Tyto novaehoollandiae castanops*), and Little Owls (*Athene noctua*), were all introduced to control rodent populations or to protect orchards from small-bird infestations (Moors and Atkinson 1984). European hedgehogs (*Erinaceus europaeus*), Wekas (*Gallirallus australis*), Marsh Harriers (*Circus aeruginosus*), Cattle Egrets (*Bubulcus ibis*), Indian Mynas (*Acridotheres tristis*), Musk shrews (*Suncus murinus*), Raccoons (*Procyon lotor*), Macaque monkeys (*Macaca fascicularis*), and Monitor lizards (*Varanus indica*) are introduced predators of known or suspected concern (Moors and Atkinson 1984).

Habitat Change and Disturbance

Humans not only exploit seabirds directly by hunting and egging, and affect seabird populations indirectly through the introduction of non-native species, but also threaten the continued existence of seabird populations through habitat alteration. Globally, habitat degradation and loss is credited as the single most important conservation issue facing the world (Meffe and Carroll 1994), and seabirds are no exception. Intensive use of coastal resources, including land, water, and biota, has not come without costs. Although coastal areas are the intentional centers of commerce, shipping, fishing, recreation, and living, coastal and adjacent nearshore areas are the unintentional centers of land-use conversion, pollution, loss of marine biodiversity, and alteration of marine ecosystems.

Broadly speaking, seabird habitat can be divided into nesting habitat, foraging habitat during the breeding season, and at-sea habitat during the nonbreeding season. While the first can be affected by development and land-use conversion, the last can be affected by increasing pollution and overexploitation of the world's fish stocks. Because we have dealt with fisheries issues elsewhere, this section will concentrate on effects to nesting habitat.

For colonial species, habitat alteration may be intentional and pronounced, as in guano mining (Frost, Siegfried, and Cooper 1976; Cline, Wentworth, and Barry 1979; Hsu and Melville 1994). Unrestrained removal of guano during the mid- to late 1800s led to the demise of Peruvian Diving Petrels (*Pelecanoides garnoti*) and Inca Terns (*Larosterna inca*), and reductions in the size of "guano bird" populations (Guanay Cormorants [*Phalacrocorax bougainvillii*] and Peruvian

Boobies [*Sula variegata*]) due to extensive excavation on the Peruvian guano islands (Duffy 1994b). Habitat alteration is also an unintended consequence of development. On tropical islands, conversion of native forest to coconut plantation affected nesting of several species (de Korte 1984). Commercial logging and land clearing for agriculture have reduced available habitat for several species (e.g., old-growth forest habitat in the Pacific Northwest for Marbled Murrelets [Sealy and Carter 1984]; hardwood forest habitat in New Zealand for Yellow-eyed Penguin [*Megadyptes antipodes*] [Seddon and Davis 1989]). Development of beach and dune habitats has reduced nesting habitat of terns (e.g., California Least Terns [*Sterna antillarum browni*]).

Nesting habitats can also be effectively degraded by increasing human disturbance. Noise from aircraft (Bunnel et al. 1981; Burger and Gochfeld 1990), especially helicopters (McKnight and Knoder 1979), presence of boats, and presence of people (Anderson, Volg, and Keith 1980), including researchers (Burger and Gochfeld 1994), can all cause a "rain of birds" with associated loss of eggs and chicks (e.g., murres and kittiwakes [*Rissa* spp.] [Sowls and Bartonek 1974]). Persistent disturbance may cause depressed reproductive success or even colony abandonment (Burger and Gochfeld 1994).

Exploitation of wildlife does not necessarily mean killing; softer impacts such as ecotourism are an example. World tourism revenue is expected to crest three trillion dollars by 1999, 5–15 percent of which could be defined as ecotourism (Miller 1993; Giannecchini 1993). Over 100,000 people visit Bonaventure Island, Quebec, each year to see the Northern Gannet colonies, without apparent impact to the birds (Nettleship and Chapdelaine 1988). Magellanic Penguins nesting along a tourist trail at Punta Tombo, Argentina, appear to habituate to human presence (Yorio and Boersma 1992). Although tourism can be a sustainable way to protect seabirds while educating a wider public than most academics, agencies, and nongovernmental organizations might otherwise reach, care must be taken to insure that effects on habitat and disruption of breeding cycles are minimized (Burger and Gochfeld 1994; Duffy 1994a).

If humans have degraded seabird habitat, they have also created it. Pilings, docks, and rooftops are used by gulls, kittiwakes, and cormorants, often explosively expanding these populations along coastal centers of human development (Furness and Monaghan 1987; Carter, Gilmer et al. 1995; Carter, Sowls et al. 1995). The incidence of Herring Gulls nesting on rooftops doubled in a six-year period to three thousand pairs in Britain (Monaghan and Coulson 1977). Abandoned buildings on islands and along coasts and islands are quickly recolonized by a variety of species (e.g., albatross on Midway Island; Nishimoto 1996). On a larger scale, nesting habitat for entire seabird colonies has been created to facilitate continued harvest of seabird products. Artificial ledges for murres have been built to facilitate egging in Russia. In Peru, the Guano Administration Company adapted headlands such that guano birds quickly colonized, raising Peruvian populations of Guanay Comorants and Peruvian Boobies to 16–20 million (Duffy 1983; Duffy 1994b). Offshore platforms in South Africa have also been built to facilitate guano harvesting (John Cooper, University of Cape Town, personal communication).

Interactive and Indirect Effects

Human influences on an ecosystem are rarely confined to a single stressor. For instance, Yellow-eyed Penguins in New Zealand and surrounding islands have declined in population size by 75 percent over the last forty years as a consequence of human disturbance, habitat destruction, introduced predators (ferrets, stoats, and cats), introduced habitat changers (cattle), and fisheries by-catch (T.D. Williams 1995). This is not a rare case but the norm and is the natural consequence of initial contact by humans and the cascade of ensuing events (i.e., hunting/egging, introductions, habitat degradation, more introductions, fishery interactions, pollution). Multiple stressors, even of a single type (e.g., multiple persistent pollutants) may interact additively, synergistically, or antagonistically (Livingstone, Donkin, and Walker 1992). It is difficult to clearly sort out relative magnitude, and, in fact, this may not even be the appropriate strategy if interactions are nonadditive. An additional layer of complexity is the effects of natural physical forces. During El Niño–Southern Oscillation (ENSO) years, food-stressed populations of Common Murres may be more prone to entanglement in fishing gear (Melvin and Conquest 1996). In a nonintuitive example, bioaccumulation of DDE was significantly lower in shags during harmful algal bloom years (Coulson et al. 1972). Finally, as population demographics for affected species change, energy flow through the ecosystem will also change, altering basic species interactions with potentially serious conservation consequences to seabirds. Paine, Wootton, and Boersma (1990) suggested that an increasing population of Peregrine Falcons (*Falco peregrinus*), as a result of successful management and conservation efforts, had a negative effect on some species of seabirds (i.e., falcon prey) but a positive effect on others because the falcons also hunted Northwestern Crows (*Corvus caurinus*), an egg predator in the system.

Features of Research That Improved Conservation

We have been able to identify issues of conservation concern to seabirds because scientists and concerned citizens (e.g., the Audubon Society) have conducted studies linking persistent declines in seabird abundance to an increase in some human-mediated occurrence (e.g., California Gulls [*Larus californicus*] at Mono Lake; Hart 1996). With the development of conservation biology as a discipline and the spread of directed research aimed at teasing out cause and effect, our collective knowledge of problems affecting seabirds has grown. Concomitant with this knowledge base is our realization that the forces affecting a single species within an ecosystem are complex, a function of natural physical, biological, and anthropogenic factors, interactive in nonintuitive ways, and almost impossible to verify statistically in the field. Several aspects of research to date have facilitated conservation efforts, including long-term and/or directed studies, cross-disciplinary studies, and review and overview publications synthesizing knowledge on taxa or conservation issues.

Much of our most powerful information has been collected in the course of long-term and/or directed studies on seabird population biology. We define long-term studies as multiyear stretching to multidecadal research on a single taxa at a single location, and directed studies of long-term effort on a single taxa regardless of whether all research is accomplished at the same colony. Longitudinal studies allow correlation with a variety of subtly changing variables (e.g., climate change, persistent pollutant loads, etc.) not necessarily detectable on shorter (i.e., several-year) time scales. Long-term datasets have also provided the baseline from which we assess the health of populations and species. In some cases, multiple studies on a single species breeding in different locations have allowed us to contrast the effect of multiple stressors across the species' range (e.g., seabird status reports). Finally, long-term studies of seabird communities allow us to contrast multispecies response to the same stressors (e.g., Farallon Islands seabird colony; Ainley and Boekelheide 1990). This is extremely useful when attempting to predict which species are sensitive to what aspects of the environment.

Cross-disciplinary studies are typified by research products that become more than the sum of the parts because of their ability to explain phenomena that no single directed study could (e.g., interaction between climate, fisheries, and seabird demography Aebischer, Coulson, and Colebrook 1990). Issues in conservation often require multidisciplinary work, to both identify as well as solve problems. Fishery by-catch, persistent pollutant uptake and effect, and interaction with introduced species are all examples. As scientists and managers become more aware of the nonlinearities of interaction between human stressors as they affect seabird species, cross-disciplinary research will become a central tenet of conservation.

In a single month, thousands of peer-reviewed publications are published in avian taxa and discipline-oriented journals around the world. Keeping up is extremely difficult, and starting from scratch is virtually impossible. Although perhaps obvious, review and overview publications either in journals or in book form allow us to keep abreast of a wide range of recent research on a taxa (e.g., *The Atlantic Alcidae*, Nettleship and Birkhead 1985; *The Petrels*, Warham 1990; *The Penguins*, Williams 1995) or issue (e.g., *Status and Conservation of the World's Seabirds*, Croxall, Evans, and Schreiber 1984; this book) while providing us with access to the primary literature if we choose. For nonseabird biologists and natural resource managers trying to become competent on issues affecting a certain species or geographic area, this type of synthesis is a must.

Previous research has also played a central role in problem identification leading to the development of laws to protect wildlife. Although academic research has contributed to this effort—for example, in determining what species should be listed as endangered—most of the drafting and implementation of laws has not been driven by seabird considerations. Nevertheless, seabirds do benefit. The four most important laws providing a framework for long-term conservation of seabirds in the United States are the Migratory Bird Treaty Act (MBTA) of 1916, the Endangered Species Act (ESA) of 1973, the Magnuson Fishery Conservation and Management Act of 1976, and the Fish and Wildlife Conservation Act of 1980.

Passage of the MBTA changed the focus of seabird exploitation from solely a local concern to a national and international issue. Although the act is difficult to enforce, it forbids the killing of any migratory bird within signatory countries out to three miles (state waters). At present, enforcement extension to cover the exclusive economic zone (two hundred miles) as well as U.S. flagged vessels worldwide is under consideration (V. Mendenhall, U.S. Fish and Wildlife Service, personal communication 1997). The "zero-tolerance" provision, if enforced, would be the strictest legal protection seabirds have. Within the United States, Canada, Japan, Mexico, and the former USSR, the MBTA has done much to regulate the exploitation of seabirds. Four exceptions may limit its effectiveness, however: seabirds can be taken for scientific and educational purposes; hunting can be done as long as a time of year is designated; subsistence hunting is allowed; and protection against injury to person or property remains a right. For example, Canada enacted regulations in 1958 that allow rural Newfoundlanders a subsistence hunt for murres from September 1 to March 31, despite the fact that current hunting levels may exceed sustainable mortality rates (Montevecchi and Tuck 1987).

The ESA currently lists ten seabirds and two seaducks as threatened or endangered. The U.S. Fish and Wildlife Service is required to publish a recovery plan for all listed species based on the best available scientific information. Although provisions are included for habitat protection and could be interpreted as ecosystem management, the ESA is primarily implemented on a species-by-species approach and the recovery process has a variety of problems (Tear et al. 1995). The Magnuson Fishery Conservation and Management Act mandates the conservation and management of fishery resources including avoiding long-term irreversible effects on species, restoring stocks, and minimizing by-catch. Although seabird research was not used in drafting or implementing the Magnuson Fishery Conservation and Management Act, seabirds stand to benefit from the implementation of Fishery Management Plans, which recognize natural sources of mortality, including predation by seabirds, and identification and protection of critical habitats for fish. Likewise, the Fish and Wildlife Conservation Act of 1980, and as amended in 1988, recognizes the ecological, educational, aesthetic, cultural, recreational, economic, and scientific value of nongame migratory species including seabirds.

Many new pieces of legislation are direct responses to problem identification by researchers. For example, in the mid-1980s, reports of hundreds of seabirds, turtles, and marine mammals being caught in driftnets began to garner widespread public attention in the United States. The Driftnet Act of 1987 encouraged agencies to gather data on the impacts of driftnets on marine life on the high seas. By 1990, data collected suggested that nontarget finfish, seabird, and marine mammal by-catch was high. As a result, the United States passed the High Seas Driftnet Fisheries Enforcement Act of 1992, discouraging the use of driftnets. At a global level, the passage of U.N. General Assembly Resolution 44/225 in 1989 called for a ban on high-seas driftnet fishing in the Pacific.

The trend of global recommendations for conservation of seabirds is relatively new and illuminates the importance of working on a global scale to address seabird problems. Several conventions and treaties address seabird conservation worldwide, including: the International Convention for the Prevention of Pollution from Ships (MARPOL, 1978), the Convention on Conservation of Antarctic Marine Living Resources (CCAMLR, 1984), the Convention on International Trade in Endangered Species of Wild Flora and Fauna (CITES, 1987), and the Biodiversity Convention (1992). MARPOL prohibits the dumping of plastics anywhere in the world's oceans and navigable waters of signatory countries. Oil discharge and a variety of other problems, including release of ballast water, are regulated under this convention. CITES prevents trade of listed species in whole or in part for all signatory countries. CCAMLR addresses concerns about biodiversity loss in Antarctica. Using the concept of "do no long-term harm" the convention has guidelines for the issuing of permits that address sustainability and ecosystem functioning. Conservation measures include control of pollution, catch and gear restrictions, and area and seasonal closures; and special conservation measures, including the collection of data on seabird injury and mortality, protect seabirds. The Biodiversity Convention endorses protection and conservation of the world's species.

Research Needed to Further Conservation

The term *applied research* recognizes the need to tie research to management questions. Few studies have addressed and quantified how management changes influence resource use and sustainability. Although problems of seabirds are mainly global and demand global solutions, these solutions must be built locally to take into consideration differences in the physical, biological, and social environments that seabirds share with humans.

The most necessary pillars to support seabird conservation while fostering changes in human behavior and environmental use are education and public support. Seabird biologists and the institutions and organizations that support them will have to work at unprecedented levels of cooperation across national and international boundaries and between institutional structures. Interest groups with divergent and sometimes conflicting priorities must cooperate in educating the public and decision makers on seabird problems and solutions in the broader context of sustained use of the marine environment. Without truly integrated studies involving a range of physical and biological scientists, cause-and-effect linkages will not be made (Marzluff and Sallabanks, this volume). Without serious participation by social scientists, economists, and managers, physical and biological data, no matter how revealing, may never be translated into action (Young and Varland, this volume).

We have identified four main research objectives for seabird conservation: (1) understanding how marine systems function and change over time, (2) understanding how seabirds respond to these eco-environmental dynamics, (3) determining how humans are changing systems that seabirds use, and (4) designing and testing modifications in law, policy, human behavior, and use of the environment to positively affect seabird survival and abundance.

Marine System Function

Long before humans began to influence seabird population demographics, natural physical forces were causing large fluctuations in seabird prey (Baumgartner, Soutar, and Ferreira-Bartrina 1992), marine habitat quality (e.g., regime shifts; Steele 1996), and breeding habitat availability (Kaiser and Forbes 1992). Understanding the links between atmospheric and oceanic circulation, primary production, and upper-trophic-level response is crucial to seabird research. For instance, Veit et al. (1997) present data on massive Sooty Shearwater declines off the coasts of the western United States over the last decade. Although these data are no doubt negatively correlated with human population size in western coastal counties and related development and pollution pressures (percent beach armoring, sewage output, number of poor air-quality days, etc.), a broader examination of the regional marine environment provided a more reasonable explanation: decreased upwelling (measured as rising water temperature; Roemmich 1992) and an ensuing decline in primary and secondary productivity (Roemmich and McGowan 1995). Less food meant fewer birds. This conclusion would not have been possible without long-term multidisciplinary monitoring of the California Current system under the auspices of the California Cooperative Oceanic Fisheries Investigations.

Seabird Response to Environmental Change

Management of seabirds requires an understanding of how seabirds respond to natural environment perturbations of varying length and amplitude (Boersma 1987; Aebischer, Coulson, and Colebrook 1990). The only way to gain understanding is through long-term studies of select seabird communities. Long-term studies are rare, yet they are the cornerstones for much of our current understanding of seabirds. There are three ways that long-term studies seem to occur. Occasionally, one or two individuals study an area for several decades, often at great personal cost. This route is less accessible to new generations of seabird biologists as academic science focuses on short-term experimentation, funding for long-term monitoring becomes scarce, and travel costs to field sites (ever further from human centers) become more expensive. However, Marzluff and Sallabanks (this volume) suggest that academic advisers may be able to accomplish some types of long-term research by continually recruiting short-term graduate student projects to the same study. The second way is an organizational structure formed to support long-term research (Point Reyes Bird Observatory, Farallon Islands; British Antarctic Survey, Southern Ocean). Finally, fish and wildlife agencies at regional and national levels are tasked with monitoring populations of stewarded species, including seabirds.

Human Activities

We have come a long way toward understanding how human activities are affecting seabird populations. This chapter touches on some of these major conservation issues. The discipline of conservation biology is a reaction to negative human-environment interactions. As studies documenting human interactions with seabird populations continue, we suggest two caveats. First, we must assess the effects of all human activities, including our own, on the seabird populations we study. Management decisions are only as good as the science on which they are based. Second, we should be careful to avoid the "Chicken Little" complex. Human activities are not detrimental by default, and we should not rush to lay blame before we have evidence (Parrish and Boersma 1995). Otherwise, science will lose its most valuable asset: credibility.

Seabird Management

Lastly, we need research to quantify how management changes modify seabird interactions and species abundance. To reduce people pressure on seabirds, large-scale changes in how the marine environment is used are needed. Remoteness and isolation are rapidly becoming descriptors of seabird colonies of the past. Duffy (1994c) suggests a 10 percent rule for seabird preservation: Colonies exceeding 10 percent of a species abundance determined regionally or globally should be protected. Large monetary settlements from natural resources damage assessment cases, such as that resulting from the *Exxon Valdez* oil spill, could be used to purchase and preserve nesting habitat of affected seabird species. Nongovernmental organizations specializing in habitat protection through purchase (e.g., The Nature Conservancy) should be encouraged to consider seabird habitat. On an international scale, debt-for-nature swaps could be used to preserve nesting habitat and perhaps migratory routes.

Seabirds require the protection of their most important breeding colonies and key feeding areas both inside and outside the breeding season. Although we know where many of the key breeding colonies are, we know little about foraging areas, particularly outside the breeding season. Research is needed to delineate and prioritize areas for preservation. The equivalent of terrestrial Habitat Conservation

Areas should be a part of marine conservation efforts. Various authors have suggested that 20–30 percent of the global marine environment be set aside as reserves (Ballantine 1997; Schmidt 1997). Zoning of ocean uses to separate the most detrimental human activities from major concentrations of seabirds is likely to become increasingly important. Evaluations of these proposals as well as current and future fisheries and ecosystem management strategies are needed.

Beyond preservation is restoration, in which degraded habitat or reduced populations are recovered through intervention, with the goals of increasing recruitment to the colony, increasing reproductive success of colony members, and decreasing adult mortality. Parrish and Paine (1996)—after Temple (1977)—define two types of restoration activity: *active,* in which organisms are intentionally added or removed (e.g., translocation, predator or competitor removal); and *passive,* in which organisms are not handled directly, but specific aspects of habitat are altered (e.g., social facilitation, provision of additional nest sites). Kress (this volume) reviews both types in the context of restoration of extirpated seabird colonies along the Maine coast.

Habitat loss and adult mortality can act in concert to create smaller, more fragmented populations with potentially altered sex ratios, age distributions, or skill levels (Pearl 1992), any of which could result in a breakdown of normal social function (e.g., the Allee effect [Allee et al. 1949], the Darling effect [Darling 1938]). Furthermore, declining species are enmeshed within a dynamic biological matrix: the community. Human activities will have not only direct effects on community members, but indirect effects as a consequence of altered species interactions (Parrish 1995; Parrish and Paine 1996). At an extreme, loss of community members will lead to a decline in ecosystem function as the interaction links between resident species are broken (Willson 1995). Without an understanding of the patterns of community and behavioral ecology, accurate predictions of population demographics and viability may be seriously flawed (Caughley 1994; Caughley and Gunn 1996), leading to incorrect or inconsequential attempts to restore the affected population (e.g., Bermudian Petrel; Wingate 1977, 1985).

Seabird restoration is in its infancy. Although the number of restoration techniques is growing, a serious research effort is needed to address the efficacy of these as well as future techniques in the context of community interactions and ecosystem dynamics.

Making Research Effective for Conservation

Sustaining the world's seabird populations is a global management task that at a minimum will require modifying human behavior and human use of the environment. Since the 1950s the world's economy has grown fivefold, while the world's human population has more than doubled from 2.6 billion to 5.5 billion (Brown 1993). Concomitant with these trends is the widening gap in the distribution of income that strongly favors growth and ever rising consumption over equity and poverty alleviation, a prescription for increasing production with little or no attention to sound use of the earth's natural resources. Seabirds like other forms of life are caught in the squeeze.

Scientific knowledge and research can: (1) identify problems and threats to seabirds, (2) highlight conservation options, and (3) assess the success of management changes. Most biologists are well aware of the declining state of natural resources, especially as this loss of biodiversity and ecological function affects their own research. Increasing numbers of researchers are becoming involved in conservation efforts, and some are creating restoration options (Parrish and Paine 1996; Kress, this volume). However, research alone is insufficient for legislative, administrative, or social action unless closely tied to political forces for change, which include education, public support, and an appreciation of the economic value of natural systems that are healthy, resilient, and functional. We believe that science, in and of itself, does not need to change to become a more effective conservation tool. Therefore, we will confine our comments to the crucial and at present mostly missing steps: realization of the consequences of status quo human activities and frameworks for change.

To protect seabirds, human activities affecting the marine environment must be managed (Ludwig, Hilborn, and Walters 1993; Duffy 1994a). The changes necessary to mitigate the extraordinary demands being placed on world resources, other living species, and the life-support system sustaining the increasing human numbers and human consumption, will require fundamental changes. The loss of biodiversity is largely a function of human actions (Soulé 1991; World Resources Institute 1992). We suggest four areas in which progress in reducing human impacts on the environment should be gauged: education, energy use, ecological function, and ethics (the 4 Es).

Education

Education should encourage humans to make "better" environmental decisions and to reduce resource consumption. Education illuminates problems, produces solutions, and fosters implementation of change. People need to be educated about environmental and conservation crises but more fundamentally about what species are where, how natural communities and ecosystems function, and what the benefits of continued ecosystem health can be. If public support for seabird conservation is lacking, perhaps the education system responsible for instilling knowledge and critical thinking skills is lacking as well. Relevant measures of education might include: number of primary, secondary, and higher education texts that discuss seabird biology, management, and conservation; public interest in volunteer birding groups; income spent on seabird-related tourism; public participation in legislation reauthorization affecting seabirds and the marine environment; and public willingness to accept up-front costs in return for long-term conservation gain.

Recognition and support for institutions and organizations that facilitate student training, public outreach, and volunteer efforts directed toward seabird conservation are needed. Local efforts should be linked into regional, national, and international efforts (e.g., Greenwood et al. 1993). The IUCN Seabird Specialist Group (SSG) and recent proposals by Birdlife International to create an international seabird conservation network comprised of regional seabird organizations are cases in point. Organizations in developed countries need to be linked to organizations in developing countries to enhance conservation efforts at local as well as global scales, as well as facilitate the social and cultural exchange necessary for truly integrated research and management efforts. For example, the SSG has recommended the creation of feral animal "hit teams" to eradicate introduced species on seabird colonies worldwide (Duffy 1994a). In developing countries, hit teams would train local personnel in eradication techniques so that future response would be local, vigilance would be possible, and local experts could export the techniques to additional islands.

Internet and other electronic means are becoming increasing useful for facilitating communication, sharing information, and generating the public support necessary for policy change. A good example showing the benefits of linking scientists, environmental advocates, and international decision makers electronically is the ongoing debate about seabird by-catch in longline fisheries. Public debate on the Internet has prompted wider acknowledgment and adoption of recommendations by CCAMLR to prevent the incidental mortality of seabirds during fishing operations. One of the results has been an IUCN Resolution (CGR1.69), adopted in 1996, outlining multiple measures to reduce or eliminate seabird by-catch in longline fisheries. Several countries have passed similar laws or resolutions regulating their fisheries.

Energy Use

Energy use is another area that directly impacts seabird conservation. People in industrialized countries consume a disproportionate share of the total energy produced. For example, the United States has only 5 percent of the world's population but consumes approximately 25 percent of the world's energy (Botkin and Keller 1995). Because of the growth in world population, most of it in developing countries, increases in energy use in these countries will have profound impacts on biological resources. Energy consumption in India has more than tripled since 1970, and China's consumption has increased twenty-two times since 1952 (Carr 1994; Lenssen 1992). Energy subsidies, status quo, and reluctance to face the external costs of energy use have resulted in local and regional marine pollution, overcapitalization and declining catch per unit of energy in the world's commercial fishing fleet (Weber 1995), global warming, and the continued push to explore for new petroleum sources in the coastal environment. All of these actions have seabird conservation consequences. If energy and other products are used more wisely, less may be needed and wastes can be minimized. Relevant measures of conservation-oriented energy use might include: per capita gross national product to energy consumption index; tons of carbon, nitrogen, and sulfur discharged; total and target catch per unit energy consumed in the world's fisheries; and assessment of chronic oiling from beached bird surveys.

Ecological Function

Ecological function is vital to our continued quality of life and perhaps to life support as well. Interactions among species are complex. Disruption of natural systems by introductions of non-native species or removal of keystone predators demonstrates the importance of interactions in ecosystem structure, function, and community resilience. Humans affect seabird populations directly via hunting, fishery by-catch, competition with fisheries, and egging and indirectly through species introductions, habitat modification, fishery-related ecosystem change, and pollution. Both types of effects can result in trophic cascades with unknown and perhaps nonintuitive consequences. We also need to identify strong and weak interactions and pay attention to them in resource management. Ecological degradation sometimes results in huge economic costs: Loss of beneficial insect pollinators through inadvertent pesticide poisoning could threaten agriculture; deforestation results in changes in local weather patterns and water quality; wetland loss to development results in increased flooding.

Interactions in the marine environment, although less well studied, may be as important. For instance, continued overexploitation of the groundfish fishery on Georges Banks resulted in catastrophic declines of commercial species and apparently persistent dominance of nontarget species despite the relaxation of fishing pressure (Murawski 1991). When detrimental ecological effects as a consequence of human actions are mitigated or reduced, diversity and the healthy functioning of the system should become sustainable. With respect to seabird conservation, relevant measures of ecological efficiency might include: assessment of the degree of native and endemic biodiversity within coastal marine ecosystems; variability in species abundance as a function of human activities such as fishing, land-use conversion, and pollutant output; alpha, beta, and gamma biodiversity; and foraging efficiency and chick growth as a function of local fishery and pollution pressures.

Ethics

Ethics are the system of social norms and societal contracts we use to govern our behavior. Legislation is the stick we use to enforce an ethical code. Were we all seabird biologists, ethics, laws, and their enforcement would reflect society's desire to maintain species and systems for the longterm. Unfortunately, short-term interests that tend to degrade natural systems are far more prevalent in society. If steps are taken to pursue conservation-minded goals in education, energy use, and ecological functioning, the result should be a new ethic incorporating the long-term as well as the short-term value of natural systems to humans. Local, regional, and national policies should not subsidize the degradation of ecosystems. Treaties, conventions, and recommendations by international organizations should foster sustained use and protection of natural systems. Legislation designed to attain management and conservation goals should be implemented and enforced. Measures of an ethical code incorporating ecological health and sustainable use of marine systems relevant to seabirds might include: number of ESA exemption consultations or CITES violations involving seabirds; number of national and international policies passed to protect and conserve seabirds and their ecosystems; public involvement in action-oriented non-governmental organizations; use of net economic welfare or index of sustainable economic welfare as a replacement for gross national product; and public willingness to accept up-front increases in costs and/or decreases in services in return for long-term guarantees of ecosystem viability and sustained use.

Adequate management and protection for seabirds will depend on our ability to collect additional information about seabird biology in the context of the marine systems they inhabit. However, conservation will depend in large part on whether humans pay attention to the 4E's. If humans increase their awareness through education, reduce energy consumption, allow for ecosystem functioning, and adopt an ethical framework for the long-term, the future for seabirds is likely to be brighter than if current trends continue.

Acknowledgments

We thank Ann Edwards for library research, initial synthesis, and stimulating discussion. Karen Jensen and Brian Walker compiled endless references. Jen Piel and Jessica Tam tracked down especially hard-to-find references. John Marzluff, Rex Sallabanks, and Steve Kress improved earlier drafts of the manuscript.

CHAPTER 17

Avian Conservation Research Needs in Western Shrublands: Exotic Invaders and the Alteration of Ecosystem Processes

John T. Rotenberry

Conservation Issues and Previous Research

Much of the western United States, particularly that in the Great Basin and throughout the Intermountain Region, originally consisted of vast tracts of native shrublands. Although a variety of shrub species occur throughout the West, individuals of the genera *Artemisia* (especially big sagebrush, *A. tridentata*) and *Atriplex* (e.g., shadscale, *A. confertifolia*) frequently dominate (Franklin and Dyrness 1973). Compared to sagebrush, shadscale is more southerly in its distribution, occurring at lower elevation on shallower soils and in relatively hotter, drier places. When undisturbed, these shrublands usually contain an understory of caespitose bunchgrasses, primarily from the genera *Agropyron*, *Poa*, *Sitanion*, *Vulpes* (=*Festuca*), *Stipa*, and *Oryzopsis* (Daubenmire 1970). Areas where grass coverage equals or exceeds that of shrubs are often referred to as shrub steppe. In years of normal or high precipitation, forbs of many species can be abundant. Because bunchgrasses do not provide continuous cover of fuel, fires do not spread easily, and thus large fires in pristine native shrublands were relatively rare (Whisenant 1990). Originally, bunchgrasses did not support large populations of native ungulate grazers, and bison (*Bison bison*) and other ungulates were mostly absent from Great Basin shrublands (Osborne 1957; Rickard, Hedlund, and Fitzner 1977; Mack and Thompson 1982). Over the last century, however, both grazing and fire have become major disturbance agents in the shrub steppe and together with agriculture have set the stage for many of the conservation issues discussed below.

Compared to more structurally and floristically diverse ecosystems, shrubland habitats generally support few bird species. The dominant species breeding in sagebrush shrublands are Brewer's and Sage Sparrows (*Spizella breweri* and *Amphispiza belli*) and Sage Thrashers (*Oreoscoptes montanus*), with Black-throated Sparrows (*Amphispiza bilineata*) more abundant in shadscale (Wiens and Rotenberry 1981; Ryser 1985; Medin 1990). Loggerhead Shrikes (*Lanius ludovicianus*), Rock Wrens (*Salpinctes obsoletus*), Green-tailed Towhees (*Pipilo chorula*), and Gray Flycatchers (*Empidonax wrighti*) may be locally abundant (Wiens and Rotenberry 1981; Ryser 1985). Common Ravens (*Corvus corax*) are ubiquitous. In areas where grasses become abundant in the understory and where grasslands become common in the landscape, several species more typical of grasslands to the east, such as Horned Larks (*Eremophila alpestris*), Western Meadowlarks (*Sturnella neglecta*), Vesper Sparrows (*Pooecetes gramineus*), and Lark Sparrows (*Chondestes grammacus*), may be common. Nonpasserines that may occur with regularity in western shrublands include Golden Eagles (*Aquila chrysaetos*), Swainson's and Ferruginous Hawks (*Buteo swainsoni* and *B. regalis*), Prairie Falcons (*Falco mexicanus*), and Sage Grouse (*Centrocercus urophasianus*), and, where shrubs have been reduced and grasses become more common, Long-billed Curlews (*Numenius americanus*) and Burrowing Owls (*Speotyto cunicularia*).

The most pressing issue relating to western shrublands is the loss and concomitant fragmentation of native shrubland habitats. Species perhaps most in need of conservation attention are those most typical of undisturbed shrub steppe: Brewer's, Sage, and Black-throated Sparrows and Sage Thrashers (the "obligate shrub steppe" species of Braun et al. 1976 and Wiens and Rotenberry 1981). It is these passerines on which I will focus.

Distribution and abundance of these species are closely tied to the distribution and coverage of native shrub species (Rotenberry 1986; Wiens and Rotenberry 1981; Dobler 1994; Knick and Rotenberry 1995). Moreover, these species, unlike more typical grassland species, appear to be especially sensitive to the negative effects of habitat fragmentation and are often absent from tracts of native habitat that otherwise appear suitable (Knick and Rotenberry 1995). For example, continuous blocks of sagebrush-dominated habitat as large as 150 hectares (sufficient for up to seventy pairs based on average territory sizes and local cover of sagebrush; Wiens, Rotenberry, and Van Horne 1985) may contain no breeding Sage Sparrows (M. Vander Haegen, personal communication). Insufficient data exist to derive "incidence functions" (proportion of habitat blocks occupied as a function of block size; Diamond 1975) to ascertain if a minimum threshold of block size exists. However, it is clear from these observations that relatively small patches may not be suitable, from a conservation perspective, for even small numbers of birds.

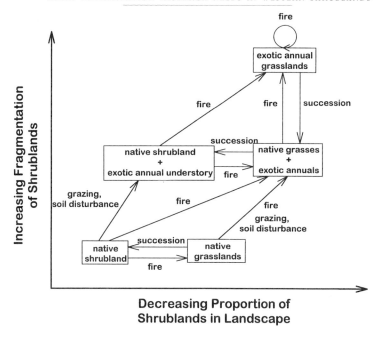

Figure 17.1. Landscape dynamics of native shrubland habitats under current ecological conditions. Boxes represent major habitat types, arrows represent transitions between habitat types, and arrow labels denote ecosystem processes that drive transitions. Distance from the origin represents increasing levels of disturbance.

Understanding the Role of Disturbances

Understanding how the principal sources of disturbance interact to produce fragmentation and loss of native shrublands is crucial to developing strategies for reversing habitat declines. These interactions are complex (figure 17.1), and repairing the problems they have generated will require integrated solutions.

Agriculture

The predominant impact of agriculture throughout the Great Basin has been straightforward: the conversion of shrubland and shrub steppe habitat to grassland and cropland. This loss of habitat has not necessarily been associated with increasing fragmentation, and in some areas remaining shrubland habitat may continue to exist in relatively large blocks of several hundred hectares (Knick and Rotenberry, in press). However, beyond the obvious effects leading to reduction of shrub steppe area, agricultural activities have also been associated with the introduction and spread of alien plants, which may now be the most serious threat to these ecosystems (see below). Agricultural areas adjacent to natural habitats may serve as continuously renewable sources for immigrant species (Janzen 1986; Alberts et al. 1993). Likewise, although agricultural areas may provide suitable wintering habitat (through the provision of waste grain) for species such as

Horned Larks, they may also support Brown-headed Cowbirds (*Molothrus ater*), a brood parasite with a relatively recent history in the Great Basin (Rothstein 1994). Agricultural areas also appear to extend the landscape-level distributions of corvids, particularly Common Ravens and Black-billed Magpies (*Pica pica*) (Marzluff, Boone, and Cox 1994), which can be major predators of the nests of passerines.

Livestock Grazing

Livestock grazing has been a significant feature of western shrublands and shrub steppe ("rangelands") since the early 1800s (Yensen 1982). Unlike the grasslands of the midwestern prairies (see, for example, Herkert and Knopf, this volume), however, Great Basin shrub steppe had not evolved under grazing pressure from large ungulates such as bison; thus, most of the impact of livestock grazing has been detrimental to native vegetation (Mack and Thompson 1982). Grazing by sheep and cattle disturbs the surface of the soil and results in selective removal of plant biomass, which alters competitive relationships among plant species and may lead to the increase of unpalatable species. The former impact destroys the microbiotic crust that usually forms on the soil surface (Cole 1990), which then adversely influences water infiltration, erosion, and nitrogen fixation (Harper and Marble 1988). Soil disturbance also provides "safe sites" for the germination of annual plant seeds, and thus promotes the invasion of these species into otherwise undisturbed habitats. Severe overgrazing, frequently associated with watering sites in an otherwise arid landscape, can lead to the removal of virtually all above-ground plant biomass.

Exotic Annual Plants

A variety of exotic annual plants has been introduced into western landscapes over the last century and a half, and many are now apparently permanent features of western shrublands. Although "tumbleweeds" (Russian thistle, *Salsola kali*) may be the most popularly known, the most pernicious is likely downy brome, or cheatgrass (*Bromus tectorum*). The principal impact of cheatgrass has been to alter the fire ecology of shrub-steppe ecosystems. Unlike native bunchgrasses, cheatgrass provides a continuous surface cover of relatively fine fuel that carries fires into and over much larger areas than likely occurred historically (Whisenant 1990). Furthermore, cheatgrass suppresses the post-fire regeneration of many native species, especially shrubs such as sagebrush or shadscale, which are killed by fire and do not resprout from crowns (Blaisdell 1953; Wright, Neuenschwander, and Britton 1979).

Synthesis of Disturbance Processes

Under the current disturbance regime, the increasing loss of native shrublands and their subsequent fragmentation seems inexorable (e.g., Young and Evans 1978; Allen 1988; Knick and Rotenberry in press) (figure 17.1). Prehistorically, large fires were relatively infrequent in native shrublands. Burned areas quickly recovered to native grasses, then underwent natural succession to shrublands. Based on historical descriptions (e.g., Frémont 1845; Vale 1975) and analytical models based on fire-return intervals of around seventy years (Rotenberry, unpublished data), at this time Great Basin landscapes likely consisted of 80–90 percent shrublands. Cheatgrass first appeared in the early to mid 1800s, probably as a contaminant in grain seed (Mack 1981), and quickly spread among agricultural areas and along roads and railroads. However, it was under the influence of widespread livestock grazing, through disturbance of the soil's surface, that cheatgrass spread into otherwise undisturbed shrub steppe (Mack 1981). Shrublands and other habitats infested with cheatgrass are twenty times more likely to burn than those without (Stewart and Hull 1949; Whisenant 1990). The establishment of cheatgrass reduces average fire-return intervals to less than five years, which is too short for most shrub regeneration to mature (Whisenant 1990). Even if an area escapes rapid reburning, recovery of many shrub and perennial grass species is greatly retarded by the presence of cheatgrass, which suppresses germination and seedling growth (Blaisdell 1949; Allen and Knight 1984; Bunting 1985).

A second fire also can destroy any viable sagebrush seeds in the seedbank if it burns within five to eight years. Thus, subsequent recovery of sagebrush can come only from existing seed sources (i.e., mature plants in adjacent unburned areas). Because sagebrush seeds are dispersed primarily by wind, total dispersal distance is about 30 m from a seed source. If fires are small and burn patchily, burned areas may fill in quickly. With large fires, this limited dispersal distance coupled with a time lag while infilling shrubs reach maturity means that decades may elapse before the central areas of large burned patches are revegetated with sagebrush.

An important feature of this disturbance cycle is its near irreversibility. Cheatgrass is autosuccessive after fire. Because of its phenology (a winter annual), cheatgrass has usually matured and set seed and its seed heads shattered by the onset of the summer fire season. Thus, post-fire survivorship of cheatgrass is high, whereas that of native perennial grasses, whose seeds are often still maturing in summer, may be quite low. Thus, once *Bromus* becomes part of an ecosystem, it is highly likely to remain a part of it.

The principal implication of the altered landscape dynamics depicted in figure 17.1 is that, under current conditions, the system is tending toward a complete loss of "pristine" native shrublands and grasslands, fragmented or otherwise (Young and Evans 1978; Allen 1988). In essence, all transitions (arrows representing the direction of change between states in figure 17.1) ultimately lead away from native habitats and through contaminated habitats, culminating in a landscape dominated by exotic annual grasslands. Even if fire frequencies can be reduced through presuppression or suppression activities, only the rate of transition, not the outcome, is affected. Analytical models suggest that, given current interfire intervals and rates of recovery, a landscape at equilibrium would likely consist of less than 40 percent shrublands (Rotenberry, unpublished data).

Loss of shrublands due to fire is also coupled with increased levels of their fragmentation, above that associated with reduction in areal extent alone (Knick and Rotenberry, in press). This fragmentation affects landscape configuration of shrubland patches at spatial scales (500–5000 ha) that are associated with negative impacts on shrub-steppe obligate bird species (Knick and Rotenberry 1995).

In summary, the synergistic processes of fragmentation of shrublands by disturbance, invasion and subsequent dominance by exotic annuals, and fire are converting landscapes to a new state dominated by exotic annual grasslands and high fire frequencies (Knick and Rotenberry, in press). Coupled with what is already known about bird species habitat relationships in this system (see above), we expect to see a change in avian composition from communities composed mainly of shrubland obligates to those consisting mainly of grassland species (figure 17.2).

Features of Research That Improved Conservation

Recognition of the scope of avian conservation problems in western shrublands is relatively recent (Braun et al. 1976). Thus, there has been relatively little research explicitly focused on the conservation of birds in these habitats. Mostly, relevant shrub-steppe avian research has examined patterns of habitat associations across a variety of spatial and temporal scales (e.g., Rotenberry and Wiens 1980; Wiens and Rotenberry 1981, 1985; Rotenberry 1986; Wiens, Rotenberry, and Van Horne 1986; Petersen and Best 1987; Dobler 1994; Knick and Rotenberry 1995), then used this information (with variable degrees of success; Rotenberry 1986) to project population changes through time and space as habitat changed under the influence of the disturbance processes outlined above.

One example of how integration of shrub-steppe bird habitat associations at landscape levels with an understanding of the processes contributing to shrubland loss and fragmentation has improved conservation prospects for birds comes from the Snake River Birds of Prey Natural Conservation Area (NCA) in southwestern Idaho. The NCA consists of approximately 400,000 hectares of formerly sagebrush and shadscale shrub steppe, now about 40 percent converted to exotic annual grasslands and other disturbed habitats and heavily fragmented (Knick and Rotenberry, in press; Knick, Rotenberry, and Zarriello 1997). During the 1980s about half of the then existing shrublands burned (Kochert and Pellant 1986). Appropriately concerned about further loss of shrublands, the Bureau of Land Management (BLM) instituted "green-stripping" in an attempt to reduce the spread of fires (Pellant 1990). Green-stripping

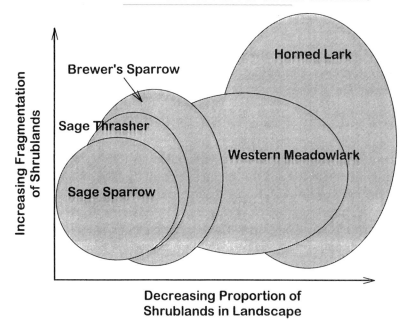

Figure 17.2. Distribution of common shrub-steppe bird species as a function of the abundance and fragmentation of native shrubland habitat in a landscape (Knick and Rotenberry 1995). Axes and scale the same as in figure 17.1. Distance from the origin represents increasing levels of disturbance.

is the strategic placement of 150–300-meter-wide strips of fire-resistant vegetation on fire-prone landscapes. Such strips may be several to tens of kilometers in length. Strips invariably consist of exotic species; forage kochia (*Kochia prostrata*) is a favorite because it has a relatively high water content (which increases its resistance to fire) and it competes well with weedy species (which increases its likelihood of establishment).

Although well intentioned, green-stripping creates numerous landscape-level effects that may adversely affect shrub–steppe birds. First, it can contribute to the further loss of existing shrublands by replacing them with a green-strip; a modest-sized strip 5 km long and 400 m wide displaces 200 ha of native shrubland. Second, it can exacerbate fragmentation; a strip through the heart of one large patch immediately creates two smaller ones. Third, it intentionally introduces alien plant species into an ecosystem that already suffers from exotics; green-strip species are selected for their competitive abilities, and thus their chance for establishment outside strips is high. Fourth, the planting process itself invariably requires heavy disturbance to the soil (often via disking), which promotes conditions suitable for the spread of other exotics, particularly cheatgrass and Russian thistle. Based on these and other considerations, BLM reevaluated its commitment to green-stripping and has now suspended the creation of any new strips within the management unit containing the NCA (S. Saether-Blair, Area Manager, BLM, personal communication).

Research Needed to Further Conservation

Although much remains to be learned about the response of shrub-steppe birds to the various threats confronting them, it seems that the most urgent research needs involve Great Basin landscapes and the plant communities they contain.

Removal of Exotic Annuals

It has become apparent (e.g., figure 17.1) that reduction of fire frequencies via green-stripping or any other form of presuppression or suppression is only a short-term solution to the loss and fragmentation of shrublands. Successful long-term preservation of shrublands will require removal of the exotic annuals that have, in essence, become self-perpetuating. However, such eradication is not easy—these species have clearly been very successful in their invasion and are well adapted to current ecological conditions throughout western rangelands. Nonetheless, finding ways to eliminate such species as cheatgrass should be of the highest priority. The ideal "magic bullet" would affect only the target species and might include a genetically engineered plant virus or other species-specific pathogen, a highly selective herbivore, or a chemical that alters some metabolic pathway unique to the species or interferes with DNA replication (also unique to the species). Alternatively, there may be some combination of grazing management and manipulation of competing plant species density (perhaps through selective application of either soil nutrients or pesticides) that may reduce levels of cheatgrass in the field.

Attention should also be paid to the identification of species that are likely to be troublesome in the future. For example, the ecological characteristics of cheatgrass that render it undesirable are also present in medusahead (*Taeniatherum caput-medusae*), which is also becoming a major pest species in regions of the West. Identification of such species well in advance of them becoming widespread could ease considerably the task of controlling them.

Recovery of Native Shrublands

Another research priority should focus on speeding the recovery of shrublands. Although ordinarily succession (in the absence of flammable exotics) should eventually achieve that goal, many ecosystems have been so perturbed that their return to anything approximating a predisturbance state is unlikely (e.g., Allen 1988). Even under unrealistically optimal conditions of appropriate remaining plant species, plentiful precipitation, and cessation of disturbance, a full century may be required for a landscape to recover fully (S. Knick, B. Hoover, and J. Rotenberry, unpublished data).

Currently, much restoration research in rangeland habitats places emphasis on recovering potential plant production biomass, not native plant species composition. If a system remains too long in grassland, soil properties may be altered so that any shrub recovery becomes unlikely (Allen 1988). Thus, it may be necessary to begin restoration efforts with exotics, just to restore some shrub component to the system and to restore nitrogen balances in the soil. However, because of the floristic specificity of most of the characteristic shrub–steppe bird species (e.g., Wiens and Rotenberry 1981), plant compositional recovery is paramount, and any use of exotic shrub species must be considered on an interim basis and only when absolutely necessary.

Grazing Impacts

Whatever its negative impacts from an ecosystem perspective might be (e.g., Fleishner 1994), because of political and economic considerations grazing will likely remain a major use of western rangelands. Because direct impacts of grazing on birds in these systems has not been sufficiently documented (Saab et al. 1995), newly defined research to assess the effects of livestock on birds is important (see also Herkert and Knopf, this volume). Unfortunately, many previous studies that attempted to couple grazing with birds suffered either from poor (or no) experimental design (especially involving pseudoreplication), or were conducted over inappropriate spatial scales (i.e., too small an area; see Saab et al. 1995). New, properly designed and replicated experiments involving a variety of alternative grazing treatments (including no grazing at all) must be implemented across the spectrum of major shrub–steppe habitat types. Because of inherent time lags in the response of both vegetation and birds (e.g., Wiens and Rotenberry 1985), these experiments must be conducted over multiple years.

Mechanisms Associated with Fragmentation Effects on Birds

Although a thorough solution to conservation problems faced by shrub–steppe birds will involve, in essence, a reconstruction of the landscape, such a solution (assuming one can be found) will take decades to change habitats at a regional scale. In the interim, it will be important to understand the mechanisms that account for the negative impacts of fragmentation on shrubland obligate birds.

The effects of fragmentation have been most thoroughly studied in formerly forested landscapes (e.g., Walters, this volume; Faaborg et al., this volume). In these studies, a sharp structural contrast exists between remnant forest patches adjacent to anthropogenic habitat, primarily agriculture and urban/suburban development. This structural contrast and adjacency produces edge effects by facilitating the penetration across the boundary and into the interior of the patch of external influences, particularly brood parasites (cowbirds), predators, and changed microclimate. Both predators and parasites can depress bird population productivity and abundance within forest patches, and altered microclimates can alter plant community composition and structure, potentially creating unsuitable habitat.

Although shrub–steppe birds are also subject to predation and brood parasitism at rates that may produce population-level effects (e.g., Rich 1978; Rotenberry and Wiens 1989) and most species also show strong habitat affinities, it seems likely that none of the mechanisms responsible for producing fragmentation effects in forested landscapes vary as strongly across patch boundaries in shrub-steppe habitats. The structural difference between a sagebrush shrub steppe and an adjacent *Bromus*-dominated grassland is insufficient to produce edge-related variation in microclimate that persists for more than a few meters. Most predators of shrub-steppe birds and their nests are generalists (corvids, snakes, small mammals such as ground squirrels and chipmunks) that move easily throughout both habitat types. Likewise, cowbirds can be found in abundance in the interior of sagebrush shrublands kilometers from a grassland edge. In essence, the demographic and ecological problems that would beset shrub-steppe birds at the edges of shrubland patches (at least those processes that have been identified as important in other systems) are almost equally likely to occur in their interiors. Thus, identification of mechanisms that produce fragmentation sensitivity in shrub-steppe birds (Knick and Rotenberry 1995) remains a pressing problem for research.

Making Research Effective for Conservation

Implementation of the results of research identified above will, to a considerable degree, depend on the types of recommendations forthcoming. Most western shrublands and grasslands, disturbed or not, fall in the public domain, particularly under the aegis of the Bureau of Land Management and the Forest Service. The presence of relatively few land management entities makes the widespread implementation of any research recommendations that are ultimately adopted much more likely.

Perhaps the easiest management technique that western land managers could implement (ignoring political considerations) will be alterations in the timing and intensity of livestock grazing, as regulation of grazing is already a major focus of their work. Because of this potential, one of the most hopeful scenarios would be to find a livestock management mechanism for ecosystem restoration. This means, however, that research on livestock impacts (both negative and, potentially, positive) should be conducted at scales and in units appropriate to current livestock management practices. Exclosures and enclosures, for example, must be commensurate with the scale of, say, BLM livestock allotments in a region. Therefore, they most often must be large, on the order of hundreds of hectares. This is not trivial, for it is also important that enclosures and exclosures be independently replicated, so as to be consistent with the design of a reliable scientific experiment. These experiments should also be implemented using measurements familiar to managers, for example, expressing stocking rates in an experiment by AUMs (animal unit months) as well as the equivalent herbivore biomass per meter squared per day. To the extent that these large-scale experiments involve manipulation of public use of grazing allotments, strict enforcement of stocking rates, however measured, is absolutely necessary.

One important research question will be the extent to which reestablishing native grasses has a potentially positive impact on livestock. The economic reality is that exotic annuals, such as cheatgrass, may also be valuable. For example, the presence of cheatgrass may extend the grazing season. During the early part of the year, when native bunchgrasses are not yet available, cheatgrass is already green and palatable to livestock. Pastoralists can use this species early, resting bunchgrass areas for use later in the season. This raises the further question of whether commercial livestock grazing is truly compatible with the return of native habitats. As noted above, the prehistoric landscape did not support extensive herds of large native herbivores; can we restore a modern landscape to those same conditions and still support extensive herds of large exotic herbivores?

Conservation research relating to restoration (and various sorts of livestock impacts as well) should also be conducted in the context of adaptive management (e.g., Walters 1986). In adaptive management, any management decision (whether it involves a specified activity or the absence of any activity) is treated as an experiment in progress. Important system variables are identified and monitored (i.e., data are collected on a continuing basis). If a variable attains a particular value (specified previously), then a new management decision is implemented and another experiment is now in progress. The value of this approach is that it involves the land manager directly in the research process, using variables and techniques with which he or she is familiar.

A final point concerns what has come to be called "technology transfer." This basically involves the flow of information, specifically management recommendations that arise from research activities, from the basic and applied scientists who perform the research to the land managers responsible for implementing those recommendations. Because the careers of academic and, increasingly, government scientists are advanced by publishing their results and conclusions in the primary literature, much of this information may not get transferred to implementation in the field in a timely fashion (see Huenneke 1994 for a good description of this problem). One way to increase the rate of transfer is the development of symposia specific to particular research problems, followed by peer-reviewed publication of proceedings. In addition to this volume, two excellent models that relate specifically to issues raised here are Monsen and Kitchen (1994), dealing with ecology and management of "annual rangelands" (primarily rangelands dominated by cheatgrass), and Martin and Finch (1995), which focuses on management of Neotropical migrant birds. It is through forums such as these that reliable, scientifically based recommendations can be most effectively communicated to the managers who need them the most. Finally, it is imperative that land managers be given, and take advantage of, the opportunity to read the appropriate scientific literature before implementing their management activities. Without their interest and willing participation in the process, all conservation research remains an intellectual exercise.

Acknowledgments

I thank John Wiens and Steve Knick for companionship and discussion during long hours in the field, at the computer, and over beer; my interaction with them has helped form the basis of the research and opinions reported here. Both contributed significant reviews of a previous version of this paper. I also thank the staff of the Snake River Field Station of the Biological Resources Division of the U.S. Geological Survey, Boise, Idaho, particularly Mike Kochert and Karen Steenhof. Much of the research reported here has been supported by the National Science Foundation, the National Biological Service, the Bureau of Land Management, the U.S. Global Climate Change Program, and the University of California, Riverside.

CHAPTER 18

Research Needs for Grassland Bird Conservation

James R. Herkert and Fritz L. Knopf

Conservation Issues and Previous Research

An increase in the level of concern for grassland birds has been prompted by recent indications that many North American grassland bird species are undergoing widespread population declines (Knopf 1994). In fact, declines among grassland birds have been exceptionally consistent among geographic areas (Knopf 1994), in contrast to the more variable declines exhibited by many other birds such as Neotropical migrant forest species (James, Wiedenfeld, and McCulloch 1992). While there is near universal agreement on the fact that populations of grassland birds are declining, much remains to be learned regarding the mechanisms causing these declines.

The purpose of this chapter is not to provide a comprehensive review of the problems faced by grassland birds, such as those provided in recent papers by Askins (1993); Bock et al. (1993); Bollinger and Gavin (1992); Saab et al. (1995); Knopf (1994, 1996b); Herkert, Sample, and Warner (1996), and Johnson (1996). Instead, we will emphasize what we believe to be some of the most important conservation problems of grassland birds, their possible management needs, and some promising research directions.

Issues on the Breeding Grounds

There is compelling evidence that problems on the North American breeding grounds have influenced (and probably still are influencing) population declines for some grassland birds. Loss of native grassland habitats has been severe in North America (Samson and Knopf 1994), and these losses have undoubtedly had a profound impact on grassland bird populations.

Declines in the acreage of agricultural grasslands have also strongly influenced recent grassland bird declines (Rodenhouse et al. 1995; Warner 1994; Herkert, Sample, and Warner 1996). Additionally, disturbances in agricultural lands, such as tilling and mowing, are likely contributing to the decline of some grassland bird populations (Bollinger, Bollinger, and Gavin 1990; Herkert 1997; Knopf and Rupert in press; Newton, this volume).

There are few data by which to assess how grassland bird populations are faring within remaining grasslands. For forest birds, evidence of population declines within specific woodlots (Briggs and Criswell 1978; Robbins 1979) sparked conservation concern that, for some species, has been only weakly supported by landscape-scale surveys such as the Breeding Bird Survey (James, McCulloch, and Wiedenfeld 1996). In contrast, concern for grassland birds has been fueled primarily by declines detected by landscape-scale surveys such as the Breeding Bird Survey (Sauer et al. 1996) and the Christmas Bird Count (Butcher and Lowe 1990). Most long-term studies of grassland birds reporting population declines within specific areas have attributed declines to successional changes within particular study areas (Sample 1989; Bernstein Baker and Wilmot 1990). Very few studies have looked at population trends of grassland birds within specific study areas not undergoing successional change. The few that have, however, suggest that grassland bird numbers within these areas may be stable (Sample 1989; Zimmerman 1993; Herkert and Glass, unpublished data). It is important to determine if grassland bird population declines are due to loss of habitat, habitat degradation (usually due to woody encroachment), or declines of birds within remaining habitat patches not undergoing succession, because conservation actions attempting to reverse declines would differ under these three scenarios.

Predation rates on grassland birds are generally high (Martin 1993). Several studies have suggested that reproductive success for grassland birds was so low that it was unlikely that populations were maintaining themselves, and that these populations seemed dependent on immigration from other areas for population maintenance (Wray, Strait, and Whitmore 1982; Johnson and Temple 1986). However, nest predation rates are geographically variable (R.J. Greenwood et al. 1995), and the local and landscape features that influence grassland bird nest predation rates remain poorly understood. Some studies have found predation rates to be significantly related to local habitat features such as degree of grassland fragmentation (Johnson and Temple 1990; Burger, Burger, and Faaborg 1994) whereas others have not (Vickery, Hunter, and Wells 1992a; Davis 1994). At present, the degree to which large-scale, landscape parameters influence nest predation rates on grassland birds is also poorly known. Few studies have looked at landscape features surrounding grasslands as a possible source of variability in grassland bird nesting studies. However, studies do indicate that grassland bird abundance is sensitive to landscape context of suitable habitats (Herkert 1994b; Vickery, Hunter, and Melvin 1994), even if causative factors are not yet clear.

There is some indication that landscape features do significantly influence grass-land bird productivity. For example, R.J. Greenwood et al. (1987, 1995) have shown that waterfowl nesting in grassland areas have higher rates of nest success in areas that have a high percentage of pasture and a low percentage of cropland in the landscape.

Issues on the Wintering Grounds

Limitations in winter resources or habitats have also been proposed as reasons for population declines of some grassland species (Fretwell 1986; Lymn and Temple 1991). Fretwell (1972, 1986) proposed that populations of Dickcissels (*Spiza americana*) were limited by the supply of seeds on their tropical winter areas. In the southwestern United States, local winter grassland bird populations also may be limited by the abundance and availability of seeds (Dunning and Brown 1982), especially in years of low seed production (Pulliam and Dunning 1987).

Recent efforts to control Dickcissels on their South American wintering grounds, where they are perceived as crop pests, may be contributing to recent population declines for this species (Basili and Temple 1995). Additionally, in the pampas region of Argentina, exposure to acutely toxic chemicals used by local farmers to control grasshoppers is likely having a major effect on Swainson's Hawk (*Buteo swainsoni*) populations that concentrate in large numbers in this region during the austral summer (Line 1996). In the southeastern United States, human conversions of grasslands to cropland and pine plantations and woody encroachment have led to a decrease in the amount of grassland habitats available for many wintering grassland birds (Hunter 1990; Lymn and Temple 1991). This loss of grassland habitat could be contributing to population declines since many declining grassland birds winter primarily or partially in the southeastern Gulf Coast region (Herkert 1994b). Similar changes are also taking place in major South American wintering areas used by grassland birds such as Upland Sand-pipers (*Bartramia longicauda*) and Bobolinks (*Dolichonyx oryzivorus*) (Bucher and Nores 1988; White 1988). Although we know that grasslands used by wintering birds are declining, so little is known about where, and in many cases what types of habitats particular species are using, that it is hard to determine what impacts these changes may be having on grassland bird populations (but see Knopf and Rupert 1995).

Features of Research That Improved Conservation

Similar to the case for western shrublands (Rotenberry, this volume), awareness of the need for grassland bird conservation has been relatively recent. As a result, research focused primarily on grassland bird conservation has been fairly limited to date. For example, only recently have conservation issues such as the effects of grassland fragmentation on breeding birds received research attention (Herkert 1994b; Vickery, Hunter, and Melvin 1994).

Grasslands are dynamic systems that are dependent on drought and periodic disturbances such as fire and grazing for habitat maintenance (Knopf and Samson 1997). Understanding the impacts of these factors (drought, fire, grazing) on grassland birds is essential to grassland bird conservation efforts. Fire, grazing, and precipitation cycles can all have substantial impacts on some grassland bird species in some areas (Zimmerman 1988; George et al. 1992; Herkert 1994c). Most research on the effects of these factors on grassland birds, however, has looked at only a single factor, such as the effects of fire or grazing. Studies that look at combinations of management practices, such as fire and grazing (Zimmerman 1997), or the interaction of factors, such as management (e.g., fire or grazing) and fragment size (Johnson and Temple 1990; Herkert 1994a) or management and precipitation patterns (Zimmerman 1992), are more useful to managers than work focusing on single factors because they show how species responses to management may differ in different ecological settings. For example, in Kansas, Zimmerman (1997) found that a combination of burning and grazing resulted in lower grassland bird abundance, poorer nest survival, and fewer young in successful nests than did grazing in unburned prairie. And in Illinois prairies, habitat area was found to have a stronger influence on grassland bird communities than did prescribed burning (Herkert 1994a). Johnson and Temple's (1986, 1990) study of the effects of prairie size, burn status, and proximity to woody edge on grassland bird nest success is the type of multifactorial study needed to provide guidance for effective grassland bird conservation. Johnson and Temple (1990) showed that grassland bird nest success was greatest for nests located in recently burned sections of large prairies that were also far from a woody edge. This type of research is very useful to managers, but there are, at present, too few of these kinds of studies. In addition, too little is known about how variable these patterns are for results of individual studies to be of widespread conservation use. For example, there are indications that results such as those reported by Johnson and Temple (1990) cannot be universally applied to all grasslands. Some studies have reported results consistent with aspects of Johnson and Temple's data (Burger, Burger, and Faaborg 1994); however, other studies have found partially conflicting results. For example: in Maine, Vickery, Hunter, and Wells (1992a) did not find nest success to be significantly influenced by proximity to woody edges; Davis' (1994) study of prairie fragments in Saskatchewan did not find nest predation to be significantly different among small and large prairie fragments; and Hendricks and Reinking (1994) did not find nesting success to be consistently higher in recently burned prairies in Oklahoma. This variability indicates that more research is needed in this important area.

Research Needed to Further Conservation

Demographic Studies

Additional data on the demographics of grassland birds is probably the most important research need on the breeding grounds. Until more is known regarding season-long fecundity of females, adult and juvenile survival rates, and adult and juvenile dispersal patterns, it will be difficult to predict the effects that various rates of nest predation and nest parasitism will have on grassland bird populations. These types of data are difficult to obtain, but they are nonetheless important. Additional studies of nesting success, site fidelity, and adult survivorship, such as those of Gavin and Bollinger (1988), Bollinger and Gavin (1989), Wittenberger (1978), and Bedard and Lapointe (1984), will be needed to enable future researchers to begin modeling the effects of different management options and conservation actions on grassland bird populations.

Efforts to estimate seasonal fecundity (Pease and Grzybowski 1995) for grassland birds are also needed in order to more fully understand the influence of high nest predation rates in some areas. One way that grassland birds may be able to offset high nest predation rates is by persistent renesting attempts, but few studies of marked populations of birds have been conducted to produce the necessary data. Renesting frequency and number of broods are key life-history traits influencing annual fecundity that have been poorly studied (Martin 1995). Although data on grassland bird nest success are needed, these data will not allow researchers or managers to differentiate population sources from population sinks unless accompanied by information on adult and juvenile survival and dispersal rates. It is critical to achieve a better understanding of grassland bird source-sink dynamics. In particular there is a need to identify where grassland bird sources are and describe their characteristics (at both a local and a landscape scale). Without more detailed information on grassland bird source-sink dynamics it will be difficult to identify or target desirable habitat conditions for grassland bird conservation efforts.

The use of behavioral indices (Vickery, Hunter, and Wells 1992c) has the potential to provide cost-effective data on grassland bird reproductive success. However, before behavioral studies are widely applied, it would be very useful to conduct studies that compare these behavioral indices with estimates of nesting success using conventional techniques. Behavioral indices can also be used to examine other important conservation issues such as the relationship between pairing success and fragment size (Gibbs and Faaborg 1990) or to study the impacts that edges have on grassland bird nest success (Delisle 1995). Researchers presently censusing grassland areas without collecting nesting success data should consider using behavioral observations as a way to supplement their census data.

Some other fundamental questions regarding grassland bird productivity also remain little studied. For example, how does burning or grazing influence grassland bird nest success? Results of existing studies appear contradictory. Several studies have shown that nest success for a variety of ground-nesting grassland birds is generally higher in recently burned grasslands than it is in unburned grasslands (Kirsch and Kruse 1972; Fritzell 1975; Johnson and Temple 1986, 1990). Toland (1986), however, found that nesting success for Northern Harriers (*Circus cyaneus*) in Missouri prairies was higher in unburned areas than in recently burned areas; and Hendricks and Reinking (1994) found nest success in burned tallgrass prairie areas to be variable from year to year and not consistently higher than unburned areas. At present, there are insufficient data to determine what impacts commonly employed grassland management activities have on bird populations or to make robust predictions regarding optimal management strategies under varying conditions and in varying landscapes. New studies utilizing an adaptive management approach (Walters 1986), in which grassland bird responses to different frequencies and combinations of management practices are monitored and used to guide future research and management activities, are needed.

The fate of fledglings once they leave the nest also needs to be determined. Most nesting studies of birds monitor nests until the young fledge and then stop. There is very little information regarding habitat use or survival for grassland birds after fledging but before their southern migration. For some species this may be an important portion of the annual life cycle, possibly constituting a period of a few months. However, work on Mountain Plovers (*Charadrius montanus*) at the Pawnee National Grasslands (Knopf and Rupert 1996) and Prairie Falcons (*Falco mexicanus*) in Idaho (McFadzen and Marzluff 1996) has shown that predation risk for grassland birds continues, and for Prairie Falcons may even increase, once chicks leave the nest. Recent work on forest birds also has shown that predation rates on recently fledged young may be very high (Anders et al. 1997).

Research in grassland areas that examines whether population size is linked with reproductive success the previous year (Sherry and Holmes 1992) also would enhance our understanding of grassland bird population dynamics and the factors responsible for grassland bird declines. Although intuitive, this relationship has rarely been documented (Robinson 1993) and has yet to be documented in grassland birds. Field experiments designed to test whether purported causes of breeding bird population declines actually affect the size of the breeding population are also needed (James and McCulloch 1995).

Landscape-Scale Studies

Research in forest systems has shown that the landscape context of a particular site exerts a strong influence on bird populations, influencing both species sensitivity to habitat fragmentation and nest predation and parasitism rates (Freemark and Collins 1992; Robinson et al. 1995). To date most studies of grassland birds have focused primarily on features within sites to study area-sensitivity (Herkert 1994b; Vickery, Hunter, and Melvin 1994) or nesting success (Johnson and Temple 1990; Burger, Burger, and Faaborg 1994). It is important to examine these aspects of breeding ecology from a broader landscape perspective. For example, there is little research data to guide conservation planners in deciding how large grassland conservation areas should be, how they should be spatially arranged, and into what types of landscapes they should be placed to have the greatest impact on grassland bird populations.

Understanding the influence of landscape context on grassland birds is an important step in making management decisions and creating conservation strategies. Frequently, there is not enough variation in landscapes at the scale typically studied by a single investigator. In order to capture a wide range of variability among landscapes and improve our understanding of the factors influencing bird distributions and nest success across these broader landscapes, it may be necessary for researchers working in different landscapes to collaborate. Studies examining grassland bird community structure and nesting success modeled after the forest work of Robinson et al. (1995) would help answer some of these basic questions and would enhance grassland bird conservation efforts.

Winter Ecology

Little is known about the winter ecology of grassland birds, and there is a need for more detailed research on nearly all aspects of this topic. Additional studies that address the influence of grassland management (i.e., grazing intensity) on grassland birds across broad geographical areas, such as the work of Grzybowski (1982), will be useful to future grassland bird conservation efforts. For some species, there is a need to identify where populations are wintering and to identify the kind of habitats used in winter. For other species, analyses of habitat trends and disturbance patterns coupled with research regarding how these changes influence wintering birds are needed. Recent indications that high concentrations of grassland birds may be particularly vulnerable to catastrophic events on their wintering areas (Basili and Temple 1995; Line 1996) highlight the need for additional studies on this subject. Evidence that major land-use changes are occurring on both the North and South American wintering grounds of many other grassland birds (Knopf and Rupert 1995) also underscores the need for additional work on winter ecology.

Additional work is needed on the effects of food-limitation on winter grassland bird populations. Dunning and Brown (1982) and Pulliam and Dunning (1987) have shown that local winter sparrow populations in southern Arizona can be food-limited, especially in infrequent years of low seed production. Additional research is needed to determine the extent of food-limitation in other areas and also regarding the scale at which food-limitation occurs in arid systems.

Long-Term Response to Management

Most studies of the effects of grassland management on breeding birds have been short term (usually two to three years, but see Johnson 1997). There are virtually no data to allow managers to predict how these short-term responses might interact to create long-term patterns of response to different management regimes. Longer-term studies of effects of grassland management are needed. These studies should also take an adaptive management approach and monitor bird response to different management frequencies, intensities, and combinations of management practices and then refine management decisions and research questions in response to research results from previous experimental treatments.

Response to Management under Varying Environmental Conditions

Precipitation patterns influence grassland bird abundance (Igl and Johnson 1996) and response to management (Zimmerman 1992), with individual species responses varying in relation to fluctuations in yearly precipitation. Factors such as the intensity and completeness of the burn and availability of adjacent refuge habitats also may influence grassland bird response to burning (Ryan 1986), but these have received little attention to date. Additionally, timing of the prescribed fire (spring vs. fall) may affect grassland bird response to burning. Higgins (1986) suggested that rates of nest success may be greater in grassland areas burned in the fall than in areas burned in the spring. Grassland bird response to grazing also is variable and influenced by many factors including stocking density and duration, soil types, soil moisture, plant species composition, and weather (Kantrud and Kologiski 1982). These results suggest that management is only one of a variety of factors that influence grassland bird distribution and abundance patterns within grasslands. As a result, some sites may be poorly suited for grassland bird conservation efforts regardless of the management attention focused on them due to their soils, small size, or proximity to woody edges.

Grazing

There is a need for additional grazing studies in the western United States (Rotenberry, this volume). Endemic grassland birds evolved in landscapes that experienced differential grazing pressures (Knopf 1996b). In the western United States the feasibility of alternative grazing strategies that allow for more variation in grazing intensities within larger areas than presently used needs to be evaluated. Larger allotments and different grazing rotation schemes would help to restore the natural variability in grazing that has been lost with the current approach of smaller, fenced parcels (Knopf 1996a).

Grassland Restoration

More research aimed at evaluating different techniques for grassland creation, restoration, and enhancement as it benefits grassland birds is needed (Vickery et al., in press; Rotenberry, this volume). In the past, development of grassland restoration techniques has focused primarily on plants, with little concern for birds or other wildlife. As more conservation attention is focused on creating habitat for declining grassland birds, a better understanding of the effectiveness of different methods of restoring grasslands will be necessary.

Making Research Effective for Conservation

There is evidence that declines in the availability of grassland habitat, reproductive failure (due primarily to high rates of nest predation and occasionally nest parasitism), and problems on the North and South American wintering grounds are all influencing grassland bird population declines. Efforts to identify which factors are most important in limiting grassland bird populations are hindered by the limited data on several aspects of their ecology. Until more information identifying specific problem areas for particular species becomes available, grassland bird research and conservation efforts should address all three of the major potential limiting factors: breeding-season habitat availability, reproductive failure, and winter ecology.

The demographic work outlined here will be essential for determining the extent to which reproductive failure may be influencing grassland bird population declines. Once there is a better understanding of the factors that influence grassland bird nest predation rates, and more is learned regarding adult and juvenile survival rates and dispersal, population sources and sinks can be identified. Once these population sources are identified, grassland bird conservation activities can be properly focused on areas where productivity is highest. Further research, including some long-term studies, is needed on the effects of common grassland management activities so managers can make more informed decisions regarding the effects of various management practices (such as the interval between successive treatments) on the grassland areas they are managing. This research should use an adaptive management design (Walters 1986), in which species response to different management frequencies, intensities, and combinations are monitored and used to refine future management activities.

Further research on effects of landscape composition and structure on grassland birds will provide much needed direction in deciding where and in what configurations grassland conservation areas should be located to achieve the greatest population benefits. Further research regarding the benefits of different grassland restoration techniques will lend guidance in making these new areas attractive and productive for grassland birds.

More research is needed on the winter ecology of most grassland birds. Until more detailed information regarding the winter ecology of most species becomes available, the extent to which winter resources may limit grassland bird populations cannot be properly evaluated. If population problems for some grassland bird species are the result of winter season phenomena, then conservation activities focused on the breeding grounds may be ineffective in stemming pop-ulation declines. Although there is reason to believe that problems associated with the winter grounds may be affecting some grassland bird species, in some cases too little is known to determine if these problems are limiting populations.

Acknowledgments

We are grateful to Carl Bock, Richard Knight, Peter Vickery, Janet Ruth, John Marzluff, and Rex Sallabanks, who made many helpful comments on an earlier draft of this manuscript.

CHAPTER 19

Urban Environments: Influences on Avifauna and Challenges for the Avian Conservationist

John M. Marzluff, Frederick R. Gehlbach, and

David A. Manuwal

Conservation Issues and Previous Research

The world's population has exceeded 6 billion people and shows no sign of slowing (Horiuchi 1992; Fleischer, this volume). Our burgeoning population, increasingly industrial society, and associated natural resource use are the fundamental reasons that much of the natural world in general, and avian biodiversity specifically, is in crisis (Mangel et al. 1996; Boersma and Parrish, this volume). Perhaps nowhere is the influence of population growth on bird life so obvious as it is when traveling from a large city to the wildlands. Such a trip in North America and much of Europe produces a transition from flocks of exotic Rock Doves (*Columba livia*), European Starlings (*Sturnus vulgaris*), House Sparrows (*Passer domesticus*), and corvids to less abundant, but more diverse native avifaunas. Despite the obvious impact of urbanization on birds, this phenomenon has attracted relatively little research interest (we conservation biologists seem too focused on getting to the wildlands rather than on the changes in avifaunal diversity encountered along the way). Here we review some impacts of urbanization on birds and suggest important research topics necessary to improve understanding about how our selection and use of living space affects avian diversity.

Urban, suburban, and *rural* have different meanings in different cultures and are nodes in an environmental continuum that can be smooth and/or patchy depending on the developmental complexity of private and public land ownerships.

Probably there are no operational definitions for these terms, but study sites can be described in a uniform manner that will aid in making comparisons. Thus, Gehlbach (1996) asked authors of urban bird studies to provide, at minimum, human population density and percentage of habitat (green) space. Kinds and amounts of industrial, commercial, domestic, and rural (e.g., agricultural, recreational) structures are also useful measures. The rural end of the urbanization continuum has the fewest humans and least human influence and is the most natural; while suburbia has more human residents, some commerce, little industry, and, in one studied example, 26 percent green space (Gehlbach 1996). The continuum's opposite or urban end is most cultural with the most humans, commerce, and industry and least green space. An intermediate point on the continuum, in which humans live in rural areas with some connection to a nearby urban area, is referred to as an "exurban" area (Davis, Nelson, and Dueker 1994). The exurban zone, usually surrounding an urban area with charismatic hinterlands (e.g., Portland, Oregon, USA; Davis, Nelson, and Dueker 1994), may have continuous residential development, but houses are on considerable land with a majority of green space (e.g., 63 percent; Gehlbach 1994, 1996).

Humans began dominating their habitats to the detriment of birds many centuries ago. Native forest habitat in England was noticeably reduced in the eleventh century and nearly entirely removed by the seventeenth century (Wilcove, McClellan, and Dobson 1986). Clearing and fragmenting native forests in the eastern United States began three hundred years ago and has resulted in the decimation of 80 percent or more of this habitat in some areas (Whitcomb et al. 1981). Such habitat destruction occurred as urban areas sprawled to contain more people and rural areas were cultivated for resources.

Many urban areas have increased in population size exponentially during the second half of the twentieth century as the world's population more than doubled from 2.5 to over 6 billion people and society became increasingly industrialized. The recent boom in communication technology is allowing us to live farther from our workplaces. This has resulted in emigration from urban areas and conversion of nearby rural areas to exurban areas (Riebsame, Gosnell, and Theobald 1996). The American Rocky Mountain region provides an illustration (Gersh 1996; Riebsame, Gosnell, and Theobald 1996; Theobald, Gosnell, and Riebsame 1996). There, permanent European-like settlements began to replace temporary Native American camps in the mid 1880s. Settlement for the next century and a half was fueled by booms and busts in timber, mining, and grazing industries. This produced a few large cities (Denver, Colorado; Albuquerque, New Mexico; Salt Lake City, Utah; and Boise, Idaho) and an extensive rural area. Large blocks of wild land remained untouched. Growth in the region accelerated (2–4 percent per year) in the 1980s and 1990s as Americans emigrated from large northeastern and West Coast cities to the Intermountain West and Southwest (Raish, Yong, and Marzluff 1997). This "second conquest" of the West (Gersh

1996) has been characterized by faster growth in suburban and exurban areas than in large cities.

Most notable is growth in resort areas, sprawling "bedroom communities" of urban areas, and subdivisions of formerly extensive ranches (10–50-hectare "ranchettes"). Growth has been most rampant (20–65 percent from 1990 to 1995; U.S. Census Bureau) in counties with significant wildlands (Rudzitis and Johansen 1989; Gersh 1996). Changes in these areas have been dramatic and foreshadow changes to be expected throughout much of the world where development near wildlands is coveted for its amenity value and possible for an increasing number of people because of telecommuting. As an example, consider the East River Valley of Colorado. Theobald, Gosnell, and Riebsame (1996) calculated that 20 percent of this 35,000-hectare peninsula of private land surrounded by public wild land is now divided into ranchettes 18 hectares or smaller. During the last thirty years, 100 km of road have been added to the area, primarily to serve the 171 new ranchettes and 1,371 new homes in high-density subdivisions. Thus, the current process of urbanization suggests that birds will be affected in two distinct settings: (1) large urban areas and their associated suburbs; and (2) exurban areas and their interface with rural areas and wildlands.

Urbanization affects birds directly and indirectly (Marzluff 1997). It directly changes ecosystem processes, habitat, and food supply. Indirectly, it affects birds' predators, competitors, and disease organisms. These effects lead to significant changes in the population biology of birds in urban areas with resulting effects on bird communities. Species able to exploit urban environments have dense and stable populations because ameliorated climate, abundant food and water, reduced predators, and increased nest sites allow for lengthened breeding seasons, increased survival, and increased productivity (Gehlbach 1994, 1996). Many native species do not attain dense and stable populations in urban areas because of the scarcity of natural habitat and intolerance of human cultural activity. Therefore, urban bird communities are simple in structure compared to nearby rural communities that more closely approximate the richer avifaunas in habitats with minimal cultural influence. Urbanized habitats typically support larger (measured by biomass) and sometimes richer (more species) but always less evenly distributed avian communities that are dominated by a few, very abundant species (Pitelka 1942; Emlen 1974; DeGraff and Wentworth 1981; Rosenberg, Terrill, and Rosenberg 1987; Mills, Dunning, and Bates 1989). Urbanization favors some species but selects against others, so that the composition of urban avian communities differs from those found in native environments (Beissinger and Osborne 1982; Rosenberg, Terrill, and Rosenberg 1987; Mills, Dunning, and Bates 1989; Blair 1996; Bock et al., in press). Urban avifaunas are typified by invaders, often exotics, that benefit from human cultural resources (Knight 1990). For example, bird species richness and diversity decreased with increased urbanization in the Seattle, Washington, area in both spring and winter (figure 19.1; Penland

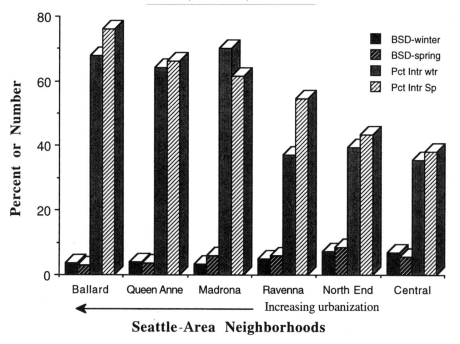

Figure 19.1. Changes in bird diversity (BSD) and percent exotic bird species (Pct Intr) over a gradient of urbanization in the Seattle, Washington, area during winter and spring. Diversity index is the reciprocal of Simpson's diversity index. Data from Penland (1984).

1984). Exotics were nearly 80 percent of the avifauna in the most urbanized areas. Native birds, such as the Black-capped Chickadee (*Parus atricapillus*) and Rufous-sided Towhee (*Pipilo erythrophthalmus*), were generally associated with natural vegetation features, whereas exotics such as Rock Doves, House Sparrows, and European Starlings were associated with the human-modified environment.

Changes in Ecosystem Processes

The effects likely to have the greatest impact on birds are changes in the basic functioning of their ecosystems, namely reduction in natural processes such as fire, altered nutrient cycling, and disrupted water cycling. These changes are accentuated by urbanization and can lead to long-term degradation of habitats.

Concern for human dwellings has prompted intense efforts to suppress fire near urban areas, which disrupts natural fire regimes with which many birds have evolved (Finch et al. 1997; Moir et al. 1997) and also disrupts nutrient cycling. The most important effect of urbanization on nutrient cycling is probably a lengthening of cycles. The lack of fire causes living and dead biomass to accumulate much faster than it degrades and recycles (Covington and Moore 1994). Prescriptions to reduce the future fire threat, such as thinning forests (Edminster and Olsen 1996), are then used rather than prescribed burning. As a result, although the urban forest accumulates energy and nutrients, it typically exports rather than recycles them. This nutrient and energy loss may steadily degrade forest growth, with long-lasting effects on birds.

Water is an important, and often limiting, resource in arid forests and grasslands. Patterns of runoff are affected by urbanization as native soils are replaced by impermeable concrete and surrounding forest substrates are compacted by vehicular and foot traffic. Urban centers have tremendous water requirements, and the resultant use of water affects the distribution and cycling of water in surrounding areas (Knight 1990). Water tables are lowered as aquifers are used at greater than replacement rates (Thorn, McAda, and Kernodle 1993). Springs and seeps important to wildlife may then dry up, causing reductions or redistributions of nonurban birds. Urban water supplies expand as pools from reservoir size to backyard birdbaths are created, benefiting urban birds.

Urban areas, especially large cities, produce heat, reflect and absorb solar radiation, and gradually release absorbed solar energy throughout the night. This results in higher average temperatures and less variable daily and annual temperatures in urbanized areas relative to rural and wild areas. Rainfall may also be increased in mature (30-year-old or more) subdivisions relative to rural areas or recent subdivisions (Gehlbach 1994). This ameliorating effect of urbanization may reduce prey fluctuation, increase prey biomass, and reduce bird energy requirements, which enables urban birds to attain higher densities, breed for longer periods of the year, produce more young, and have higher annual survival than birds living outside of urban areas (Balda and Bateman 1972; Tatner 1982; Plesnik 1990; Gehlbach 1996; Rosenfield et al. 1996). Also, the urban heat island effect (Landsberg 1981) may promote certain colormorphs such as the rufous Eastern Screech Owl (*Otus asio*; Gehlbach 1988, 1994).

Changes in Habitat

Vegetation in moderately urban environments is typically more fragmented, includes less coverage at mid and upper levels, and has more coverage at ground level than vegetation in natural environments (Beissinger and Osborne 1982; Blair 1996). Not only are patches of vegetation isolated in urban environments, they also rarely include the full complement of vertical strata found in natural forests (Beissinger and Osborne 1982; Knight 1990). Native plant species are often removed from urban environments and replaced by exotic ornamentals (Beissinger and Osborne 1982; Rosenberg, Terrill, and Rosenberg 1987). Even moderately urban environments contain few standing or downed dead trees that provide nest and foraging sites for cavity nesters and timber drillers. Extreme urbanization leads to decreases in vegetation at all levels as human-made structures replace vegetation (Blair 1996). In heavily urbanized areas, native habitat may be modified well beyond the city boundaries (Kamada and Nakagoshi 1993).

Direct habitat modification by urbanization likely: (1) benefits ground-gleaning and probing birds that are tolerant of human activity (e.g., American Robin [*Turdus migratorius*] and American Crow [*Corvus brachyrhynchos*]); (2) benefits species that nest in human-made structures or ornamental vegetation (e.g., some raptors, Rock Dove, and Barn Swallow [*Hirundo rustica*]); (3) reduces shrub and canopy nesters and foragers (many warblers, vireos, tanagers, and grosbeaks); and (4) reduces burn specialists, cavity nesters, and bark drillers (some flycatchers, tits, and woodpeckers) (Bird, Varland, and Negro 1996; Marzluff 1997).

Changes in Food

Urban centers provide food directly to birds at feeders and indirectly to them at areas of waste treatment, collection, and transfer. Seed eaters and nectarivores may obtain substantial portions of their daily energy intake from feeders (Brittingham and Temple 1992), which may increase their survival (Brittingham and Temple 1988). Scavenging omnivores (gulls, corvids, blackbirds, and European Starlings) benefit from spilled waste (Robbins, Bystrak, and Geissler 1986; Marzluff, Boone, and Cox 1994).

Food resources are also affected indirectly by the changes in vegetation and amelioration of climate discussed above. Exotic plants have fewer insects than native plants, but urban lawns provide rich and consistent feeding grounds (Rosenberg, Terrill, and Rosenberg 1987). These changes favor ground foragers and granivores while selecting against shrub and mid-canopy foliage gleaners. Certain moths and beetles may fluctuate in density less in suburban than nearby rural areas, thereby allowing insectivores that tolerate human activity to attain denser and more productive populations in suburbia (Gehlbach 1994, 1996).

Raptors may change their diet in urban relative to rural areas, reflecting differences in prey availability. Urban Tawny Owls (*Strix aluco*) throughout Europe (Galeotti, Morimando, and Violani 1991) and eastern Screech Owls in Texas (Gehlbach 1994) switched from the typical rural diet of small mammals to avian prey regardless of season. Foods and their usage surveyed concurrently in suburban and rural Texas indicated that avian prey was denser overall and the common permanent resident species were more continuously available in suburbia. Eastern Screech Owls fed opportunistically despite prey population cycles and had no long-term influence on songbird populations or composition of the suburban avian community (Gehlbach 1994).

Changes in Predators

Urbanization causes major changes in predator communities. Introduced predators (cats, dogs) are more abundant in urban areas than native forest, and they may have substantial effects on the avifauna (Churcher and Lawton 1987; Soulé et al. 1988). Other avian predators (e.g., small owls, falcons, and accipiters) may attain high densities in urban areas and concentrate their activities at feeders where prey are abundant (Rosenfield et al. 1996). Large predators are usually eliminated from areas of human habitation, which results in competitive release of, and hence greater effects from, smaller predators ("mesopredator release"; Soulé et al. 1988). In suburban central Texas, for example, Eastern Screech Owl biomass exceeds that of other larger owls, hawks, and corvids by 1.2 times/km², whereas at a nearby rural site these predators sustain 10 times more biomass/km² than screech owls (Gehlbach 1994). Recent increases in coyotes (*Canis latrans*) and mountain lions (*Felis concolor*) in and around urban areas provide another example. These predators may benefit many songbirds by reducing mammalian nest predators, especially cats and foxes (Soulé et al. 1988; Quinn 1992). Humans are predators, as well, that can seriously deplete local songbirds, even if only for "sport." An example of this occurred in Flagstaff, Arizona, when a teenager shooting Pinyon Jays (*Gymnorhinus cyanocephalus*) one breeding season was the major source of nest failure and mortality experienced by the flock that year (Marzluff and Balda 1992).

Urbanization may reduce nesting productivity by subsidizing predators that destroy bird nests (Tomialojc 1979; Soulé et al. 1988, Engels and Sexton 1994). Some nest predators are more abundant in urban areas than native habitats, and their numbers increase with increasing urbanization (figure 19.2; Wilcove 1985; Marzluff, Boone, and Cox 1994). Moreover, the greatest densities of nest predators in rural areas are found at sites near urban areas (in figure 19.2 Mormon Lake is only 25 km from Flagstaff). Population viability may be compromised by increasing nest predators. For example, nearly half of all Pinyon Jay nests in Flagstaff, Arizona failed from predation in the 1980s. This was a significant increase over predation in the 1970s and was closely correlated with increasing raven populations in the city. Reduced jay productivity led to a decrease in population size and an increased reliance on immigration to sustain the Flagstaff population. Thus, the population functioned as a "sink" during the 1980s although it was probably a "source" population in the 1970s (Marzluff and Balda 1992).

The effect of predation may have been accentuated in Pinyon Jays, which nest colonially and build conspicuous nests. In other studies, predation on open (Tarvin and Smith 1995) and cavity (Gehlbach 1994) nests in cities was inversely correlated with proximity of human dwellings, suggesting that some nesting birds are protected in cities, which leads to high urban productivity (Tomialojc and Gehlbach 1988). Dispersed urbanization where residential and industrial areas are interspersed with natural habitats, in which harassment of nest predators is minimal and availability of supplemental food maximal, is the setting most likely to inflate nest predation on urban birds. This is the case in many small cities and exurban areas like the one (Flagstaff, Arizona) where the Pinyon Jay study was conducted.

Disease

Disease rarely regulates temperate bird populations, but urban populations are likely to be more susceptible to disease than populations in wildlands because

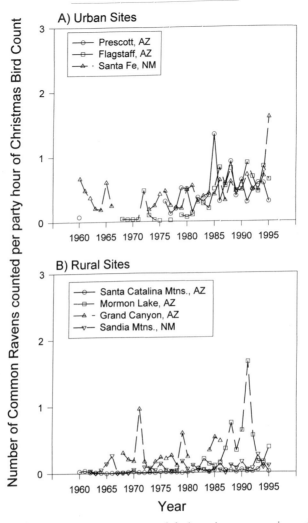

Figure 19.2. Number of Common Ravens counted during winter surveys in southwestern pon-derosa pine areas. Birds were counted each winter at the same location as part of the National Audubon Society's Christmas Bird Count. Counts were standardized by observer effort, which varies annually, by dividing the total number of birds counted by total observation time (party hours). All sites in Arizona and New Mexico that had relatively complete counts from 1960 to 1995 were used. Sites inside the city limits of urban centers are plotted in the top panel, and those outside city limits are plotted in the bottom panel. Redrawn from Marzluff (1997).

artificial feeders concentrate birds and increase the ability of disease to spread among individuals. Moreover, some urban species, such as Rock Doves and blackbirds, may function as reservoirs for disease (Garner 1978). Rock Doves are more common in urban environments than in natural environments (Marzluff 1997) and are known to carry diseases such as *Trichomonas*. This protozoan may be able to survive in urban settings better than rural ones because of the large Rock Dove population. When environmental conditions (warm springs) favor its growth, it can be quickly transmitted through urban bird populations at communal feeding sites like backyard feeders. Seed eaters and nectarivores would be most susceptible to such diseases because they frequent urban feeders.

Competition

Competition may be reduced or increased by urbanization. Reductions in numbers of predators may release those predators that tolerate human activity from competition and allow them to attain higher densities and increased reproduction (e.g., Eastern Screech Owls; Gehlbach 1994). Competition is heightened by urbanization in forested areas where availability of nest sites affects the population density of cavity nesting birds (Brawn and Balda 1988). An unnatural feature of urban and suburban areas is the relative scarcity of tree cavities because of cosmetic pruning, cavity filling, snag removal, and tree thinning. Gehlbach (1994), for example, found an average of only three cavities per nesting pair of suburban Eastern Screech Owls compared to twice that number at a rural site. The shortage may intensify interspecific competition for nesting and roosting space. Gehlbach and McCollough (unpublished data) noted that successful Screech Owl nests in boxes were inversely correlated with the number of box-nesting European Starlings in similar suburban habitat, but not with fox squirrel (*Sciurus niger*) nests, because the squirrels and owls select similar habitats. Starlings prefer parklike vegetation and thus are segregated spatially from owls, which select more densely wooded nest sites.

Behavioral Adjustment to Urbanization

Individuals may adjust their behavior in response to novel features of urban environments. Such adjustments may enhance a species' ability to persist in the urban environment and may have ramifications beyond the urban setting. The best documented changes involve nesting behavior. Songbirds that are repeatedly disturbed at their nest may increase their aggressiveness (Knight and Temple 1986). Persecution by humans may also select for reduced aggressiveness (Knight et al. 1989). Species that live closely with humans, such as American Crows, may habituate to human presence in urban areas (Knight, Grout, and Temple 1987). Increased predation on nests in urban sites may be met with changing nest placement to minimize subsequent losses by some birds (Knight and Fitzner 1985; Marzluff 1988).

Foraging behavior also may be modified in urban environments. Species that use feeders or other anthropogenic sources of food may reduce their use of natural foods and change their temporal and spatial foraging behavior to include provisioning sites (Brittingham and Temple 1992). Reduced reliance on natural foods may interrupt seed dispersal and pollination far beyond the urban area. Clark's Nutcrackers (*Nucifraga columbiana*), for example, are important dispersal agents for whitebark pine in Colorado. Nutcracker reliance on human handouts at Rocky Mountain National Park may have decreased the dispersal of whitebark pine in the region (Tomback and Taylor 1986).

Legal Protection

Many of the factors discussed above (increased nesting and feeding opportunities, reduced competition and predation, and ameliorated climate) interact with the legal protection afforded birds in metropolitan areas to enable dense, nuisance populations to develop. This is most obvious for several species of waterfowl that find ideal conditions in urban environments. Wild birds such as the Mallard (*Anas platyrhynchos*), Canada Goose (*Branta canadensis*), and American Coot (*Fulica americana*) reach nuisance population levels in cities bordering or containing lakes, ponds, and rivers. Urban waterfowl cause problems in both terrestrial and aquatic habitats. On land, high concentrations of birds may denude the vegetation or change plant composition. Fecal material makes areas unsuitable for recreational use. This is particularly a problem in waterfront parks and golf courses, which have been impacted by geese in the eastern United States since the late 1970s (Conover and Chasko 1985). In water, fecal concentrations greatly lower water quality, increase eutrophication, and may cause human health risks.

During the past sixty years there has been a large increase in Canada Geese in North America, which has resulted in significant management problems (Conover and Chasko 1985; Manuwal and Ettl 1990; Cooper 1991; VanDruff, Bolen, and San Julian 1994). Many of the nuisance goose populations are the result of transplants. The large increase in Canada Geese in Seattle, Washington, since the 1960s (figure 19.3) when a small number of birds were transplanted from nesting islands on the Columbia River is typical. With its mild winters and abundant water, the Puget Sound region of western Washington provides overwinter habitat for these geese as well as large populations of other species of waterfowl. American Coots, American Wigeon (*Anas americana*), and resident Canada Geese are the primary winter nuisance species in urban areas of Puget Sound (Manuwal and Ettl 1990).

Several management strategies exist to control nuisance waterfowl, but local situations will dictate which ones are likely to be implemented. Possible control alternatives include (from Conover and Chasko 1985; Manuwal and Ettl 1990; and VanDruff, Bolen, and San Julian 1994): (1) fear-provoking stimuli; (2) nonlethal chemical repellents; (3) government trapping; (4) fences and enclosures; (5) hunters; (6) lethal poisons; (7) chasing birds from specific locations with trained dogs; (8) trapping and transplanting problem birds to other areas; and (9) addling eggs. Discouraging humans from feeding urban waterfowl will help prevent birds from causing local damage. More research is needed to develop nonlethal control measures and educate the public on the problems associated with too many waterfowl in the urban environment.

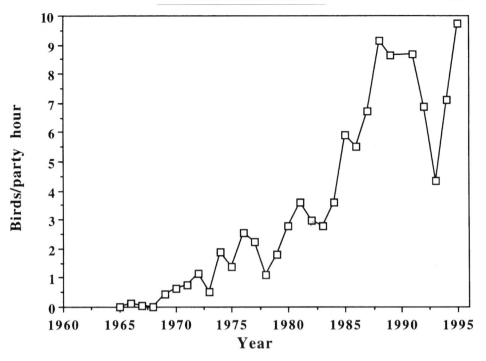

Figure 19.3. Number of Canada Geese counted each winter in Seattle, Washington, as part of the National Audubon Society's Christmas Bird Count. Counts were standardized by observer effort, which varies annually, by dividing the total number of birds counted by total observation time (party hours). Geese were introduced into Seattle in the mid 1960s.

Features of Research That Improved Conservation

There has been no drastic reduction in urbanization or modification of how we settle the land in response to studies of urbanization's effects on birds. Rather, research has served to document the effects of human beings. Some approaches have provided salient understanding, and a few have discovered opportunities to provide for birds in people's dominion. Fruitful approaches share some combination of the following features: (1) long-term investigation; (2) use of rigorous experimental design allowing simultaneous comparison of population viability along several points of the urban-wildland continuum; (3) identification of mechanisms responsible for the effects of urbanization on birds; and (4) quantification of avian demography.

The first important contribution of research to conservation was to heighten our awareness of impacts. Studies that quantified avian diversity and abundance along gradients of urbanization identified the types of species most susceptible to urbanization and the types able to benefit from human development. Studies that included replicate study areas in various types of urban, rural, and wild areas were most effective at documenting avian responses to urbanization (Beissinger and Osborne 1982; Rosenberg, Terrill, and Rosenberg 1987; Gehlbach 1994; Blair 1996). Recent collaborations among avian ecologists, geographers, and social scientists have provided spatially explicit details of the rate and pattern of human land use (Knight, Wallace, and Riebsame 1995; Riebsame, Gosnell, and Theobald 1996). These collaborative efforts have been especially important in suggesting that urbanization has far-reaching effects that may transcend boundaries into rural or wild ecosystems (Knight and Clark, in press; Buechner and Sauvajot 1996).

Research has increased our understanding of the mechanisms by which urbanization affects avian abundance and diversity. Brittingham and Temple's (1986, 1988, 1992) experimental investigation of the effects of bird feeders on Black-capped Chickadee behavior and demography stands out as an excellent, but unfortunately singular, example of applying rigorous scientific methodology to understanding one of the most common ways humans affect birds. A few long-term studies of demography have shown that species able to exploit humans do so because urbanization ameliorates climate, reduces some predators, and provides rich and consistent sources of food, all of which combine to lengthen breeding seasons, increase productivity, and improve survival of some birds (Tatner 1982; Plesnik 1990; Gehlbach 1994). Experimental and demographic studies of urban populations have shown that nest predation rates are lower than in rural areas (Snow 1958; Gehlbach 1994), may increase in the city but are lower generally than in rural areas (Tomialojc 1979), or can be higher than in rural habitat (Marzluff and Balda 1992); hence, urban populations may be either sources or sinks with respect to the surrounding countryside.

Properly designed experiments allow us to investigate the relative importance of the many landscape changes associated with urbanization. This is critical for understanding what features of urbanization actually affect birds. Friesen, Eagles, and MacKay (1995) studied a large number of study plots in each of two years to relate bird abundance to woodlot size and degree of human development. By having five to seven replicate plots in each of twelve combinations of size and development classes, they were able to statistically assess the importance of woodlot size, degree of development, and interaction of size and development. Their rigorous experimental design allowed them to quantify the relative importance of two aspects of urbanization: forest fragmentation and human density. They concluded that housing density was more strongly correlated with avian diversity than was forest fragmentation.

Occasionally, research has provided management recommendations to improve the suitability of urban areas for relatively sensitive species. Realization that nest sites, especially cavities, are reduced with urbanization stimulated placement of nest boxes, which allow many tits, wrens, raptors, and bluebirds to better exploit urban areas. This can initiate (Gehlbach 1994) and help restore populations, but restoration may take time, as initial colonists are relatively unproductive yearlings (Petty, Shaw, and Anderson 1994). Such yearlings characterized a new nest-box population of suburban Eastern Screech Owls studied concurrently with an older, more productive, and denser suburban population 50 km away (C. McCollough, Baylor University, unpublished data). Thus, both urban and rural populations augmented by artificial nests may be sinks initially, but their productivity could improve as the breeders mature. When they become more productive than new urban or long-established rural populations, old urban populations could be the sources of emigrant colonists (Gehlbach 1994). Research on the hunting and nesting habits of Peregrine Falcons suggested that they could exploit even the most densely populated urban centers if given a chance. Captive propagation and reintroduction provided the chance and has been very successful (Cade et al. 1996).

Research Needed to Further Conservation

Despite the conspicuousness of urbanization and its obvious effects on bird population ecology and community composition, our understanding of the impacts of urbanization on birds is based primarily on few experimental or controlled comparative studies. In order to understand urban bird ecology, it is critical to do simultaneous studies in different parts of the urban-wildland continuum. Gehlbach (1994, 1996) demonstrated the utility of making such comparisons that include several environmental and population features and monitor several avian generations and environmental fluctuations. Future investigations need to use carefully designed experiments and long-term monitoring so that causal relationships between human activity and bird population viability can be established (Gutzwiller 1995). Experiments need to account for natural factors, such as weather and food availability, by being designed to control for them, or better yet, test their interactive effects with urbanization. Long-term monitoring needs to determine fitness, which can be estimated as fledglings per breeding pair (Marzluff and McFadzen 1996) or more accurately by population recruits per breeding pair. Ultimately, though, fitness is judged as recruits per breeding pair relative to the pair's age-related survival, hence reproductive potential. Henny, Overton, and Wight (1970) provided a basic formula for this, which Gehlbach (1994) used along with other measures in an urbanization continuum from rural through a ten-year old suburban patch to a thirty-year old suburb joined to a medium-size city.

Critical evaluation of human effects will require long-term monitoring of abundance, distribution, and fitness of uniquely marked birds. Long-term demographic studies that allow us to compare avian population viability in urban, suburban, exurban, rural, and wild areas are especially needed. However, such intensive studies cause us to reduce the scale of investigation and study only one or a few species. We suggest monitoring abundance of entire avifaunas in demographic study areas while focusing detailed fitness measurements on easily studied representatives of various guilds or keystone species sensitive to human activity (James, Hess, and Kufrin 1997; Marzluff and Sallabanks, this volume). Important predators, nest parasites, and competitors (species such as American Crows, Common Ravens [*Corvus corax*], European Starlings, and House Wrens [*Troglodytes aedon*]) that appear to benefit from our activities should be carefully monitored as they can secondarily affect many birds that may not decline from the direct actions of human beings. A "serial" or "adaptive" management approach (Walters 1986) should be employed during the extrapolation of results obtained at small spatial or temporal scales to the larger scales relevant to management of avian diversity (Marzluff and Sallabanks, this volume).

Some of the most important specific research questions concerning urbanization are:

1. *How do we characterize urbanization?* Standard measures of urbanization need to be developed and reported in studies of urban effects. There is a continuum of intensity of human use from urban city centers to rural and wild lands. Avian study sites must be accurately quantified as to their position on this gradient. We suggest metrics such as percent natural habitat, green space, housing density, population density, road density, and industrial development. However, we need detailed study of how these and other variables relate to bird population and community ecology so the most influential variables can be identified and used to characterize the degree of urbanization.

2. *How do bird populations and communities change with increasing urbanization?* As human density increases, do species nest progressively earlier with greater success, and do the populations stabilize with longer-lived breeders as in Gehlbach's (1996) model? Long-term studies of changing urban sites are needed to answer such questions. Standardized surveys, such as the American Christmas Bird Counts and Breeding Bird Surveys, should include urban-rural continua, so changes within and between sites over time are assessed. The decline and disappearance of certain species and colonization and population growth of other species should be studied. Are niches emptied by intolerant species and filled by others more readily habituated to human activity? What are the pre-adaptive differences of such birds and their consequences for successful urban life?

3. *How does the type of urban development affect birds?* Comparisons of bird abundance and productivity among different intensities of urbanization are rare. In particular, the effects of dispersed housing, which is growing rapidly, should be compared with the more traditional clustered housing developments.

4. *What types of urban developments are most compatible with native birds?* We need to know if and how urban development can be made more "bird friendly." Research on the benefits of landscaping with native plants, using alternative energy sources, and educating homeowners on providing food, water, and shelter to birds would help identify long-term adjustments that city, county, and state governments could mandate to minimize our impacts on avifauna.

5. *How do nest predators influence nesting success in urban areas relative to rural areas?* Nest predators commonly increase in urban areas, but their general effect on other birds is unknown. Nest predation rates are typically high even in rural areas (Martin 1993) where bird communities are diverse and populations are viable, and may be higher still or lower in cities (see above). Nest predation may be mitigated by successful renesting in urban (Gehlbach 1994) and rural (Faaborg et al., this volume) areas, although increased predation in the city can be associated with the increased failure of renestings as well as the failure of first nests (Marzluff and Balda 1992). We need to know more about how productivity is influenced by nest predation at urban and nearby rural sites concurrently and how the different types of nests (open versus cavity, ground versus canopy, clustered versus solitary) are affected.

6. *How do birds respond to the urban-rural interface?* Rural and wild areas are not immune to the effects of urbanization (Buechner and Sauvajot 1996). We need to know how far from urban centers the negative and positive effects of urbanization extend into surrounding wildlands and develop ways to buffer the urban influences. House cats and perhaps other subsidized predators range from urban centers to hunt in surrounding rural areas, but how important this impact is at varying distances from urban sources is unknown.

7. *How does the interspersion of urban, suburban, exurban, rural, and wild land affect avian diversity and population viability?* We need to provide land-use planners with information enabling them to understand how entire landscapes affect birds. Detailed knowledge about bird diversity and viability in urban, suburban, exurban, rural, and wild areas of various size will be needed. Additionally, we need to know how the juxtaposition of these habitats affects birds in each habitat. It will be especially important to understand how edges between various types of habitats affect birds at varying distances from the edge. Knowledge of the dynamics of birds in each type of habitat can then be combined with knowledge of dynamics across boundaries between habitats to model expected effects of landscape composition (amount and juxtaposition of urban, suburban, exurban, rural, and wild habitat) on avifaunas. Such models would be especially useful for providing hypotheses for empirical testing and validation.

8. *How will projected human growth in an area affect birds?* Many rural and wild areas will likely become urban, suburban, or exurban areas in our lifetimes. We need to anticipate these changes, learn where they are most likely to occur, and proactively gather research results that can be used to objectively guide the pattern, pace, and location of future development.

Making Research Effective for Conservation

How do we increase the chances that the results of our research will be incorporated into policy to reduce the impact of urbanization? General prescriptions are given in Young and Varland (this volume) and developed further in the chapters by Hejl and Granillo, Kochert and Collopy, and Ganey and Dargan. A recurring theme from these authors is the need to relay results in less technical formats to appropriate managers and policy makers. Success in this area is crucial to successful implementation of research. With respect to managing urbanization, this will require conservation biologists to work with geographers, social scientists, and land-use planners so that important suggestions for minimizing urbanization effects can be incorporated into long-term, spatially extensive, land-use plans. We must work with urban planners and developers to appreciate what spatial arrangements of housing, roads, and natural areas are financially feasible. We will need convincing data and presentation of results when wildlife needs do not appear financially viable. We must be able to suggest ways to minimize impacts during development and recognize special areas where development is not compatible with wildlife. The costs of providing for wildlife need to be incorporated into the costs of development *before* construction begins, and this will occur only when developers and politicians understand our research and are presented with ways to provide for humans *and* wildlife. Our needs must be made known to the general public so they will be prepared for increased costs. Failure to make an effective case for wildlife to developers, politicians, and the general public will doom efforts to minimize the impacts of urbanization on wildlife.

For research to be effective we must consider the cumulative effects of humans in urban and exurban areas. Although cities provide habitat for relatively few native birds, their value for birds needs to be recognized and enhanced. Increasing the natural diversity of cities not only directly provides for birds, but also increases the quality of human life. Increasing the quality of human life in cities is central to the reduction of urban impacts on birds because it encourages dense human settlement thereby reducing sprawl and stemming the "amenity migration" (Riebsame, Gosnell, and Theobald 1996) to the exurbs. Truly effective research will provide information that allows land-use planners to enhance the attractiveness of urban areas, spatially arrange development in exurban areas to minimize impacts on birds, and minimize the flow of urban effects from private land to public reserves. This will require researchers to collaborate as teams investigating urban–rural–wild landscapes. Teams should include researchers working on non-avian fauna and flora so that the needs of biodiversity in general are met.

Human population growth and our reliance on technology are unlikely to decline in the near future. Therefore, it is important to realize that urbanization will continue and that currently rural and wild lands may soon be urban and exurban lands. Effective research will anticipate these changes and provide managers and planners with the information they need to decide how proposed development will impact avian diversity. Identification of important source areas for each species is needed so that future urbanization has minimal effects on native birds.

Acknowledgments

Rick Knight provided helpful review of our ideas. Cheri McCollough and Steve Penland provided unpublished data.

Global Variation in Conservation Needs

The conservation issues discussed in the previous chapters are broadly applicable. However, most were derived from studies conducted in North America by North American scientists. Part VI serves to broaden the perspective by providing assessments of conservation needs from other continents. Geographically, we learn about problems and research needs from the European continent (Britain and Russia), the interface of Europe with the paleotropics (Israel), the Neotropics, tropical islands (Marianas), and the Southern Hemisphere (Australia). Although this variation determines the richness and uniqueness of the avifaunas discussed, the basic conservation problems and research needs are rarely derived from ecological or geographical differences. Rather, they are derived in large part from the unique political surroundings and human influences that characterize each area.

Britain has been occupied by a large, industrial society far longer than North America and as such can be used as a crystal ball to forecast future human-caused changes in habitats and resultant effects on birds on other less peopled continents. Newton (chapter 20) warns us that future threats are unpredictable but suggests that agricultural intensification is currently the most pressing conservation issue in Britain and much of temperate Europe. Almost all species characteristic of arable land and grassland have declined in recent decades, and in Britain, where populations have been monitored, some species have fallen by more than 80 percent in the last twenty years. Rare species have been maintained by sympathetic management of some small areas where they remain. But to prevent further declines in currently widespread species, wide-scale changes in current land-use practices are required, which in turn depend on central policy changes. The main requirement for the future is for conservation needs to be more firmly rooted in national and European-wide agricultural, forestry, and fisheries policies and regulations.

Perhaps nowhere is the threat to avian diversity greater than in the Neotropics. This region holds tremendous diversity (one-third of all bird species breed there, 40 percent of all species reside there for some portion of the year, and 20 percent of all families are endemic to the Neotropics) and is under increasing pressure to produce resources for nations increasing their industrial output and economic activity. Brawn et al. (chapter 21) note that estimates for the tropics overall predict that between five hundred and thirteen hundred bird species may be "committed to extinction" by the year 2040. Losses of endemic species that are habitat restricted will be especially severe. Grajal and Stenquist (chapter 22) suggest that many of the conservation threats to birds are related to poor understanding of causes and consequences of human alterations to natural habitats, and to the lack of timely and appropriate information for decision making at many levels, from governments to individual landowners. Both sets of authors argue that unknowns about the basic natural history and evolutionary ecology of most tropical species hamper the design of conservation strategies in situations where wholesale preservation of habitats or ecosystems is not feasible. Subjects in which research is particularly needed include: patterns of dispersal and movements, effects of fragmentation and creation of edge, the spatial structuring of populations, the interplay of ecosystem processes and birds, and the genetic structure of tropical bird populations.

These types of studies have a greater conservation impact under particular circumstances, such as when: (1) they involve local researchers and natural resource managers; (2) results are included in action plans, priority-setting exercises, and red data books that help to galvanize a sense of urgency to conservation action; (3) results are disseminated in popular media as well as in dedicated journals that provide national or regional scientific outlets in Spanish, Portuguese, and English languages; (4) they are accompanied by local professional training at many levels; and (5) they include long-term monitoring to provide a time-series analysis of trends in populations, species dynamics, or whole communities. Grajal and Stenquist reason that one of the biggest challenges to bird conservation research in the Neotropics will be the integration of human needs with conservation actions. Complex issues in conservation will require a sophisticated and flexible assessment of the political, social, and economic factors that induce conservation action.

The lack of basic natural history information on birds in remote or lightly populated regions of the world is a recurring theme throughout this section. Yosef and Malka (chapter 23) and Galushin and Zubakin (chapter 24) make this point for Israel and Russia, respectively. However, reasons for lack of information in these two countries differ from those for the tropics, which serves to emphasize the overriding effects that politics can have on conservation. Yosef and Malka note that Israel has been subject to rapid human population growth; industrial, economic, and agricultural development; and political strife and violence since the beginning of the twentieth century. Since its inception as a nation, Israel has always been in a state of war. Thus, environmental issues have never been a priority to the people (including many scientists) of Israel. However, it is the only Middle Eastern country where conservation is practiced to any degree comparable with North America. Many of the diverse habitats that exist in the country have been degraded or extinguished. Environmental problems have been exacerbated because of the recent peace treaty with Jordan. Since the treaty was signed, the number of foreign investors has grown exponentially and the number of planned development projects is extensive. A blanket program for conservation is direly needed in Israel.

The dissolution of the Soviet Union has greatly affected avian conservation in Russia. Past ornithological research in Russia was federally funded and well organized, albeit restricted. Since the dissolution, socioeconomic changes have given Russians the freedom to conduct research where they want and on which topics they select. However, this potential is not being realized because funds are limited, salaries are low, and travel is expensive. A major change has been the reduction of volunteers and activists available to conduct research and promote avian conservation. In the past their contribution was enormous; now it is reduced because of increased expenses, reduced free time, and reduced salaries. Conditions in the large network of protected areas are growing worse due to lack of funds for research and protection. One positive note for the environment is that agricultural intensification and environmental pollution have declined due to the recent disastrous economic situation.

The Mariana islands are a showcase for illustrating the fragile nature of small, endemic, isolated avifaunas. Their largest island, Guam, no longer has a dawn chorus of birdsong. It has been silenced by the actions of introduced species, notably humans. Humans disrupted this ecosystem not in the usual way of destroying habitat (although we have done our share of that), but by inadvertently introducing a host of predators, especially the nocturnal, bird-eating, brown tree snake. Rodda, Campbell, and Derrickson (chapter 25) note that the loss of virtually all forest species on Guam has left many species globally endangered and a variety of forms extinct. The most urgent research need is to determine the causes of ongoing and prospective declines among the remaining species. The task of preserving remaining avian diversity in the Marianas is made more difficult because: (1) basic ecological and natural history data are lacking for most species; (2) techniques for controlling exotic predators are either unknown or are of unknown practicality; and (3) if in situ preservation fails, there are numerous uncertainties about the number of bird species that can be artificially packed onto small islands, which may limit the use of translocation as a tool for protecting birds by transporting them to "ark" islands. The Marianas offer a great challenge to avian conservationists, namely to re-create an avifauna. This will require us to understand how to maintain the constituent parts in captivity, fix the problem on the islands, and successfully reintroduce the captives.

Relative to most regions of the world, avian conservation in Australia is a shining beacon. Here, conservation has been simplified by the restriction of most species to a single nation. Management has been aided by the development of a series of national strategies and initiatives, which generally translate the available research results to management actions. Woinarski (chapter 26) points out that bird conservation in Australia has been mostly reactive, not proactive. While research and management of threatened species over the last two to three decades have ensured that the populations of the most endangered birds have generally been maintained or increased (only one extinction has occurred in the last two hundred years), many currently widespread and reasonably abundant species appear to be declining, presumably in association with pervasive land-use changes. Relatively few resources have been directed to research on, and amelioration of, the impacts of most land uses or threatening processes generally.

The broad views of avian conservation presented in this part are unified in four important messages:

1. Future threats to avian diversity in a given region are unpredictable. There-fore, we need to study many species and many human activities proactively to understand current agents of decline and detect new agents. This is crit-ically dependent on long-term studies that identify causal relationships between human activity and avian population viability.

2. To determine the cumulative impacts of human activities, we need a wide-scale, long-term monitoring system. Those in place in Europe and North America, despite their shortcomings, are sorely missed by researchers else-where in the world.

3. Conservation research is effective only when it is used to guide natural resource use policy. Avian researchers must get their message to the general public and policy makers as well as to other scientists. Ways to improve this transfer of information are detailed in part VII.

4. Conservation biologists in North America and parts of Europe need to col-laborate with colleagues in regions with less information on the effects of human impacts on birds. We can learn much from each other and together more efficiently improve the conservation of our avifauna.

CHAPTER 20

Bird Conservation Problems Resulting from Agricultural Intensification in Europe

Ian Newton

Conservation Issues and Previous Research

Broadly speaking, the same major conservation problems that exist in North America, and form the subjects of several other chapters in this volume, exist in Europe. But if you asked which were the most important, the answer would no doubt differ among biologists, depending on their particular interests and perceptions and where on the continent they happen to live. In this chapter, I shall be concerned mainly with one major widespread problem in Europe (and Great Britain especially), not discussed at length by others in this volume, but which over the last forty years has been responsible for bigger percentage declines in a wider range of bird species than any other factor, namely agricultural intensification. This has affected not only rare species, such as the Great Bustard (*Otis tarda*) of extensive grassland, but many of the commonest and most familiar songbirds of the countryside whose declines are evident not just to ornithologists, but to anyone with even the most casual interest in natural history.

Because Europe has been peopled at high density for much longer than North America, it has been more extensively modified, and few wholly natural areas remain. About 42 percent of the continental land area is now classed as agricultural, either arable or grazing land, but this proportion varies from as little as 10 percent in Nordic countries to more than 70 percent in Britain and some other mid-latitude countries, that are blessed with reasonable rainfall (Stanners and Bordeau 1994). Forest that once covered more than 80 percent of the land surface is now reduced to 33 percent, and most is in a highly modified state. Over the past forty years, agricultural changes have been most marked in Britain and other west European countries, but modern procedures are now spreading increasingly to eastern and southern countries. Most of the examples of impacts on bird populations given here are based on research in Britain, where extensive long-term data sets are available. However, trends in bird populations that can be quantified in Britain are paralleled in continental Europe (Tucker and Heath 1994), not least because many of the policies that affect birds and their habitats relate to the European Union.

Trends in Farmland Birds

Two major schemes have given wide-scale baseline information on breeding bird populations in Britain: (1) The Common Birds Census (CBC), involving annual counts of singing males on particular census plots, has been conducted annually since 1963 and has provided information on long-term trends in breeding densities (Marchant et al. 1990). (2) Two "Atlas" projects, conducted about twenty years apart, in 1968-72 and 1989-92, mapped breeding bird distributions throughout the country as presence or absence in 10-km squares (Sharrock 1976; Gibbons Reid, and Chapman 1993). In both types of survey, the information was collected mainly by amateur observers whose efforts were coordinated by the British Trust for Ornithology (for an up-to-date review, see Fuller et al. 1995).

In the twenty years between the two Atlas projects, declines in breeding range were much more prevalent among birds of farmland than among birds of other habitats (table 20.1). In fact, virtually all species of farmland declined in distribution in this time, some by more than 80 percent (figure 20.1). Although count data were available for a smaller range of species, declines in local density were evident in most species, again in some exceeding 80 percent. Declines occurred in species characteristic of grassland or arable land. Hence, the main findings to

Table 20.1. Distributional changes in British birds, 1970–90 (from Gibbons, Reid, and Chapman 1993).

	Numbers of species*	% increased	% decreased
Farmland	26	12	88
Woodland	47	43	57
Upland	38	50	50
Wetland	29	52	48
Coastal	24	46	54
Urban	5	60	40
General	20	35	65
Overall	189	41	59

*Excluding thirteen introduced species, of which eleven increased and two decreased.

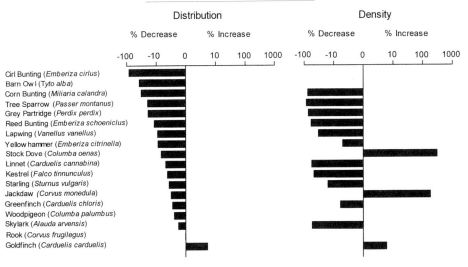

Figure 20.1. Changes in the distributions and densities of farmland birds in Britain, 1970–90. Modified from Fuller et al. (1995).

emerge from these monitoring programs were that a substantial part of the farmland bird fauna has declined over this period, and that in some species the declines have been catastrophic. By restricting the comparison to resident species (figure 20.1), and excluding summer and winter migrants, we can be sure that the causal factors lie within Britain rather than elsewhere.

Changes in Agricultural Land

The main problem in finding causes of these declines is that agricultural intensification is not a single process but has several different components, each of which can affect different species. Moreover, these various components of change have occurred more or less simultaneously, which makes the role of any one change hard to separate from the confounding effects of others. Third, the problems must be studied retrospectively, not when they began, but only after the agricultural changes and bird population declines have been underway for up to several decades.

The main changes that have occurred in European agricultural procedures over the last forty years can be listed as follows:

1. Massive increases in the use of chemicals, both pesticides and fertilizers. Rising pesticide use is reflected in increases in the range of chemicals used, the acreage treated, and the numbers of applications per year. Some such chemicals, notably the organochlorines, have had direct effects on birds, causing reproductive failures or enhanced mortality (Ratcliffe 1980; Newton 1986), while others have affected birds indirectly by reducing their food supplies (Newton 1995).

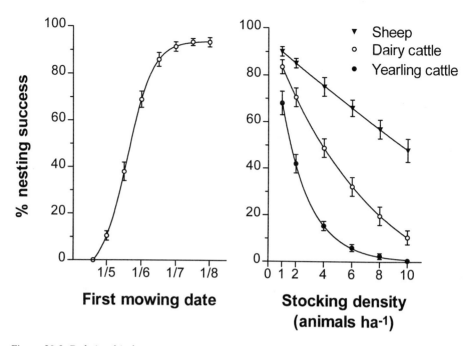

Figure 20.2. Relationship between nesting success of Lapwings (*Vanellus vanellus*), mowing dates, and stocking densities in Dutch meadows. Yearling cattle (which move around a lot) crushed more clutches than older cattle, and cattle in turn crushed more than sheep at the same density. However, because sheep are usually kept at higher densities than cattle, their total impact is not dissimilar to that of cattle. Modified from Beintema and Muskens (1987).

2. Removal of hedges and other uncultivated areas to produce larger fields that are more suited to large-scale mechanization. This has greatly reduced the amount of seminatural habitat that existed within the agricultural matrix. Some bird species both nest and feed in such habitat, while others nest there but feed on neighboring fields.

3. Change from spring ploughing to late summer ploughing of cereal stubbles soon after harvest. This change has removed the supply of spilled grain and other seeds on the ground, on which many seed-eaters formerly depended in winter. It has also removed the invertebrate food supply provided by fresh till in spring and the short-vegetation nesting habitat favored by Lapwing (*Vanellus vanellus*) and Skylark (*Alauda arvensis*) (autumn-sown cereals having grown too tall and dense by spring).

4. Extensive land drainage, which, through lowering the water table, has changed wet grassland to dry grassland or enabled a change to cereal culture. This has affected mainly dampland species, such as certain waders, but also other species, such as the European Starling *(Sturnus vulgaris)*, which feed on invertebrates from near the soil surface.

5. The trend from mixed farms, producing a variety of animal and plant products, to monoculture cereal production. This process of habitat simplification has reduced the diversity of bird species found on individual farms and rendered large areas poor in birds for much of the year.

6. The trend to earlier harvesting dates, caused by a combination of earlier sowing dates, earlier ripening cereal varieties, and a change from hay to silage (grass harvested green at an earlier growth stage). This trend means that more cultivation procedures (including harvests) fall within bird breeding seasons, causing destruction of nests and chicks of field-nesting species.

7. More intensive grassland management, involving use of inorganic fertilizer and more frequent reseeding. The fertilizer stimulates grass growth, making it no longer suitable for some ground-nesting bird species, and reseeding reduces invertebrate densities (which increase with age of grassland; Tucker 1992).

8. In hill areas, a massive increase in sheep-stocking densities, which changes habitat structure and sward composition and reduces the densities of phytophagous insects. Over the years, these changes have been encouraged by payment of subsidies, which has made previously uneconomic practices profitable. While all these changes have affected existing farmland, additional farmland for intensive use has continually been created from other seminatural areas, such as scrub and marsh land.

Features of Research That Improved Conservation

So far, only a small proportion of declining species have been subject to detailed study to identify the causal factors. Some of these species were already so rare that they were not represented in the Common Birds Census. The main finding to emerge was that no single causal factor was responsible for the declines in all species; different species have been affected by different aspects of agricultural change, depending on their ecology. They also began to decline at different times, although in several species, declines became apparent simultaneously from around the mid 1970s, corresponding to a time of increased pesticide use (mainly carbamates and organophosphates) and a change from spring-sown to autumn-sown cereals (Fuller et al. 1995). The following case studies provide examples of the types of problems revealed.

Grey Partridge (Perdix perdix)

This species lives mainly on arable land, and its decline has been attributed primarily to herbicide use (Potts 1986; Potts and Aebischer 1991). Herbicide destroys the food plants on which the insects eaten by Partridge chicks depend.

The chicks then survive poorly, so that insufficient young are produced to offset the normal adult mortality, and the population declines. This mechanism has been well documented, and explored in mathematical models, and each stage has been tested by experiment in the field (Rands 1985; Sotherton 1991). In several trials, the experimental withdrawal of herbicide use resulted in greater weed growth, greater insect densities, better chick survival, and increase in both the post-breeding and subsequent breeding densities of Partridges (Sotherton 1991). The same was true of the Ring-necked Pheasant *(Phasianus colchicus)*, but this species did not decline because its numbers were maintained by artificial rearing programs, funded by hunters. In both species, deaths of chicks may be due partly to starvation or chilling, and partly to increased predation, as they move around more in their attempts to find food.

Lapwing *(Vanellus vanellus)*

The Lapwing breeds mainly on damp grassland, used for summer grazing of cattle or sheep. Drainage causes the land to dry out earlier in spring, stimulating earlier grass growth. It also gives farmers earlier access with tractors, allowing grass rolling and fertilizer applications, which in turn lead to earlier grass cutting (especially for silage) or to higher stocking densities. Lapwings and other waders that nest in the fields suffer from earlier tractor access and from greater stocking densities, both of which destroy nests, and from earlier cutting, which destroys young (Beintema and Muskens 1987). In a study in the Netherlands, the proportion of nests that hatched increased with later mowing date, or with lower stocking density (figure 20.2). The same patterns held for three other wader species nesting in the same areas. Similar results were found in northern England, where drainage of upland "permanent" pasture or lowland meadows, followed by reseeding and fertilizer use, led to reduced breeding success and breeding numbers of Lapwing and other wader species (Baines 1989, 1990). In this study, the main cause of breeding failure was increased nest predation, as nests were more conspicuous against the uniform background of improved pasture. Only on unimproved pasture did productivity approach the estimated one young per pair needed, on prevailing mortality rates, to maintain population density long term. A second favored nesting habitat, spring-sown cereals adjacent to grazed grassland (Shrubb and Lack 1991), has largely disappeared with the conversion of spring-sown to autumn-sown cereals, and of mixed farms to arable monocultures. Low productivity and population decline have also been found among Lapwings nesting on arable land in several studies in different parts of Europe (Beser and von Helden-Sarnowski 1982; Matter 1982; Kooiker 1984; Galbraith 1988; Baines 1990). Analysis of British ringing results revealed no change in the survival rates of full-grown Lapwings over the period 1930–1990 (rather, a slight increase since 1960), thus implying that the widespread population decline was due to reduced production of young, as these detailed local studies had shown (Peach, Thompson, and Coulson 1994). On prevailing mortality, Lapwings would need to produce 0.83–0.97 young per pair per year to maintain the population. From a review of available literature, this level of productivity was found in only eight out of twenty-four studies (Peach, Thompson, and Coulson 1994).

Black Grouse (Tetrao tetrix)

Not a typical farmland bird, the Black Grouse breeds on rough upland grazing land, which supports low densities of domestic sheep and cattle as well as wild Red Deer (*Cervus elaphus*). Its widespread decline in areas still superficially suitable has again been attributed mainly to poor reproduction, resulting chiefly from decline in the arthropod food of chicks (Baines 1996). But for this species, decline in relevant insects (mainly caterpillars on *Vaccinium* and other low shrubs) was attributed to increased grazing pressure, which removed the young plant growth on which preferred insects depended. Comparing twenty different areas in five regions, breeding densities and breeding success of Black Grouse were correlated with insect abundance and vegetation height (Baines 1996), while all four measures were inversely correlated with densities of sheep or deer. An estimated 1.5–2.0 young per female were required to maintain the grouse population level, and such high productivity was seldom found in heavily grazed areas. In many parts of Europe, the mean number of young produced per female has declined over the past forty years, while adult mortality has increased, and these changes are held responsible for population decline, their relative importance varying from one area to another (Baines 1991).

Linnet (Carduelis cannabina)

This species feeds mainly on seeds from certain weeds of arable land (especially *Polygonum* and *Chenopodium*), picking them from the plants themselves or from the ground. In the past, the seed bank in the soil formed an important food source, as seeds were turned to the surface at every cultivation. Herbicides have thus destroyed the food plants themselves and over a period of decades have also led to progressive depletion of the seed bank. Each year seeds germinate, but the resulting plants are killed before they can seed. The decline of the Linnet and some other seed-eaters can thus be attributed to the decline in the arable weeds that once formed a large part of their diets. Seed-eaters that over the same period have maintained their numbers feed mainly or partly on tree seeds (e.g., Siskin [*Carduelis spinus*]) or from garden feeding trays in winter (e.g., Greenfinches [*Carduelis chloris*]), or are recovering from past human impact (e.g., Goldfinch [*Carduelis carduelis*]), which until recent decades was widely trapped as a cage bird). Other seed-eaters that have suffered marked declines include Corn Bunting (*Miliaria calandra*), Tree Sparrow (*Passer montanus*), and Bullfinch (*Pyrrhula pyrrhula*) (Marchant et al. 1990; Fuller et al. 1995).

A summary of purported causes of declines in these and other species is given in table 20.2. The main message is the wide range of factors involved, from

Table 20.2. Causes of population declines in some farmland bird species that have been studied in detail.

Species	Habitat	Timing of decline	Environmental cause of decline★	Demographic cause of decline	References
Starling, *Sturnus vulgaris*	Farmland, especially short-cropped pasture	Late 1960s to present	Increased conversion of pasture to arable land, and drainage of remaining pasture. Redesign of poultry and pig units to exclude birds.	Increased mortality of nestlings and full-grown birds.	Feare 1994 Tiainen et al. 1989
Linnet, *Carduelis cannabina*	Mixed farmland	1960s to present	Herbicide use, which has reduced the abundance of farmland weeds, which provided the food, and long-term depletion of the seed bank in the soil.	Unknown, but but probably increased over-winter mortality.	Newton 1986, 1995 O'Connor & Shrubb 1986
Cirl Bunting, *Emberiza cirlus*	Mixed farmland with old pasture and stubble fields	1950s to present	Loss of weedy stubble fields, through herbicide use and change from spring to autumn plowing. Loss of old pasture, which provided grasshoppers and other insects for chicks.★	Increased over-winter mortality and decreased breeding rate.	Evans & Smith 1994

Species	Habitat	Period	Cause	Effect	References
Corn Bunting, *Miliaria calandra*	Farmland, especially arable	Mid-1970s to present	Loss of winter stubble fields, through change from spring to autumn plowing, and increasing use of herbicides.	Unknown, but probably increased winter mortality.	Donald & Forrest 1995
Skylark, *Alauda arvensis*	Mixed farmland, pasture and arable	1970s to present	Use of pesticides, which has reduced arthropod abundance, and change from spring to autumn plowing of stubble fields, which has reduced winter seed supplies.	Inadequate chick production.	Jenny 1990 Schläpfer 1988
Corncrake, *Crex crex*	Hayfields	Late 19th century to present	Change from slow hand-cutting of hay to faster mechanized cutting, so that the birds cannot escape the blades. The problem is accentuated by edge-to-center cutting in ever decreasing circles and by use of nitrogen fertilizers, which permit an earlier cut (when chicks are younger) for hay and silage production.★	Increased mortality of chicks and adults.	Norris 1947 Green 1995

continued

Table 20.2.—*continued*

Species	Habitat	Timing of decline	Environmental cause of decline★	Demographic cause of decline	References
Lapwing, *Vanellus vanellus*	Farmland, damp pasture or arable with short vegetation	1980s to present	Widespread land drainage, which dries the surface soil, making invertebrate food less available and giving access to machinery earlier in the year; earlier grass cutting and higher stocking densities. Also, changes from spring-sown to autumn-sown cereals, and from mixed farming to monoculture cereal crops.	Decreased breeding rate. No change in annual mortality.	Beintema & Muskens 1987 Baines 1990 Peach, Thompson, and Coulson 1994
Partridge, *Perdix perdix*	Farmland, especially arable	Late 1950s to present	Increasing use of herbicides, which destroys the broad-leaved weeds that support the insects eaten by Partridge chicks. Also, removal of hedgerows and other field boundaries that provide nesting cover, making remaining nests easier for predators to find.★	Inadequate chick production.	Rands 1985 Potts 1986 Potts & Aebischer 1991

Black Grouse, *Tetrao tetrix*	Heathland and other rough grazing		Overgrazing and conversion of heath-dominated to grass-dominated swards; associated reduction of arthropod abundance.	Inadequate chick production.	Baines & Hudson 1995 Baines 1996
Stone Curlew, *Burhinus oedicnemus*	Short turf dry pasture and heathland	Late 19th century to present	Conversion of dry downland pasture and heathland to arable. Use of machinery on arable land, which destroys eggs and chicks.★	Inadequate chick production.	Green 1988
White Stork, *Ciconia ciconia*	Damp grassland	Late 19th century to present	Widespread land drainage, and intensification of grassland management, leading to loss of frogs, grasshoppers, and other prey species. Also, drought and reduction of food supplies in African winter quarters.	Inadequate chick production and increased adult mortality.	Bairlein 1996, Kanyamibwa, Bairlein, and Schierer 1993

habitat changes, such as massive reduction in food supply (mostly consequent on increased grazing or pesticide use), to excessive destruction of adults or nests due to procedural changes, such as earlier grass cutting. For the most part, the declines in all the declining species are continuing.

In particular species, traditional natural history studies have proved vital in assessing causes of declines and in developing effective conservation measures. In some studies, conservation actions were treated as experiments, and techniques were improved progressively. The most effective approaches involved: (1) comparison of population and habitat parameters in different areas or different time periods with different trends (as in Lapwing, Black Grouse); (2) the experimental testing of hypotheses developed from such comparisons (as in Grey Partridge); and (3) examination of historical trends in nest success or survival rates (from ringing) which can identify the problem as breeding or mortality (as in Lapwing). In contrast, nonexperimental studies in single areas of uniform habitat have been generally ineffective in revealing causal factors.

Research Needed to Further Conservation

With any declining species, a prime requirement is for continued monitoring of distributions and densities, for only then can we find whether declines continue or "bottom out," with populations stabilizing at a new lower level. We can also assess the impact of any further changes in land use or other factors that might occur and of any conservation measures that might be taken.

So far, only a few species out of the many declining ones have been studied sufficiently to define the factors causing their decline. Because so many different factors are involved, extrapolations from one species to another are unwise. A major research need, then, is to extend research to a wider range of declining species. While priority should obviously be given to the rarest ones, research is often easier on species that are still common enough to be studied in a range of habitat types. For each species, the main questions are which aspects of the agricultural environment have changed so as to cause declines, and whether they act on breeding or mortality rates. In many European countries, ringing recoveries now cover several decades, including periods of predecline and decline. As for the Lapwing and others, these data could be examined to find whether annual survival rates have declined over the years. If not, population decline is likely to be due to reduced breeding success, in which case the problem is in spring-summer on breeding areas rather than in winter. In Britain, nest records are also available over several decades and can be examined for changes in nest success (see Crick 1994 for various buntings).

Uncertainty hangs over the role of predators in population declines. Several predatory species, especially corvids, have increased in recent years and account for many nest failures. It is unclear, however, to what extent their predation is predisposed by habitat changes, such as reduced cover resulting from intensive grazing, or by reduced food supply, which forces birds to expose themselves for longer each day. Hence, more research is needed on the role of predators and particularly on the interaction between habitat and predation and between food supply and predation.

So far, most progress has been made by examining species individually, making use of existing variation in agricultural patterns from one region to another to formulate hypotheses that can be tested experimentally. More could be learned from making use of schemes designed to take farmland temporarily out of production, and more particularly from experiments that involve changing one aspect of agricultural procedure at a time and monitoring which species (if any) respond. Attempts at this latter approach have already been made by reducing pesticide impacts (Sotherton 1991), and by switching from late summer to spring ploughing of cereal stubbles (Evans and Smith 1994).

Making Research Effective for Conservation

Some of the species discussed above are now so restricted in distribution that they can be maintained only in particular localities by managing farmland in a sympathetic way. As this is usually less profitable for farmers than conventional management, such management usually entails buying the land for the purpose or paying farmers compensation for lost revenue. Thus, local Corncrake (*Crex crex*) populations have been maintained by paying farmers to forgo intensive grass management, cutting later in the year (hay rather than silage), and cutting from the center of the field outwards. This last procedure forces birds to the edges of the fields, from which they can escape unharmed on foot to neighboring fields. Local Stone Curlew (*Burhinus oedicnemus*) populations have been maintained by paying observers to find and mark the nests, which can then be avoided by the farmer during rolling and harrowing. Similarly, in parts of France and Spain, observers find and mark Montagu's Harrier (*Circus pygargus*) nests, so that the young can be saved during cereal harvesting. The effectiveness of such measures depends, of course, on correct diagnosis of the factors causing decline.

Such costly and time-consuming methods are acceptable in the short term for rare and restricted species. But different approaches are needed to halt the general declines of species that are still widespread. For these, the findings from research have been used to lobby government for changes in agricultural policy or subsidy structure, in an attempt to lessen their damaging activities. The most conspicuous success in this approach was the reduction and eventual banning in the use of organochlorine pesticides, which facilitated recoveries in the numbers of certain raptors and other affected species (Ratcliffe 1980; Newton 1986). Another success of this approach has been the withdrawal of subsidies for hedge removal, but otherwise this approach has not been spectacularly successful, partly because of the powerful vested interests against it.

Agricultural policy over much of Europe still encourages intensification and its associated practices. Until this central policy is changed, or applied less widely, we can expect that other bird and wildlife populations over much of the continent will continue to decline. Since the policy is encouraged by payment of subsidies, much could be achieved in conservation terms by minor changes in the products for which specific grants are paid. The fact that some subsidized activities have other negative consequences, such as overgrazing, soil erosion, and pollution of drinking water, means that arguments against specific subsidies need not be made on conservation grounds alone, as such subsidies jeopardize the whole agricultural enterprise in the longer term.

Meanwhile research has a role to play in monitoring wildlife populations and in identifying causal factors in the declines of particular species. The availability of good data on long-term trends in bird numbers has been largely responsible for raising the level of public awareness and debate about agriculture and environment. Almost certainly, some other types of organisms have declined on farmland even more than birds, but we have insufficient data to quantify the trends. Fortunately, birds are popular creatures, the general public demands their conservation, and voluntary data collection has proceeded on a wide scale over many years. In this way, ornithology has made a major contribution to nature conservation, and the value of birds as environmental indicators is well established.

Although I have concentrated on farmland birds, in parts of Europe, similar problems occur in forestry as old-growth mixed stands give way to short-rotation conifer monocultures. This mainly affects species dependent on cavities and dead wood, or on rich ground vegetation (such as woodland grouse), all of which are scarce in modern conifer plantations. Many studies have demonstrated the role of nest-site shortages in limiting the breeding densities of hole-nesting species, and after provision of nest boxes, increases of up to twentyfold in breeding density have been recorded in some such species (Newton 1991). As on agricultural land, the main obstacle to progress is not research, but changes in forestry management that would produce the desired conservation benefits. For some seabirds, overfishing could begin to play an increasing role in reducing population levels (see also Boersma and Parrish, this volume). This is becoming of greater importance, with the development of "industrial fishing" based on small fish species (such as sandeels) that are important to birds. Added to this reduction in food supply is the extra mortality to seabird populations imposed by oil spills, drift netting, and other fisheries procedures.

If present population declines continue, some landbird species will be increasingly confined to reserves and other restricted localities that provide favorable conditions. This will not only reduce their overall numbers and distributions, but also can be expected to bring other problems associated with the demography and genetics of small populations (Caughley 1994).

The potential value of birds for monitoring wider environmental conditions was recognized by the establishment of the Common Birds Census in 1963, together with repeated distributional surveys. These showed the successful recovery of some bird populations from agrochemical poisoning, but also revealed the disastrous impact in conservation terms of changes in agricultural practice in the past thirty years. Widespread monitoring has led to the recognition that conservation legislation and policy must address the wider countryside, as well as protected areas.

An important lesson from past experience is that future threats are unpredictable, even at a gross scale. In the 1950s, the greatest perceived threat to nature was urban spread, and few predicted the major damage that would be inflicted by agricultural practice, afforestation with alien conifer monocultures, land drainage, and coastal developments. For the marine environment few could have predicted the damage to other fauna that would be caused by driftnetting and other fishing procedures. Thus, long-term monitoring needs to be extended to as wide a range of species and habitats as possible, and to be cast beyond today's problems or species of concern.

The most significant current threats for nature conservation arise from national (and European Union) rural land-use policy, with its emphasis on intensive production systems. Habitat loss and degradation are the main current causes of species declines, even though other factors may contribute to the final stages of extinction. Such policy could provide opportunities for the enhancement of nature conservation if only ecologists could influence it. Action could be taken at two scales: (1) a "top-down" approach employing government policy and support of the kind used to promote the intensification of arable agriculture and livestock production; and (2) a "bottom-up" program of grant aid for individual farmers to improve their local farm environments for wildlife. Influencing central land-use policy in these ways is the main challenge for the immediate future.

It is not that wide-scale intensificiation of agriculture is required to support the present human population of Europe. In the past forty years, growth in agricultural production has greatly outstripped the growth in human numbers, and the last twenty years have seen the production of enormous food surpluses, which have been stored or destroyed at great expense. More importantly, recent farming methods have resulted in other environmental problems that will be hard to rectify, such as wide-scale soil erosion and pollution of water resources. Perhaps the main beneficiaries of agricultural intensification have been the farmers themselves and the associated agrochemical and other industries, which collectively have become a powerful lobby in favor of the status quo.

Acknowledgments

This chapter could not have been written without the efforts over many years of hundreds of bird-watchers, who have provided the data necessary to quantify widespread trends in bird numbers. I am grateful to them, and to R.J. Fuller and R.E. Kenward, who (in their capacity as referees) made helpful comments on the manuscript.

CHAPTER 21

Research Needs for the Conservation of Neotropical Birds

Jeffrey D. Brawn, Scott K. Robinson, Douglas F. Stotz, and W. Douglas Robinson

Conservation Issues

Two themes emerge from an assessment of research needs for Neotropical birds: urgency and unknowns. Here, we review important threats to avian diversity in the tropics and identify specific needs for avian research in the Neotropics in the twenty-first century. Importantly, strategies for Neotropical birds will largely fail if the immediate threats to many regions are not alleviated—an action that cannot wait for detailed ecological information.

Pending habitat loss and threats to an extraordinarily high number of species are issues that motivate a complicated research agenda for birds in the Neotropics (here defined from northern Mexico to southern Argentina, including the Caribbean). The major human causes of habitat loss or disturbance are logging, agricultural development, mining (petroleum and minerals), and urbanization (Canaday 1997). Some species are further threatened by unregulated, direct exploitation (Robinson and Redford 1991). Reported rates of habitat loss vary widely depending on the technology used to estimate them and the region or habitat under consideration (Fearnside 1990, 1993; Skole and Tucker 1993). Nonetheless, these rates are increasing in several geographic areas of important conservation concern. For the tropics overall, the United Nations Food and Agriculture Organization estimates that 15.4×10^3 km^2 per year have been deforested since 1980 (cited in Dale et al. 1994). In the Brazilian Amazon Basin, rates of deforestation have increased from an average of 13,600 km^2 per year in the 1970s to anywhere from 50,000 to 80,000 km^2 per year in the 1980s (Skole and Tucker 1993). The latter figure represents a loss of about 2 percent per year. This estimate and others illustrate that much attention has been devoted to deforestation in the wet or moist lowland forests of Amazonia. Certain habitats or ecosystems outside Amazonia are being lost at much faster rates, however, and include the Atlantic Forests of southeastern Brazil and the Cerrado region of dry forest. Protection of remaining natural habitat in these and other areas is critical for conserving many endemic species (Stotz et al. 1996).

Another issue is that simple rates of deforestation or habitat disturbance underestimate the extent of habitat alteration. Remaining habitat becomes isolated, and the proportion of edge to area inevitably increases. Estimates from forest habitat in the Brazilian Amazon indicate that the total area affected by fragmentation and edge effects is more than twice that actually cleared (Skole and Tucker 1993). Adverse edge effects on the viability of avian populations in North America are well known (e.g., Robinson et al. 1995; Brawn and Robinson 1996), but information about the ecology of edges within tropical ecosystems has only recently been presented (Murcia 1995; Schelhas and Greenberg 1996).

Information on these issues is important because the contribution of the Neotropics to global avian biodiversity is immense. About 3,750 species of birds breed in the Neotropics (Stotz et al. 1996). About 60 percent of *all* bird families are represented in the Neotropics, and nearly a third of these are endemic. Within habitat and between habitat diversity are extraordinarily high by temperate-zone standards. For example, a country as small as Panama contains more bird species than all of North America above the Mexican border (Ridgely and Gwynne 1989).

The number of known extinctions of birds in the Neotropics since European contact is surprisingly low. Twenty-six Neotropical species are thought to have gone extinct since the 1600s and about seventy species are currently on the brink of extinction in the wild (Stotz et al. 1996). In the tropics worldwide (barring oceanic islands), about 113 species have possibly gone extinct since 1600 (Reid 1992). Nearly all those on the brink in the Neotropics have been endangered due to habitat loss, but several have also been exploited by hunting or the pet trade (Robinson and Redford 1991; Stotz et al. 1996)

Whereas these numbers suggest that avian biodiversity in the Neotropics is largely intact, loss of habitats and ecosystems may lead to unprecedented rates of local extirpation or global extinction within the next fifty years. Estimates of extinction rates are typically based on anticipated losses of habitat and species-area relationships; that is, How many species will be supported within the resulting habitat islands (Simberloff 1986; Reid 1992)? Reid (1992) derived estimates of future losses of plants, birds, and mammals within closed-canopy forests under three scenarios of habitat destruction and estimated that one species will be "committed to extinction" (following Simberloff 1986) every one to two days. For birds, under the scenario of high rates of habitat loss, 10–32 percent of species may be lost. Even low rates of loss may lead to a 2–4 percent loss of species. The time frame for these losses is difficult to characterize.

Of more immediate concern are those species that are habitat restricted and endemic to a specific region. Areas with comparatively high rates of habitat loss and a comparatively large proportion of habitat-restricted endemics include the Atlantic Forests of Brazil, the northern and central Andes, central South America, and the Greater Antilles (fig. 21.1; see Stotz et al. 1996 for specific delineation of these areas). Habitat losses in these areas are generally higher than current losses in the lowland forests of Amazonia. Continued habitat destruction in these regions may lead to losses of hundreds of species. In the Northern Andes region alone, there are 111 habitat-restricted endemics. In the Atlantic Forests of eastern Brazil, 30 percent of the 170 species of forest birds endemic to the region are considered endangered (Whitney and Pacheco 1995). Deforestation is more advanced there than elsewhere in the Neotropics (as much as 93 percent of the forest was already destroyed by the 1970s; Mori, Boom, and Prance 1981). The

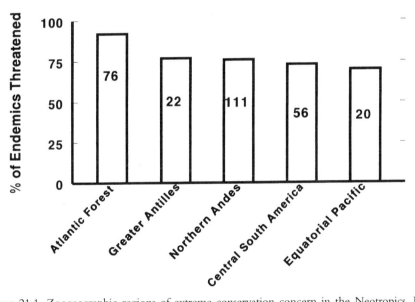

Figure 21.1. Zoogeographic regions of extreme conservation concern in the Neotropics. Bars show proportion of habitat-restricted endemic species that are threatened. Numbers in bars indicate number of endemic species within region (from Stotz et al. 1996). "Atlantic Forest" refers to the humid coastal region of eastern Brazil; "Northern Andes" refers to montane regions from the coastal mountains of Venezuela south to northern Peru; "Central South America" refers to the lowland, open habitats separating humid Amazonian forests and humid forests of coastal eastern Brazil, south through Paraguay and parts of central Argentina; "Equatorial Pacific" refers to the arid and semiarid lowlands of western Equador and Peru. See Stotz et al. (1996) for expanded definition of zoogeographic regions.

Atlantic Forests are the only lowland forest with threats comparable to those in montane forests.

Species that breed in the Neotropics have generally suffered more from habitat alteration than species of Nearctic or Austral migrants. While some migrants have been—and will be—seriously threatened by habitat destruction, many use secondary habitats readily and typically have broader wintering ranges and broader habitat tolerances than do the resident species that occupy the same habitats and regions (Stotz et al. 1996).

Features of Research That Improved Conservation

Conservation has long been an element of ornithological research in the Neotropics (see Buckley et al. 1985; Karr, Robinson, et al. 1990), but explicit conservation research has proliferated in the last decade (e.g., see the 1995 and 1996 volumes of *Bird Conservation International*). One long-standing approach that reflects the "frontier" aspect of much Neotropical research has been to describe avifaunas in new habitats or regions. Typically, these surveys combine scientific collection of specimens (sometimes for alpha taxonomy) with natural history study (Remsen 1995). A fundamental objective of these studies has been to assign priorities for preservation and conservation to regions that are critical for maintaining avian biodiversity (Hernández-Baños et al. 1995; Parker et al. 1995; Stotz et al. 1996).

A recent approach has been to study directly changes in avifaunas owing to human induced disturbance (e.g., Johns 1991; Kattan, Alvarez-Lopez, and Giraldo 1994; Thiollay 1996; Greenberg 1996; Canaday 1997). This approach, now common in North America, has proved informative, but more studies are needed in the Neotropics in diverse biogeographic and ecological settings.

Two specific studies deserve special mention: changes in the avifauna on Barro Colorado Island (BCI), Panama, and the Biological Dynamics of Forest Fragments Project near Manaus, Brazil. Studies of birds on BCI have established the case history of long-term (i.e., over eighty years) changes in species composition subsequent to isolation (Willis 1974; Karr 1982). These studies have been correlative and experimental (Sieving 1992; Sieving and Karr, in press; W. D. Robinson, University of Illinois, unpublished data). The forest fragments project is the first (and, to our knowledge, only) large-scale experimental study of fragmentation in the Neotropics (Stouffer and Bierregaard 1995b). This experiment along with the longer-term information from BCI has given conservationists the foundation for predicting and explaining the effects of fragmentation.

Research Needed to Further Conservation

We believe a three-tiered approach is useful for identifying conservation research needs for Neotropical birds. This approach is forced by the immediacy of threats to a large number of species and by the reality that much habitat will be disturbed or lost. Therefore, we offer three broad categories for research questions and priorities: preservation, conservation, and restoration. The first category refers to the protection of large areas or ecosystems that are unique and will likely not tolerate large-scale anthropogenic disturbance. Continued faunistic studies and development of methods to protect preserves are priorities in this category. The second category "conservation," includes issues pertaining to fragmentation, dispersal, metapopulations, and small populations. The final category, "restoration," stems from the reality that major disturbances and habitat losses have occurred and are inevitable. Owing to limitations on the scope of this chapter, our emphasis is on terrestrial forest birds in Central and South America.

We also believe that a program for monitoring avian populations and communities in the tropics is urgently needed. At present, few baseline data on long-term variation (spatial or temporal) in avian abundances or community structure are available in relatively intact habitats, let alone in disturbed situations.

Preservation Research

The primary research needs for preservation of habitats and ecosystems are to conduct faunistic surveys and develop methods for assuring that preserves actually do "preserve." Surveys have been completed for many important areas in the Neotropics, and we briefly summarize major patterns below.

Diversity and endemism among birds in the Neotropics are highest in humid forests. However, diversity reaches its peak in the lowlands, while endemism is substantially higher in montane regions (Stotz et al. 1996). The humid Andes from Venezuela south to Bolivia contain the greatest concentrations of endemic birds in the Neotropics. Other montane regions contain fewer endemic birds than the Andes, but all have more endemic birds than the humid lowlands at their bases. Many of these endemics also occupy a relatively narrow elevational range, with almost half having an elevational range less than 1,000 m. Since birds with restricted ranges are much more likely to be threatened than widespread species, this has resulted in a much larger level of threat to avifaunas in humid montane regions than in the humid lowlands.

The West Indies, especially the large islands of the Greater Antilles, is another area with significant endemism. Although avian diversity is much lower on these islands than in continental regions, over half of the resident landbirds are endemic to a single island or group of islands (Stotz et al. 1996).

Despite fewer species of birds in open and dry habitats, significant pockets of endemic birds inhabit these habitats throughout the Neotropics. Most important are the mosaics of grasslands, arid scrubs, and deciduous forests in the Cerrado and Caatinga regions of central and northeastern Brazil; the tropical deciduous forests of western Ecuador and northwestern Peru; and the scrub and deciduous forests along the Pacific coast of Middle America and in the Yucatán Peninsula. The rather patchy distribution of these habitats across the Neotropical region has resulted in a higher degree of endemism. Endemic habitat specialists in tropical deciduous forests are among the most threatened sets of birds in the

Neotropics and substantially more likely to be threatened than are endemic habitat specialists in humid forests (Stotz et al. 1996).

We believe that detailed faunistic surveys must continue in areas that are comparatively undersampled. This priority emphasizes the need for continued support of museum collections and systematic research in the Neotropics (Remsen 1995). The basic "currency" of biodiversity cannot be quantified without these studies. Systematic studies of the Neotropical avifaunas are needed with diverse methods and sampling techniques. Particular attention should be given to areas such as the northern Andes that are thought to be "species pumps" for the Neotropics (see Fjeldså 1994). More analyses are needed, but biogeographical evidence suggests that these areas may be where a high proportion of species originate. If so, these regions should receive high priority for wholesale preservation of ecosystems.

Continued research on the factors that enhance the effectiveness of parks and preserves is vital. This subject will need to combine ecological, socioeconomic, and political expertise. An important nonbiological problem with preserves in the Neotropics is that they are understaffed and underfunded (Peres and Terborgh 1995). This problem leads to little or no enforcement, and the preserve's wildlife and habitats are essentially undefended against exploitation. Peres and Terborgh suggest that accessibility to exploitation by roads or rivers should be an important factor in the location and design of reserves. How the criteria of defendability would work in important areas such as the mid-elevational forests of the Andes requires study. A comparatively cool climate often makes middle-elevation areas preferable for human settlement, which increases the demand for wood and other resources offered by forest habitats.

Conservation Research

We identified five general (and nonexclusive) areas for conservation research in the Neotropics. The general theme linking these areas is to develop the means to conserve biodiversity in situations where total preservation is not possible.

Dispersal and Movements

The natural history of dispersal and movements by most species of Neotropical birds is unclear. Some species undergo seasonal migrations over a large geographic area in the tropics (Chesser 1995; Levey and Stiles 1992), while other species may move on smaller regional scales. Altitudinal migration, for example, is common among nectarivorous and frugivorous species in Costa Rica (Stiles 1988; Loiselle and Blake 1991, 1992; Levey and Stiles 1992). Similar movements may occur in many areas of the Andes (Terborgh et al. 1990; Robinson, Terborgh, and Munn 1990), but such migrations are virtually unstudied. Regional movements may also occur within lowland forests. In central Panama, for instance, frugivorous manakins (*Pipra* spp.) appear to track food resources that vary with moisture conditions (Martin and Karr 1986). Some evidence even indicates the possibility that small-scale daily movements within the forest are driven by daily changes in humidity (Karr and Freemark 1983; Karr and Brawn 1990).

All these behaviors require much more study. Identifying the habitats needed by tropical birds for all seasonal and daily movements is particularly important. Do species that move altitudinally, for example, require continuous tracts of forest from lowland to highland areas (Loiselle and Blake 1991)? The destruction of forest at middle elevations may restrict the movements of some species, as has been suggested for central Panama (Karr 1990).

The vagility of tropical species appears to vary greatly depending on habitat and life history. Some species are nearly ubiquitous and apparently excellent colonizers. Widespread species in the West Indies, for example, tend to be widespread on the mainland of Central and northern South America as well (Faaborg 1979). Many of these species occupy second growth and may be efficient at dispersing and finding new patches of disturbed or early successional habitat. Some birds of mature tropical forest, however, may only rarely cross water or light gaps as short at 100 m (Terborgh and Weske 1969; Willis 1974; Terborgh 1975; Faaborg 1979; Harper 1987; Stouffer and Bierregaard 1995b; W. D. Robinson, unpublished data). Populations of those species are more likely to become locally extinct than those of more vagile species that periodically recolonize or receive immigrants (Schoener and Spiller 1987; Pulliam 1988). We discuss the subject of spatially structured populations in more detail below.

The evolutionary and ecological processes leading to interspecific variation in vagility merit detailed study (Levey and Stiles 1992). Are sedentary species of forest habitat less likely to move across habitat discontinuities because of physiological limitations, such as extreme sensitivity to light or inability to sustain flight for moderate distances, or perhaps because of ecological limitations, such as increased predation risk?

The influence of different types and sizes of barriers to dispersal also requires study. Evidence supporting the importance of barriers such as water gaps in restricting immigration includes Capparella's (1988, 1992) work on genetic differences among populations of certain forest understory birds in Amazonia. On landbridge islands off the coast of Panama, overwater distances of only 5 km appear to restrict dispersal—even in relatively common, second-growth species (Brawn, Karr, and Nichols 1995). In addition to water gaps, deforested areas such as pastures can act as effective barriers against dispersal by some forest birds. Stouffer and Bierregaard (1995b) recently showed that such gaps separating forest fragments in Amazonia significantly limit the ability of forest undergrowth insectivores to move between forest fragments. Finally, species that have become extirpated on BCI are significantly less likely to cross open areas than species that are still present on BCI (W. D. Robinson, unpublished data).

The high sensitivity of certain species to disruptions in habitat emphasizes that much work is urgently required to determine if habitat corridors can ameliorate the effects of fragmentation on tropical birds. Some woodcreepers and wrens at a middle-elevation site in Costa Rica, for example, are more likely to move between fragments if the fragments are connected by a narrow forested corridor (R. Borgella, Cornell University, personal communication). Alternatively, corridors may have negative ecological effects such as the movement of pests, predators, or vector-borne diseases (Simberloff and Cox 1987).

Fragmentation

In contrast to the virtual explosion of interest and data on the effects of fragmentation and disturbance in the temperate zone (see review by Faaborg et al. 1995; Walters, this volume), little information is available from the tropics. The most extensive tropical data are available from the aforementioned study near Manaus, Brazil (Bierregaard 1990; Stouffer and Bierregaard 1995a,b), where several 1-, 10-, and 100-hectare fragments were created. Severe reductions in species richness were documented, especially in the smaller fragments, and species of understory insectivores were particularly sensitive to fragmentation. Several understory insectivores were also lost from BCI after its isolation from the mainland (Willis 1974; Karr 1982). One possible mechanism for the sensitivity of these species is strong stereotypy in their diets and foraging behavior (Sherry 1984). Reductions in species richness have also been noted from isolated habitat patches in Ecuador and southeastern Brazil (Leck 1979; Willis 1979).

Most investigations of fragmentation effects have been limited to lowland forest. An exception is a study at middle-elevation sites in San Antonio, in the Andes of Colombia (Kattan, Alvarez-Lopez, and Giraldo 1994). After eighty years, more than 30 percent of the original species composition was lost due to fragmentation effects. Particularly vulnerable species were canopy frugivores and understory insectivores. As pointed out by Kattan et al., the almost complete absence of baseline data from faunistic surveys of middle-elevation sites severely hampers efforts to document the effects of fragmentation.

Aside from the expected reductions in species richness, almost nothing is known about other ecological factors associated with habitat fragmentation. Changes in rates of nest predation so frequently reported from landscapes in temperate latitudes fragmented by urban or agricultural intrusion (Robinson et al. 1995; Brawn and Robinson 1996) remain virtually unstudied (see Loiselle and Hoppes 1983 and Sieving 1992 for results from artificial nest experiments). Indeed, we still lack data from community-wide studies of reproductive success in forest-interior sites (but see Oniki 1979; Skutch 1985). Furthermore, effects of edge on nest predation rates, bird species' occupancy, and predator populations have received little attention (Martin 1996).

Spatial Structuring of Populations

This subject follows from studies of dispersal and fragmentation. A fundamental uncertainty is the extent to which spatially explicit population dynamics are applicable to Neotropical birds. Do tropical species have metapopulations (here defined as a set of local populations linked by dispersal [Lindenmayer and Lacy 1995]) or isolated single populations? The complexity of ecological and evolutionary factors with tropical birds will not likely yield an "either/or" answer. If so, what are the intrinsic and extrinsic factors that account for variation in the spatial structuring of tropical species?

Recent developments suggest great promise in the utility of spatially explicit population models for conservation (Dunning et al. 1995; Conroy et al. 1995; Villard, Schmidt, and Maurer, this volume). The demographic and population parameters needed to apply spatially explicit population models prescribe detailed and long-term demographic studies and include survival rate, dispersal rate, annual fecundity, population size, recruitment rate, and turnover rate. Field data needed to derive these parameter estimates are difficult to attain, and, to our knowledge, few studies using mark-recapture and modern estimation procedures such as Jolly-Seber stochastic models have been carried out (see Karr, Nichols, et al. 1990; Faaborg and Arendt 1995; Brawn, Karr, and Nichols 1995). Therefore, we know little about the demographic parameters or life-history traits of tropical species. To illustrate, the importance of density-dependent variation in the demography of tropical birds has long been assumed, but, notwithstanding *Geospizid* finches (Grant and Grant 1989), this assumption has virtually no direct empirical support. Long-term demographic studies in undisturbed and disturbed landscapes are a fundamental need, as are the resources to sustain these efforts.

A topic related to metapopulations and spatially explicit modeling is sources and sinks (Pulliam 1988; Pulliam and Danielson 1991). Do tropical species naturally have source and sink habitats, and, if so, what is the scale of movements among them? Will fragmentation create sink habitats in the tropics as it does in North America (Robinson et al. 1995; Brawn and Robinson 1996)? And, if so, what can be done to lessen these effects? The viability of populations within fragments was typically ignored in the single large versus several small reserves debate (see Gotelli and Graves 1996), and it is now clear that species richness is a poor/naive index for the success or failure of a system of preserves. Again, parameter estimates from diverse landscapes are needed for "sensitive" species such as understory for insectivores and relatively resistant species such as those inhabiting second growth.

Effects of Ecosystem Disturbances on Bird Population and Communities

A striking aspect of many tropical ecosystems is that they are much more intact than those of North America. For example, every river system in North America is dammed, straightjacketed with levees, or otherwise altered by agricultural runoff and siltation. Thus, we have little understanding of the role of floods in creating habitats for disturbance-dependent, floodplain birds. Our opportunities to study other disturbances such as fire are similarly constrained by changes in fire regimes that followed the arrival of Europeans in North America. In much of the Neotropics, especially in South America, there are still enormous tracts of "pristine" floodplains, grasslands, and savannas. As a result, we have the opportunity to study the role of natural disturbances in maintaining a diverse array of habitats. By understanding how birds respond to successional gradients created by disturbance, we can improve our ability to predict which species are most vulnerable to anthropogenic disturbances, and suppression of disturbance, and the potential roles played by birds in restoring degraded lands (see "Restoration Research" below).

The extensive river systems of Amazonia offer opportunities to study bird communities along primary successional gradients created by the meandering of rivers (e.g., Salo et al. 1986; Rasanen, Salo, and Kalliola 1987; Kalliola et al. 1991). As rivers meander back and forth across floodplains, they erode forest on the outside of meander loops and deposit new soil (silt) on the inside (the meander tongue). The meander tongue is the site of primary succession on which tree species diversity and complexity of vertical structure increase progressively through time for hundreds of years (Terborgh and Petren 1991). Bird community structure changes dramatically along these gradients, with many species confined to the early and later ends (Terborgh and Weske 1969; Remsen and Parker 1983; Dyrcz 1990; Rosenberg 1990; Robinson and Terborgh 1997). Oxbow lakes formed when meander loops are pinched off from the rest of the river create further habitat heterogeneity as they fill in with silt and various successional stages (Robinson 1997). The existence of multiple successional stages along rivers and the associated backwater habitats (swamps, marches, oxbows, palm swamps) in close proximity with mature forest have been invoked as one contribution to increased richness of Amazonian bird communities relative to the rest of the world (Remsen and Parker 1983; Salo et al. 1986; Karr, Robinson et al. 1990).

In general, species that live along river-disturbed habitats are most tolerant of human disturbances (Terborgh and Weske 1969; Remsen and Parker 1983). Likewise, those species restricted to the oldest forests are likely to be most vulnerable; roughly seventy species do not appear until the forest is likely to be at least three hundred years old in the Manu area (Robinson and Terborgh 1997). Nevertheless, not all species that occupy successional vegetation along rivers also thrive in anthropogenically created secondary succession (Robinson and Terborgh 1997). Similarly, we know little about the extent to which disturbed habitats along rivers act as "source" habitat for species that live at very low population densities in other habitats (Robinson, Terborgh, and Munn 1990).

The continued existence of many "top" predators may also be important in maintaining ecosystem function (Robinson 1995). Large eagles and cats may control populations of mesopredators that, in turn, prey upon bird nests. The extent to which this top–down control of ecosystem function operates remains an open question. Some mesopredators such as monkeys remain abundant even in habitats where all of their predators remain (Terborgh 1983).

Population Structure and Conservation Genetics

The relationship between population size and genetic structure is a complicated function of ongoing ecological processes and historical events. This relationship is especially unclear for Neotropical birds. Data on associations between population structure and components of fitness are especially few. Recent studies do suggest that isolation will lead to loss of genetic diversity (Brawn et al. 1996; Sieving and Karr, in press), but the conservation relevance of these changes remains to be established. For example, in a study of island populations, Brawn et al. found clear evidence of isolation and subsequent drift in the diversity of mtDNA haplotypes. Yet the study species (Crimson-backed Tanager, *Ramphoceleus dimidiatus*) is abundant with no apparent fitness consequences. Sieving and Karr report inbreeding in some terrestrial insectivores within small, but not large, island populations in Gatun Lake, Panama, with unknown consequences on population viability. Much research on the "genetic natural history" of Neotropical bird populations is needed in diverse biogeographic and ecological settings. An informative approach would be to assess short-term changes in the genetic structure of populations before and after isolation or disturbance. Again, information on dispersal and immigration is critical. Comparisons of already isolated populations with those in contiguous habitats are also needed (e.g., Brawn et al. 1996). Priority locations for these studies would be areas where species have deep phylogenetic branches and may serve as "species pumps" for the Neotropics, such as the montane regions of the northern Andes (Brooks, Mayden, and McLennan 1992; Fjeldså 1994). We believe that conservation in the Neotropics should help maintain extant biodiversity and the evolutionary processes that could lead to the formation of future geographic variation or species.

Restoration Research

In situations where habitat has been severely degraded, much effort has gone into restoration techniques such as reforestation (e.g., see Uhl 1988). How these techniques affect bird populations requires study. A few case histories have been presented (da Silva, Uhl, and Murray 1996), but analyses from Central America, the Amazon Basin, and other areas are not yet feasible. How successional processes affect bird communities and our efforts to restore bird populations needs to be answered.

Conversely, birds must play an especially important function in the reforestation process owing to their role (along with bats) as seed dispersers (da Silva, Uhl, and Murray 1996); therefore, birds may drive successional patterns at restoration sites. Restoration techniques that could enhance the ecological services provided by birds require additional study. Will increasing isolation diminish the probability that certain species of frugivorous birds will visit an area? Will an abandoned pasture surrounded by second growth be likely be visited by a different compliment of frugivorous species than one surrounded by more mature forest (da Silva,

Uhl, and Murray 1996)? What effect will this have on the pattern and pace of succession?

The Need for Comprehensive Monitoring in the Neotropics

Our survey of research priorities closes with a plea for ecological monitoring in the Neotropics. The success of various conservation measures cannot be established without reliable information on trends in populations or communities and changes in local landscapes (Soulé and Simberloff 1986; Kremen, Merenlender, and Murphy 1994). For Neotropical birds, we believe that monitoring should be modeled after programs established in North America and emphasize three key areas: population and community dynamics, demographic parameters, and reproductive ecology. The political and logistical barriers to establishing these programs in the Neotropics are formidable and will require close cooperation of diverse governments, nongovernmental agencies, and international organizations (see Grajal and Stenquist, this volume).

The first area involves monitoring abundances and species composition through techniques such as point counts or spot mapping. The scale of censusing may be very important for such studies. For example, W. D. Robinson (unpublished data) has found in Panama that a census of a small plot (about two hectares) with extrapolation to larger areas versus a census of a large area (about one hundred hectares) can lead to significantly different density estimates. The use of mist nets and mark–recapture data for estimating population size (as opposed to density) will also be useful. The second area involves intensive use of mist nests, mark–recapture or mark–resighting data, and stochastic models for open populations (Karr, Nichols, et al. 1990). As discussed above, vital parameters that could be estimated over time are survival rates, recruitment rates, and rates of population turnover (Brawn, Karr, and Nichols 1995). The third area involves monitoring the reproductive success of birds in disturbed and undisturbed habitats or landscapes.

Making Research Effective for Conservation

The Neotropics present an especially complex set of challenges to the goal of effective conservation research. Diverse political, socioeconomic, and biological considerations reduce the effectiveness of a single approach. The success of conservation research may therefore be at the local scale. For example, long-term monitoring and research on populations and local movements of birds within the former Panama Canal area (Karr 1990) have led to specific recommendations about the need to retain habitat over the distinctive moisture gradient that spans the isthmus from the dry Pacific slope to the much wetter Atlantic coast (J. Hautzenroder, U.S. Department of Defense, personal communication). Research demonstrating that it is in the long-term interests of local peoples and governments to conserve wildlife and preserve habitat will be particularly effective (see Grajal and Stenquist, this volume). Outreach programs for environmental education at the local scale are an essential complement of field research. Sustainable exploitation of wildlife populations or natural habitats will add incentives for preventing wholesale habitat loss (Robinson and Redford 1991).

Finally, inclusion and training of local biologists is another essential approach. Longterm projects that train a series of local biologists will enhance the effectiveness of resource agencies where these personnel will eventually be recruited.

Acknowledgments

We thank D. Niven, T. Sherry, and K. Sieving for constructive comments on earlier drafts. John Marzluff and Rex Sallabanks were more than helpful editors.

CHAPTER 22

Research Applications for Bird Conservation in the Neotropics

Alejandro Grajal and Susan Stenquist

Conservation Issues and Previous Research

With nearly 4,130 species, the Neotropics harbor almost 43 percent of the world's avifauna (Wege and Long 1995). Proportionally, it is also one of the least-known avifaunas. New species are described at a rate of almost two to three per year (Stotz et al. 1996). Nearly 327 species have been classified as endangered or threatened, while several species have become extinct in recent years, and a handful exist only in captivity (Collar et al. 1992; Wege and Long 1995). However, this is only part of the picture, as many lesser-known cryptic species are probably in great danger of extinction, with little or nothing known of their basic biology. The main threat to Neotropical bird species is habitat loss, although other factors, such as pollution, hunting, and the live-bird trade, are very negative for the conservation of a number of species. Species of large biomass such as cracids, tinamous, anseriformes, raptors, and parrots are particularly sensitive to hunting pressure (Redford and Robinson 1987; Silva and Strahl 1991; Strahl and Grajal 1991), while many species of parrots (e.g., *Anodorhynchus leari, Cyanopsitta spixii*) and certain emberizine finches (e.g., *Carduelis cucullata*) face imminent extinction due to the live-bird trade for the pet markets (Thomsen and Mulliken 1992; Coats and Phelps 1985).

The ever growing threat of habitat loss in the Neotropics is a direct consequence of human population growth and its associated economic development, which are then reflected in rates of deforestation, draining or alterations of wetlands, marine and terrestrial pollution, land conversion to agriculture, and direct resource competition with birds. This is nothing new. However, it is important to analyze other indicators, such as gross domestic product, industrial output, and urbanization trends. In general (in a very diverse socioeconomic context), most Neotropical countries are decreasing their net population growth rates but increasing industrial output and gross economic activity, which have a larger impact on natural habitats. Increased urbanization in most countries has not diminished the expansion of the agricultural frontier, as large agro-industrial conglomerates have consumed most agricultural land and continue to displace landless peasants into marginal or "frontier" areas (World Bank 1994; World Resources Institute 1994). All these human activities have resulted in fragmentation of habitats. As a result, many parts of the Neotropics are classified now as areas with endangered habitats (ICBP 1992; Wege and Long 1995), which in turn contain the largest share of endangered bird species (Collar et al. 1992).

Compounding this overwhelming wave of habitat destruction is a clear dearth of trained human resources in conservation in general. The shortage of conservation professionals is particularly acute in the fields of research applied to conservation, such as natural resource management. There is an urgent need for further training at many levels, from parabiologists and community park rangers, to protected-area managers and graduate-level researchers.

Previous Research

The growing trend of knowledge of Neotropical ornithology started with the initial interest of field ornithologists and ornithological collections during the 1950s that in turn produced the basic information to print the first field guides (e.g., Howell 1985; Meyer de Schauensee 1966; Meyer de Schauensee and Phelps 1978; Ridgely 1976; Ridgely and Tudor 1994; Willis and Eisenmann 1979). The publication of these field guides allowed new research in areas where the avian fauna was known only to specialists. As a result, birds are probably the best-known vertebrate group in the Neotropics in terms of basic natural history, distribution, and ranges (e.g., Stotz et al. 1996). Greater knowledge of Neotropical avifaunas has allowed new analyses that highlight biogeographic conservation priorities (Dinerstein et al. 1995; World Bank 1994; ICBP 1992; Biodiversity Support Program et al. 1995) as well as conservation needs of individual bird species (Collar et al. 1992; Stotz et al. 1996; Wege and Long 1995). But much remains to be known about the basic natural history of Neotropical birds (Brawn et al., this volume). For example, basic life-history traits such as fine-scale geographic distribution, population size and dynamics, migratory patterns, diet, reproduction, and nesting behaviors are unknown for the vast majority of Neotropical bird species (Short 1984).

Research Activities Needed to Further Conservation

Local Involvement

One way to increase the conservation value of any scientific work is to involve local researchers. This involvement should go beyond hiring local field assistants or field guides. Ornithologists in northern countries should foster and develop formal and informal relationships with their counterparts in Neotropical countries to encourage conservation-oriented studies and publications (Duffy 1988; Foster 1993; Jenkinson 1993; Rosenberg and Wiedenfeld 1993; Strahl 1992). Furthermore, ornithologists should help to produce research results that are replicable and useful to decision makers. Direct interactions between researchers and natural resource managers are critical in this respect (Kochert and Collopy, this volume). Of particular success have been studies that are directly applicable to the creation or management of protected areas (Beebe 1947; Parker and Bailey 1991; Remsen and Parker 1995; Roca 1994).

Research Results That Influence Policy

Whenever research results can be included in action plans, priority-setting exercises, and red data books, they provide information needed to impart a sense of urgency to conservation actions (Biodiversity Support Program et al. 1995; Morales et al. 1994; Rodriguez and Rojas-Suárez 1995; Saavedra and Freese 1986; Stotz et al. 1996; World Bank 1994). Furthermore, some of the most influential information for conservation decision making has been the publication of relatively basic studies with wide audiences—such as field guides, species lists, environmental impact assessments, trip reports, long-term studies, and rapid ecological assessments (e.g., Andrade 1992; Boersma, Stokes, and Yorio 1990; Carrillo and Vaughan 1994; Kattan, Alvarez-Lopez, and Giraldo 1994; Miller 1995; Meyer de Schauensee and Phelps 1978; Parker and Bailey 1991; Parker, Parker, and Plenge 1982; Ridgely and Tudor 1994; Willis and Eisenmann 1979).

Scientific Publications and "Gray" Literature

Contrary to trends in developed countries, scientific publications in refereed journals are of relatively limited impact in Neotropical countries, because most international journals have extremely limited distribution and are generally not available to the nonspecialists. This may explain the large number of "gray literature" publications in the Neotropics that never make it to publication in international journals (Duffy 1988). The reason for the lack of dissemination of research results is that many of the necessary studies for conservation and management of Neotropical birds involve basic natural history, which is in turn disdained by "hard" scientific journals and colleagues that see these descriptive studies as only of secondary value. This turns into a vicious cycle that destroys incentives for publication of studies by Neotropical ornithologists (Gibbs 1995).

For example, we undertook a small survey of all ornithological literature as found in eight issues of the *Recent Ornithological Literature* supplement (ROL) from 1991 to 1994. We further separated the Nearctic and Neotropical listings into studies with direct application to conservation and with no immediate application to conservation. Even though ROL has a section on conservation and management, additional studies were included if they followed other conservation criteria, such as population and species abundance estimates, endangered species, management of species and populations, reintroductions, transcontinental migrations and habitat use by migratory species outside their breeding range, descriptions of new species, and the effects of harvesting, hunting, and exploiting of species groups or bird communities by humans. Of 8,880 article titles from all regions of the world, 3,317 (37 percent) were from the Nearctic region and 815 (9 percent) from the Neotropical region. These numbers reveal the already known deficiency of scientific publications on Neotropical ornithology in countries with the greatest avian diversity. Even though the absolute number of ornithological publications is much greater in the Nearctic region, the proportion of articles directly related to conservation is significantly greater in the Neotropical region (40 percent) than in the Nearctic region (26 percent).

This situation is beginning to reverse, thanks to efforts to increase the availability of ornithological periodical journals (Foster 1993), as well as the appearance of refereed journals that provide outlets for locally valuable scientific information in Spanish, Portuguese, and English languages. Among these valuable publishing resources are the journals *Vida Silvestre Neotropical, Bird Conservation International,* and *Ornitología Neotropical.* Another important development in scientific publications is the strengthening of national or regional journals that strive to maintain publication stability, such as *El Hornero, El Pitirre, Revista Brasilera de Ornitología,* or *Eco Trópicos.* While most Neotropical ornithology research remains to be done, supporting these local and regional scientific outlets is crucial for the advance of Neotropical bird conservation (Gibbs 1995; Mares 1986). The existing lack of basic scientific information for Neotropical birds is a major negative factor affecting the decision-making process in conservation.

Long-Term Monitoring Studies

One of the major weaknesses in biodiversity management in the Neotropics is the paucity of analyses of trends in populations, species dynamics, and whole communities. The few studies that have taken a repeated look at a bird community over several years have found dramatic changes in community assemblages (Boersma, Stokes, and Yorio 1990; Gandini et al. 1994; Kattan, Alverez-Lopez, and Giraldo 1994; Willis 1974). These long-term monitoring studies provide crucial information that can help to interpret how human development affects bird dynamics (Brawn et al., this volume). Given the enormous value of long-term data for conservation, and the fact that few researchers have the time or funds to dedicate to long-term studies, ornithologists need new and creative ways to further monitoring approaches. Some budding approaches with volunteers and parabiologists have recently started in the Neotropics, like monitoring of bird populations by parabiologists in Ecuador (SUBIR project), in Costa Rica (Inbio), or simultaneously around the world (BirdLife International's World Festival of the Birds). These types of participatory, long-term studies provide some of the very basic information that conservation requires.

Training

The dearth of trained professionals is, and will continue to be, one of the main weaknesses of research on Neotropical bird conservation. Therefore, professional training at many levels will remain one of the highest priorities for bird conservation. A number of graduate schools in wildlife management have sprouted recently in the Neotropics, covering important areas. Examples include dedicated master's programs in wildlife management and conservation like the Mesoamerican Program at the National University in Costa Rica, the wildlife management program at the University of the Llanos (UNELLEZ) in Venezuela, and the program at the University of Córdoba in Argentina. Many other universities in the region are producing the future cohorts of ornithologists with a decidedly conservation slant. In addition to university-level training, courses in conservation biology, experimental design, environmental education, and protected-area management have an enormous impact on the maturity and progress of conservation leadership (e.g., Organization for Tropical Studies course for Latin Americans, University of Colorado Course in Protected Area Management, Buffer Zone Management Course by Wildlife Conservation Society and University for Peace in Costa Rica). Further integration and support by ornithologists from developed countries can have a great impact (Jenkinson 1993). Several authors have dealt with ways to improve the role that highly trained ornithologists from developed countries have on their Neotropical colleagues (Duffy 1988; Foster 1993; Rosenberg and Wiedenfeld 1993; Strahl 1992). But the sad fact is that conservation needs several orders of magnitude more researchers and managers (Duffy 1988; Gibbs 1995; James 1987; Short 1984; Soulé and Kohm 1989)

Biological Conservation and Human Needs

One of the biggest challenges to research applied to bird conservation in the Neotropics is how to integrate human and conservation needs. A significant development in conservation thinking has been the relatively recent increase in literature advocating "sustainable development" (IUCN, UNEP, WWF 1991). This approach has taken a hard look at ways to accommodate human needs into conservation action, although it has often relied on unproven biological paradigms, such as the possibility of human "development" without biodiversity losses (Robinson 1992). Moreover, the poor definition of what sustainable development entails or what can be the "allowable" losses to biodiversity has made sustainable development a cause for everybody and a way to justify everything from blatant attacks on biological conservation to protective conservation measures. The sobering conclusion is that the needs of humans and biodiversity will in most cases be in direct conflict, and therefore will require a serious appraisal of the level of biodiversity erosion at the expense of human development. From the research standpoint, the resolution of conflicts between biodiversity conservation and human development will require creative thinking and equally imaginative research hypotheses.

Mainstream ornithological research for the most part has missed the momentous changes in global biological diversity policy (Fjeldså 1995). For example, the global Convention on Biological Diversity is already having major repercussions on funding for conservation research, such as the Global Environmental Facility (GEF) and other multilateral funding sources. These funding sources are pouring billions of conservation dollars into extremely data-poor decisions, whereas ornithologists have captured just a minimal amount of these funds, even though ornithological information is readily available and can provide timely answers to biodiversity action plans and conservation priorities (ICBP 1992; Wege and Long 1995; Biodiversity Support Program et al. 1995; Dinerstein et al. 1995).

Similarly, the Convention on Biological Diversity is affecting ways in which biological research will be conducted in developing countries, and particularly in Neotropical countries. Issues of access to genetic resources, intellectual property rights over biodiversity information sources, and new developments in biotechnology are rapidly changing the landscape of priorities and funding for biodiversity conservation (Ruiz Muller 1996; Simpson, Sedjo, and Reid 1996). Given that Neotropical birds are relatively better known than any other taxonomic class, Neotropical ornithology has the opportunity to address some of the most urgent needs imposed by the Convention on Biological Diversity, such as national biodiversity strategies and national action plans. Otherwise, ornithologists will miss one of the greatest opportunities to insert conservation priorities into national development policies.

Making Research Effective for Conservation

Translating research findings into actual conservation action has been one of the most puzzling issues for scientists. In some instances it is easy to identify the main culprit as a lack of political will by the decision makers, even if this is just a simplistic assumption. In most cases, translating research findings into conservation action requires educating the decision makers in local communities or in central or local governments of the potential benefits of conservation actions. Sometimes creating a constituency for the conservation issue adds enough weight to the political will of decision makers (e.g., Butler 1992). In many cases, however, there will be real or apparent contradictory needs between the economic development of local communities and conservation priorities. Mena and Suárez (1993) assigned this disparity to the fact that biological research is usually ahead of social, economic, and political research on conservation issues. As a corollary, we can say that complex conservation problems require a sophisticated assessment of local issues and variables: Anyone advocating a simple recipe for translating biological research into action is probably not aware of the complex political, social, and economic factors that influence conservation action.

Finally, we would like to emphasize the important role of the international scientific community in conducting ornithological studies that provide relevant information for decisions in conservation. Research should focus on results that can directly help conservation action, which also depends not only on scientific publications and good information, but also on political will. The challenge for the conservation of Neotropical birds will be to provide solutions that extend benefits of conservation to local human communities and decision makers. These solutions should include biodiversity conservation concerns in resource use plans, such as in tropical commercial forestry (Mason 1996) or coffee plantations (Wunderle and Latta 1996).

Acknowledgments

S. Stenquist's contribution was possible by an internship at the Wildlife Conservation Society. The manuscript benefited from comments by Jeffrey Brawn, Richard Hutto, and Steve Beissinger.

CHAPTER 23

Avian Conservation in Israel

Reuven Yosef and Rony Malka

Conservation Issues and Previous Research

Israel (28,000 square kilometers), critically located at the only land junction of Europe, Asia, and Africa, is influenced by the biogeographical regions of these continents. Viewed from any direction, Israel constitutes a series of transition zones—species of the Mediterranean regions have their southern or eastern range limits in Israel, and African and Asian species reach either their northern or western limits in these zones. Thus, Israel has extremely high levels of biodiversity that are restricted to very small areas (Paz 1987; Danin 1992a,b). The avian checklist for Israel shows that 204 bird species have been recorded breeding, 185 regularly; the rest are either occasional or former breeders (Shirihai 1996). A latitudinal cline, from the deserts to the temperate north is evident—75 species breed south of 30°N, 91 species south of 31°N, 129 south of 32°N, 137 south of 33°N, and 143 south of the northern borders with Lebanon and Syria. In addition, 216 species are winter visitors (late September–late April). Israel's location at the only landbridge connecting Europe, Asia, and Africa is also strategic from the point of avian migrations (Safriel 1968; Dovrat 1980; Christiansen et al. 1981; Shirihai and Christie 1992). A total of 283 species migrate through Israel during autumn and spring. An additional 130 species (as of 1994) have been observed and are categorized as "vagrants." Included in the above are several globally threatened species that either breed in or migrate regularly through the country (table 23.1; Collar, Crosby, and Stattersfield 1994). Israel is one of the few countries of the region wherein all raptors, as well as most other migratory species, are protected by law.

Israel has been subject to rapid human population growth (in 1997 its population was 5.8 million); rapid industrial, economic, and agricultural development;

and political strife and violence since the beginning of the twentieth century. Thus, environmental issues have never been a priority to the peoples living in the region. Political slogans such as "Make the deserts bloom" or "Drying swamps for human settlement" reflect the lack of understanding by political leaders that have for decades influenced the policies of the region (e.g., Ben–Gurion 1971; Gradus 1996). The excuse has always been the need to protect the country from the enemies that surround it. This has led to degradation, and sometimes even extinctions, of many of the diverse habitats that exist in the country.

In Israel, the Wild Animals Protection Law was passed in 1955, and it decreed that "wild animal means a mammal, bird, reptile, amphibian which does not by its nature live in association with man." Further, an additional law of National Parks and Nature Reserves was passed in 1963 and proclaimed that a "nature reserve means an area in which animals, plants, soil, caves or water of scientific or educational interest are preserved from undesirable changes in their appearance, biological composition or process of development and which the Minister of the Interior, upon the proposal of, or after consultation with, the Minister of Agriculture, has declared a nature reserve." Israel has 159 nature reserves with a total area of approximately 575 million acres. An additional 373 reserves of an additional 1.3 million acres have been proposed. Only 3 percent of the Mediterranean regions in Israel are protected as nature reserves. Nearly 20 percent of the deserts are protected (Dr. E. Frankenberg, chief scientist, Israel Natural Resources Administration, unpublished report), but a large proportion of these overlap with army training grounds. The conflicts that result between conservation and army training, agriculture, and human settlements make it almost impossible to conserve the small populations of extremely diverse flora and fauna. Nathan, Safriel, and Shirihai (1996) found that of the 185 avian species that bred regularly between 1863 and 1993, 14 species have become extinct and 58 species are threatened. They further report that most extinct species are raptors and waterfowl, and their extinction is related to human pressure (table 23.1). Based on data from Shirihai (1996), Nathan (1996) found that the Syrio-African rift (including the Arava or Great Rift Valley) is the most important for conservation of avian habitats. He reports that the valley has the largest biodiversity of breeding passerines and also the largest number of listed species.

Agriculture

Israel has made its mark on the world's agricultural market because of its technological innovations, genetic manipulations, and, subsequently, better crop-yielding plants and animals. As in Europe, agricultural intensification has major effects on birds (Newton, this volume). A unique and important effect of agriculture in Israel is its increasing dependence on plastic. The biggest drawback to this lies in the fact that Israel does not have the recycling capability to remove most of the plastic that is used annually by the farmers. Thus, this plastic is usu-

Table 23.1. Listed species that either breed in or migrate regularly through Israel (data from Collar, Crosby, and Stattersfield 1994; Evans 1994).

Breed	Migrate/Winter
Globally threatened:	
Marbled Teal (*Marmaronetta angustirostris*)	Greater Spotted Eagle (*Aquila clanga*)
Lesser Kestrel (*Falco naumanni*)	Imperial Eagle (*Aquila heliaca*)
	Corncrake (*Crex crex*)
	Great Bustard (*Otis tarda*)
	Red-breasted Goose (*Branta ruficollis*)
	White-eyed Gull (*Larus leucopthalmus*)
Near-threatened:	
★White-tailed Eagle (*Haliaeetus albicilla*)	Pallid Harrier (*Circus macrourus*)
	Cinereous Bunting (*Emberiza cineracea*)
	Pygmy Cormorant (*Phalacrocorax pygmeus*)
	Little Bustard (*Tetrax tetrax*)
	Black-winged Pratincole (*Glareola nordmanni*)
Vulnerable:	
Ferruginous Duck (*Aythya nyroca*)	White-headed Duck (*Oxyura leucocephala*)
	Sociable Lapwing (*Vanellus gregarius*)
Regionally threatened or declining:	
★Griffon Vulture (*Gyps fulvus*)	Honey Buzzard (*Pernis apivorus*)
★Lappet-faced Vulture (*Torgos tracheliotos*)	Levant Sparrowhawk (*Accipiter brevipes*)
Egyptian Vulture (*Neophron percnopterus*)	Lesser Spotted Eagle (*A. pomarina*)
Grey Francolin (*Francolinus francolinus*)	Bittern (*Botaurus stellaris*)
★Lanner Falcon (*Falco biarmicus*)	Saker Falcon (*F. cherrug*)
Sinai Rosefinch (*Carpodacus synoicus*)	Great Snipe (*Gallinago media*)

★ reintroduction program currently in place

ally buried in the vicinity of the fields, left as large mountains of garbage that disperse with strong winds, or burned.

Advanced technology also allows farmers to grow different crops throughout the year. As a result, populations of open-habitat species have been drastically reduced. These include Collared Pratincole (*Glareola pratincola*), Scops Owl (*Otus scops*), European Bee-eater (*Merops apiaster*), Calandra Lark (*Melanocorypha calandra*), and Short-toed Lark (*Calandrella brachydactyla*).

Farming in deserts means extensive use of biocides, whether pesticides or fertilizers. Pesticides are used to prevent the resourceful desert species that are adapted to exploit minimal resources, or to overcome exotic species brought in with the crop to be grown and proliferate, from causing damage (also termed "pest-control"). The low carrying capacity of the desert forces the farmers either to fertilize heavily in order to reap a satisfactory crop, or alternatively, to use other technological advances like hydroponic cultures. Mendelssohn and Leshem

(1983) reported that by the early 1980s over 600 different pesticides had been approved and were used in Israel. Mendelssohn (1972) believes that the main factor that affected raptor populations after 1950 was the "exaggerated application" of thallium sulfate (used legally until the 1970s) as a rodenticide. A wide range of raptors were found dead owing to secondary poisoning following the spraying of thallium, including Griffon Vulture (*Gyps fulvus*), Lappet-faced Vulture (*Torgos tracheliotos*), Cinereous Vulture (*Aegypius monachus*), Egyptian Vulture (*Neophron percnopterus*), Peregrine (*Falco peregrinus*), Lanner (*F. biarmicus*), Black Kite (*Milvus migrans*), White-tailed Sea Eagle (*Haliaeetus albicilla*), Marsh Harrier (*Circus aeruginosus*), and Brown Fish Owl (*Ketupa zeylonensis*)(Mendelssohn 1972; Shirihai 1996).

In Israel, as in most westernized nations, DDT was outlawed officially in the 1960s; however, to date many government agencies recommend the continued use of this deadly agent for a wide range of purposes. One recommendation is to spray the vicinity of human habitations in the deserts with 4 percent DDT in an attempt to eradicate the sand fly vector of the single-celled parasite, leishmania. Yosef (unpublished data) documented the die-out of several birds at a local kibbutz following such a spraying. How many more such incidents go unnoticed or unreported, and how well the farmers practice the restriction of DDT to human settlements is a question that no one is able to answer or supervise. We are aware that extensive use of fluoroacetamides, azodrin, and other chlorinated hydrocarbons and organophosphorus insecticides (including endosulfan, methomyl, monocrotophos, parathion, etc.) continues (Shirihai 1996; Yosef, personal observation). In addition, smuggling of biocides between Israel and the neighboring Arab countries is well developed. It has reached a level that has forced the Israeli Ministry of the Environment and the Nature Reserves Authority to set up a special unit to try and stop the biocide trafficking to and from Israel.

In contrast, the farming practices, and conservation efforts, have resulted in proliferation of specific species. These and some migratory species that exploit crops have become pests for farmers. In many cases, owing to political pressures, farmers have been allowed to kill these pests, even if the species is red-listed globally. A glaring example is the shooting of several tens of White Pelicans (*Pelecanus onocrotalus*) in northern Israel because of their feeding in commercial fish ponds. Other species considered pests are Common Cranes (*Grus grus*) and Pygmy Cormorants (*Phalacrocorax pygmeus*), and at present a strategy for the management of their wintering populations is being formulated in order to try and prevent further damage to agricultural fields.

Human Settlements

Israel's policy for "making the deserts bloom" (Ben-Gurion 1971) has resulted in a serious disturbance of the environment. Building settlements in the heart of the desert, gardening with exotic species, and creating human-related landfills have allowed tolerant species to exploit these introduced resources and to invade and displace indigenous species (Marzluff, Gehlbach, and Manuwal, this volume). Thus, the indigenous species are either relegated to shrinking areas or become extinct (Nathan, Safriel, and Shirihai 1996). An example of such a case is the settlement of the Arava Valley. Species like the Spur-winged Plover (*Hoplopterus spinosus*), Rufous Bushchat (*Cercotrichas galactotes*), Indian House Crow (*Corvus splendens*), and Graceful Warbler (*Prinia gracilis*) are colonizing the environs disturbed by humans (Shirihai 1996). Conversely, species like the Lappet-faced Vulture, Hoopoe Lark (*Alaemon alaudipes*), Bar-tailed Desert Lark *(A. cincturus)*, and Scrub Warbler (*Scotocerca inquieta*) are either already extinct in the wild or their population levels are very low. Shirihai (1996) found these changes to have occurred extensively for all the known major habitats of Israel. He found that in the Mediterranean garrigue and batha habitats heavy grazing, new human settlements, military training zones, and excessive planting of pines has led to the extinction of those species that breed in open or semi-open areas. These include Lesser Kestrel (*Falco naumanni*), Eagle Owl (*Bubo bubo*), Little Owl (*Athene noctua*), Long-billed Pipit (*Anthus similis*), Tawny Pipit (*A. campestris*), Black-eared Wheatear *(Oenanthe hispanica)*, Blue Rock Thrush (*Monticola solitarius*), Upcher's Warbler (*Hippolais languida*), Spectacled Warbler (*Sylvia conspicillata*), Orphean Warbler (*S. hortensis*), Whitethroat *(S. communis)*, Lesser Whitethroat (*S. curruca*), Great Grey Shrike (*Lanius excubitor*), Woodchat Shrike (*L. senator*), Cretzschmar's Bunting (*Emberiza caesia*), and Black-headed Bunting *(E. melanocephala)*. Additional examples resulted from the draining of marshes, wetlands, and riverine/riparian habitats in northern and central Israel during the 1950s. On the one hand, over twenty avian species became regionally extinct because of these actions. These include Grey Heron (*Ardea cinerea*), Purple Heron (*A. purpurea*), Great Crested Grebe *(Podiceps cristatus)*, White-tailed Eagle, Ruddy Duck (*Oxyura jamaicensis*), White-headed Duck (*O. leucocephala*), Spotted Eagle (*Aquila clanga*), Marsh Harrier, Brown Fish Owl, and Black Tern *(Chlidonia niger)*. On the other hand, several species established themselves well in these artificial habitats, and many are expanding their range. These include Little Egret (*Egretta garzetta*), Night Heron (*Nycticorax nycticorax*), Squacco Heron (*Ardeola ralloides*), Cattle Egret (*Bubulcus ibis*), Glossy Ibis (*Plegadis falcinellus*), Mallard (*Anas platyrhynchos*), Coot (*Fulica atra*), Spur-winged Plover, and White-breasted (or Smyrna) Kingfisher (*Halcyon smyrnensis*).

Pollution

Israel has few natural water sources. This has led to most water sources being captured and exploited for human purposes. Those that are not exploited are heavily polluted by the discharge of the industries and sewage of the towns and settlements of the region, which includes polychlorinated biphenyls (PCBs) and mercury (Shirihai 1996). This pollution has killed many species, including amphibians, molluscs, fish, and plants, seriously affecting the waterfowl and wader populations that breed in Israel (Paz 1987). No study on the effects of the contaminants on breeding birds has been carried out.

Groundwater Exploitation

More than one-half of Israel is desert. Exploitation of water sources for human purposes is considered a national priority, and groundwater is one of the major sources in the desert regions (Adar 1996). Along the Mediterranean coastline, overexploitation has led to the sea water advancing and replacing the brackish or sweet groundwaters. This affects the above-ground vegetation, which in its turn affects the fauna that can subsist in these disturbed areas (Paz 1987). In the deserts, overextraction of groundwater has led to desiccation of surface vegetation such as acacia stands and the drying up of natural water holes and oases on which the wildlife depend for their survival (Shirihai 1996). Many of these animals are now forced to drink or feed in agricultural fields. This is worrisome because we remain in the dark as to what chemical loads the animals pick up during their foraging activities.

Tourism and Recreation

Tourism is one of the largest industries in Israel and brings in a lot of much needed foreign currency. Israel is considered a good place for recreational activities, especially by Europeans during the winter. But it is not only the foreigners who have influenced the environment because hiking is considered an important element in the education of younger generations, and for decades priority has been given to outdoor activities with the aim of instilling a "love of the land through the feet." However, the organizers and participants consider these eco-tourist activities as something to do on holidays or as a patriotic duty and do not take into account the effects of their activities on the environment. Modern technical advancements have created additional problems of even greater magnitudes because better field vehicles (all-terrain vehicles, cheaper 4x4s, ultralight aircraft, etc.) allow increased access to areas previously undisturbed (Bury and Luckenbach 1983; Yosef 1997).

Sports such as rappeling (abseiling) and mountain climbing are known to disturb breeding raptors (Knight and Gutzweiller 1995) because rock climbers choose routes that are commonly used by breeding, foraging, and roosting wildlife. Climbers often disturb and on occasion cause desertion of nests of Bonelli's Eagle (*Hieraeetus fasciatus*), Griffon Vulture, Egyptian Vulture, Short-toed Eagle (*Circaetus gallicus*), Long-legged Buzzard (*Buteo rufinus*), Lanner, and Barbary Falcon (*F. pelegrinoides*). Rock climbers also cause other unintentional damage by removing what little vegetation grows along cracks in the rocks because it hinders their climbing (Knight and Gutzweiller 1995). This vegetation is important for wildlife, which use the rock face as a perch for viewing, feeding, and nesting. The Nature Reserves Authority is now regulating the sites at which rock climbing is allowed and preventing the use of known raptor breeding sites during the breeding season.

Trade in wildlife as pets is also a problem. Birds can be bought at pet stores, and many eventually escape into the wild or are released by the owners. An extreme example is that of the Ring-necked Parakeet (*Psittacula krameri*) that now breeds in several colonies in Israel and has displaced other hole-nesting species (e.g., Syrian Woodpecker, *Dendrocopos syriacus*). Recently, an aviary was opened to the public in central Israel, and since its construction species from all over the world have been reported in the environs of the Tel-Aviv region. The number of these escapees is unknown, as is their impact on the indigenous species.

It is essential that official agencies, such as the Ministry of Tourism, recognize the potential effects of their activities. In parallel to the marketing of Israel as the Middle Eastern Riviera, they should actively advocate conservation, so that they will have natural wonders left to market in the future.

Military Training

Since it received independence in 1947, Israel has always been in a state of war. This necessitates a well-trained army, which has resulted in most of the open areas being set aside for army training bases; these areas include very sensitive habitats. However, because of Israel's small dimensions, the army is unable to remain in the confines of the training areas, and exercises are also carried out in natural areas. Often this leads to a conflict between the training troops and the environment. Cases of soldiers shooting at wildlife have been documented (Ilani 1981).

Pilots often disregard flight safety regulations and fly extremely low through canyons. This can lead to the desertion of breeding raptors such as Griffon Vultures, Egyptian Vultures, and Bonelli's Eagle (e.g., Court et al., in press). It is probable that these disturbances also disrupt the breeding cycle of nocturnal raptors (e.g., Hume's Tawny Owl, *Strix butleri*) that breed in the desert canyons; however, this remains to be studied. Studies from North America are not encouraging (Andersen, Rongstad, and Mytton 1989; Awbrey and Bowles 1990; Ellis, Ellis, and Mindell 1991). At present the army coordinates all flight plans and training exercises with the Israeli Nature Reserves Authority, and all officers-in-training have to undergo courses in conservation; however, breach of regulations occurs occasionally (Court et al. in press).

The Peace Process

The above-mentioned problems have been exacerbated because of the success of the recent peace process among the nations in the region. Since the signing of the peace treaty between Israel and Jordan, the numbers of foreign investors has grown exponentially, and the number of development projects planned for the region is extensive. If all the proposed plans are implemented within the next decade, there will be no piece of the desert that will not have been "developed" for human purposes (Vardi, Benvenisti, and Seroussi 1995; Vardi et al. 1996).

Israel's Academia

Although none of the above-mentioned problems are new to Israeli scientists and environmental organizations (e.g., Yom-Tov and Mendelssohn 1988; Yom-Tov and Tchernov 1988; Nathan, Safriel, and Shirihai 1996), no serious effort to evaluate the status of the various groups of birds has been made. The exception is the raptors; however, no conservation implementation perspectives were ever applied to the breeding raptor surveys conducted (e.g., Bahat 1986; Frumkin 1986), the soaring bird surveys carried out annually (e.g., Dovrat 1980, Shirihai and Christie 1992), or the work with the Israeli Air Force implemented with raptors as the major conservation focus (e.g., Leshem and Yom-Tov 1996a,b). Only recently was a steering committee for raptors established and an evaluation of the breeding raptors of Israel carried out.

Thus, although many master's and Ph.D. dissertations are submitted annually in Israel's universities, little is known about the status of the passerines, near-passerines, waders, waterfowl, and pelagic birds. However, Israel's academia have shown that they care and that their actions can affect conservation-related issues. One such incident occurred in the 1950s, when the Hulah swamp was drained for human purposes. The objections raised by academia, and the subsequent creation of the first nongovernment, conservation-oriented organization, called the Society for the Protection of Nature in Israel (SPNI), forced the government to set aside a small area (approximately 5 percent) as a nature reserve. The SPNI has since been active in education, conservation, and ecotourism.

Hunting

Although hunting is not as extensive in Israel as in the neighboring countries (Woldhek 1980; Bildstein 1993), Israel's fauna has been hunted extensively in the past by its conquerors. Thus, for example, the Middle Eastern race of the Ostrich (*Struthio camelus syriacus*), which still existed at the beginning of the twentieth century in the Negev Desert, was hunted into extinction (Paz 1987). Today, hunting is regulated and licensed by the Israel Nature Reserves Authority. At present there are approximately six thousand registered and licensed hunters in Israel. None are allowed to hunt in nature reserves or on army training grounds, areas that comprise a large proportion of Israel.

Research Needed to Further Conservation

An important lesson from experience is that future threats are unpredictable, even at a gross scale (Moser et al. 1995; Newton, this volume). The greatest perceived threat to nature is human settlements, and few predicted the damage that has been inflicted by modern agricultural practices; reforestation with alien species in even-aged, monocultural stands; land drainage; water-table lowering; habitat fragmentation; and coastal developments. Birds are vital bioindicators of changing conditions in the natural world. Like ourselves, birds are at the top of many food webs, and their high metabolic rates make them particularly sensitive to stress and habitat degradation. Thus, long-term monitoring and environmental impact studies need to be emphasized in the immediate future if we are to preserve any of the remaining biodiversity in the remaining, patchy landscape of Israel. The most significant current threat for nature conservation is from rural land-use policy and development plans. There is potential to change this, but conservation biologists have to first clarify to themselves the extent to which each of the habitats and wildlife populations need protection. The conservation organizations need to keep on top of the range of threats, and the academics need to emphasize and encourage more applied research that will further our understanding of the existing flora and fauna and help to recommend to the authorities the compromises possible between human and environmental requirements and ways of implementing conservation measures.

A wide-ranging and scientifically rigorous monitoring program is essential for following changes in bird populations (Moser et al. 1995). Israeli conservationists must realize that it is important to maintain the full range of species and habitats in order to avoid species becoming vulnerable (Nathan, Safriel, and Shirihai 1996), and they should shoulder the responsibility of addressing the causes of changes and declines in populations at an early stage while the chance of recovery is high. The need to be proactive is stressed by the fact that many breeding species in Israel are at the periphery of their distribution. The species that are extinct in Israel were significantly more peripheral than others, and the 113 nonthreatened species are significantly less peripheral than others (Nathan, Safriel, and Shirihai 1996). Nathan et al. also report that most extinctions are related to human pressure. The same opinion is voiced by Paz (1987), who also reports that no species is confined exclusively to Israel. Current theoretical conservation biology is as yet incapable of predicting habitat conservation and species occupancy. This necessitates fieldwork, which can be used directly to influence the construction of land-use policies and development plans.

A blanket program for conservation is direly needed in Israel. Its importance is underlined by the fact that Israel is the only country in the Middle East wherein conservation is implemented and practiced to an extent comparable to North America and Europe. The aims of the conservation program need to be clarified and renewed based on present reality. The different habitats need to be mapped, and their relative importance needs to be established. It is also important to develop criteria for establishing the conservation status of each of the habitats and the species that inhabit them. Only when this is done can we evaluate the impact of the previously mentioned parameters on existing habitats and the need for establishing either short- or long-term nature reserves, or maybe even areas in which human activities will have to be kept to a minimum.

In the future, ecological studies based on sound natural history will be needed to suggest possible population changes and to try to understand the implications of these changes. It is the Israeli biologist of the year 2000 who will have to make the choice between becoming an ivory-tower ecologist, who may care for the environment but will not voice his or her opinion in a public forum, and a conservation biologist, who will make a stand, and hence also make a difference, against goverment and regional policies in an effort to save the remaining wildlife of the Holy Land.

Making Research Effective for Conservation

Research pertaining to the effects of the wide range of biocides used in agriculture, the compromises needed between human pressures and environmental requirements, and the effects of human encroachment in the various regions are direly needed. Further, this information needs to be divulged to the public and the policy makers in their language. The prevailing attitude of not publishing scientific data in lay journals and not getting the information out to the general public is probably one of the greatest problems facing Israel's scientific community. Its involvement in policy making, conservation implementation, and education of the general public is lacking and is badly needed.

Studies in the use of biocides by Israel's well-developed and extensive agricultural community are critical. This is because Israel's location at the junction of the three continents, and its function as a major migratory bottleneck, means that chemicals are taken up not only by the breeding populations but also by migratory species that breed in Europe and Asia. The results should be incorporated into the regional development planning and the relevant authorities alerted. Until the agricultural intensification policy is changed, we can expect the local and migratory bird populations to be negatively affected (Newton, this volume).

Research into the effects of conifer monocultures is also required. The creation of monocultural forests in the deserts will almost certainly have affected the breeding populations and their distributions. One would expect woodland species to penetrate into arid regions and desert species to retreat with the change in habitat quality. Also, planting of same-aged monocultures affects the breeding capabilities of hole-nesting species (Newton 1991). Researchers need to address the effects on avian populations and bring their study results to the attention of the agency that conducts and is responsible for these activities.

If present land-use policies are allowed to continue in their present form and state without taking environmental requirements into consideration, natural and undisturbed avian populations will exist only in nature reserves or not at all.

Acknowledgments

We thank the editors, Professor Berry Pinshow, and Richard Porter for their constructive criticism of earlier drafts of the manuscript.

CHAPTER 24

Research Priorities for Bird Conservation in Russia

Vladimir M. Galushin and Victor A. Zubakin

Conservation Issues and Previous Research

Occupying 17 million square kilometers (about 11 percent of earth's land area), a major part of which is still wild, with natural or close-to-natural bird habitats, Russia supports 732 bird species (Flint 1995). About twenty of these nest exclusively (or almost so) within Russia. Thirty-five are globally threatened, and twenty-one more are included in lists of near-threatened and conservation-dependent species (Collar, Crosby, and Stattersfield 1994). Russia is the population center for the great majority of waterfowl and waders in the Eastern Hemisphere, similar to Canada and the United States in the Western Hemisphere. Many waterfowl and waders nest in Russia and migrate to winter in western Europe, south and Southeast Asia, and Australia, thus playing a vitally important role in natural ecosystems and hunting economies of the many countries covered by their nesting and wintering ranges and the migratory pathways between. Healthy bird-nesting habitats in Russia means healthy migratory and wintering populations in western Europe, Africa, and southern Asia.

With the drastic changes in the socioeconomic situation in Russia over the last five to seven years, governmental supervision and regulation of nature conservation and environmental education were suddenly weakened. In some areas nothing replaced the previous methods of management and conservation. As a result, the general state of bird conservation and study deteriorated. Consequently, priorities for research and activity in the field of bird conservation need to be changed in accordance with the new reality of life in post-USSR Russia.

To better understand the impact of socioeconomic changes in current conservation priorities, a short review of past avian conservation and research efforts is presented.

Coordination of Research and Conservation Organizations

Bird conservation research in the former USSR was characterized by comprehensive coordination, planning, and funding through a well-developed, broad system of scientific, educational, and conservation organizations (Ilyichev and Flint 1982). Bearing in mind the large size of the country, ornithologists themselves sought such coordination in order to avoid duplication and concentrate efforts on important priorities. One of the most effective tools for coordination was regular conferences, workshops, and other scientific and conservation meetings.

The most significant of these meetings were the USSR Ornithological Conferences, from the first in Leningrad in 1956 to the tenth (and last) in Belorussia in September 1991, just three months prior to the disintegration of the Soviet Union. All conference expenses were paid from the state budget. Thanks to such support, many hundreds of ornithologists from all regions participated in meetings to discuss scientific findings and problems, or simply to meet each other regularly. The XVIII International Ornithological Congress in Moscow in August 1982 also offered a unique chance for Soviet ornithologists to exchange views and data with their foreign colleagues.

Since 1962, coordination of ornithological conferences, bird research, and conservation has been performed on a permanent basis by the USSR Ornithological Committee. The committee was included in the highest scientific body in the country, namely the USSR Academy of Sciences. After 1991, however, the committee ceased to exist. At the end of the 1980s, State Committees on Nature Conservancy functioned in the USSR and all republics. These were later transformed into ministries, but quite recently the aforementioned ministry in Russia was divided into the Ministry of Natural Resources and the State Committee on Environmental Conservation (GosComEcologia). Now, endless bureaucratic games will certainly prevent them from effective conservation of nature.

At the beginning of the 1990s a key position, the president's adviser on environmental protection, was instituted. Well-known zoologist and popular conservationist Professor Alexey Yablokov was appointed to that position, and he began pressing the government for solutions to environmental problems, including protection of wildlife and biodiversity as a whole. When his important position was simply terminated, the high officials that remained in position paid little attention to solutions of urgent environmental problems, despite a number of bilateral agreements and international conventions of environmental character signed by the Russian Federation.

Three major nongovernmental organizations have been addressing avian conservation in Russia for the last three decades: the All-Russian Society for Conservation of Nature, the USSR Ornithological Society, and the Movement of Student Brigades for Nature Defense.

The All-Russian Society for Conservation of Nature was formed in 1924. At the end of the 1980s its membership exceeded 30 million—over 20 percent of the population of the country. At present the membership has decreased 80 percent. Initially, the society was quite effective, particularly in the field of bird protection, thanks to well-known ornithologists, ecologists, and conservationists. Despite a general decrease in participation, some regional and local parts of the society continue to struggle for protection of birds and their habitats.

The USSR Ornithological Society was established in 1983 with the support of the Academy of Sciences. At the end of the 1980s its membership was over 2,500 ornithologists from all republics of the USSR. Its priorities are bird study and conservation, organization of conferences, and publications. Working groups on cranes, waders, bustards, raptors, and some other birds began in the society. Publication of popular books and papers on birds and their protection has been widely supported by the society through a contest named after the late naturalist and author Professor Alexander Formozov. Regional divisions of the society are active in protection of birds and bird habitat and in environmental education. A significant shortcoming of the society, though, is its failure to promote widespread bird-watching as the most reliable way to introduce people to bird study and conservation. The main cause of this was and is a shortage of inexpensive, popular field guides on the birds of Russia. After the USSR split, the society was renamed the Ornithological Society.

The Movement of Student Brigades for Nature Conservation appeared in the 1960s mostly in universities and other institutions of higher education. The first brigade was formed at Moscow State University in 1960. During the 1970s and 1980s, student brigades were the major force in the field of nature conservation. At that time, over one hundred brigades formed throughout the country. Their major activities were: (1) fighting violations of nature-protective legislation; (2) environmental education; and (3) scientific research. The movement was fully independent of governmental administrations and self-supported in both funding and structure. At the end of the 1970s, the movement established the program Fauna to promote bird study and conservation among young people. The program was successful in collecting data for red lists of rare birds, widespread environmental education, and preparation of proposals for protected areas.

An important product of the student movement was activists. Many now work in various capacities to benefit conservation of nature in Russia. Volunteers from nongovernmental environmental organizations used weekends and out-of-work periods in the field pursuing nature protection. Thanks to cheap transportation, volunteers were mobile and effective, and impressive results were achieved from small expenses but great enthusiasm.

Unfortunately, at present, most such enthusiasts lack the opportunities to continue in conservation activities. The majority are students, teachers, and young scientists who are forced to search for extra jobs to support their very existence; their "basic" salaries seldom exceed $50—80 per month and are often paid with three-to-five-month delays. Thus, volunteers lost both their free time and their ability to pay reasonable membership fees. Now expensive transportation has practically stopped travel to the field. These new circumstances, aggravated by cool attitudes of administrators toward the conservation movement, significantly affect its activity at the present time. Traditional nongovernmental environmental organizations failed to adapt to recent social and economic conditions, which significantly diminished their effectiveness. There is an urgent need to establish a functional network of new nongovernmental environment organizations to readdress vitally important research and conservation priorities. Yet fee-based volunteer organizations are not possible under present economic conditions in Russia. In western countries, the most effective environmental organizations are operated by a paid staff. Well-known examples are the National Audubon Society (USA), Royal Society for the Protection of Birds (UK), Vogelbescherming Nederland, Naturschutzbund (Germany), and the Lega Italiana Protezione Uccelli (Italy), which have thousands of members and multimillion dollar annual budgets. Therefore, our only chance in Russia is to establish organizations with initial support from friendly foreign and international societies. The first such effective new environmental organizations in Russia were the Socio-Ecological Union, Biodiversity Conservation Centre, and several smaller regional groups and societies.

One new organization, the Russian Bird Conservation Union (RBCU), was established in 1993; essential financial support and technical assistance was issued from Vogelbescherming Nederland in 1995. RBCU joins the partnership network of BirdLife International as its Partner Designate. RBCU has sixteen regional branches and about one thousand individual members. The RBCU goals are to organize, coordinate, and support bird conservation in the country. Its first priorities are site and species conservation, mostly covered by two large programs, Important Bird Areas in European Russia, and Globally Threatened Species. The primary activities of the former program are inventory of existing areas and identification of new areas important to birds, and collection of data for a new issue of the European catalog *Important Bird Areas in Europe*. As for the latter program, twenty-three globally threatened bird species occur in Russia, for which National Action Plans for conservation are being developed. The RBCU also publishes bulletins, leaflets, and newsletters and is preparing to publish a *Field Guide on Birds of Russia*. Various activities like Bird Days, Bird of the Year, and participation in World Bird Watching Days are current objectives of RBCU in the fields of ornithological education and public awareness of bird protection. Almost fourteen thousand people from 297 regions participated in the latter events in October 1995 (Lebedeva 1996).

Habitat Conservation

By the beginning of the 1990s, the USSR had established a dense network of Strictly Protected Areas at various levels: Strict Nature Reserves (Zapovedniks), National Parks, State and Local Sanctuaries (Zakazniks), and Nature Monuments (Boreiko 1995, 1996). In 1991 the USSR had 173 Zapovedniks (Krever et al. 1994; Volkov 1996). Five years later there were 93 Zapovedniks and 31 National Parks in Russia alone (over 360,000 square kilometers, nearly 2 percent of the country's total area). There are now 233 such areas in the Moscow region, 216 in Nizhny Novgorod region, and 120 in Buryat region southeast of Lake Baikal.

Conditions in Protected Areas are growing worse, however, due to lack of funds for wildlife research and conservation and the attempts of local people and administrations in some regions to privatize the lands of local Zakazniks. Nevertheless, contrary to pessimistic expectations, the existing system of Protected Areas still functions and is even growing, thanks to the enthusiasm of their staff, authorities within GosComEcologia, and grants from international organizations. Almost all of the Protected Areas conserve some bird habitat; hence, proposals for their continued development remain a research priority for ornithologists.

Rare Species

Before the end of the 1960s, birds in Russia were generally partitioned into "useful" (game, insectivorous, and traditionally respected species such as swans, cranes, white storks, and others) and "harmful" species (granivores, raptors, piscivores, corvids). The former were protected, while the latter were persecuted. After it was proved that such a partition was unnatural, however, attitudes toward birds were changed, thanks to extensive ecological research (Galushin 1963, 1982). After wide and comprehensive discussion, almost all species except the Hooded Crow *(Corvus corone cornix)* have been declared as ecologically valuable and protected.

For threatened and rare species, the *USSR Red Data Book* was formally established by the government in 1974; its two editions were published in 1978 and 1984. The *Red Data Book of Russia,* published in 1983, included 107 of the 732 species recorded in the country. The second edition is under preparation; its larger red list includes 125 bird species or subspecies. By a special decision of the Russian government in 1996, statuses of species in the *Red Data Books* were confirmed by legislative acts: all cited species and their habitats are now strictly protected by federal laws.

Regional *Red Data Books* were subsequently developed that necessarily included species from the USSR and Russian books plus locally rare species. Unfortunately, their quality widely varies from region to region, subject to data collection by local researchers. Preparation and publication of scientifically valid *Red Data Books* for every administrative region (republic, kray, and oblast) are current priorities of local zoologists and conservationists.

Publication of *Red Data Books* stimulated both bird study and conservation. *Red Data* species have been included in federal and local research plans and budgets as priorities, and both work time and money have been allocated for intensive studies. As a result, scientific data on rare bird species have significantly increased. Conservation has also been successful for many rare species.

Population trends of raptors are exemplary of changes seen subsequent to legislative protection and conservation research prioritization (Galushin 1996). Of the thirty-four raptors known to breed in European Russia, two rare species, the Booted Eagle *(Hieraaetus pennatus)* and Red-footed Falcon *(Falco vespertinus)*, are recommended for inclusion in the second edition of the *Red Data Book* of the Russian Federation as species of conservation concern. Regarding population trends, all raptors could be distributed into three main categories: (1) species with decreasing populations; (2) species with relatively stable or poorly known population trends; and (3) species with generally increasing populations. These divisions are derived from tentative population trends that are subject to further monitoring. Causes of the above trends are not clear for some of the species, and so thorough studies of raptor populations are also a conservation research priority. Proposals have been developed for national and joint international efforts to study and protect raptors. Breeding ranges and populations of certain threatened and rare European raptors are located almost entirely within European Russia (80–100 percent), yet their distributions and population trends are much less well known here than in western and central Europe. Conservation measures are consequently less advanced in Russia. Likewise, the number of ornithologists and the technical and financial potential for research in the east are hundreds of times smaller than in the west.

To develop scientific bases for their protection, large-scale population studies of raptors in Russia may best be achieved by close cooperation with ornithological and conservation societies, organizations, and foundations from western Europe and North America. An excellent example of such fruitful collaboration is the long-term study of pesticide contamination of raptor eggs in various regions of European Russia by Dr. C. J. Henny from the United States, with many Russian colleagues (Henny et al. 1994, 1995, 1996).

Recent Changes in Bird Conservation and Research

Consequences of the Present Socioeconomic Situation

To summarize the state of bird conservation and research beginning in the 1990s, some achievements and shortcomings have to be mentioned. The major advance in the immediate past has been the administrative and financial support for a strong system of research and conservation by official scientific and environmental structures on both federal and local levels. Such support was strictly planned in advance and thus is far from perfect concerning response to actual day-to-day situations, including international developments in conservation. In order to influence research and conservation directions, both nongovernmental organizations and individual scientists had to be prepared to prove their findings and arguments to high officials in the Academy of Sciences, various ministries, and other governmental and party structures. When such proposals were approved, they immediately became governmental targets and were constrained due to limited personnel, planning, and budgeting.

The breakup of the USSR also resulted in weakening of communications among ornithologists. Coordination between basic science and conservation fell apart. After the tenth USSR Ornithological Conference in 1991, there were no such gatherings in any of the post-USSR countries. Fund allocations to field research in federal and local budgets tended toward zero. Administrative support for strict observation of conservation legislation virtually ceased, and poaching of game and commercially valuable birds such as large falcons became more and more common. With decentralization, regional and local authorities changed hunting regulations to the detriment of sustainable use of game populations. Poor observance of hunting and conservation regulations was only slightly countered by the increased expense of hunting. Illegal capture and smuggling of raptors for falconry or private zoos abroad still weigh heavily on populations of rare birds. In Kazakhstan alone, Saker Falcons (*Falco cherrug*) suffer severe losses from nest robbing and illegal removal (Kenward et al. 1995, 1996); about one thousand sakers were taken from the country in 1994 (Sklyarenko 1995). Curbing illegal removal and smuggling of valuable and rare birds is a vitally important priority of bird conservation in all post-USSR countries.

Nevertheless, some consequences of socioeconomic changes have been positive. Ornithologists are now free to pursue any kind of research. Joint international projects are now possible, including field studies in remote regions of the country. National and international foundations provide grants to support bird research and conservation. Administrative obstacles to information exchange with foreign colleagues and participation in international conferences have been waived. Before the 1990s, only three to five ornithologists from the USSR, funded from the central budget, were allowed to participate in International Ornithological Congresses; in contrast, over fifty participants from Russia, Ukraine, Belorussia, and other former USSR countries took part in the 1994 International Ornithological Congress in Vienna. Their participation, though, was almost entirely supported by the Organizing Committee and various international foundations. With similar support, ten to twelve specialists from the former USSR participated in international conferences on raptors in Germany in 1992, Spain in 1995, and Italy in 1996. For the great majority of ornithologists, however, this remains a very new and yet unresolved dilemma: now they have the rights to go anywhere but lack the money to do so.

Consequences of the economic disaster in post-USSR countries that are negative for the great majority of people are at the same time positive for wildlife. Recent drastic changes in the economy will certainly modify bird habitats and eventually bird populations. Millions of hectares of agricultural lands are being neglected, haymaking has stopped on meadows, and forest clearings are being overgrown. Industrial and agricultural pressures on bird habitats have softened, as have human disturbances. Environmental pollution has sharply decreased. Relief from industrial and agricultural (particularly pesticide) pollution may also be reflected in bird populations and is likewise worthy of thorough research. This unique "natural experiment" across huge areas of eastern Europe and northern Asia should not go unnoticed by ecologists and conservationists.

Research Needed to Further Conservation

In the coming decades, the crucial priority must be a full-scale restoration of the total amount of avian research and conservation to their levels at the beginning of the 1990s. Concomitant to this goal, coordination among research and conservation and greater adaptability to present and potential socioeconomic conditions are essential.

As for avian conservation research, three top priorities are: (1) assessment and prediction of the impact of present socioeconomic conditions on bird diversity and population dynamics; (2) survey and monitoring of bird fauna, populations, and habitats in various regions; and (3) implementation of conservation biology practices, including development of appropriate measures for protection of birds and their habitats. Consequences of the large scale, reduction in human impact on landscapes for bird species diversity, distribution, and population changes also warrant thorough scientific investigation.

Finally, changes in human attitudes to various birds (game, rare, so-called "useful" or "harmful," etc.) under present social, economic, cultural, and educational conditions also merit study. Biologically appropriate recommendations for adequate response to population changes must be developed by both governmental and nongovernmental conservation organizations.

Research Needed to Assess Bird Populations

The huge size of Russia and its rich biodiversity mean that populations of the same species in different regions are unevenly known. Generally speaking, rare and game bird numbers are better studied everywhere; the reliability of assessment of other bird populations varies from region to region. In some cases original data are widely scattered in the literature and hidden in often obscure local publications. Hence, one priority is population assessment supplementing available data with further surveys. This is vitally important in the context of the

BirdLife International Project for determination of Important Bird Areas. The European Bird Atlas approach of using standard 10 x 10- or even 50 x 50-km plots is impossible in the vast territory of Russia, with its relatively few ornithologists and inadequate transportation. Therefore, the only means of preliminary population assessment for many bird species is via extrapolation of available data from limited areas. To make such extrapolations more reliable, surveys in key areas must be performed.

Such projects could be more efficiently carried out jointly with international or foreign support. For example, the Important Bird Areas (IBA) Project in European Russia is currently being undertaken by the RBCU with substantial support of both Vogelbescherming Nederland and BirdLife International. Data obtained through this project could later be used for better bird conservation within IBAs and as nuclei for threatened species population assessment by extrapolation to large areas. Another example is a computer database at the Novosibirsk Institute of Animal Systematics and Ecology (Siberian Division of Russian Academy of Sciences), where data on bird numbers are collected from key areas and used for extrapolation of passerine distribution across central Siberia (Ravkin 1984).

Monitoring bird populations is also an important research priority. Monitoring of selected species has been performed within Strict Nature Reserves for a long time, as well as by Regional Game Management Offices (for game birds) and by the Ministry (State Committee) on Environmental Conservation (for rare birds). Common birds and potentially rare species are not covered by regular monitoring. The first steps in development of a bird-monitoring network should be selection of typical key areas within all natural zones and large ecosystems, appointment of ornithologists to carry out regular surveys every three to five years at each site, and use of a unified methodology. Dozens of professionals and hundreds of experienced bird-watchers are needed to make this survey system efficient. Building up a large network of bird-watching through the whole country is an important goal for RBCU and ornithological and conservation organizations.

In light of the extensive use of the continental shelf for oil excavation and other activities, the compilation of a seabird catalog (similar to that for Alaska) has become a particular priority. Through a joint project of the Institute for Biological Problems of the North (Russian Academy of Sciences) and the United States Fish and Wildlife Service, a catalog of seabirds of northeastern Russia is under preparation. Another one for the Barentz Sea is being prepared by the Institute of Nature Conservation. New seabird colonies have been recently described on Novaya Zemlya, Severnaya Zemlya, and Franz Josef Islands (Pokrovskaya and Tertitsky 1993; Gavrilo et al. 1994; Krasnov 1995). However, a seabird catalog for all aquatories in Russia remains a task for the future.

Monitoring of rare bird populations on both federal and regional levels continues to be an important priority for the preparation of valid *Red Data Books* for each of the eighty-nine administrative regions of Russia, over twenty of which are already published. At the same time, species from the new federal list of 125 threatened and rare birds are not adequately known. Outlines for conservation of 5 especially rare species appear below:

The Slender-billed Curlew (*Numenius tenuirostris*) is the most mysterious bird of Russia. The species is obviously on the brink of extinction. From time to time, individuals and small groups have been recorded wintering in Morocco and along migratory routes in Bulgaria, Rumania, Hungary, Greece, Italy, Turkey, and the Ukraine. Based on these records, the world population is roughly estimated at 50–270 pairs (Gretton 1991; Tucker and Heath 1994; Heredia, Rose, and Painter 1996). However, despite many years of intensive searching, nesting sites of the species have been found neither in Russia nor in Kazakhstan. Ideas on potential breeding locales of the species have recently appeared, which may help to reveal nesting sites of this secretive bird (Danilenko, Boere, and Lebedeva 1996). Sophisticated techniques such as miniature satellite transmitters could also solve the problem. To find and save this species is the most urgent priority of Russian ornithologists and conservationists.

The Lesser White-fronted Goose (*Anser erythropus*) nests almost entirely in Russian tundra. Its breeding range and population have decreased for the last few decades. The species was included in the *Red Data Book* of Russia (Elyseev 1983). The breeding range has become extremely fragmented; assessments of population totals vary widely from 30,000 to 160,000 individuals (Krivenko 1991; Morozov 1995). To work out appropriate conservation measures, this species' range and population size have to be refined.

The Baikal Teal (*Anas formosa*) breeds only in eastern Siberia. Its population is estimated at 75,000 individuals (Krivenko 1991; Rose and Scott 1994) but has decreased for the last twenty years. Particular causes of that decline are still unknown.

The Pallid Harrier (*Circus macrourus*) is disappearing from Russia. Its traditional breeding range and population size have shrunk considerably over the last twenty years (Galushin 1994). At the same time, quite unexpected nesting sites have been recorded in the taiga of the Perm region, far beyond this species' known breeding range (Lapushkin et al. 1995). This phenomenon may hint at a redistribution of the species and warrants more precise study.

The Spotted Greenshank (*Tringa guttifer*) probably nests only within Russia. About forty pairs have been counted on the Sakhalin Island (Nechaev 1991). The world population is assessed at less than one thousand individuals (Rose and Scott 1994). Therefore, searching for still unknown nesting sites is important.

Research Needed to Assess Bird Habitats

Purposeful bird habitat surveys are carried out in two major directions. From the middle of the 1970s, under the Ramsar Convention on Wetlands of International Importance as Wildlife Habitats, the USSR government established twelve internationally important wetlands, including three in Russia. In 1994, thirty-five such wetland habitats were designated within Russia. At present, wetland habitats of both international and national importance continue to be described and cataloged.

In 1987, the USSR Ornithological Society joined a Programme of International Bird Areas (IBAs) in Europe. In two years 75 IBAs have been revealed within European Russia (Grimmett and Jones 1989). At present, RBCU coordinates this work under the auspices of the BirdLife International IBA Project, supported within European Russia by BirdLife International and Vogelbescherming Nederland. At the end of 1996, about 150 IBAs within European Russia were described by RBCU. Within the next decade, RBCU plans to complete a catalog of IBAs and key ornithological areas of federal and regional importance. These areas, together with wetland habitats, need to be protected and monitored as Special Protection Areas.

Natural History and Education

Successful bird conservation depends on knowledge of each species' biology. Particular characteristics of ecology and behavior of various species and their adaptability to changing environment are vitally important for their survival and well-being (Soulé and Wilcox 1980; Soulé 1987). A single change of environment can lead to quite different responses even by related species, resulting either in a population increase or in a sudden population crash, as with the Baikal Teal and Ferruginous Duck (*Aythya nyroca*) in Russia. In this respect, conservation priorities in Russia are to determine specific biological, ecological, and behavioral characteristics for species of particular concern that may be under direct and indirect anthropogenic pressure. This approach seems promising regarding comparative study within model groups of birds: (1) game birds under hunting pressure, (2) former "harmful" raptors, (3) cranes, divers, woodpeckers, and the majority of small birds toward which human attitude is traditionally neutral.

The exceptionally important and mandatory component of any conservation strategy and crucial long-term priority is overall environmental education to advocate human awareness of birds.

Making Research Effective for Conservation

To be effective, conservation of birds has to be developed on the results of sound ornithological research and widely supported by a great number of experienced bird-watchers. Therefore, bird conservation studies are aimed in two general directions: professional and amateur. Ideally, both must be equally prioritized. However, in Russia the latter is in fact lagging in comparison to western countries.

Scientific research on birds has always been in progress in Russia. Current avian research and conservation priorities are based on the following convictions. First, Russia has the richest diversity of bird populations in need of research and protection. Russia also has hundreds of enthusiastic ornithologists and thousands of activists in conservation organizations doing bird research and conservation. Finally, Russia is open to further collaboration with foreign and international persons and organizations to the benefit of Russian birds, as both national and global wealth.

Acknowledgments

The authors are thankful for everybody who helped them in collection of data and preparation of this paper. Most of all, we are grateful to the enthusiastic people who still continue bird research and conservation in Russia, as well as to those who support these vitally important priorities. We thank Lauren Gilson, Mark Fuller, and John Marzluff for comments on earlier drafts and help in anglicizing our prose.

CHAPTER 25

Avian Conservation Research in the Mariana Islands, Western Pacific Ocean

Gordon H. Rodda, Earl W. Campbell III,

and Scott R. Derrickson

Conservation Issues and Previous Research

Of the three island regions of the Pacific (Polynesia, Melanesia, and Micronesia), Micronesia has the smallest land area. Although spread over an area roughly as expansive as the contiguous United States, the aggregate land area is less than 2,000 square kilometers, a little larger than the Hawaiian island of Maui (Engbring and Pratt 1985). Micronesia consists of the archipelagos of the Marianas, Marshalls, Gilberts, and Carolines. The Mariana Islands constitute over half (1,019 square kilometers) of the Micronesian land area and include Micronesia's largest single island, Guam. The four largest islands of the Marianas are at the archipelago's southern end and house virtually all of the human population: Guam (53 percent of Mariana Islands' land area), Saipan (12 percent of area), Tinian (10 percent), and Rota (8 percent). None of the thirteen other major islands in the Marianas constitutes more than 5 percent of the total area. Within this remote, fragmented land area are a high proportion of endemic land birds (table 25.1). Of the twenty historically resident species, fourteen to seventeen (uncertainty due to questionable taxa) are endemic to the Marianas at either the species or the subspecies level. Endemism is especially high among the sixteen forest birds, twelve to fifteen of which (75–94 percent) have unique forms in the Marianas (eight endemic spp.). Only three of the forest bird species range beyond Micronesia. In contrast, the nearshore and pelagic seabirds are all wide ranging

(Reichel 1991). Due to the absence of local endemism, research work on seabirds in the Marianas has been limited to status surveys, and conservation work has focused on the resident landbirds, as will this chapter. However, as for many oceanic island birds worldwide (Atkinson 1989; Burger and Gochfeld 1994), introduced species represent the primary conservation threat to seabirds (and landbirds) in the Marianas (table 25.2). There are no native resident predators on birds except other resident birds, primarily crows. Subfossil accumulations of bird bones in eyries suggest the prehistoric presence of a bird-eating raptor such as the Peregrine Falcon (*Falco peregrinus*) (Steadman 1992). Native mammalian or reptilian predators have always been absent.

With such a small land area, one would suppose that habitat destruction would be a major threat to resident birds. The potential for large-scale habitat loss is pervasive and has already occurred in some areas. Guam, for example, has approximately the same human population density as India, with the concomitant fragmentation and loss of natural habitats. However, the extant birds appear relatively tolerant of habitat modifications. The episodically limited resources of the islands and the passage of extremely destructive super-typhoons (sustained winds of more then 240 km/h; see also Wiley and Wunderle 1993) seem to have produced recurring resource bottlenecks that over evolutionary time pre-adapted the birds to dietary and habitat opportunism (Engbring and Pratt 1985). Such tolerance was notable following the Second World War. Several islands, including Guam, Tinian, and Saipan experienced extensive destruction during that war. Guam was subjected to the most severe naval bombardment in history, which destroyed over 80 percent of the island's structures (Morison 1953). Prewar habitat loss from shore-to-shore sugar cane production on Tinian, Saipan, and other islands exacerbated the loss of habitat due to the fighting. Other islands in Micronesia (e.g., Beliliou) experienced even more intense habitat alteration, yet no bird forms were lost (Engbring and Pratt 1985). Much of the habitat temporarily destroyed in the war was permanently replaced by stands of the exotic leguminous tree tangentangen, (*Leucaena leucocephala*), which was spread over much of Micronesia to curb soil erosion after the hostilities ended. This tree maintains monotypic stands and in most areas has not yielded to succession by native trees in the half century since the war (Craig 1994). Nonetheless, most of the native forest birds have done well in this and other modified habitats, although often not as well as in native forest (Craig 1990, 1996; U.S. Department of the Interior 1996). As more data are gathered, it seems likely that habitat requirements will be found to have greater importance in conserving some species, but present information indicates that introduced predators are the immediate threat to vertebrate welfare in the Marianas.

Thus, unlike many mainland areas, retention of pristine habitat may not be a priority for many species in the Marianas. Indeed, restoration of large amounts of habitat, to the extent that this requires eradication of all exotic predators, is not possible with current technology, and may never be. Instead, it may be

Table 25.1. Native resident forest birds of the Marianas (following Pratt, Bruner, and Berrett 1987).

		Endemism	
		Marianas	Micronesia
Native forest birds			
Bridled White-eye	*Zosterops conspicillatus*	e	
Guam	*Z. c. conspicillatus*	x	
Rota	*Z. c. rotensis*	e	
Saipan, Tinian	*Z. c. saypani*	e	
Collared Kingfisher	*Halcyon chloris*		
Rota	*H. c. orii*	e	
Tinian, Northern Islands	*H. c. albicilla*	e	
Golden White-eye	*Cleptornis marchei*	e	
Guam Flycatcher	*Myiagra freycineti*	x	
Guam Rail	*Rallus owstoni*	x	
Island Swiftlet	*Aerodramus vanikorensis*	ssp?	
Mariana Crow	*Corvus kubaryi*	e	
Mariana Fruit-dove	*Ptilinopus roseicapilla*	e	
Micronesian Honeyeater	*Myzomela rubrata*		e
Marianas	*M. r. saffordi*	e	
Micronesian Kingfisher	*Halcyon cinnamomina*		e
Guam	*H. c. cinnamomina*	x	
Micronesian Megapode	*Megapodius laperouse*	ssp?	e
Micronesian Starling	*Aplonis opaca*	ssp?	e
Nightingale Reed-warbler	*Acrocephalus luscinia*	e	
Aguiguan, Saipan	*A. l. nijoi*	e	
Alamagan	*A. l. yamashinae*	e	
Guam	*A. l. luscinia*	x	
Rufous Fantail	*Rhipidura rufifrons*		e
Aguiguan, Saipan, Tinian	*R. r. saipanensis*	e	
Guam	*R. r. uraniae*	x	
Rota	*R. r. mariae*	e	
Tinian Monarch	*Monarcha takatsukasae*	e	
White-throated Ground-dove	*Gallicolumba xanthonura*		e
Native nonforest landbirds			
Common Moorhen	*Gallinula chloropus*		
Marianas	*G. c. guami*	e	
Mariana Mallard	*Anas platyrhynchos oustaleti*	x	
White-browed Crake	*Porzana cinerea*		
Yellow Bittern	*Ixobrychus sinensis*		

e = endemic and extant.
x = endemic but extinct in the wild (only rail and Micronesian Kingfisher in captive propagation).
Many observers recognize endemic subspecies not recognized by Pratt et al.; these are denoted "ssp?" Questionable occurrences (e.g., Common Buzzard on Anatahan) are omitted.

Table 25.2. Introduced species of potential significance to birds.

Invertebrates	cockroach	*Periplaneta americana*	H
	various mosquitos		
	(none native)	Cuculidae	D
	African giant snail	*Achatina fulica*	F
	flatworm	*Platydemus manokwari*	C
Amphibians	treefrog	*Litoria fallax*	F, C
	marine toad	*Bufo marinus*	D, C
Reptiles	house gecko	*Hemidactylus frenatus*	F, C
	oceanic gecko	*Gehyra oceanica*	F, C
	mutilating gecko	*Gehyra mutilata*	F, C
	green anole	*Anolis carolinensis*	F, C
	skink	*Carlia fusca*	F, C
	monitor	*Varanus indicus*	P, C
	brown tree snake	*Boiga irregularis*	P, F
Birds	Black Francolin	*Francolinus francolinus*	C
	Blue-breasted Quail	*Coturnix chinensis*	C
	domestic fowl	*Gallus gallus*	C
	Rock Dove	*Columba livia*	D,C
	Philippine Turtle-dove	*Streptopelia bitorquata*	C
	Black Drongo	*Dicrurus macrocercus*	C
	Eurasian Tree Sparrow	*Passer montanus*	H, D, C
	Chestnut Mannikin	*Lonchura malacca*	C
Mammals	house mouse	*Mus musculus*	F, P
	roof rat	*Rattus tanezumi*	P, C
	Polynesian rat	*Rattus exulans*	P, C
	Norwegian rat	*Rattus norvegicus*	P
	musk shrew	*Suncus murinus*	P, F, C
	dog	*Canis familiaris*	P
	cat	*Felis catus*	P
	Philippine deer	*Cervus mariannus*	H
	pig	*Sus scrofa*	H, P
	goat	*Capra hirtus*	H
	cow	*Bos taurus*	H
	water buffalo	*Bubalus bubalis*	H
	human	*Homo sapiens*	P, C, H

The potential roles are coded as C = competitor, D = disease vector or poison, F = food, H = habitat modification, P = predator.

necessary to find a completely new path to avian conservation. The approach of promoting avian conservation without the assistance of the environmental conditions under which the birds evolved carries great risks and even greater uncertainty. If one is going to remake an ecosystem, it is necessary to have a commanding knowledge of ecosystem design constraints. Given the current paucity of research on birds and other key species of the Marianas, this is a daunting requirement.

Although many species can tolerate changes in habitat composition, few forest birds can thrive in the absence of forest; native bird populations will diminish in response to urbanization. Naturally, the protection of suitable habitat is always desirable and will ultimately set the limit on the preservation or restoration of avian populations. For example, wetland birds cannot be maintained if wetlands are eliminated. Prior to human habitat modifications, permanent natural wetlands numbered fewer than a dozen on each of the Mariana Islands, and many of these have since been reduced or eliminated. Of the four wetland bird species, one, the Mariana Mallard (*Anas platyrhynchos oustaleti*), is extinct (Reichel and Lemke 1994); another, the White-browed Crake (*Porzana cinerea*), is extirpated; and a third, the Mariana Common Moorhen (*Gallinula chloropus guami*), is endangered (Stinson, Ritter, and Reichel 1991; Wiles et al. 1995). Qualitatively, the research and conservation activities needed to address this loss of habitat do not differ in the Marianas from that required elsewhere, and this concern, though vital, will not be addressed further in this chapter.

Of the introduced species that have influenced the avifauna of the Marianas (table 25.2), the most notable have been humans (Steadman 1992) and the brown tree snake (*Boiga irregularis*) (Savidge 1987; Engbring and Fritts 1988). Since the arrival of humans and their commensals, thirteen of twenty-two (59 percent) avian species on the island of Rota have disappeared (one shearwater, one tern, one duck, one megapode, three rails, two pigeons, one parrot, one swift, one monarch flycatcher, and one parrot finch; Steadman 1992, 1995a,b). The losses experienced on other islands are less well documented but will undoubtedly be found to be quantitatively and qualitatively similar to those on Rota and other Pacific islands (Steadman 1995b).

The snake was accidently introduced to Guam among derelict war matériel retrieved from the New Guinea area following World War II (Rodda, Fritts, and Conry 1992). In addition to causing about one hundred power outages per year (Fritts, Scott, and Savidge 1987), consuming innumerable pets and domestic fowl (Fritts and McCoid 1991), threatening the survival of human babies (Fritts, McCoid, and Haddock 1994), aiding or causing the loss of the majority of the native mammals (all bats; Wiles 1987) and lizards (Rodda and Fritts 1992), the snake bears primary responsibility for the loss of virtually all of Guam's native forest birds (Savidge 1987; Engbring and Fritts 1988). As of mid 1996, only three native forest bird species survived on Guam: The Mariana Crow (*Corvus kubaryi*) numbers less than twenty individuals, and no pairs persisted through the last breeding season (C. F. Aguon, Guam Department of Fish and Wildlife, personal communication); the Micronesian Starling (*Aplonis opaca*) has only one viable population, of fifty to one hundred birds (Wiles et al. 1995); and the Island Swiftlet (*Aerodramus vanikorensis*) is reduced to a remnant population of several hundred roosting primarily in a single cave (G. J. Wiles, personal communication).

One full species, the Guam Rail (*Rallus owstoni*), and one arguably full species, the Guam form of Micronesian Kingfisher (*Halcyon c. cinnamonmina*), are being maintained in captivity. The population of the latter has declined from sixty-five to forty-eight birds over the last six years, and there exists some doubt as to whether the population can be sustained (Hutchins, Paul, and Bahner 1996). The rail has done much better, providing sufficient numbers of individuals to allow release of selected individuals on the presumed snake-free island of Rota. To date, the seven releases have not been successful, although a wild offspring was seen (Witteman and Beck 1991; M.K. Brock, personal communication). No attempts have yet been made to restore birds to Guam, as this effort would seem unlikely to succeed until snake-free or snake-reduced areas have been created.

As has been widely noted (Pimm 1987; Jaffe 1994), the forests on Guam are now qualitatively different from those in Hawaii, despite the absence of native forest birds in most lowlands of both areas; in Hawaii the forests support a variety of introduced birds, whereas in Guam there is silence. Introduced Black Drongos (*Dicrurus macrocercus*) and Philippine Turtle-doves, (*Streptopelia bitorquata*) are present in many forested areas of Guam, but they are sparse in comparison to their densities prior to the irruption of the snake.

This avifaunal catastrophe has not, to date, spread to other islands. However, the other Mariana Islands receive virtually all of their supplies via Guam, and brown tree snakes are frequent stowaways within cargo (Fritts and Rodda 1995). Saipan has experienced nearly forty snake sightings in the last six years, including several in which the snake was captured and positively identified as a brown tree snake. Rota has experienced poorly understood declines in the populations of its four smallest bird species, leading to speculation that the snake has colonized that island as well. If the snake were to colonize Rota, Saipan, and (as seems likely as a consequence) their intervening neighbor, Tinian, and if the loss of bird species experienced by Guam were to be repeated, occupied habitat would be reduced to less than 10 square kilometers for all of the Mariana endemic forest birds except for the Nightingale Reed-warbler (*Acrocephalus luscinia*), which would have less than 20 square kilometers.

Although the snake-caused extirpations on Guam have understandably received the most attention, other bird populations have been lost for a variety of reasons: the Golden White-eye (*Cleptornis marchei*), from Tinian; the Island Swiftlet from Rota and Tinian; the Mariana Mallard from its entire range (Guam, Tinian, Saipan); the Micronesian Megapode (*Megapodius laperouse*) from Rota, Tinian (possibly reestablished), Saipan (since repatriated), and Guam; the Nightingale Reed-warbler from Tinian and Pagan; and the White-browed Crake from Guam (Wiles, Beck, and Amerson 1987; Reichel and Glass 1991; Stinson and Glass 1992). None of these losses resulted in the extinction of a currently recognized taxon, but their cumulative effect adds urgency to the avifauna diversity crisis in the Marianas.

Due to the magnitude of the biodiversity problem, there is ample justification for exhaustive research on Mariana birds. However, fieldwork is costly and difficult due to the remoteness of the archipelago. Prior to about the 1960s, research focused on either taxonomic or distributional questions. Since the early 1970s, formal status surveys have been conducted on a variety of species (Engbring and Ramsey 1984; Engbring, Ramsey, and Wildman 1986; Grout, Lusk, and Fancy 1996). Valuable natural history anecdotes have been compiled by several systematists and visitors (e.g., Baker 1951; Marshall 1949; Stophlet 1946), but focused natural history and ecology studies have been published only in the last two decades. We are aware of only eighteen publications focused on ecology or natural history, exclusive of gray literature and status and distribution surveys. Of these, over half are notes, and only four full papers provide data on the ecology of birds in the wild. White-eyes have been the subject of several papers (Bruce 1978; Craig 1989, 1990, 1996; Stinson and Stinson 1994), but some species have been overlooked entirely. Eleven of the sixteen Mariana native forest species have not yet been the subject of a single, focused natural history or ecological investigation (notes excluded). Fifteen of the eighteen natural history papers were published in the last decade. Regrettably, the recent increase in research was too late to materially influence the extinction crisis on Guam. Toward the end of the Guam extinctions, a major research program was initiated on the brown tree snake, but comparable programs have not been started on other potentially important species introductions.

Thus, the features of avian conservation and research in the Marianas that distinguish this area from that of continental areas and other archipelagos such as Hawaii are that: (1) forest birds, whether introduced or native, are almost totally absent from the largest island, Guam; (2) the introduced species that are largely responsible for this biodiversity loss, especially the brown tree snake, appear to be spreading through the archipelago; and (3) the natural histories of the species at risk are exceptionally poorly known.

Features of Research That Improved Conservation

It is important to distinguish between *management* actions that could have been implemented to conserve avian populations and *research* initiatives that could have been completed to formulate sound conservation policy. By either standard it is difficult to be optimistic about avifauna conservation in an area where native birds have been extirpated from the majority of the habitat. However, putting the management outcome aside for the moment, did research supply crucial information in a timely fashion? For Guam, the crucial decisions were made in the immediate postwar period, when local avian knowledge was confined to anecdotal natural history and morphology-based taxonomy. At least initially, little consideration was given to the potential ecological impacts caused by the introduction of a snake. This focus reflected prevailing attitudes in avian conservation. There were earlier examples of snakes causing the loss of native vertebrates on islands (on Mauritius, Cheke 1987; on Menorca, Eisentraut 1950), but these non-Mariana examples are even now poorly understood and in any event did not affect a large number of bird species. By the time the cause of Guam's avifauna collapse was identified and published (Savidge 1987), most of the species had been extirpated. This is not a criticism of the quality of avian research in the Marianas but rather an acknowledgment that the resources available for conservation research were not commensurate with the magnitude of the environmental turmoil created by political, military, and commercial change.

Two classes of especially productive avian research were status surveys (numerous authors) and snake predation studies (Savidge 1987; Fritts 1988; Conry 1988). Wetland bird status surveys have been invaluable for identifying the habitats most vital to the conservation of these birds (Stinson, Ritter, and Reichel 1991). Status surveys of the forest birds on Guam revealed the decline and disappearance of local bird populations from the 1960s onward (Engbring 1983). Researchers publicized these findings, so that additional research resources could be directed to determining the factors underlying the population declines. Though not concluded in time to avert avian extirpations, the resulting snake research did provide clear evidence that the snake was the principal cause of recent avian extinctions on Guam, and extensions of that work have begun to identify control measures that can be used to avoid or mitigate snake-induced conservation problems. Other islands in the Marianas have also experienced avian population declines or extinctions, but unequivocal explanations remain elusive.

Quantitative natural history studies (e.g., Craig 1990, 1992) have been especially useful for understanding the relationship between endemic species and their habitat requirements. For identifying the key units of biodiversity, modern molecular taxonomic techniques have been especially fruitful (e.g., Tarr and Fleischer 1995). Avian paleontological studies (Steadman 1992, 1995a,b) shed new light on the integrity of extant avian communities in the Marianas.

Research Needed to Further Conservation

Future research activities are predicated on three relatively firmly established facts: (1) extant avian communities on many of the Mariana Islands are now depauperate in relation to their prehistoric condition; (2) exotic predators, especially the brown tree snake, present the most immediate threat to birds of the Mariana Islands; and (3) once firmly established, many exotic predators cannot be eradicated (eliminated island wide) with current or foreseeably available resources and technology. Thus, much research is being directed at possible translocation of threatened species to islands least at risk from introduced predators. To date, research has taken the form primarily of status surveys, including surveys to identify the populations most immediately at risk (Engbring, Ramsey, and Wildman 1986; Craig and Taisacan 1994; Ramsey and Harrod 1995) and community surveys of potential recipient ("ark") islands, to identify factors that might preclude successful translocation (Craig et al. 1992).

Among ecologists, there is a wide range of opinion on the merits of relying on selected translocations to preserve threatened species within the Mariana archipelago. Rather than focusing narrowly on translocation, we recommend a broad approach to avian conservation in the Pacific. Below, we discuss a series of questions that highlight the important areas to include in this approach.

How Do We Obtain Replicable Population Estimates?

Population sizes for birds of most of the islands in the Marianas have been estimated twice, once in the early 1980s (Engbring, Ramsey, and Wildman 1986) and once in the mid 1990s (Ramsey and Harrod 1995; Craig 1996). Variable circular plots were used for population estimation. For most samples, the number of detections declined from the earlier to the later survey, while the estimated density increased (due to lowered detection distance estimates). This has inspired some skepticism about the technique, which has not been validated on any local system. Moreover, no estimates of interobserver consistency are available for the Mariana observers, as no two teams have sampled the same population concurrently. Are there reliable and cost-effective alternatives to variable circular plots that would be better adapted to local conditions? How often are the true population densities estimated accurately by surveys?

What Avian Biodiversity Is There to Protect?

The taxonomy of most species was determined using traditional morphometric techniques (Baker 1951; Pratt Bruner, and Berrett 1987). While this was appropriate at the time, modern molecular techniques and the emergence of the evolutionary or phylogenetic species concept as a dominant paradigm creates uncertainty about most earlier taxonomic renderings (Hazevoet 1996). A case can be made that each island population is on a separate evolutionary trajectory, but modern molecular techniques have shown that such is not always the case (e.g., Tarr and Fleischer 1995). In addition to identifying the key entities for biodiversity preservation, information on genetic variability within populations will be useful in identifying minimal propagule sizes for translocation or captive propagation (Fleischer, this volume).

What Is the Ecological Role of Each Species in Its Ecosystem?

Basic biological facts are needed for almost all species. For example, for some species we do not know if both sexes defend territories. We have no survivorship data, and if we were to obtain a single set, we would not know if it represented natural conditions or some gross distortion brought on by habitat or community perturbations. Basic biological knowledge is a necessary precursor for manipulations.

What Effect Do Each of the
Introduced Species Have on Native Birds?

While the evidence is now convincing that the brown tree snake played a large role in Guam's avian extinctions, the mechanisms for these extinctions remain largely speculative. Were the population declines a result of increased adult mortality, lowered recruitment as a result of egg or nestling predation, or a combination of both factors? Do the birds lack the behavioral traits responsible for nest protection in the parts of the world with natural snake predators? To what extent can learned behaviors play a role in predator defense (Maloney and McLean 1995)? Is selection capable of producing adequate predator defense traits on a time scale of relevance to management? What valuable or unique traits would be lost by such selection?

This line of inquiry can be followed with regard to predation by the brown tree snake, but what about the many other exotic species that may be responsible for loss of population viability? Does nearly complete elimination of ground-level vegetation affect birds on the islands where goats have irrupted? What role have rats played in the decline of megapodes and other native birds? What is the present distribution of rat species on the various islands within the archipelago? Do juvenile birds of some species die from ingesting young marine toads (*Bufo marinus*), the introduced species that has been deemed responsible for population declines in a variety of Australian vertebrates (Covacevich and Archer 1975)? Several ornithologists have suggested that the introduced Black Drongo is responsible for the precipitous decline of the Bridled White-eye (*Zosterops conspicillatus*) on Rota (e.g., Craig and Taisacan 1994). Under what circumstances does the interaction influence white-eye population structure or abundance?

In addition, we need to be prepared for many new introductions and be capable of rendering a rapid judgment as to whether they constitute a threat to the native species. For example, several snakes arrived on Guam in a 1995 shipment of Christmas trees. Those collected were destroyed by wildlife personnel before they could be identified, but if others escaped and became established, would we be able to predict their ecological significance to birds? A more pressing example is the several species of parrots that are feral in urban areas of Guam. Some of these are sufficiently numerous to form flocks and breed. Do these constitute a threat to restoration of the natives? How long will the window of opportunity to eradicate these incipient colonizations remain open? Or should exotic bird species be retained on an island largely devoid of bird life as a surrogate for the lost species?

What about accidental introductions of other taxa from other parts of the world? Are there key vectors of avian disease that are likely to arrive inadvertently with shipments? What are the higher-order interactions among introduced species? Would elimination of the brown tree snake result in unacceptable irruptions of rats or shrews? What introduced species might elevate the population densities of introduced predators and thereby tip the balance of survival against an indigenous bird, as the introduction of rabbits increased the abundance of cats on Macquarie Island, leading to the demise of the Macquarie Island Parakeet (*Cyanoramphus novazelandiae erythrotis*; Taylor 1979)?

To What Extent Can the Introduced Species Be Controlled in Practice?

Despite a decade of concentrated work, herpetologists are just beginning to probe the limits of what can be done to manage the brown tree snake. Several promising approaches include broadcast toxicants and large-scale snake exclosures. Modest-size areas can be made brown tree snake free using barrier fences and trapping (Campbell 1996). However, can areas as large as a thousand hectares be rendered snake free?

Is absolute elimination of the brown tree snake necessary for the success of a large-scale snake exclosure? Most introduced species irrupt after initial colonization, reaching population densities not subsequently maintained. The brown tree snake reached densities of about 100/ha, roughly four times that of all avian species combined (Rodda et al., in press). Perhaps the birds can survive if snake densities are held below 15/ha. Is it possible to maintain native birds by creating vast areas that are snake-reduced although not snake free? These brown tree snake questions are under active investigation, but little is being done with regard to the other introduced species. Can drongos be controlled in practice? Can one eliminate monitor lizards from small islands? New Zealanders have had notable success in controlling or eradicating introduced mammals (Veitch 1994; Innes et al. 1995), leading to enhanced reproductive success for native birds (Clout et al. 1995). Feral animal control will be an issue for avian conservation in all of the prospective ark islands (Aguijan and far northern islands).

How Much Natural or Seminatural Habitat Is Needed to Maintain Population Viability?

While little effort has been directed at habitat preservation to date, it is clear that many species will not survive in the urban areas that are progressively replacing natural habitat. Long-term viability will require knowledge of the spatial and habitat requirements of each species. These requirements need to be ascertained not only for quasi-natural areas, such as wildlife refuges, but also for possible predator exclosures, which may be the only recourse for protecting some species. Unfortunately, basic information on movements, dispersal, home range, and territory size is lacking for most species. Do territories remain constant in size if the number of species in a guild is changed? For example, if we translocate another insectivorous bird to Aguijan Island (the 720 hectare island that is targeted to be the primary ark by most translocation advocates), will resident insectivores require larger territories? What are the effects of rare but predictable catastrophic events (e.g., super-typhoons) on population viability? Was the prehistoric loss of Guam Rails from Aguijan due to a natural event such as a super-typhoon?

How Many Species Can You Pack into a Small Area?

While the emphasis of the preceding section was on the natural history attributes of individual species, the issue must also be addressed at the community level. Aguijan Island, the postulated ark island, appears to have exceptional species richness for an island of its size. (See figure 25.1; the number of species shown for each island is the number of native birds present historically, not the current or prehistoric species richness.) Is this an artifact of the other Mariana Islands being artificially depauperate as a result of human presence? Island biogeographic data and experimental results of translocations in other archipelagos should be brought to bear on this question (Moulton 1993; Armstrong and McLean 1995). To what extent do niches tend to widen on very small islands? Can we use this information to predict the number of species that can be packed onto Aguijan? Or a wildlife refuge of comparable size? Does the number of potential cohabiting species increase (or decrease) if one removes exotic pest species, such as the goats and monitors that have dramatically altered Aguijan Island? How are the small insectivores partitioning resources (c.f. Wheeler and Calver 1995) on Aguijan as compared to other islands?

One especially fruitful approach to the question of species packing is the work of David Steadman and colleagues on subfossil remains that can sometimes be dated to prehuman times (Steadman 1995a,b). Clearly, the Mariana Islands have been the home of many species that are now absent. Many extinctions seem to coincide with the arrival of humans. However, there is some uncertainty as to the temporal resolution of the fossil record: Were all the prehuman species found concurrently? The postulated island taxon cycle (MacArthur and Wilson 1967) suggests that species should turn over fairly rapidly on small islands. On Pacific Islands, the prehuman rate of turnover does not seem to have been high in comparison to more recent ecological changes (Steadman 1995b). However, it may not be enough to know that Rota once held x number of bird species, if extirpated species vacated niches that are not of importance to the species proposed for translocation. Were there keystone species, now absent, that were needed for coexistence of proposed translocatees?

Under What Conditions Is Captive Propagation Viable and Desirable?

To date, captive propagation has been undertaken both because no alternatives existed in the short run (mallard, rail, kingfisher), and because zoos expressed interest in developing techniques for anticipated future needs (white-eye, fruit-dove, crow). However, the approach is expensive and has not yet brought about hoped-for restorations. The available zoo space is being subjected to increased

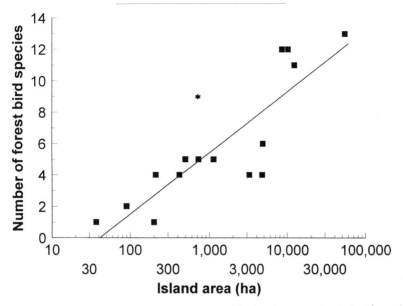

Figure 25.1. Species area relationship for Mariana Island birds. The regression is significant ($F_{(1,14)}$ = 31.5, P = 0.001; Bird species = 3.8154*$\log_{10}$[island area in ha] −5.767) and implies that the island indicated by the star symbol, Aguijan, which has nine species and is frequently mentioned as a possible destination for translocated species (ark island), has an expected richness of 5.1 species based on its area (720 ha).

demands (Snyder et al. 1996), and these may result in the loss of zoo services now being provided for Mariana birds. If this captive propagation service is to be called upon in the future, which species make the most appropriate candidates? Under which circumstances? What natural history information is needed as a precursor to successful captive propagation? How can we obtain the behavioral or avicultural information that is needed to convert the declining Micronesian Kingfisher colony into a growing population?

A related set of research is needed to address the vexing issue of what is required to reestablish wild populations of individual birds that have lived their entire lives in a cage (Soderquist 1994; Snyder et al. 1996). Both of the captive breeding programs (rail and kingfisher) that have succeeded in at least temporarily maintaining captive colonies of birds now extinct in the wild (the Mariana Mallard program failed: Engbring and Pratt 1985) have nonetheless experienced great difficulty in releasing the captive-reared birds (rails: Witteman and Beck 1991; M.K. Brock, personal communication) or even maintaining normal parenting behaviors (kingfishers: B. Bahner, personal communication). To correct these problems, a substantial body of research may be needed to determine the causes of failure and identify pre-release protocols (captive predator trials, socialization training, etc.) to prepare the birds for life in the wild (McLean, Lundie-Jinkins, and Jarman 1994).

What Services Are Lost When Conservation Depends on Remote ex Situ Populations?

Environmentalists are fond of enumerating the various ecological services that are provided by wildlife. Birds pollinate trees, disperse seeds, limit agricultural pests, provide an aesthetic resource for island residents and tourists, etc. To maintain bird species exclusively by translocation to remote islands or by establishing ex situ captive populations eliminates these ecological benefits. How important are they? Under what conditions would they tip the balance in favor of in situ preservation?

To these sources of biological uncertainty is added the unresolved question of whether it is better to expend extreme and potentially futile effort to preserve a species in situ or to quickly translocate the species to distant and uninhabited islands where they may survive with minimal intervention but also provide minimal ecological services to humans.

In addition to the ecological and tactical questions there are important administrative and ethical questions. Is translocation good governmental policy? Is it fair to subsequent generations? Does it represent the will of the current generation? While ornithologists may be no better than anyone else in probing these ethical and policy considerations, they can contribute to the debate by sharpening understanding of the trade-offs and reducing the inevitable uncertainties regarding the likelihood of success of alternate courses of action. For example, if ornithologists can determine that the addition of another small insectivore to an ark island will result in niche shifts and the eventual loss of one of the competing insectivores, policy makers will be better positioned to evaluate the wisdom of translocating the at-risk insectivore to the ark island.

What Are the Key Parameters for the Success of Translocation?

The majority of endangered species translocations fail (Griffith et al. 1989; Ounsted 1991). Why? If one is going to create avian communities at will, how important is the order of assembly? If crows are added before potential prey birds such as white-eyes, will the crows starve, or begin preying on their own young? What is the genetically or numerically minimum propagule size for each species? Is the season of release important (Armstrong and McLean 1995)? Hypotheses can be based on natural history studies, but experimental manipulations and studies of appropriate surrogate species will be needed to have any confidence in the highly artificial procedure of human-caused translocations.

How Can the Infrastructure of Science in the Marianas Be Improved?

The vast number of questions raised explicitly or by inference in the above sections requires vastly more research than has been accomplished to date. Moreover, the answers to all of these questions would be helpful for decisions being made this year; thus, the timeline for completion of this research is not generous. In the absence of any nearby research universities, most research has been funded directly or indirectly from federal wildlife management sources, primarily the U.S. Fish and Wildlife Service. The current political climate suggests that this support is not likely to increase in the near future. What can be done to expand the pool of financial resources? What can be done to insure that all resources are expended in the most professional manner possible? Is there a need for greater reliance on resident scientists than on short-term visitors? What management activities should be excluded from consideration because scientific uncertainty precludes a reasonable assessment of the chances for success?

Making Research Effective for Conservation

Given the pace of human population increase and the priority given to development of an industrial economy, it is difficult to envision how conservation research can be funded at a level sufficient to address even the major issues implied by the current avian extinction crisis in the Marianas. The highest-priority tasks include: (1) identification of the factors responsible for the decline of various bird populations; (2) systematic compilation of basic natural history data on the species at risk; (3) determination of the limits of practicality of control technologies applied to introduced species that are a threat to birds, and assessment of the likely effectiveness of these controls; (4) if in situ preservation is impractical for a given species, careful determination of the circumstances under which the species might be preserved through translocation to an island less at risk from the factors causing the population decline; and (5) enlistment of new allies and new sources of support, so that the unique avian biodiversity of remote Pacific Islands is not lost through ignorance.

Avian conservation in the Marianas has not yet advanced to the point where tactics have been chosen and now must only be refined. Instead, the reasons for species declines are still a mystery in many cases; in other cases, the appropriate type of response has yet to be identified. For these reasons, research will play a central role in Mariana avifauna conservation in the coming years.

Acknowledgments

We greatly appreciate the improvements to the manuscript suggested by Kathy Dean-Bradley, Steve Fancy, Thomas Fritts, Scott Johnston, Annie Marshall, John Morton, and John Marzluff.

CHAPTER 26

Research and Management for the Conservation of Birds in Australia

John C. Z. Woinarski

Conservation Issues and Previous Research

Australia now holds a moderately rich (approximately 770 species) but highly distinctive bird fauna. This fauna is the residue of a far richer assemblage, much of whose most spectacular components (including flamingos and the giant flightless mihirungs) has been lost during the 60,000–plus years of Aboriginal management of the continent (Rich and Baird 1986). The explanation of these losses has been sharply contested (Flannery 1994) but undoubtedly includes some contributions from imposed fire regimes, hunting, and rapidly fluctuating climates. As an aside, the birds themselves have provided strong evidence of climate change, with temperature-dependent amino acid racemization reaction in radiocarbon-dated Emu (*Dromaius novaehollandiae*) eggshell fragments, suggesting temperatures in central Australia that averaged at least 9°C lower in the period 45,000–16,000 years B.P. than in the present.

In the two centuries since European colonization, only one bird species, the Paradise Parrot (*Psephotus pulcherrimus*), has disappeared from the Australian mainland (Male 1995). This low rate of extinctions contrasts to the staggering losses (extinction of seventeen of the ca. two hundred terrestrial species) suffered by Australian mammals during this period. The disparity in these figures suggests that very different causal factors are operating, and/or that the recent dismal fate of Australian mammals may be a precursor to that awaiting Australian birds (Recher and Lim 1990). The limited mainland loss of birds also contrasts to the far higher rate of bird extinctions on Australian islands (eight species; Male 1995) and the nearby islands of New Zealand.

Threatened Species

The bulk of the research and management for conservation of Australian birds has been directed at threatened species (Male 1995). The status of all Australian species was assessed largely by consensus of experienced ornithologists (Garnett 1992a), preceding the more rigorous and quantitative assessment methodology now recommended by the International Union for the Conservation of Nature. Including subspecies, 23 taxa were regarded as extinct, 26 as endangered, 40 as vulnerable, 32 as rare, 29 as insufficiently known but possibly threatened, and 924 as not threatened. This categorization was used to outline and estimate costs of research and management actions for threatened species, and to prioritize these within and between species (Garnett 1992b). A framework and support for this program was established, with subsequent federal endangered species legislation, although the resources directed toward threatened species continue to fall well short of those required to undertake the research and management actions specified in Garnett (1992a).

For only a few species, intensive research and management have involved long-term population studies. Typically, these population monitoring studies have alerted managers to issues with far wider ramifications than the single-subject species. The documentation of gradual decline in numbers of albatrosses nesting and foraging around Tasmania established the significance of bird mortalities associated with long-line fisheries and led to changes in the operation of that industry (Brothers 1991). Regular monitoring of the gradual recovery of Noisy Scrub-birds (*Atrichornis clamosus*) from a single critically small population demonstrated the significance of fire regime for this species and two co-occurring threatened birds (Smith 1985). Along with population-monitoring studies of other species in other environments (notably Brooker and Brooker 1994), this research suggested a profound significance for fire management more generally for a broad range of Australian birds. Monitoring of the fate of juvenile Mallee Fowl (*Leipoa ocellata*) demonstrated the significant impact of predation by feral foxes (Priddel 1990; Priddel and Wheeler 1994), contributing to the implementation of predator reduction campaigns that have benefited a broad sweep of Australian wildlife.

For a few Australian bird species, numbers have fallen so low that captive breeding has been used for security and/or to bolster population size (e.g., Helmeted Honeyeater [*Lichenostomus melanops cassidix*], Lord Howe Island Woodhen [*Tricholimnas sylvestris*], Orange-bellied Parrot [*Neophema chrysogaster*]). In most cases, this intensive intervention has been successful—in some cases, spectacularly so (Miller and Mullette 1985). Although adequate data are available for relatively few species, population viability analyses have been used successfully for some threatened species in order to aid the prioritization of research and management efforts (Venn and Fisher 1993).

Communities, Habitats, and Threatening Processes

Recognizing that designated threatened species were perhaps simply sympto-matic of a pervasive malaise, Recher and Lim (1990), Robinson (1991), and Robinson and Traill (1996) argued that contemporary land management prac-tices across Australia will cause substantial declines of many bird species that are currently relatively abundant. Among the declining species, there is a dispropor-tionate representation of species with particular ecological traits (including feeding or nesting on the ground, and breeding in hollows) and/or occurring in particular habitats (especially woodlands and heaths).

Australia has only very limited areas of fertile lands suitable for intensive agri-culture, and the native vegetation formerly occurring in these areas has largely been fragmented and cleared (Braithwaite, Belbin, and Austin 1993). This has spawned a considerable body of research on the impacts of habitat fragmentation upon wildlife in general and birds in particular (Saunders et al. 1987; Saunders and Hobbs 1991; Walters, this volume). Much of this research has been devel-oped into management guidelines (Saunders, Hobbs, and Ehrlich 1993)—although the clearance of native vegetation continues to be a major problem for the conservation of Australian wildlife, particularly in Queensland (where the current clearing rate of forests and woodlands is 300,000 hectares per year). A notable feature of forest and woodland fragmentation is that its effects are often magnified by the usurpation of remnant patches by aggressive open forest and ecotonal bird species, including some introduced species, but more commonly by large honeyeaters (typically *Manorina* spp.), which exclude most other insec-tivorous birds and hence contribute to decline in the health of the remnant veg-etation itself (Loyn et al. 1983; Loyn 1985).

Eucalypt open forests occupy a relatively small proportion of Australia and have been substantially modified by timber harvesting (Norton 1997). The localized impacts of forestry operations upon birds have been documented in a number of recent studies (Loyn 1980; Recher, in press). As with forestry impacts on other continents, the costs appear greatest for the many bird species, such as the Sooty Owl (*Tyto tenebricosa*), that are reliant upon naturally occurring tree hollows and hence old-growth forests (Scotts 1991; Gibbons and Lindenmayer 1996). To some extent, the conservation message from this research has been incorporated into a coordinated national strategy for Australian forests (Commonwealth of Australia 1992).

Far less extensive than eucalypt forests, Australian rain forests are now relatively well reserved and researched, in part reflecting the richness and distinctiveness of their bird fauna. However, fragmentation during the last two centuries has led to declines for some of the largest and/or most specialized rain forest bird species, such as the Southern Cassowary (*Casuarius casuarius*) and the Coxen's Fig-parrot (*Cyclopsitta diophthalma coxeni*) (Crome and Moore 1990).

Arid and semi-arid shrublands and grasslands dominate much of Australia and have received a disproportionately small research effort. Much of these habitats is devoted to extensive pastoralism, whose impacts upon birds remain very poorly known (Woinarski 1993). A recent national strategy (National Rangeland Management Working Group 1996) aims to include some conservation goals for these rangelands.

Temperate grasslands have been almost entirely alienated and modified by agriculture, intensive grazing, and/or urban development. Few bird species were associated primarily with these habitats, but at least one of these species, the Plains-wanderer (*Pedionomus torquatus*), is now threatened by these habitat changes (Baker-Gabb, Benshemesh, and Maher 1990). The relatively small areas of coastal heathlands of southern Australia have a distinctive bird fauna. These habitats have been substantially altered by grazing and clearing for coastal developments. The fragments remaining are now often subjected to fire regimes that are unsuitable for much of their constituent bird fauna (McFarland 1988), and the combination of fragmentation and fire has constrained dispersal and management options.

Wetlands are relatively restricted in Australia, but while their conservation values have been reasonably well documented (Australian Nature Conservation Agency 1996), they are not managed under any coherent national conservation strategy.

Superimposed on the modification of Australian environments arising from their utilization and development are the pervasive impacts of feral animals, weeds, and changed fire regimes. Feral herbivores, most notably rabbits, have had a devastating impact across much of Australia, with consequent landscape change adversely affecting many Australian birds (e.g., Pedler and Burbidge 1995), although also notably becoming an important resource for raptors (Baker-Gabb 1984), whose populations may be expected to decline with effective means of rabbit eradication. Feral predators, particularly foxes and cats, have played a key role in the extinction of many Australian mammals and have been implicated in the decline of some Australian birds, such as the Bush Stone-curlew (*Burhinus grallarius*) (Garnett 1992b) and Mallee-fowl (Priddel 1990). Introduced plants have transformed many areas of Australia with resultant changes in the ecology and status of many bird species (e.g., Whitehead, Wilson, and Saalfeld 1992; Smyth and Young 1996). Research and management of feral pests and weeds have been very uneven, with relatively little attention directed at their impacts upon birds. In some contrast, there has been a considerable body of research directed at the effects of fire upon bird conservation (Woinarski and Recher, in press). Many Australian environments appear to be extremely fire prone, and the destructive effects of wildfire upon human life and property have led to the imposition of a fire management regime unlike that occurring before European colonization. Unsuitable fire regimes are now one of the main threatening factors for Australian birds (Garnett 1992b). Many threatened species—such as the Noisy

Scrub-bird—are reliant upon long-unburnt vegetation, a rapidly diminishing environment (Woinarski and Recher, in press). In a contrasting example, change from a frequent fire regime previously imposed by Aboriginal people to a current fire exclusion policy is leading to the expansion of fire-sensitive rain forests at the expense of some tropical eucalypt tall open forest, and thus threatening some bird taxa particularly associated with this latter vegetation type (Harrington and Sanderson 1994).

There is little information on the diseases and parasites of Australian birds (except where birds are vectors for diseases that subsequently affect humans) and the extent to which this generally isolated bird fauna has been colonized by recently introduced parasites (McOrist 1989). However, the very limited research suggests that decline in some Australian bird species may be associated with newly established parasites (Tidemann et al. 1992).

Some Australian environments, and their associated fauna, are expected to be particularly susceptible to rapid climate change. There have been some attempts to model the impacts of climate change on a range of Australian wildlife species and to include these predicted impacts in management priorities and strategies (e.g., Bennett et al. 1991).

Monitoring and Movement Patterns

Broadscale monitoring programs and studies of dispersal, two key components in the conservation of birds elsewhere (J. J. D. Greenwood et al. 1995), have been hampered in Australia by low human population density. Although a large proportion of Australian ornithologists and bird-watchers contributed to a broad distributional atlas (Blakers, Davies, and Reilly 1984) and a more limited population monitoring program (the Australian Bird Count; Ambrose 1994), there is no ongoing national or regional population-monitoring program for Australian birds. Despite a bird-banding program that has extended over at least four decades and 3 million marked birds (G. B. Baker et al. 1995), recoveries are few and movement patterns remain largely unknown for most Australian birds. This is particularly so as the climatic vagaries characteristic of much of Australia force birds to move with limited predictability. For a few threatened species, radio-tracking has been used successfully to identify causes of mortality, resource requirements, movement patterns, and habitat preferences and hence to then guide management actions (Benshemesh 1990).

Documentation of the Bird Fauna

Dedicated (fanatical) collectors of the nineteenth and early twentieth century fleshed out the species-level taxonomy of Australian birds, leaving only a few species from very remote locations to be discovered in the second half of this century. However, the last few decades have seen a remarkable revolution in higher-order taxonomic opinion about Australian birds (Christidis and Boles 1994), firmly establishing the distinctiveness of most of the Australian bird fauna and to some extent recasting conservation priorities. Intensive genetic research has also been directed at resolving specific and intraspecific boundaries in some groups, notably those with highly disjunct populations (Joseph and Moritz 1994), although, in general, subspecific categorization remains poorly resolved or documented. Some of this research has had notable conservation repercussions, with the resolution of the taxonomic status of the Black-eared Miner (*Manorina melanotis*) being most notable. This species was described from a small area of mallee vegetation (stunted eucalypt trees typically with many trunks, growing in semi-arid areas). With massive clearing and fragmentation of this environment over the course of this century, its dominant congener from surrounding open woodlands and farmlands, the Yellow-throated Miner (*M. flavigula*), moved into the mallee fragments and interbred with Black-eared Miner, genetically "swamping" it. With the number of phenotypically "pure" wild individuals reduced to a handful during the last two decades, debate polarized between those who considered that widespread hybridization demonstrated that the Black-eared Miner and Yellow-throated Miner were conspecific and hence few resources should be directed toward conserving the few distinctive Black-eared Miner genes (Silveira 1995), and those who considered that the Black-eared Miner had become Australia's most endangered bird species (Schodde 1990). Recent detailed genetic analysis (L. Christidis, unpublished data) has apparently supported the latter case. Lack of management action during the debate may have now compromised the survival chances of the taxon.

Exploited Species

The overwhelming majority of native birds are legislatively protected throughout their Australian range. Excluded from this protection are a few species that have become agricultural pests, a small number of species that are taken commercially mainly for consumption—most notably, the Short-tailed Shearwater (*Puffinus tenuirostris*) (Skira 1996)—and species permitted to be taken by recreational hunters. The latter are mostly waterfowl but also include smaller numbers of quail. Hunting regulations vary between states, and bag limits are usually set following population surveys. Monitoring suggests that this take is generally sustainable, although special protection has to be afforded to some of the rarer species (most notably Freckled Duck [*Stictonetta naevosa*]), and recent research has indicated that high levels of lead in the sediment of some wetlands used by shooters have caused substantial mortality in foraging waterfowl (Whitehead and Tschirner 1991). The amount of game hunting in Australia is appreciably less than in Europe or North America, with consequently less dominance of the wildlife research and management agenda, although (as elsewhere) the protection of some wetlands has been promoted by game clubs (King 1995).

More recently, some state conservation agencies have advocated sustainable harvest for aviculture of parrots and other birds, especially for species that may have some agricultural impacts (Parks and Wildlife Commission of the Northern

Territory 1995). Although there have been some limited attempts to examine the economics of such commercial use, there has been little research directed at the capacity of the target species to withstand harvesting, and no detailed assessment of the conservation benefits and costs of this approach. The issue remains contentious.

Features of Research That Improved Conservation

There are two threads evident in the above account of conservation of birds in Australia: relative success with conservation programs concerned with the recovery of individual threatened species, but relatively little progress with broader-scale programs that consider population trends of bird communities and the impacts of threatening processes more generally.

The former approach has characteristically attracted relatively high levels of funding and public interest. The approach has been rooted in a generally well-considered identification of the taxa requiring most urgent conservation attention, providing a framework for the research and management of those taxa (Garnett 1992a,b). Success with threatened species has been greatest for localized sedentary species whose preferred environments had been modified by management practices imposed over the last two centuries, and where management of the limited areas in which the species persisted could subsequently be readily changed to favor the species. Success has not been so forthcoming for threatened species with wider ranges and/or those affected by a more complex array of factors, partly due to the greater research difficulties inevitable in teasing apart multiple and sometimes interactive influences, and partly due to the more formidable obstacles to conservation management across extensive ranges. These problems are probably most pronounced for highly "nomadic" species—such as the Regent Honeyeater (*Xanthomyza phrygia*) and Swift Parrot (*Lathamus discolor*), which track rich, but spatially and temporally variable, nectar sources in the eucalypt woodlands and forests of southeastern Australia (Franklin, Menkhorst, and Robinson 1989). As these environments have been greatly diminished, fragmented, and modified over the last two centuries, these species have rapidly lost dispersal options and consequently have undergone massive and continuing decline. Such species cannot be maintained without extensive land protection strategies. Their maintenance is further jeopardized by loss of opportunity to document "natural" dispersal patterns or the factors underlying them.

One feature that has undoubtedly aided the conservation of Australian birds is the restriction of a very high proportion of the fauna to a single nation, which allows an unusually coherent approach to their conservation. This coherence has been maintained for species that occur across several Australian states by the establishment of national strategies for threatened species, developed cooperatively by all state and federal governments, and well-developed mechanisms for involvement of all relevant and responsible parties in recovery teams (Male 1995).

In contrast to the substantial rescue effort directed at designated threatened species, there has been little attempt to assess population trends for whole communities or responses of birds generally to some pervasive land-use practices. The few cases that have considered changes in bird faunas have generally been based on qualitative or very imprecise quantitative assessments (Saunders and Curry 1990; Smith and Smith 1994), due to the historical and continuing lack of national, or even regional, bird census programs. Nonetheless, these studies suggest an ongoing decline for many species whose status is not yet officially designated threatened.

Research into the impact of land uses has been notably patchy. The response of birds to forestry operations has been relatively well considered, perhaps due to widespread public concern for forests, coupled with the concentration of forestry activities on public lands (which ensures some governmental accountability for impact assessment). However, the research on forestry impacts has generally been localized and concentrated on successional responses. Despite some notable recent research, landscape-scale forestry issues (such as patch size and age composition of retained forests, number and dispersion of cavities required to maintain viable populations of owls and other hollow-dependent or old-growth species, and attributes of connective strips) remain poorly known and sharply contested (Recher et al. 1987; Norton and May 1994), in parallel with that for boreal forests. The lack of adequate information on broadscale impacts is even more pronounced for the far more extensive pastoral and agricultural industries, at least partly because these are concentrated on privately owned or leased lands.

Research Needed to Further Conservation

The management of Australian wildlife has been handicapped by its very distinctiveness. Research and management assumptions made from European or North American vantage points have proven insufficient or misleading (Yom-Tov 1987; Rowley and Russell 1991). Consequently, the ecological characteristics (and hence conservation requirements) of the Australian fauna have to be described de novo. This is clearly a time-consuming and resource-hungry demand, especially critical when conservation actions may be required urgently, and especially onerous when there are relatively few bird researchers in Australia compared to most developed countries. The depth of research required may also be unusual, as for much of Australia the rationale for particular ecological traits is apparent only during the critical, characteristic but irregular periods of drought or especially wet seasons (Friedel, Foran, and Stafford Smith 1990). Continental-scale responses to such climate-induced landscape conditions render long-term monitoring difficult to plan and interpret. As an aside, some of the ecological distinctiveness of the Australian passerine fauna, most notably the characteristically low reproductive rates (Rowley and Russell 1991), may often exacerbate conservation problems, especially in environments subject to rapid modification.

As elsewhere (Wiens 1995), conservation research for Australian birds also suffers from the profound difficulty of scaling up from localized intensive research to broader landscape management. While this is a particularly critical problem for highly mobile species (Woinarski et al. 1992), it also hampers conservation of species with fragmented populations, species whose maintenance depends on the retention of a range of dissimilar patches or resources, species restricted to particular seral stages, and species affected by processes whose management requires a broadscale response (such as control of feral predators). Research addressing these issues must extend beyond short studies at single sites to focus instead on the temporal and spatial patterning of resources, how these have been modified by land use, and how birds respond to these patterns. To conserve these species, research must then be linked to integrated regional planning.

The health and future of Australia's birds is not measurable only by the number of designated threatened species and the success of management directed to them. It requires also information on trends in the distribution or status of the entire bird fauna. The lack of such monitoring is probably the greatest single deficiency in conservation research for Australian birds, particularly as the limited information available on population trends for a range of regions suggests widespread declines for many species (Reid and Fleming 1992; Smith and Smith 1994; Robinson and Traill 1996). Where such monitoring detects consistent patterns of decline, research attention may then be more economically redirected from localized single-species studies to identifying, investigating, and managing more pervasive threats affecting whole suites of environments and species. A notable example of this approach is that outlined by Robinson and Traill (1996) for one of the environments in which birds are declining most rapidly, the temperate woodlands of southern Australia.

I've argued here that researchers and managers have done reasonably well with the heroic surgery required by endangered species, but that this effort may have come at the expense of the diminishment of wildlife values more pervasively across the landscape. Perhaps this is a false dichotomy, as both strands of research may enlighten management and provide complementary insight into conservation problems. Research and management on endangered birds are still very necessary, as some of these species would undoubtedly become extinct without the support that they are now afforded. However, an increased priority should be given to research and management on the impacts of some threatening processes generally (notably fire, feral animals, agriculture, pastoralism), and to some environments that have been relatively neglected.

Largely due to logistic convenience (and perhaps more generously, reflecting magnitude of land modification), research on the conservation of birds has been concentrated on the narrow, relatively fertile belts of temperate eastern and southwestern Australia. The neglect of arid and semi-arid areas, hummock grasslands, chenopod shrublands, *Acacia* shrublands, and tropical grasslands (all relatively remote from most population centers) may have compromised our ability to detect change or implement management aimed at conserving birds in these environments. These regions and environments merit a more equitable share of the research and management resources.

Making Research Effective for Conservation

Extension to the General Public

Extension of research results from the very small corps of active ornithologists to the interested general public has been a main commitment of the organizations catering to Australian bird-watchers and research scientists (the Royal Australasian Ornithologists Union, Australian Bird Study Association, Bird Observers Club of Australia, Queensland Ornithological Society, and South Australian Ornithological Association), although the effectiveness of this transfer has probably been hampered by the multiplicity of those organizations. The involvement of interested volunteers has also been instrumental in many projects dealing with threatened birds, and has been explicitly sought in the composition of most recovery teams supervising such research (e.g., Smales, Menkhorst, and Horrocks 1995). This has greatly assisted the conservation of birds on private lands (Bennett, Backhouse, and Clark 1995), as has a national Landcare program in which private landholders (mostly farmers) receive resources for voluntarily revegetating and rehabilitating degraded lands.

Translation to Management Strategies

On a species-specific basis, the Recovery Team approach has proven a generally effective forum for promoting relevant research and translating research results to the resolution of practical management problems, largely because of its inclusion of all stakeholders and relevant expertise.

On a broader level, individual research efforts are gradually being filtered into the development of regional and national management strategies. For example, research into the impacts of forestry were collated and assessed in a series of public enquiries, from which a national forest policy was developed that sought to provide a framework for the maintenance of conservation values and a logging industry. There have been comparable national strategies or initiatives seeking to provide for protection of threatened species and ecological communities, a comprehensive conservation reserve network, ecological sustainability of all main industry sectors, wise use of rangelands, and adequate control of feral pests and weeds, etc. The last few years have also seen a series of major symposia (and subsequent publications) bringing researchers and managers together to consider habitat fragmentation (Saunders et al. 1987), the role of corridors (Saunders and Hobbs 1991), habitat restoration (Saunders, Hobbs, and Ehrlich 1993), conservation off reserves (Hale and Lamb, in press), the recovery process (Stephens and Maxwell 1996), the role of captive breeding (Serena 1994), and conservation through sustainable use of wildlife (Grigg et al. 1995).

Research on the movement patterns of birds migrating beyond Australia has also been pivotal to the development of bilateral agreements with Japan and with China seeking cooperation in the conservation of migrant birds, especially shorebirds (Watkins 1995).

Realigning Research and Management Priorities

Partly spurred by the framework provided by the World Conservation Strategy, there has been a productive (if inconclusive) debate about research and management priorities for the conservation of Australian biodiversity (e.g., McIntyre et al. 1992; Burbidge 1993; Cogger 1993). Largely due to the popular appeal of birds and the lamentable conservation status of mammals, these two groups have monopolized funds available for conservation of Australia's biota. Advocates of the plight of other organisms seek to redress this taxonomic imbalance. Prioritization of research and management should probably sidestep that parochial argument and instead advance projects that provide the greatest insight into factors responsible for the most rapid, irreversible, and/or substantial modifications of environments, and the most opportunity to protect those environments from such factors (Marzluff and Sallabanks, this volume). In some cases, research into individual endangered bird species may provide such a key. However, in many more cases, research and management that focuses more broadly on environments or threatening processes is likely to produce greater long-term benefits. It is timely that the research and management effort for the conservation of Australian birds (and especially threatened species) was woven more tightly into the overall protection of Australian environments and their biota.

Acknowledgments

I thank Peter Whitehead, Don Franklin, David Baker-Gabb, Hugh Ford, Harry Recher, Ian Rowley, and the editors for helpful comments on earlier drafts.

PART VII

Relevance of Conservation Research to Land Managers

Research is of little use to avian conservation if it cannot be applied to real problems. Previous sections discuss specific research needs aimed at helping us manage birds. Here, the authors do not discuss specific research areas. Rather, they discuss general approaches aimed at making research more meaningful to managers. They stress ways that researchers can improve the likelihood that their results get incorporated into recommendations and actions to conserve avian diversity. The authors of these chapters draw upon their experiences in the research and management realms of North America. They are either researchers that work closely and effectively with managers or managers that work closely with researchers. Their suggestions are relevant around the world, and their insights provide a novel look at typically forgotten or unknown properties of research that profoundly affect the utility of our work.

Arnett and Sallabanks (chapter 27) conducted a simple and thought-provoking survey whose results suggest that research and monitoring are not meeting managers' goals. Most managers do not feel that current information is sufficient to understand how birds respond to common management activities, a point also noted by Ganey and Dargan (chapter 31). Given the wide variety of research discussed in the previous chapters, why don't managers understand how land-use activities affect birds? In this section we learn that the answer has two parts: (1) important types of research remain to be done, and (2) the important task of relating research results to managers is often not completed.

The types of research that remain to be done are generally those called for repeatedly in this volume: long-term, large-scale, experimental assessments of the effects of land management activities on birds and their habitats. Managers want information on the causal link between activities and birds. This link needs to include an assessment of how activities affect avian demographics, and should indicate how human activities affect birds at multiple scales.

How one approaches a research project so that results are used by managers is really what sets this section apart from others in this book. Researchers need to understand that managers rarely are able to make decisions about a land-use practice solely on the basis of how that practice may affect birds. However, there are many, often obvious, ways that researchers can increase the palatability of their work to managers. Young and Varland (chapter 28) discuss this point in general terms by laying out twenty-one principles that make research meaningful to managers. Kochert and Collopy (chapter 29) build upon that chapter by stressing how adhering to six of the principles (addressing management needs, promoting interaction between scientists and managers, establishing appropriate reference frames, clearly describing desired research products, producing unambiguous results, and conducting forthright research) aided two research projects they were involved in. Hejl and Granillo (chapter 30) and Ganey and Dargan (chapter 31) offer two additional perspectives on how they have succeeded in making research results applicable to management goals.

The important point made repeatedly by the authors is that research must be highly integrated with management needs to be meaningful to managers. This needs to be an ongoing, standard practice. Researchers should ask managers what information they need and design studies that explicitly address those needs. This is rarely done but very effective when accomplished. Research that is used to make management decisions will typically result from long-term, integrated involvement of researchers and management agencies.

Integrating research and management is the key to producing scientifically based management decisions. This is facilitated by communication between researchers and managers. Lack of communication is at the heart of the delay to implement scientifically sound management. Researchers need to get their ideas to managers, and this requires participating in nontraditional activities such as preparing general reports, attending meetings of management societies in addition to meetings of professional research societies, and participating in workshops with managers. Researchers that meet regularly with local managers likely to be interested in their work will usually be rewarded by having their findings quickly incorporated into management policy. Arnett and Sallabanks's survey quantifies the rarity with which researchers and managers interact to develop applied research projects. It also suggests that when researchers and managers do work together, research results are more easily implemented into management actions.

This is a must-read section for avian researchers interested in getting their results used. Follow the general guidelines in chapter 28 carefully and always ask managers what they need to know in addition to suggesting important information that you think they should know. Be aware that there are at least two distinct types of managers: resource specialists and decision makers (chapter 30). Both types require different sorts of research information, and both must be reached by researchers if our results are to be used. This requires publishing in a variety of venues and participating in formal and informal meetings, workshops, and briefings.

CHAPTER 27

Land Manager Perceptions of Avian Research and Information Needs: A Case Study

Edward B. Arnett and Rex Sallabanks

Conservation Issues and Previous Research

To conserve avian populations, land-use managers need to know how to manage various landscape attributes that affect bird species' habitat and demographic characteristics; such information typically comes from the research community. Unfortunately, this learning process has been hampered by a gap between research and management (MacLaren 1992; Finch and Patton-Mallory 1993). Numerous reasons for this gap may exist, including the following: (1) results from research projects are not readily accessible to managers; (2) research findings are too narrow in their focus to be effectively used by managers; (3) terminology differences and philosophical barriers prevent open communication between researchers and managers; and (4) researchers and managers rarely design and implement wildlife studies together or maintain dialogue throughout the course of a project to ensure that results are relevant to management. Managers continually deal with political and economic realities, and public opinion and concern over resource issues (Lautenschlager and Bowyer 1985). These conditions are not always appreciated or understood by researchers, who may criticize managers for not confronting resistance or embracing new ideas quickly enough (Finch and Patton-Mallory 1993). Perhaps the most fundamental reasons for the research-management gap, however, are that researchers have not done a good job of listening to managers' needs (e.g., Montrey 1991), and managers have not done a good job explaining their needs or soliciting research help (Finch and Patton-Mallory 1993). Within government agencies, efforts to assess managers' information needs for directing research programs are an ongoing process (see Bureau of Land Management 1996; USGS 1996; Smith et al. 1997 for recent examples). These efforts, however, are generally broad in scope (relative to avian conservation) and usually involve only administrative-level managers, while rarely including input from field personnel who are responsible for day-to-day management implementation. Thus, to effectively bridge the gap between research and management for improving avian conservation, an important first step is for researchers to identify the needs of managers at all levels within a given organization. Despite this rather obvious conclusion, we know of few previously published efforts to address this potentially important conservation issue specifically for management of birds and their habitat.

One working example of an existing effort to improve the conservation and management of birds and their habitats in North America (and to some extent, Central and South America, also) is the Partners in Flight—Aves de las Americas (PIF) nongame landbird conservation program. This program provides researchers and managers with unprecedented opportunities to interact and, together, improve avian conservation (see Finch and Stangel 1993 and Martin and Finch 1995 for recent products). PIF is based on cooperative partnerships among an array of federal and state agencies, nongovernmental organizations, private industry, and academic institutions. It fosters coordinated avian conservation efforts through its framework of working groups that provide a network for communication, data sharing, and technology transfer. Agency managers and wildlife biologists regularly attend PIF meetings together, especially at the state and local levels.

One of the PIF technical working groups, the National Research Working Group, has provided the most thorough evaluation of avian research needs to date (Martin et al., unpublished report); such work exemplifies the kind of product that an integrated program like PIF can generate. In summary, a questionnaire presenting a wide variety of research topics relating to the conservation of Neotropical migratory birds was sent to researchers and land managers throughout the United States. Completed questionnaires were received from 256 respondents from 17 agencies including academic institutions, federal and state land management agencies (e.g., U.S. Fish and Wildlife Service [USFWS], U.S. Forest Service [USFS], U.S. Bureau of Land Management [BLM]), National Aeronautics and Space Administration (NASA), Audubon, The Nature Conservancy (TNC), and private industry. Martin et al. (unpublished report) summarized results of the questionnaire by first grouping similar questions into four main subject areas. Within each of these areas, those research questions that received the highest mean rank scores were then restated more succinctly.

Identifying and addressing causes of avian population change were ranked highly by questionnaire respondents, and included both proximate (e.g., nest predation, brood parasitism) and ultimate (e.g., grazing, fragmentation) causes (Martin et al., unpublished report). Understanding fitness parameters (e.g., clutch size, nesting success, nest predation rate) required for self-sustaining populations, as well as how population patterns are influenced by metapopulation processes also ranked high. Information on ecological conditions required for populations to remain viable and on which conditions create source and sink populations are also important areas on which research is needed. Finally, identifying which species are currently at risk and what characteristics define their vulnerability were considered to be of high priority (see Martin et al., unpublished report, for more details). Many of these research priorities are similar to those identified by others (Soulé and Kohm 1989; Martin and Finch 1995; Faaborg et al., this volume; Haufler, this volume; Marzluff and Sallabanks, this volume).

A similar, but more narrowly focused questionnaire (developed independently by the USFS Rocky Mountain Forest and Range Experiment Station) was sent to district biologists of the USFS in the western United States asking them to list two general research needs that would help them better manage sensitive species (Squires, Hayward, and Gore, this volume). Responses received from thirty-five wildlife biologists resulted in the identification of eighteen research needs (table 11.1). Most biologists agreed that a definition of the range of natural variability for species and data on species-specific habitat requirements were their most important information needs.

Information-needs questionnaires are potentially an important tool for better directing research to ensure improvement in avian conservation, but other than the two questionnaires described here, we are unaware of other similar efforts to specifically identify avian information needs most important to land managers and how research should be prioritized. The pressing issue then, and on which this chapter focuses, is how managers' information needs can best be identified and how we should conduct studies and present results such that effective management and conservation of avifauna can follow.

Features of Research That Improved Conservation

In general, research that improves conservation will be that which addresses managers' tangible needs and is based upon well-replicated experimental studies that measure population health, not just species' distribution and abundance (Marzluff and Sallabanks, this volume; Young and Varland, this volume). Measures of density alone can be misleading indicators of a species' viability (Van Horne 1983; Vickery et al. 1992b), and management plans based solely upon such measures are risky. One particular topic, the development of bird-habitat relationship models, has been an important component of previous research on birds. Indeed, effective conservation may require specific, accurate, predictive models of wildlife-habitat relationships (O'Neil and Carey 1986). This is one area of research that may be particularly difficult for managers to interpret, however, and there are many instances in which habitat measurements have failed to accurately predict response by bird populations (Toth and Baglien 1986).

All too often, management recommendations resulting from research are not effectively implemented and fail to improve avian conservation. Reasons for this failure are numerous: (1) research that has been conducted is not meaningful to managers (Young and Varland, this volume); (2) communication between researchers and managers is poor (Finch and Patton-Mallory 1993; Ganey and Dargan, this volume); (3) results from research have not been published in the most appropriate forum for effective management and conservation to ensue (Hejl and Granillo, this volume); (4) most studies are correlative and observational rather than experimental and fail to address synergistic affects of multiple land-use activities (Marzluff and Sallabanks, this volume; Ganey and Dargan, this volume); and (5) if declines or deleterious effects of a specific land-use activity are found, the mechanisms driving such effects are rarely identified (Marzluff and Sallabanks, this volume). The most obvious exceptions include species that are threatened and endangered (e.g., Northern Spotted Owl [*Strix occidentalis*], Florida Scrub Jay [*Aphelocoma coerulescens*], Red-cockaded Woodpecker [*Picoides borealis*], Piping Plover [*Charadrius melodus*], and Bald Eagle [*Haliaeetus leucocephalus*]). Threatened and endangered species are intensively studied, projects often have large research budgets, and managers often are involved at the inception.

Snag management, modified by research on cavity-nesting birds, also has led to improved conservation of this important avian guild (Davis, Goodwin, and Ockenfels 1983). Perrins, Lebreton, and Hirons (1991) perhaps offer the best examples of other species and guilds for which avian research has had significant relevance to conservation and management. With a plethora of studies on songbirds, including Neotropical migrants (Hagan and Johnston 1992; Martin and Finch 1995), we are certainly more aware of their habitat needs and factors that influence their population dynamics now than two decades ago. Several excellent examples of management-oriented information on the needs of bird species have been published (Bushman and Therres 1988; DeGraaf et al. 1992; Hamel 1992), but how often have results from the majority of avian studies actually been translated into improved conservation? For the reasons outlined above and elsewhere in this volume, examples of research on songbirds leading to improved conservation are less common than perhaps initially perceived.

To improve the relevance of science, PIF has undertaken the important task of identifying research questions that would improve the conservation of songbirds. In fall 1996, the Oregon/Washington (Arnett) and Idaho (Sallabanks) chapters of PIF collaborated to assess avian research priorities in the Pacific Northwest using a questionnaire, following previous efforts at the national (Martin et al., unpublished report) and state (Andelman and Stock 1994) levels and by the Western Working Group of PIF. Our approach differed somewhat from previous efforts, because we specifically targeted land managers and management biologists to assess their needs for making day-to-day decisions that affect bird populations and habitat. The remainder of this chapter describes our efforts and questionnaire results.

Research Needed to Further Conservation

The research needs presented in this section are those directly identified by the results of our survey. Research needs questionnaires were sent not only to organizations that owned and managed land in the states of Idaho, Oregon, and Washington, but also to state wildlife agency personnel who oversee wildlife management on properties owned by others. Recipients ($n = 426$) of the questionnaire represented diverse groups and organizations, including BLM, Bureau of Reclamation, National Park Service (NPS), Native American tribes, nongovernmental organizations, private industry, U.S. Army Corps of Engineers, U.S. Department of Defense, U.S. Department of Navy, U.S. Fish and Wildlife Service, and U.S. Forest Service. We attempted to census each affiliation within the three states by sending questionnaires to all representative management units (e.g., all districts for each national forest, all forest products companies).

In developing the questionnaire, we utilized several concepts and questions presented by Martin et al. (unpublished report), position statements drafted by the PIF Western Working Group Research Subcommittee, and our own perceptions of important management issues in the Pacific Northwest. The questionnaire was divided into three parts: background and general information, research needs, and information dissemination. In the background section, we asked about each recipient's organizational affiliation, position, and major habitat managed. Also in this section, recipients were asked to respond to general questions about the types of management activities they implemented and whether research was addressing their specific needs relative to avian conservation and management. In the research needs section, recipients were: (1) given the opportunity to write in suggestions for their most pertinent information needs; and (2) asked to rank a series of questions from low (1 = no value for successful management implementation) to high (5 = critical for successful management implementation). These questions were categorized into the following sections: monitoring and research methods, population dynamics, habitat ecology, habitat management, and landscape ecology. Mean ranked scores (and standard errors) were calculated and summarized by each major habitat managed by respondents, regardless of affiliation. In the last section, we asked recipients several questions about communications with the research community, suggestions for improving communication and technology transfer, and usable formats for disseminating research results.

We received 185 completed questionnaires (43 percent return) from 13 different groups and organizations: USFS (36 percent, $n = 68$); state wildlife and natural resources agencies (23 percent, $n = 42$); BLM (14 percent, $n = 27$); USFWS (7 percent, $n = 13$); forest products industry (7 percent, $n = 13$); Native American tribes (4 percent, $n = 7$); U.S. Department of Defense (2 percent, $n = 3$); NPS (2 percent, $n = 3$); utilities industry (2 percent, $n = 3$); U.S. Army Corps of Engineers (1 percent, $n = 2$); U.S. Bureau of Reclamation (1 percent, $n = 2$); and TNC (1 percent, $n = 2$). Seventy-one percent ($n = 131$) of the respondents were management biologists, while 19 and 10 percent ($n = 35$ and 19, respectively) were managers (e.g., district ranger, refuge manager) or staff biologists, respectively. Forest, range, and wetland habitat managers and biologists (69 percent, $n = 128$; 19 percent, $n = 35$; and 6 percent, $n = 12$, respectively) were the primary respondents to our questionnaire, while those managing coastal, agricultural, and urban areas composed the remaining responses ($n = 10$).

In the background section, we attempted to assess perceptions of managers relative to the adequacy of research and availability of scientific information that assists them with making day-to-day decisions that affect bird populations and habitat. When asked "Do you feel that existing methods and protocols are adequate to meet your objectives for monitoring birds?" 44 percent said yes, while 56 percent said no or were unsure. Of those who answered no or were unsure, only 37 percent believed that there was enough scientific information available to develop methods and protocols for monitoring birds. We then asked participants whether they felt, in general, that current bird monitoring and research were adequate to address issues important for day-to-day management implementation. Only 18 percent responded yes to this question, while 52 percent said no and 30 percent were unsure. We then requested that participants list the most common management activities performed on their ownership and indicate whether they believed enough scientific information was available to make decisions relative to bird response to these activities. Consistently, respondents indicated that there was not enough information to make informed decisions about bird response to all management activities listed in this survey (figure 27.1).

Specific information needs were addressed by asking participants to rank a series of questions among several different categories. Table 27.1 summarizes mean ranked scores and standard errors for questions in each of five categories and by each major habitat managed by respondents. In the "Monitoring and Research Methods" section, respondents consistently indicated that determining the most effective and cost-efficient methods for monitoring birds was a top priority. Methods to monitor birds in conjunction with other vertebrates and deter-

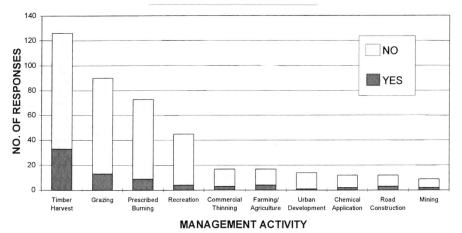

Figure 27.1. Most common management activities conducted on lands managed by questionnaire respondents, and the number of those who feel there is (yes) or is not (no) enough scientific information currently available to assist them with making day-to-day decisions about bird responses to those activities.

mine the times of year when birds are most susceptible to disturbance also were important to most respondents. Relative to population dynamics, most respondents believed that evaluating effects of human disturbance on avian productivity and survival was the major information need. Range and wetland managers ranked the determination of bird population status and causes of population declines highest in this section. Responses to questions about habitat ecology varied, but forest, rangeland, and urban managers suggested that understanding structural habitat features important to birds and how to provide them at different spatial scales was the most important research need in this section. In addition, identifying long-term trends in habitat condition was a priority for most respondents. Not surprisingly, the top-ranked research priority for the "Habitat Management" section mirrored the respondent's primary management activity. Bird response to silvicultural prescriptions, green tree/snag retention, and riparian management zone widths were ranked highest among forestland managers, while response to grazing was most important to rangeland managers. Relationships between birds and agricultural practices were ranked highest by wetland and agricultural land managers.

We pooled several questions and concepts from the previous three sections and broadened their scope into a section on "Landscape Ecology," relative to avian conservation and management. Forestland managers believed that understanding the influence of habitat fragmentation resulting from management activities and natural disturbance was of highest priority, followed by determining minimum area requirements to sustain viable bird populations and cumulative effects of management activities and habitat change. Rangeland, wetland, agricultural, and

Table 27.1. Mean ranked scores[a] for questions in each of five categories and summarized by each major habitat managed by respondents. Numbers in bold represent the highest-ranked research need for each category of questions (n = 128 forest respondents [F], 35 range [R], 12 wetland [W], 4 coastal [C], 3 ag/farm [A], and 3 urban [U]).

	Mean Score					
	F	R	W	C	A	U
A. Monitoring and Research Methods						
1. Effective and cost-efficient methods for monitoring.	**4.0**	**4.1**	**3.9**	**4.8**	3.7	**4.7**
2. Techniques to monitor birds in conjunction with other vertebrates.	3.8	3.9	3.5	3.3	**4.3**	4.0
3. Experimental design and analyses to evaluate population change and response to habitat variables and manipulations.	3.4	3.5	3.2	2.5	3.0	4.3
4. Habitat and habitat change for birds through space and time.	3.5	3.2	3.1	3.0	2.0	3.0
5. Distribution and abundance throughout the year.	3.2	3.3	3.6	2.8	4.0	4.0
6. How nonhabitat factors (e.g., weather) affect birds.	2.8	3.2	3.3	2.3	4.0	3.0
7. Local and regional conservation and management plans.	3.5	3.5	3.8	4.5	3.0	**4.7**
8. Seasonal susceptibility to disturbance.	3.6	3.5	3.5	4.5	2.7	4.3
B. Population Dynamics						
1. Nest success and survival among different habitats and management areas.	**3.9**	3.7	3.8	2.0	3.0	3.0
2. Effect of nest predation/cowbird parasitism among different habitats.	3.0	2.8	2.7	1.8	2.7	2.7
3. Limiting factors associated with different bird populations.	3.8	4.0	4.0	2.5	3.7	3.7
4. Determine whether populations are increasing, decreasing, or stable.	**3.9**	4.0	**4.1**	2.5	4.0	4.3
5. Causes of specific bird population declines.	3.8	**4.1**	3.6	2.5	**4.7**	4.3
6. Demographic characteristics and how they influence bird populations.	2.8	2.8	2.5	1.3	3.0	2.7
7. Natural history of bird species about which little is known.	3.3	3.1	2.7	1.8	3.3	3.3
8. Effects of human disturbance on productivity and survival.	**3.9**	3.9	3.8	**4.0**	**4.7**	**4.7**
C. Habitat Ecology						
1. Key habitats where birds show greatest declines.	3.8	4.0	3.6	4.0	3.0	4.3
2. Structural habitat features of greatest importance to birds and how to provide them at different spatial scales.	**4.3**	**4.2**	3.6	4.0	3.3	**4.7**
3. Abiotic habitat variables (e.g., soil, aspect) that influence bird use of habitat.	3.1	2.8	2.7	3.8	2.3	4.3
4. Location and management of stopover habitats for migrating birds.	3.4	3.9	3.8	**4.5**	3.7	4.3
5. Key wintering grounds of migratory birds and preferred habitats.	3.4	3.6	3.8	4.3	**4.0**	4.3

	(1)	(2)	(3)	(4)	(5)	(6)
6. Age and sex differences in habitat use.	2.4	2.4	2.4	1.8	2.0	2.3
7. Habitat use patterns for birds not well studied.	3.4	3.2	3.1	1.8	3.0	3.0
8. Relationships between birds and other vertebrates (e.g., ungulate grazing).	2.9	2.9	2.7	1.8	3.7	3.0
9. Long-term trend of habitats and its effect on birds in my area.	3.9	4.1	4.0	2.8	3.7	4.0
D. Habitat Management						
1. Agricultural practices.	2.5	2.9	4.2	3.0	4.0	3.3
2. Various silvicultural prescriptions.	4.5	3.3	2.8	3.0	3.0	4.0
3. Green tree and snag retention options.	4.3	3.4	2.5	3.0	3.0	3.0
4. Chemical and mechanical vegetation management.	3.3	3.9	3.6	3.3	3.3	4.0
5. Spread of noxious weeds and exotic vegetation.	3.3	4.2	4.0	2.5	3.0	3.7
6. Wildfire and prescribed burning.	3.8	4.3	3.3	2.8	2.7	3.3
7. Livestock grazing practices.	3.1	4.5	4.0	2.3	3.3	4.0
8. Oil/gas development.	2.1	2.4	2.2	2.5	2.7	3.7
9. Water management (e.g., sump rotation).	2.2	2.7	4.0	3.3	2.3	3.7
10. Riparian management zone widths and basal area requirements.	4.1	4.0	3.7	4.3	3.3	4.0
11. Habitat fragmentation.	4.2	4.2	3.6	4.0	4.0	4.0
E. Landscape Ecology						
1. Responses of bird communities to distribution of structural habitat features.	4.0	3.5	3.1	2.8	3.3	3.7
2. Minimum habitat area requirements to sustain viable populations.	4.2	4.1	3.5	3.8	4.0	4.7
3. Influence of landscape habitat fragmentation resulting from management activities and natural disturbance.	4.2	4.3	3.3	3.8	4.0	4.0
4. Cumulative effects of management activities and habitat change.	4.1	4.2	3.4	4.8	4.0	4.3
5. Effects of patch size and isolation on use of patches and landscapes.	3.9	4.0	3.2	2.8	3.0	3.7
6. Impacts of nest predation and parasitism in a landscape context.	3.1	2.9	2.6	2.0	3.0	2.3
7. Influence of natural ecosystem processes on distribution and productivity.	3.7	3.8	2.9	2.5	3.0	3.0
8. Development of management prescriptions and adaptive trials.	4.0	4.3	4.2	4.5	4.7	4.7

[a] Information needs question ranking system: 1 = not useful for successful management implementation; 2 = interesting and somewhat useful for successful management implementation; 3 = useful for successful management implementation; 4 = very useful for successful management implementation; and 5 = critical for successful management implementation.

urban managers indicated that developing adaptive management trials and evaluating bird responses to various management prescriptions were most important from a landscape perspective.

The effectiveness of technology transfer and information dissemination between researchers and managers was addressed in the last section of the questionnaire. We began by asking participants if they communicated with researchers and, if so, whether this interaction was frequent or infrequent. Thirty-two percent indicated that they communicated frequently with researchers, while 65 percent and 3 percent said infrequently or never, respectively. Of the 180 respondents who communicate with researchers, 75 percent indicated that there could be improvements in communication and information dissemination between researchers and managers. Participants were asked to suggest ways to improve communication and technology transfer between researchers and managers. The most common suggestions were: (1) establishing a contact list of specialists by species and/or specialty (16 percent, $n = 31$); (2) involving managers with development and implementation of research projects (13 percent, $n = 24$); (3) communicating more frequently and constructively (11 percent, $n = 21$); (4) improving the distribution of research findings and recommendations (11 percent, $n = 21$); (5) improving the availability of scientific literature and other information sources (11 percent, $n = 21$); and (6) conducting workshops on research findings and management implementation (10 percent, $n = 20$).

We then asked if any research organization had contacted them directly about *their* specific needs and research priorities for birds. Only 34 percent of respondents answered yes to this question, but 89 percent of those replying yes indicated that this communication had been very useful. We also asked recipients whether they hosted and/or attended cooperative meetings relative to avian research. Forty-three percent said yes, of which 90 percent suggested that these meetings were useful for information dissemination and building working relationships.

When asked which information sources were most commonly used to make decisions about bird populations and their habitat, respondents ranked U.S. government agency technical publications highest (mean $\pm$ SE rank score = 4.15 $\pm$ 0.07). Refereed scientific journals (3.89 $\pm$ 0.09), state agency technical reports (3.69 $\pm$ 0.08), staff (expert opinion) (3.48 $\pm$ 0.10), symposium and workshop proceedings (3.37 $\pm$ 0.08), and literature reviews (3.35 $\pm$ 0.09) were other important information sources identified by respondents. We also asked, "If research topics were to be synthesized into one document to assist with more efficient information dissemination, which would be most useful to you: a smaller, one- to two-page 'cookbook' summary or an in-depth report?" About half of all respondents suggested either approach, but biologists generally preferred more in-depth reports, while managers suggested "cookbook" documents would be more useful. Respondents suggested that the *Journal of Wildlife Management* and *Wildlife Society Bulletin* were scientific journals they most frequently referenced and deemed most useful when gathering information on birds and their habitat.

Making Research Effective for Conservation

Implementation of local, regional, and global management programs that promote avian conservation can be limited by a variety of factors (e.g., inadequate funding) but certainly is further impeded by philosophical and technical differences between researchers and managers. Because of the complexity of land management activities and their associated effects on wildlife, knowledge from many disciplines is required to address environmental problems and management issues (Leedy 1987; Davis, Halvorson, and Ehorn 1988; Beissinger 1990; MacLaren 1992). More importantly, as human demand on natural resources accelerates, the necessity for increased communication and cooperation between researchers and managers is even more pronounced (Finch and Patton-Mallory 1993).

Although interpretation and extrapolation from surveys using questionnaires should be viewed cautiously (McIvor and Conover 1994), several key points emerged from our evaluation of managers that have implications for avian conservation and effective management implementation. Respondents generally believed that existing scientific information is inadequate to assist them with monitoring bird populations and making day-to-day decisions about: (1) bird response to management activities; and (2) how to maintain and/or enhance habitat in actively managed landscapes. In some instances, this perception may simply be a function of poor communication or information dissemination about a particular subject that actually has been well studied. In other cases, research topics may have been addressed elsewhere, but findings and recommendations have not been evaluated locally to determine applicability or effectiveness. And, of course, there are many questions about avian ecology and relationships with management that have yet to be answered (e.g., dispersal; Walters, this volume). Regardless of the situation, the managers we surveyed considered existing scientific information to be inadequate to implement effective management for avian conservation.

Responses to questions were fairly consistent among respondents managing different habitats and seem logical within a typical land-management framework. Given the dynamic nature of funding among most organizations, it is not surprising that managers want to maximize efficiency and effectiveness of monitoring programs while minimizing costs. Researchers are challenged not only with offering a variety of monitoring tools (DeSante and Rosenberg, this volume; Hutto, this volume), but also with presenting limitations, biases, and trade-offs associated with utilization of various monitoring methods so managers can make informed choices for effective implementation to meet their objectives. Publications summarizing this information in a format usable for managers are needed, and Hamel et al. (1996) offers an excellent example for implementing point-count monitoring for birds. Ralph et al. (1993) also provides a comprehensive, easily understood review of monitoring protocols and procedures for studying landbirds.

Other results reflect issues facing managers on a daily basis. For example, evaluating the effects of human disturbance on birds is an important research need for managers because disturbance-related activities are often frequent and continual and usually have regulatory implications. In general, the effects of disturbance on birds are poorly understood, and subsequent management decisions are often based on professional opinion rather than empirical data. Relative to habitat management and landscape-scale implementation, it is clear that managers need more information about bird response to major activities that are conducted on their lands. Understanding how to best provide structural habitat features and identifying location and condition of key habitats are obviously important needs for managers, who directly influence manipulation of both structure and area for birds. Consistently, managers ranked fragmentation, cumulative effects of management activities, and developing management prescriptions and adaptive trials as their main research needs from a landscape perspective. While continuing studies of bird response to various management activities are warranted in specific locations where information is lacking, managers may benefit most from timely syntheses of research findings on bird response to a particular activity (e.g., grazing). Managers require information more quickly than researchers generally can produce it (Hejl and Granillo, this volume), and syntheses (e.g., Hejl et al. 1995; Saab et al. 1995) coupled with management recommendations could be used in the interim as new research projects are developed. Researchers and managers could then collaborate to identify research priorities and hypotheses, develop adaptive trials, and locally test prescriptions and models developed from other studies.

We anticipate that responses to questions in our survey will assist with prioritizing future avian conservation research needs, or at least stimulate discussion during research coordination. But more importantly, a key message to researchers is that hypotheses need to be developed and addressed in the context of specific management regimes and must consider operational and funding limitations if research findings are to be effectively implemented by land managers. Too often, researchers develop project objectives and sampling designs without input from managers, and only when the study is complete are managers approached for review of the products (Finch and Patton-Mallory 1993; Young and Varland, this volume). The outcome is usually frustration for both groups: managers feel left out of the process, and research objectives may not coincide with their specific needs, while researchers may find that operational constraints hamper their experimental design or data collection procedures. If managers, most importantly those "on the ground," are involved early in the development of research projects, these constraints are addressed more quickly and effectively, thus conserving valuable resources and deterring alienation among participants. Also, by being included early in the research development process, managers are more likely to accept results and implement recommendations at the end of studies (Finch and Patton-Mallory 1993).

We also believe that involving those individuals who actually are responsible for work on the ground is vital for successfully implementing management for birds and their habitat. For example, Weyerhaeuser Company has successfully implemented selective timber harvest programs in south-central Oregon to maintain and enhance nesting habitat for Bald Eagles (Anderson 1985; Weyerhaeuser Company 1992). Biologists, foresters, engineers, and entomologists from the company and several agencies collaborated to develop harvest plans, but ultimately company field personnel and forestry contractors were responsible for implementing silvicultural prescriptions. Including these individuals early in the process was important for several reasons: (1) dialogue and trust were established among researchers, managers, and field personnel; (2) wildlife and silvicultural objectives were developed mutually, clearly explained, and well understood by field personnel; (3) researchers and managers were advised of operational limitations, costs, and safety issues that influenced the development and successful integration of prescriptions with harvest operations; and (4) field personnel were given a sense of ownership in the project that undoubtedly increased the success of management implementation. We encourage researchers and managers alike to foster working relationships like that described above by actively involving all individuals at the inception of management projects.

Gaps in communication and technology transfer still exist between researchers and managers in the Pacific Northwest. Finch and Patton-Mallory (1993) provide a thorough and comprehensive discussion on why gaps exist between research and management and offer several recommendations for bridging such gaps. Results from our survey strengthen their discussion. We conclude that implementation of the recommendations made by Finch and Patton-Mallory (1993) could greatly improve effectiveness of research and management relative to avian conservation.

In particular, the way information is transferred to managers seems critical for successful management implementation to achieve avian conservation goals. Respondents to our questionnaire indicated that sources other than refereed journals are important to them, and many would prefer one- to two-page summary reports that simplify results and management recommendations. Research results are typically scattered across various publications and generally not in a form that is readily usable by managers (Finch and Patton-Mallory 1993; Hejl and Granillo, this volume). Synthesizing information into a usable form will be important in the future, but this requires funding and technical personnel. An inherent problem with research institutions, particularly academia, is the "publish or perish" paradigm, which does not appear to recognize, reward, or offer incentives for disseminating information in forums other than refereed journals. We agree with Finch and Patton-Mallory (1993) that overcoming this barrier will require that research institutions develop procedures to reward researchers for technical assistance, production of "nontechnical" publications, and other forms of problem solving and technology transfer for managers. Hamel et al. (1996)

exemplifies the opportunity for researchers and managers to collaborate on developing publications that are technically credible, yet "user friendly" for managers.

Because research results are typically scattered across various publications, improving the distribution of publications is equally important. Respondents indicated that of refereed publications, those offered through membership with The Wildlife Society were most frequently relied on for information. We speculate that most wildlife managers and management biologists do not belong to multiple professional organizations that offer refereed journals, as many researchers do, and consequently are most familiar with publications of the parent society for their profession. Therefore, managers may have limited access to avian journals where usable information is frequently published. We are not suggesting that options for publication be limited but do encourage both researchers and managers (and their institutions) to work collaboratively to develop better processes for publishing and distributing information, particularly for those managers and biologists in remote areas.

Although most respondents in our survey suggested there is room for improving communication, an overwhelming majority implied that what communication they did have with researchers was very useful. We encourage researchers to communicate more frequently with managers at all levels within a given organization to better understand: (1) political, economic, and operational issues that influence research development and implementation; (2) managers' specific information needs and how to obtain their input to research projects; and (3) what products will maximize effectiveness of management implementation for avian conservation. Managers, too, need to take a more active role in understanding science, research methodology, and terminology (e.g., sampling design considerations) and in pursuing assistance from researchers. We recommend annual cooperative meetings between principal research investigators and managers, including both administrative and field staff, for all avian research projects. These meetings should combine both field visits and office discussions of research progress, data interpretation, management implications, and future project direction.

With respect to nongame landbird conservation at least, we concur with Finch and Patton-Mallory (1993) that PIF offers a logical forum for technology transfer between research and managers. Other avian conservation groups may be able to perform a similar function for other avifauna (e.g., Ducks Unlimited, Western Hemisphere Shorebird Reserve Network, Wild Turkey Federation). Respondents in our survey offered numerous ideas for improving communication and information dissemination that could easily be developed and implemented by working groups within this organization. We suggest that the Information and Education, Monitoring, and Research Subcommittees at all levels within PIF work collaboratively to produce and distribute tools that effectively bring researchers and managers together to achieve common goals for avian conservation. We offer that future functions of the PIF Research Subcommittees could include: (1) synthesizing and publishing avian research findings of use to managers; (2) sponsoring and coordinating local and regional workshops and symposia for disseminating research findings and implementing management strategies; and (3) coordinating local cooperative research projects that link avian conservation goals with larger research programs or management planning efforts (e.g., Columbia River Basin assessment; Quigley, Haynes, and Russell 1996). While these recommendations are made specifically with respect to nongame landbirds (especially songbirds), other species (e.g., waterfowl, shorebirds, raptors) could benefit from similar efforts.

The results of our questionnaire complement those of Martin et al. (unpublished report) in a way that, when taken together, provides researchers in the field of avian conservation with critical information about the most pressing and useful issues to study. We hope that avian biologists will make use of this information recognizing both the limitations and the strengths. There are many topics about which we need additional information but only limited funds with which to study them. Given that we are ultimately challenged with improving the conservation of avifauna, this assessment provides both researchers and managers with a common link with which to move toward this important goal.

Acknowledgments

Although too numerous to mention individually, our first thanks go to the many respondents of our questionnaire. The development of the questionnaire benefited from the comments of B. Altman, J. Augsburger, K. Bettinger, S. Bouffard, J. Doremus, R. Floyd, R. Goggins, C. Paige, J. Plissner, T. Reynolds, T. Rich, J. Rochelle, D. Runde, V. Saab, and M. Vander Haegen. Special thanks also go to J. Daugherty and M. Yordy for creating the database and assisting with queries, analysis, and manuscript preparation. J. Bart, E. Fritzell, D. Runde, M. Vander Haegen, and J. Marzluff provided comments on an earlier version of this manuscript. T. Martin provided us with a draft report summarizing results of the PIF National Research Subcommittee's priority research needs questionnaire. The Oregon/Washington and Idaho chapters of PIF sponsored our efforts and provided logistical support. Weyerhaeuser Company provided financial support for production and mailing of questionnaires and other correspondence with recipients.

CHAPTER 28

Making Research Meaningful to the Manager

Leonard S. Young and Daniel E. Varland

American ornithology had its beginnings in the 1800s as early naturalists documented the life-history characteristics of newly discovered species (Wilson 1812; Audubon 1831; Nuttall 1832). Gradually, ornithology began to approach management as some formerly abundant birds such as the Passenger Pigeon (*Ectopistes migratorius*) and the Great Auk (*Pinguinis impennis*) were driven to extinction (Blockstein and Tordoff 1985; del Hoyo, Elliott, and Sargatal 1996). These calamities gave rise to an increased emphasis on conservation. Avian biology's management connections solidified in the 1930s with the publication of Aldo Leopold's *Game Management* (Leopold 1933) and the advent of scientifically based wildlife management.

Linkages between avian biology and natural resource management continued to strengthen throughout the twentieth century as an increasing awareness of the necessity of basing management programs on scientific information grew and was expressed through legislation such as the Endangered Species Act of 1973 and the National Forest Management Act of 1976. Today, one of the primary pursuits of avian biology is to provide information needed to wisely manage birds and their habitats. Much of the funding for avian research is from sources related to management.

Contemporary Management Environments

There are several ways in which avian researchers and natural resource managers interact. First, researchers identify major environmental problems and work with managers to find solutions. Examples of this type of interaction between researchers and managers include work during the 1960s to identify and correct the effects of organochlorine pesticides on reproductive success and declines in many bird populations (Keith 1966; Porter and Wiemeyer 1969), and ongoing work to identify and correct the effects of organophosphate insecticides on Swainson's Hawks (*Buteo swainsoni*) wintering in Argentina (Di Silvestro 1996).

Second, researchers help managers solve problems related to everyday management. Examples of this type of interaction between researchers and managers include ongoing efforts to integrate Northern Goshawk (*Accipiter gentilis*) habitat requirements into commercial forestry programs in the southwestern United States (Reynolds et al. 1992; see also Braun et al. 1996) and recent work to resolve conflicts between communal roosts of Common Ravens (*Corvus corax*) and electric power transmission systems (Engel et al. 1993).

Third, researchers provide information needed to design effective rules and regulations. Examples of this type of interaction between researchers and managers include research on the habitat needs of Northern Spotted Owls (*Strix occidentalis caurina*) (Lehmkuhl and Raphael 1993; Bart 1995) and on the effects of river recreation on wintering Bald Eagles (*Haliaeetus leucocephalus*) (Knight and Knight 1984; Stalmaster and Gessaman 1984).

Fourth, researchers develop tools and techniques needed for management programs. Examples of this type of interaction between researchers and managers include development of survey techniques for Marbled Murrelets (*Brachyramphus marmoratus*) (Pacific Seabird Group 1994) and captive breeding techniques for Whooping Cranes (*Grus americana)* (Kepler 1978).

Last, researchers investigate the demographics of hunted species to help managers set seasons and harvest limits. An example of this type of interaction between researchers and managers is the comprehensive program carried out by the governments of Canada, the United States, and Mexico to evaluate waterfowl population levels and reproductive success as a basis for setting hunting seasons and bag limits.

In addition to these focused interactions between researchers and managers, management programs often benefit from unexpected findings and new management questions that arise as secondary products of focused research. Such insights can be extraordinarily beneficial in guiding management programs in new directions.

The basic roles of the researcher and the manager are distinct. The manager's role is to weigh diverse information and make choices, while the researcher's role is to provide timely, high-quality information that empowers the manager's decision-making process. Generally researchers do not directly conserve birds. Managers make the decisions that affect birds and their habitats, within boundaries established by law and policy. Although researchers sometimes are asked to share managers' decision-making responsibilities (e.g., on conservation planning teams and recovery teams; Noss, O'Connell, and Murpy 1997), the researcher's work toward avian conservation is usually expressed through the manager's decision-making process.

In this chapter, we offer practical advice on how research can succeed in management environments. Much of the advice we offer may seem simple, even intuitive. However, the points we emphasize are often overlooked and have been responsible for many research projects failing to produce results that are meaningful from a management perspective. We hope that the ideas we present will help researchers more fully unlock the power and potential of their scientific work.

Principles of Meaningful Research

Research is meaningful in a management environment if it can be used to help make decisions; throughout this chapter we use the term *meaningful* in this context. As evidenced by the other chapters in this volume, researchers usually view research in terms of topical priorities. However, "meaningful" research is just as dependent on basic characteristics that allow it to successfully meet the needs of management programs. Knowledge that appears adequate to the researcher may be largely unusable by the manager, and knowledge gaps that are evident from the researcher's perspective may not impede conservation at all.

In most instances, avian conservation is one of several management goals. Management decisions usually reflect legal, policy, and economic considerations in addition to scientific input (Sinderman 1982; Thomas 1988), and managers must often consider diverse, conflicting recommendations. Rarely will a management decision strictly reflect a researcher's recommendation. Researchers should recognize the challenges that managers face and should appreciate the distinction between research and management environments.

For example, in Washington state, ecosystem conservation is the primary management goal on only 2 percent of over 3 million acres of state-managed forest and shrub-steppe lands. By statute, generation of revenue to support state institutions and local governments must be the primary management goal. The challenge for managers is to successfully integrate avian conservation into revenue-oriented management programs. Even National Wildlife Refuges must sometimes achieve goals such as oil and gas production and cattle grazing.

Managers have limited time to make decisions and put management programs into effect. Researchers should design experiments that can provide the best results possible with available resources and time, and they should realize that their results may be used sooner than they desire and be extrapolated beyond what they feel is scientifically prudent. Judging what information is needed for management requires researchers to recalibrate their eye using additional criteria. To promote meaningful research, we offer the following principles.

Research should be honest about addressing management needs. Researchers should not establish tenuous management connections as a means to garner research funds. Unless the researcher is serious about providing useful information for management, it will be serendipitous for a study to produce usable results. Superficial management recommendations create confusion for management programs and thereby do a disservice to avian conservation.

Research should be interactive. Researchers and managers must work together to set research priorities, design experiments, and review results (Finch and Patton-Mallory 1993; Christensen et al. 1996). Working relationships should be characterized by mutual respect and cooperation, and awareness of the unique challenges faced by each party. Researchers should periodically update managers on research progress while a study is underway and should exert extra effort to ensure that their work is understood by managers before, during, and after the study.

Meaning often can be enhanced through cooperative studies, whereby landowners, stakeholders, and other interested parties participate in a cooperatively designed and supported research program. Such programs have the strong advantage that all interested parties are working off the same page, using the same scientific information. Differences may still exist in terms of what management actions and choices are appropriate, but the confusion and discord that result from "my science vs. your science" (Jasanoff 1990) can be minimized. Cooperative research also helps to ensure adequate funding for large-scale problems that are expensive to investigate.

Research should be self-aware. Researchers must be actively aware of why their work is being funded. Researchers should think about the purpose of their research at every stage of the research process—objective setting, design, data collection, analysis, and write-up—and should anticipate how the information they are producing will be used by managers.

Research should set high standards of quality and objectivity. Meaningful research must produce accurate information that withstands the intense scrutiny of peer review. More and more frequently, management decisions related to avian conservation are being made in the courts, in response to lawsuits. As part of this process, scientific work is often aggressively reviewed with the specific intent to find fault and to discredit (Murphy and Noon 1991; Gutiérrez 1994).

Research should not be dogmatic. Meaningful research avoids untested presumptions that have become ambient knowledge through repetition. Research that does not proceed from a valid starting point may lead managers in the wrong direction and result in erroneous or unnecessary management actions that later have to be corrected. Researchers need to rigorously and quantitatively test many basic assumptions of conservation biology that have slipped into textbooks and common knowledge as dogma.

Research should be geographically specific. Whenever possible, research should be carried out within the ecoregion in which the related management program will be implemented. It can be risky to extrapolate the results of research carried out in one ecoregion to other regions where the same relationships may not exist.

Research should consider the entire ecosystem. Meaningful research proceeds with a fundamental understanding of the ecosystem in which it is carried out. Researchers should understand the ways in which ecosystem processes can influence research results. For example, research to determine the forest habitat relationships of Marbled Murrelets as a basis for conservation planning should consider the influence of oceanic conditions on murrelet breeding biology. Murrelet occupancy of forest stands may be influenced by currents, water temperatures, and the distribution of forage fish (Burkett 1995).

Research should establish an appropriate frame of reference. Researchers should evaluate functions and processes that exist in undisturbed ecosystems to gain an understanding of the full expression of bird populations. However, they should not assume that populations can persist only if pristine conditions are emulated in all respects. Researchers should recognize that management programs must succeed in managed environments subject to periodic disturbance.

Research should employ current technologies. Meaningful research uses the best and most appropriate contemporary research approaches. Geographic information systems, spatially explicit population models, and advanced inferential statistics are three sets of tools that are more and more frequently brought to bear on avian conservation (Villard et al., this volume). However, researchers and managers alike should realize the limitations of these technologies; the heart of a successful research project will always be the logic, intelligence, and creativity with which it is designed and carried out (Romesburg 1981).

Research should emulate management activities. Where possible, "real-world" management activities should form the replicates for experimentation. For example, research on the effects of timber harvest on forest bird populations that employs commercial timber harvest units as experimental replicates will be more powerful and readily accepted in a management environment than similar research employing small research plots. Experimental research that employs manipulations that are too small or do not emulate management activities is of limited value (Hilborn and Walters 1981; Marzluff and Sallabanks, this volume).

Research should illuminate limiting factors. Meaningful research directs management attention where it is most needed. Research that helps managers isolate the particular features of the environment that must be addressed is most effective in achieving desired outcomes. For example, some of the most promising contemporary research on Northern Spotted Owl–habitat relationships emphasizes the relationships among landscape pattern, forest structure, prey populations, and owl breeding success (Carey, Horton, and Biswell 1992; North 1993). This research is allowing forest managers to target specific habitat conditions and design management programs that have a reasonable chance of meeting avian conservation goals. Management programs that are designed without such knowledge and must simultaneously address a wide range of environmental variables often meet with limited (or indeterminate) success.

Research should clearly describe desired outcomes. Desired outcomes of management programs should be expressed as specific, measurable conditions that are not artifacts of experimental design. Researchers should suggest several approaches for obtaining the same outcome, thereby creating flexibility and enhancing the manager's ability to achieve multiple goals. Researchers must work with managers to ensure that the outcomes they specify can actually be achieved.

Research should be unambiguous. Meaningful research lends itself to consistent interpretation by all interested parties. The better the job a researcher does in clearly stating results and conclusions, the less guesswork there will be in putting the study to work. Researchers should not expect managers to be able to do much with their work if results and conclusions are muddled. Such studies often spawn antagonistic discussions among parties with differing agendas. An important part of making research results unambiguous is to integrate and interpret them within the overall context of accumulated knowledge and scientific theory. A clear statement of the study's limitations will also increase comprehension by all.

Research should be forthright. Researchers should not shrink from controversial, unpopular, or unexpected conclusions if these are clearly supported by their data. However, research findings that are markedly different from the prevailing body of science must be carefully qualified in research reports. Conclusions and recommendations should be within the scope of what research results actually say (Ratti and Garton 1996). The conclusions and recommendations sections of a report are not the place to promote personal views or popular beliefs not supported by research findings. Hagan (1995, 976) noted how hard this can be: "If we conduct our science with proper objectivity, we may even find our own science in conflict with our personal values."

Researchers should publish results in journals and as management reports. Research results should be published both in high-quality refereed journals and as easily understood management-oriented reports (see Hejland and Granillo this volume). Risser (1993) suggested that the journal *Ecological Applications* require authors to conclude their papers with a section explaining how results could be used by managers. Research institutions should provide the time and resources needed by scientists to publish research results. Universities should allow graduate students to prepare theses in manuscript format to expedite publication and should provide financial support for students until papers are in press. Researchers should resist the tendency to use technical jargon (Finch and Patton-Mallory 1993) and instead should describe their results, conclusions, and recommendations in terms that have meaning to managers. This greatly improves managers' comprehension of their work and eliminates the need for difficult, time-consuming, third-party translations.

Researchers should follow through. Researchers must work with managers to help incorporate research results into management programs after the research itself has been completed. Researchers should be available to answer managers' questions and help make mid-course corrections to management programs. This process of continually modifying management programs in response to new scientific information is termed adaptive management (Walters 1986), and most modern resource management programs are adaptive in nature. Research institutions should recognize the need for scientists to support adaptive management and should reward research scientists for successful follow-through with management programs (Finch and Patton-Mallory 1993).

Researchers should remain faithful to their role. Although we have emphasized the importance of researchers closely coordinating all aspects of their work with managers, we urge researchers to remain faithful to their role. Managers rely on researchers to report their research findings in an unadulterated manner. Researchers should not attempt to make deals or construct compromise solutions to management problems. The manager is the appropriate nexus where diverse information converges and is weighed. Premature compromise disrupts the balance of the decision-making process.

Researchers should not be activists in their own field. We believe that researchers working in support of a management problem should carefully consider their personal behavior. As Sparrowe (1995, 560) stated, "We cannot simultaneously be objective voices on behalf of science and strident lobbyists for our views." Credibility is a fragile and subjective quality that can easily be damaged (see Adams and Hairston 1996). Activist behavior can give rise to allegations that data have been cooked or experiments designed to support preconceived ideas about outcomes or actions (which is difficult enough to avoid in the best of circumstances; Lélé and Norgaard 1996). These considerations may be particularly important for scientists at the start of their careers, who do not have an extensive publication record or other evidence of quality, objective work.

To us, this is a matter of role congruence. All parties involved in a management discussion must be able to rely equally on research findings bearing on the situation. This involves a large measure of trust, which will exist only if the researcher producing the relevant information conducts her or himself with scrupulous objectivity. A scientist must therefore make a choice as to what role he or she wants to play: researcher or activist. Both are legitimate roles, but they are seldom compatible.

Having offered these cautions, it is important to recognize the vital role that scientists play in bringing new information to the attention of society at large and helping society correctly interpret it. Researchers, by definition, are frequently the first segment of society to become aware of new information bearing on society's activities. Scientists should not shrink from standing face to face with other citizens in appropriate public forums to cogently explain and discuss their findings (Ehrlich and Ehrlich 1996).

Summary

We believe that the greatest contribution that a researcher can make to avian conservation is to furnish high-quality information that is equally trusted by all segments of society. Researchers can most effectively support avian conservation if they embrace principles that enhance the meaning of their work in management settings. Recognition of the relative roles of the researcher and the manager, an appreciation of the manager's responsibilities, and an understanding of the way in which management decisions are made are essential for a scientist to establish a management perspective for her or his work. Strong, mutually supportive working relationships characterized by high levels of interaction and excellent communications are essential to effective incorporation of research results into management programs.

Now is a wonderful time to be a natural resource manager. Rote is no longer encouraged. Citizens' expectations of their natural resource managers have never been higher. Administrators are actively challenging the managers they direct to embrace new concepts and demonstrate creativity and innovation in day-to-day decision making. Managers are responding to this encouragement with renewed enthusiasm and heightened awareness of the fullness of their responsibilities.

The research community represents a source of new knowledge and new capabilities that managers badly need. Managers are looking to researchers for direction. They want to find information they can use, and they have to find guidance that will work in the real world. By heeding the principles that make research meaningful to managers, researchers can ensure that their work meets this challenge—that it delivers, not disappoints—and that their work comes alive where it counts: on the ground.

Acknowledgments

We are most grateful for thoughtful reviews of a draft manuscript by Andrew B. Carey and Bruce G. Marcot. We thank Stanley W. Biles, Carl J. Cederholm, Bryan D. Griffith, John M. Marzluff, Robert D. Meier, and Rex Sallabanks for comments and assistance.

CHAPTER 29

Relevance of Research to Resource Managers and Policy Makers

Michael N. Kochert and Michael W. Collopy

For decades, if not centuries, the study of birds and their place in the natural world was driven by the curiosity of naturalists and amateur bird-watchers. More recently, however, development of the science of ornithology has been increasingly led by formally trained researchers affiliated with educational and research institutions. In many cases, this has influenced the focus of the avian research being conducted, with expectations by managers and policy makers that the work undertaken be relevant to their needs.

Recent experiences within the U.S. Department of the Interior (USDOI) illustrate the trend toward a greater integration of effort and increased need for relevance in federally supported research. Creation of the National Biological Survey (NBS) by the U.S. Secretary of the Interior in 1993 was an attempt to consolidate and better coordinate the biological research functions within USDOI and, as the National Research Council report *A Biological Survey for the Nation* (1993) stated, "provide a better and more efficient information base from which to make planning and operational decisions, thereby strengthening the quality of such decisions and improving the management of biological resources." Although the research functions of NBS are now vested in the Biological Resources Division (BRD) of the U.S. Geological Survey, the mission of BRD continues this commitment "to work with others to provide the scientific understanding and technologies needed to support the sound management and conservation of our Nation's biological resources" (BRD 1996).

BRD's commitment to addressing the current and emerging research needs of other USDOI bureaus is best illustrated by its Bureau Information Needs program. This program annually solicits input from other USDOI bureaus on their research needs and convenes national and regional-level meetings to prioritize the issues that they would like BRD to address. At these meetings, scientists and managers collaborate to ensure that priorities are relevant at various scales and that proper coordination is occurring both within and between agencies. Although this process continues to be refined, it has already proved to be an effective mechanism to promote interagency communication and information sharing about common resource management priorities and associated research needs (Hejl and Granillo, this volume) and has given BRD the direction it needs to make its research programs more relevant (Arnett and Sallabanks, this volume).

For research to be relevant and responsive to management needs, it must provide managers and policy makers with information necessary to make informed management decisions. In the previous chapter, Young and Varland describe in detail what characteristics they believe make avian research more meaningful to managers. Their twenty-one "principles of meaningful research" focus largely on various aspects of the scope, scale, approach, and role of ornithological research. We agree with these principles but emphasize that it is especially important for researchers to recognize and understand that resource managers base decisions on more than the results of avian conservation research. Land managers and decision makers operate in a complex management environment that involves multiple issues and, often, conflicting mandates. Avian researchers need to recognize that they contribute to a larger body of knowledge and understand that decisions cannot be made on their information alone.

Avian research that is relevant to management is conducted at multiple biological, spatial, and temporal scales, and relevant research relates results at the spatial and temporal scales at which managers make decisions. In this chapter we discuss two case studies in which relevant research was applied to avian conservation efforts at different scales. The first case demonstrates how research on raptors in the Snake River Birds of Prey National Conservation Area was used to address a local issue. The second describes how Northern Spotted Owl (*Strix occidentalis*) research contributed to the development of the president's Northwest Forest Plan and a bioregional plan being implemented on federal lands throughout the Pacific Northwest. (USDA and USDOI 1994; Raphael, McKelvey, and Galleher, this volume).

Case Studies

We discuss the case studies in relation to some characteristics of relevant research presented by Young and Varland (this volume), specifically how this research: (1) honestly and directly addressed management needs; (2) relied on interaction between scientists and managers; (3) established an appropriate frame of reference; (4) clearly described desired outcomes; (5) was unambiguous; and (6) was forthright.

The Snake River Birds of Prey National Conservation Area

The conservation area is unique because it contains an unusual concentration of nesting raptors, possibly the densest nesting assemblage of noncolonial raptors in the world (USDOI 1979b). Located in southwestern Idaho, this 240,064-hectare area (196,225 hectares are public land) along a 130-kilometer stretch of the Snake River was established by Congress in 1993 (USDOI 1995). Many users vie for the habitat in the conservation area, which is managed by the BLM. The first efforts to protect the area occurred in 1971 when the secretary of the interior designated a narrow 12,980-hectare (10,686 hectares public) area along the Snake River as the Snake River Birds of Prey Natural Area under an administrative withdrawal. Early investigations revealed that the natural area protected only about half of the unique nesting population and very little of the foraging habitat, and that this habitat was threatened by agricultural development (USDOI 1979b). In response to this need, the BLM initiated a large-scale, long-term team research project in 1975 to assess the amount of area needed by the raptors and to evaluate the effects of conversion of range habitat to irrigated agriculture on the raptors (USDOI 1979b). The results of the research provided the biological basis for establishing the conservation area.

Research Addressed Management Needs at Hand

The research focused directly on the management questions mainly because the research was funded, conducted, and directed by the BLM. The Boise District BLM, which had management responsibility for the area under study, managed the research funds and employed the BLM researchers. These researchers, who were sensitive to the needs of the area, designed and conducted the research and oversaw the BLM contractors.

Research Was Based on Interaction Between Scientists and Managers

The BLM researchers were housed in the district office, and the close proximity of scientists and managers allowed for daily interaction. Managers took part in designing and establishing research priorities. They were constantly apprised of progress and problems and participated in any changes in the research approach. This interaction reinforced researchers with the reasons for their research funding and how the information would be used by the managers. However, this close proximity and ready access to managers had disadvantages. Frequently, researchers were involved in management activities as technical experts, which often detracted from time that could have been spent on research.

Research Established an Appropriate Frame of Reference

The project was one of the earlier landscape-scale research ventures. It consisted of numerous individual projects that were conducted at different scales; site- and species-specific studies were integrated with larger-scale studies to address landscape-level questions (USDOI 1979b). The research design considered that the area was subjected to many land uses and much of the habitat was greatly altered. The research was designed to assess the current conditions and to predict future changes if agricultural development continued or increased. It emulated a real-world management situation and considered management in a highly disturbed system.

Research Clearly Described Desired Outcome

The research described the ecological interactions among the habitat, prey, and nesting raptors and identified the spatial requirements of the three principal raptor species: Prairie Falcons *(Falco mexicanus)*, Golden Eagles (*Aquila chrysaetos*), and Red-tailed Hawks (*Buteo jamaicensis)* (USDOI 1979b). Using a geographic information system (Steenhof 1982) and computer simulations, responses of nesting Prairie Falcons to conversions of range habitat to irrigated agriculture were predicted. The research described the management actions necessary to ensure the stability of the raptors in the area and the biological consequences of incremental conversions of range habitat to irrigated agriculture. The secretary of the interior considered this information and weighed the political and biological consequences of excluding certain areas from protection (i.e., private land). Based on this assessment, the secretary recommended boundary modifications to the Boise district BLM manager, who in turn consulted the researchers on the biological consequences of the modifications, which were subsequently transmitted back to the secretary. This interactive process continued while the environmental impact statement was being developed and until a biologically and politically feasible solution was obtained.

Results Were Unambiguous

The final report presented straightforward and unambiguous results (USDOI 1979b) and recommendations that were implementable by managers. The environmental impact statement proposed that the conservation area be established through congressional legislation (USDOI 1979b); however, the provision to prohibit any further conversion of federal lands to irrigated agriculture was highly controversial. When Congress took no action on the legislation in 1980, the secretary of interior withdrew the area proposed for the conservation area under a twenty-year administrative withdrawal. The BLM and the secretary were subsequently sued by parties who wanted the area to be more accessible for agricultural development. Results of the research, however, withstood the court challenge.

Research Followed Through and Was Forthright

After the final report and the EIS were completed, the BLM researchers continued to work with local BLM managers to implement the results of the research. They attended public hearings, helped managers develop court testimony, and conducted briefings of congressional and state officials. BLM researchers also advised managers on the revisions of the management plan in 1985 and 1995 (USDOI 1985, 1995). Although research personnel are no longer housed in the BLM office, the research staff still advises the BLM managers on management activities in the conservation area.

Research results have been published in both scientific journals and management reports. In 1975, while the research was being conducted, the BLM initiated an annual report series for the conservation area, which provided preliminary results of the ongoing work. The final results of this research program were published in USDOI (1979b) and distributed to more than a thousand people. As of 1994, a total of sixty-eight papers had been published in the scientific literature using data collected during this program (Steenhof 1994).

Northern Spotted Owl Research and the Northwest Forest Plan

Conservation of Northern Spotted Owls in the coniferous forests of the Pacific Northwest has been an extremely contentious issue for more than twenty years (Wilcove 1994b). Since the 1970s, when the first comprehensive Northern Spotted Owl research was published (Forsman 1976; Gould 1977), concern has grown among avian researchers regarding the adverse impacts on owl populations of harvesting old-growth and late-successional forests. During the 1970s and 1980s, timber harvest levels remained high; however, environmental concerns regarding the loss of old-growth forests and declining owl populations increased. As a consequence, numerous interagency committees, teams, and panels were established in the late 1980s and early 1990s to develop conservation strategies that would provide a compromise solution to the conflict (e.g., Thomas et al. 1990; USDOI 1992). In spite of these efforts, a series of lawsuits brought by environmental groups, coupled with the listing of the owl as threatened by the U.S. Fish and Wildlife Service in 1990, essentially shut down the sale and harvest of old forests on federal lands within the range of the Northern Spotted Owl.

This gridlock persisted until the newly elected Clinton administration convened a Forest Conference in Portland, Oregon, on April 2, 1993. At this conference, the president asked that a science-based forest management plan be developed that would end this impasse. Furthermore, he directed that this plan should: (1) adhere to the nation's laws; (2) protect and enhance the environment; (3) provide a sustainable timber economy; (4) support the region's people and communities during economic transition; and (5) ensure that federal agencies work together. After a concerted interagency effort on the part of both scientists and managers, a Forest Plan was developed and approved in April 1994, incorporating over 110,000 public comments.

Research Addressed Management Needs at Hand

Concerns over the adverse effects of harvest of old forests on spotted owl populations resulted in the establishment of numerous demographic studies between 1985 and 1990 in Washington, Oregon, and northern California. These studies provided managers with a science-based monitoring tool to track the vital rates of owls. Because these studies were highly coordinated and used standard data collection protocols, greater consistency was achieved in the interpretation of the results from studies throughout the owl's range (Franklin et al. 1996). The spatial distribution and replication of these studies also made broader inferences regarding the status of the species across its range (i.e., meta-analysis) possible (Burnham, Anderson, and White 1996).

Research Was Based on Interaction between Scientists and Managers

All spotted owl research conducted on federal lands is conducted in close coordination with the agencies responsible for their management (e.g., BLM, U.S. Forest Service, National Park Service). In many cases, these demographic studies are true collaborations, with the land management agencies contributing significant resources, including staff time, to collect field data. Demographic performance data are analyzed annually, with results made available to resource managers. In addition, workshops were convened in 1991 (Anderson and Burnham 1992) and again in 1993 (Burnham, Anderson, and White 1996) to conduct meta-analyses of all available demographic data sets; it is anticipated that similar workshops will be conducted as needed in the future.

Research Established an Appropriate Frame of Reference

Establishment of demographic study areas in the protected old-growth forests of Olympic National Park and wilderness areas of national forests, as well as the highly fragmented forests managed by the BLM and U.S. Forest Service, provides a full range of ecological conditions against which to evaluate the population status and trends of spotted owls. This network of studies, coupled with the use of standardized protocols, established a powerful monitoring tool that managers could access for information relative to the vital rates of spotted owls.

Desired Outcome Was Clearly Described

Researchers provided managers with scenarios of different desirable ecological conditions based on current information on owl demographics, habitat relationships, and other ecological questions. These scenarios were updated as new information became available. Annual reporting of results from the spotted owl demographic studies throughout the Pacific Northwest, coupled with the periodic meta-analysis workshops, have provided managers and policy makers with a predictable flow of biological information with which to make management decisions.

Results Were Unambiguous

Because of the close coordination and consistency among the demographic studies, it has been possible to provide clear and consistent results and interpretations of findings. These findings have withstood controversial court challenges and have been instrumental in developing the initial Habitat Conservation Area reserve design (Thomas et al. 1990) and, more recently, the late-successional reserve network that is a major element of the president's Northwest Forest Plan.

Research Was Forthright

Subsequent to the adoption of the Northwest Forest Plan, researchers have continued their demographic monitoring throughout the range of the spotted owl. They are developing a habitat model that will link spotted owl population status and reproductive performance to key habitat attributes. If successful, this will enable managers to monitor forest habitat and perhaps key understory characteristics as surrogates for spotted owl population status and reproductive performance.

Conclusions

The Snake River Birds of Prey National Conservation Area is one of the few wildlife conservation areas where the boundary has been based on the spatial requirements of the species for which it was established and where research has been the driving force behind the effort to protect the area. This effort is a classic case of a long-term, close working relationship between researchers and managers to solve problems on a local scale. Protection of the unique resource in the area occurred incrementally and spanned more than twenty years from the time the natural area was established in 1971 until the permanent legislation created the conservation area in 1993. Each step was based on research results and the close interaction of researchers and managers. Importantly, there has been continuity during the last twenty-five years involving some of the researchers of the original research in the 1970s. The long-term interaction between researchers and managers continues to address management problems in the conservation area as demonstrated by the recent research on the effects of military activity and fire (USDOI 1996a).

With the implementation of the Northwest Forest Plan, the forest management gridlock experienced by the Pacific Northwest is loosening. It is important to recognize that avian research on the Northern Spotted Owl was instrumental, not only in bringing the significance of the loss of old-growth forests and associated species to the public's attention, but also in creating a science-based conservation strategy that provides for both endangered species conservation and resource extraction. The tradition of rigorous, coordinated research on spotted owls in the region is exceptional. Researchers, while not always sharing the worldview of resource managers charged with multiple-use mandates, have demonstrated repeatedly that their research data and interpretations are credible and relevant to the needs of informed resource management.

We considered the two case studies reviewed in this chapter to be examples of relevant avian research, as characterized by the criteria presented by Young and Varland (this volume). It should be noted that a key characteristic of both research efforts was the long-term, highly integrated involvement of avian researchers with management agencies. It is true that this level of cooperation and coordination has occasional drawbacks in terms of ready access by managers to researchers for technical assistance often unrelated to their research programs; however, this level of interaction invariably results in better communication and mutual understanding by both resource managers and scientists of the issues and constraints involved. Arnett and Sallabanks (this volume) stated that researchers and land managers must work together to effectively implement strategies for managing avian populations and their habitat. It has been our experience that avian researchers most successful at establishing these types of long-term, relevant research programs are those who work most closely with resource managers and design studies that directly and efficiently address their highest-priority research information needs.

CHAPTER 30

What Managers Really Need from Avian Researchers

Sallie J. Hejl and Kathleen Milne Granillo

Natural resource managers today deal with increasingly complex issues related to local, regional, national, and even international constituents. Most managers need to balance several competing uses of a given area or a given resource. The effects of management on birds, and managing "for" birds, are just a part of most managers' concerns. Also, there are many diverse and differing viewpoints regarding "proper management" of natural resources that include and affect birds. Several questions arise: Do managers really want to know more about birds? If so, how can they best improve their knowledge of avian conservation needs and stay current in avian research developments? What sorts of information do they need from researchers in order to manage birds effectively?

Managers do want answers to a multitude of questions about birds. Proof of this is found by examining the U.S. Geological Survey, Biological Resources Division (BRD), Bureau Information Needs (BIN) process. The BIN process seeks to determine the high-priority biological research needs of all of the Department of the Interior agencies. Each agency presents ten to fifteen high-priority research needs each year to BRD. Generally, 25–33 percent of these deal specifically with birds; in addition, a good percentage deal indirectly with birds. When examined by agency within the Department of the Interior, most of the needs identified by the U.S. Fish and Wildlife Service concern birds, as do several of the needs identified by the National Park Service and the Bureau of Land Management. Even agencies not specifically charged with wildlife as a prime concern, such as the Bureau of Reclamation and the Office of Surface Mining, have research needs that relate to birds. Other land management agencies such as the U.S. Forest Service do not have formal, annual research needs processes, but wildlife researchers work with constituent land managers to develop five-year research plans, many of which concern birds. The U.S. Forest Service also has conducted many conservation assessments (Northern Spotted Owl, *Strix occidentalis caurina* [Thomas et al. 1990]; California Spotted Owl, *S. o. occidentalis* [Verner et al. 1992]; Northern Goshawk, *Accipiter gentilis,* in the Southwest [Reynolds et al. 1992]; forest owls [Hayward and Verner 1994]; Marbled Murrelet, *Brachyramphus marmoratus* [Ralph et al. 1995]) and land management assessments (Forest Ecosystem Management [FEMAT 1993], Sierra Nevada Ecosystem Project [Sierra Nevada Ecosystem Project 1996], Interior Columbia Basin [Quigley, Haynes, and Russell 1996], and Tongass Land Management Plan [USDA 1997b]), each of which enumerate research needs specific to those topics.

What managers most need is available, usable research results on the desired topics. Many of the chapters in this volume elaborate on which topics are important and suggest how to design and conduct research so that it is more meaningful for managers. These ideas should be discussed, mulled over, digested, and incorporated by researchers. Many researchers already apply many of these ideas in their research. Even more important than relevancy of research is the accessibility of that research. The literature rarely addresses this important step, other than in passing, but information transfer is the key to getting research incorporated into management decisions. Researchers should continue publishing their results in peer-reviewed journals and the other places they have published in the past. However, most managers are not able to wade through scientific journals searching for information that could potentially help them to do their jobs better. Or, if they do read the journals, they must decide how to apply the results presented to their particular management problem. Researchers need to take steps to get their results in a usable form and into the hands of managers (Arnett and Sallabanks, this volume).

There are many avenues for this transfer of information. In most land management situations that involve potential effects on birds, there are two kinds of managers: (1) wildlife biologist or resource specialist, and (2) decision maker (often nonbiologists). Wildlife biologists or resource specialists often work with other resource specialists to create management alternatives among which the decision maker selects. Both types of managers need to be reached for research results to help guide management decisions. Information transfer will therefore require several different tactics. Presentations to managers, and not just to peers, can be highly effective. Many land managers work for federal or state agencies. These agencies usually have yearly (or more frequent) meetings where they would welcome a research presentation and interaction with researchers; both types of managers can be reached in this setting. Most managers do not attend the annual ornithological society meetings, but many, especially the biologists, do attend the yearly meetings of The Wildlife Society and the North American Wildlife and Natural Resources Conference, local chapter meetings of The Wildlife Society and similar societies and organizations, Partners in Flight–Aves de las Americas meetings, and assorted symposia. Presentations during these meetings let managers know which researchers to contact for information on certain topics. For policy formulation related to natural resource management to be based on the best available biological information, this level of scientific understanding must get into the hands of the higher-level, often nonbiologist, bureaucrats. This can be at least partially accomplished by working with groups such as The Nature Conservancy and The Wildlife Society, who meet regularly with agency leaders and who have strong constituencies. We suggest researchers consider presenting research results in these and other, less traditional forums. In addition, the Ornithological Council (1725 K Street, NW, Suite 212, Washington, D.C. 20001–1401) is a nonprofit organization formed by seven ornithological societies, whose purpose is to act as a conduit between ornithological science and legislators, managers, conservation organizations, and private industry. Those in need of credible scientific information and expert analyses on birds to make laws, regulations, policies, and management decisions can obtain such information from the council. The council also keeps ornithologists informed about policy issues affecting birds.

For research results to be directly applicable to managers and therefore used by managers, researchers need to be intimately familiar with management. Part of producing usable research results is asking the right questions. Some managers do not know what specific questions to pose about effects of their proposed management on birds. Researchers must then be able to help managers articulate their research needs. We suggest that researchers apply results to specific management situations, i.e., make results "real" to managers. To design research that will result in usable management recommendations, researchers may need to step out of their research environments, visit the managed lands, and be introduced to the management process. Many managers have experienced research at some level (e.g., many have conducted research projects for advanced degrees), but few researchers have been land managers. Researchers could learn much from participating in the alternative development process for actual projects, touring field operations (timber harvesting, prescribed fires, precommercial timber operations, grazing operations, water development projects, road building projects) and participating in reviews of various management practices.

There often is a psychological barrier between managers and researchers. Managers generally do not have Ph.D.s, while most researchers do, and this difference in education makes many managers perceive researchers as all-knowing, unapproachable, and imposing. They feel uncomfortable requesting assistance or challenging statements made by researchers about management. They also believe that most researchers are not aware of the practical constraints that limit the real choices managers must make. Many researchers are condescending with managers because they perceive them to be less educated, less sophisticated scientifically, and parochial. These stereotypes and barriers need to be broken down and will be as managers and researchers work more closely together in the future. In addition, it would be valuable if researchers understood the social, political, and economic considerations that are a part of every natural resource decision.

Managers need information relevant to the many different scales of management (Ruggiero, Hayward, and Squires 1994). Obviously, district rangers and refuge managers want to know the effects of alternative local-scale management actions on birds, such as cutting trees or building water-control structures. Many managers search for answers to these sorts of questions, and these are the questions that have been most studied. But managers are also interested in larger scales. They are interested in the bioregional level of analysis for comprehensive planning efforts. A few managers and researchers think about still larger scales. Indeed, managers work at every scale, making decisions that affect conservation of natural resources, including birds. Many of the higher-level managers are viewed by researchers (and much of the public) as bureaucrats and not really as managers of natural resources. But they make many of the decisions and develop policies that can have huge implications for birds. For example, at the larger-scale levels of management, managers need researcher input for habitat conservation planning, bioregion plans, Federal Energy Regulatory Commission licensing renewals, and comprehensive management plans. They, too, need to hear research results in such ways that they can incorporate the latest research findings into their view of the world.

Clearly, both researchers and managers need good baseline distribution data for each species and habitat (Hutto, this volume). Based on this type of information, researchers can ascertain what could be the potential threats (if any) to each bird species (e.g., loss of old-growth forests, loss of the integrity of a riparian system, fire suppression, fragmentation of habitats). Some regional reports detailing wildlife-habitat relationships are also available that help define at least the first cut on predicted responses to proposed management actions (DeGraaf et al. 1980; Dobkin 1992; and Hamel 1992). Ideally, researchers could then systematically study the key indicators of population health (e.g., pairing success, nesting productivity, juvenile and adult survivorship, dispersal) for each species within each habitat under each treatment of concern for multiple spatial and temporal scales (Martin 1992; DeSante and Rosenberg, this volume; Marzluff and Sallabanks, this volume; Walters, this volume). Using good statistical design with before-after replicated treatments to examine these indicators of population health would be ideal (Marzluff and Sallabanks, this volume); coordination between managers and researchers will be required to set up such studies. Researchers must also ensure that managers realize the nature of the research, the time requirements to complete the research, and whether replicate sites are needed; they must coordinate on management activities within the study areas so as not to jeopardize the research. It would be invaluable if researchers remained involved in the project after the initial research is complete in order to conduct or oversee follow-up monitoring. Where persistent questions arise regarding the validity of predictions, scientists should conduct research to test assumptions. Managers also need advice in designing useful monitoring systems.

Managers need results from "basic" and "applied" research and for common and rare species. Fundamental research that contributes to our basic understanding of the functioning of ecological systems truly is applied research and often is as important for management decisions as is what is typically considered applied research (Orians 1997). Many researchers study common species because they are easier to study, while managers desperately need information on uncommon species and habitats. Managers need some researchers to tackle the more difficult species and habitats, focusing on high-priority ones (as currently being identified by state working groups within Partners in Flight–Aves de las Americas), while others continue to study common species and habitats. Common species and habitats should not be avoided, however, because unfortunately, they too can quickly become rare and of concern (Newton, this volume).

Timing of the distribution of research information is critical. Knowing that their research will be used for management decisions, many researchers often wait to disseminate their results until they have many years of data and understand many parts of the system they are studying. Managers, however, need information as soon as researchers can supply it (often sooner). Researchers cannot wait until they know everything. They must share what they have as soon as they can, being sure to put it in the proper context (i.e., detail all of the limitations of the data). Managers need to allow for flexibility in their recommendations so that when new data become available new management recommendations can follow. The ideal situation would be for researchers to work with managers to create usable management recommendations.

Hopefully, general trends will be obvious with well-designed research, even in the early stages. For example, a critical review of the literature on the effects of silvicultural practices on forest birds in the Rocky Mountains (Hutto et al. 1993; Hejl et al. 1995), the first few years of an intensive study of birds in western red-cedar (*Thuja plicata*) and western hemlock (*Tsuga heterophylla*) forest (Hejl and Paige 1994; S. J. Hejl, unpublished data), and a bird-monitoring program in the northern region of the U.S. Forest Service (Hutto, this volume) all indicated that Brown Creepers (*Certhia americana*) and Winter Wrens (*Troglodytes troglodytes*) are two of the species most likely to be harmed by current logging practices in the northern Rockies, including the loss and fragmentation of old-growth forests. Each of these studies provided unique but complementary information. While we do not know all the details of how to maintain the viability of these species, researchers can confidently propose that these are species of concern in the northern Rockies and give some suggestions as to how to manage for these species (e.g., maintain existing tracts of continuous old-growth cedar-hemlock forest). Of course, the needs of many species (sometimes conflicting) must be considered in creating management plans. For suggestions on how to balance the needs of several species at the same time and maintain all of these species, we suggest mimicking "natural" or presettlement patterns and processes as a goal, which allows for a diversity of habitat components, successional stages, habitats, and landscapes (Hejl et al. 1995).

Researchers need to consider how managers use information to create useful management recommendations. Variation is an important concept in ecosystem patterns and in the dynamics of natural disturbance processes; therefore, variation is an important concept for both researchers and managers. For example, the number of snags in old-growth stands naturally varies from stand to stand. Researchers need to provide managers with information on this variation (i.e., recommend not only the mean number of snags per acre, but also provide a range), and managers need to imitate this variation rather than simply manage for minimum numbers. Because of competing needs, wildlife biologists are often forced to manage for the average or minimum value. Researchers need to express whether this is sufficient for certain birds. Often, if an average and/or minimum were followed across a landscape, variation likely would be lost, and this might not suffice for some species. One possible approach is to leave a "natural" variety of values for a given attribute across the landscape (Hejl et al. 1995). Another approach would be to manage some areas for the minimum value of some attribute and other areas for the maximum value. In a similar vein, Hutto (1995) suggested that it may be best to log trees from one part of a burn and leave another part of the burn untouched to maintain the variety of microhabitats available within a burn. This suggestion must be tempered with the realization that fire-associated bird species prefer the tree species and sizes that are often selectively removed from a burn in salvage-logging operations. Leaving the preferred tree species and sizes unlogged would probably benefit fire-dependent birds more than leaving usually unused tree species and sizes.

We see the design of a good system (or several systems) for sharing information quickly and easily between researchers and managers as the managers' biggest need (Boersma and Parrish, this volume). We need a simple, usable, well-oiled network of information flow between researchers and managers. Because managers are often managing public lands, we need a similar network with the public on the advantages and disadvantages (i.e., trade-offs) of various management alternatives as well as on the continual importance of the understanding of basic ecological concepts (as through newspaper articles and editorials, environmental education, public meetings). As Fleischer and Marzluff, Gehlbach, and Manuwal (this volume) note, the most critical issue of our time is the ever increasing human population and the concomitant ever increasing standard of living, which results in an overexploitation of resources and massive loss of habitat and biodiversity. We must convey the consequences of alternative solutions to the managers and the public. In addition to the traditional forum of publications, we envision a combination of formal and informal information-sharing networks, including but not limited to scientific or management meetings, internet connections, telephone conversations. In general, managers need to put forth the effort to invite researchers for input, and researchers need to make themselves available for consultation.

Two of the major barriers to both managers and researchers communicating as we have discussed are: (1) lack of time and (2) lack of incentives (Rensselaerville Roundtable 1995). All parties need to state their time limitations and respect each other's. Monetary awards and professional recognition could help encourage collaborative efforts. Researchers' performances are primarily evaluated on the number of publications. We suggest that agencies and universities modify performance evaluation processes to include outreach and information transfer efforts as important components.

Finally, frequent, open communication between researchers and managers is the key to researchers conducting pertinent research that will provide useful management recommendations and to managers easily obtaining useful information for land management decisions. Interactive discourse is the most valuable, such as that occurring in current U.S. Forest Service ecosystem projects (e.g., Bitterroot Ecosystem Management/Research Project [BEMRP], Yaffee et al. 1996). Scientists and managers have worked together from the design through the dissemination phase of BEMRP. Numerous meetings have been held between researchers and managers and between researchers, managers, and the public. The BEMRP is unusual in that frequent, two-way communication has occurred. Throughout this discourse, there has been a strong concern to keep managers' and researchers' roles distinct (Rensselaerville Roundtable 1995). In general, managers often listen to researchers (via presentations and publications), but researchers rarely listen to and understand managers' problems, concerns, and constraints. We think deep understanding and discourse would help enormously in creating workable land management solutions for the restoration and maintenance of all native bird species for the twenty-first century.

Acknowledgments

We benefited greatly from discussions with various managers, especially Mike Hillis, Bruce Marcot, and Helen Ulmschneider. Thanks also to Mike Hillis, Chuck Hunter, and John Marzluff for reviewing the manuscript.

CHAPTER 31

Avian Conservation on National Forest System Lands: Linking Research and Management

Joseph L. Ganey and Cecelia M. Dargan

Land managers with the U.S. Department of Agriculture Forest Service (USFS) are charged with maintaining well-distributed, viable populations of native vertebrates on USFS lands (National Forest Management Act, 1976, and enabling rules and regulations [36 CFR 219.19]). Their ability to meet this mandate is often severely hampered by a lack of ecological information on the native fauna inhabiting those lands. Research on a plethora of topics is needed to facilitate appropriate land management, yet funding for research is limited. For research to best benefit land managers, it must address their most pressing needs (Young and Varland, this volume). In this chapter, we briefly discuss some of those needs with respect to conservation of native avifauna on national forest lands, evaluate some of the research priorities presented in this volume, and describe some steps researchers can take to make their results more accessible to land managers. We recognize that conservation of some species of birds may be beyond the control of USFS land managers. For example, many species do not reside on national forest lands throughout the entire year, and their populations may be limited by factors occurring outside of national forest lands. Therefore, the following discussion pertains primarily to species for which management on national forest lands can address the most critical conservation issues.

Research Needs of Forest Managers

To effectively foster avian conservation, USFS biologists need to understand how lands under their jurisdiction function as habitat for those birds across both spatial and temporal scales, and how land-management actions influence the ability of those lands to support native birds. This understanding requires species-specific knowledge on: (1) population status and trends; (2) patterns of resource use; (3) relative fitness in different habitats; and (4) interactions with other species. It also requires better information on the effects of land-management activities on birds and their habitats at a hierarchy of spatial and temporal scales. Specific research needs are discussed below; these could be viewed as separating roughly into the categories of basic and applied research, although there are obviously interactions between categories.

Bird-Habitat Associations (Basic Research)

Agency biologists often lack even basic information on population status and habitat associations of birds. Unless a particular species is of special interest, which usually means threatened, endangered, sensitive, or hunted, detailed information on its natural history is rarely available. Consequently, there is a pressing need for basic research on life-history attributes of many native birds.

The most immediate research need is for better information on habitat associations of native birds at a variety of spatial scales. Researchers should go beyond simply documenting presence or numbers of birds in different habitats, however. Because some habitats may function as population sinks, where mortality exceeds reproduction (Gibbs and Faaborg 1990), managers also need information on survival and reproductive rates in different habitats, as emphasized in this volume by Brawn et al.; Herkert and Knopf; Faaborg et al.; and Walters. Until we can link habitat use with fitness, we will be unable to rigorously evaluate the importance of different habitats in conserving native avifauna.

Land managers often are responsible for large areas, and management activities alter the amounts and spatial configuration of different habitats across these areas. Consequently, research documenting the effects of landscape composition and pattern on bird demographics is essential. Relationships between birds and the landscapes they occupy are likely to be complex, however. Understanding these relationships will require studies that encompass long time frames and a hierarchy of spatial scales, and will require better information on dispersal patterns of birds (Faaborg et al., this volume; Walters, this volume). Recent developments in geographical information systems (GIS) technology provide a powerful tool with great potential for fostering a better understanding of the effects of landscape composition on bird communities (Villard, Schmidt, and Maurer, this volume). Accurate digital data sets on habitat variables relevant to birds are often difficult to come by, however. Therefore, effective application of GIS to avian research will require, at a minimum, the development of more and better sources of relevant digital data at the appropriate scales.

Managers also require information on population trends of birds. In some cases, population trend may be estimated based on vital rates (Burnham, Anderson, and White 1996). More commonly, however, estimating trends will require monitoring bird numbers (DeSante and Rosenberg, this volume; Hutto, this volume). Monitoring is mandated by law on National Forest System lands (e.g., Morrison and Marcot 1994) and is therefore often viewed more as a management than a research activity. Implementation of credible monitoring programs presents special problems for land management agencies, however, because most land management agencies lack the technical expertise, funding, and personnel necessary to design and accomplish rigorous monitoring. Therefore, designing credible schemes to monitor bird populations may best be accomplished through partnerships between management agencies and researchers (Hutto, this volume). Well-designed monitoring programs could benefit both researchers and managers in other ways, as well. These include identifying species or groups of species that are declining and require more detailed study (Newton, this volume) and providing important information on background levels of variation in abundance and species composition within bird communities (Brawn et al., this volume; Hutto, this volume). Finally, monitoring is a necessary component of adaptive management and can aid in: (1) determining whether management actions have the predicted results; and (2) developing appropriate modifications to treatments where desired outcomes are not being achieved (Lancia et al. 1996; Ringold et al. 1996; Grumbine 1997).

There is also a need for rigorous field testing of wildlife-habitat relationships models. These models typically describe the range of habitats occupied by a particular species and may include information on relative abundance in different habitats and suitability of those habitats for life-history aspects such as nesting or feeding (Morrison, Marcot, and Mannan 1992; Patton 1992). In the absence of more specific information on treatment effects (see below), managers must often rely on these models to evaluate the effects of land management activities on native birds. Few wildlife-habitat relationships models have been tested (Berry 1986), however, and substantial error rates have been noted in some models that have been tested (Dedon, Laymon, and Barrett 1986; Block et al. 1994).

Influence of Land Management on Birds (Applied Research)

Perhaps the primary research need facing USFS managers is the need for a better understanding of cause-and-effect relationships between land management actions, birds, and their habitats (Finch et al. 1997). Most studies on this subject have been correlative rather than experimental, and most have suffered from little or no replication of treatments (Finch et al. 1997; Marzluff and Sallabanks, this volume). Both factors limit our ability to draw inferences regarding the effect of land management on birds and their habitats. Few studies have attempted to address synergistic effects of multiple management activities, and fewer still have attempted to address the cumulative effects of management actions over time and/or space. Consequently, there is a great need for well-designed, replicated experiments that can elucidate cause-and-effect relationships between land management activities, birds, and their habitats (Finch et al. 1997; Marzluff and

Sallabanks, this volume). These studies should evaluate the effects of land management activities both singly and in combination and should consider the cumulative effects of various management actions over time (Finch et al. 1997). Understanding cause-and-effect relationships is a formidable task. It will require well-designed experimental studies conducted cooperatively with managers. The practical difficulties inherent in conducting large-scale, long-term experiments on complex systems are enormous (Carpenter et al. 1995), but there is no viable alternative.

Making Research Applicable to Managers

The above wish list of research needs is not exhaustive but is nonetheless daunting. Much of the needed information can be obtained only through long-term studies and/or large-scale experimentation, and a reality check suggests that it will be difficult or impossible to fund and conduct the necessary experiments and monitoring programs in an era of shrinking budgets. Thus, in the short term, managers will still be forced to deal with high levels of uncertainty when evaluating the effects of management activities on birds. Researchers should recognize this uncertainty and do what they can to assist managers in dealing with it. We suggest that there are at least three areas where researchers can make immediate contributions to management.

First, researchers should recognize that managers are often asked to define either targets (e.g., snag densities) or thresholds with respect to acceptable change (i.e., points beyond which further disturbance results in undesirable effects). Thresholds can relate either to single management actions or to cumulative effects across both space and time. Researchers typically shy away from defining thresholds or targets, for good reason: Uncertainty is typically great in ecological systems, and the consequences of errors are potentially enormous. Nevertheless, researchers should recognize that managers are faced with such decisions and provide assistance when they can. One possible approach might be for researchers to provide information on both mean tendency and range of variability of the variable(s) of interest, and for managers to strive to maintain that range of variability rather than managing for a single (often minimum) value.

Second, researchers should consider alternative means of sharing current research results with managers. As the ecological literature grows more voluminous, it becomes more difficult for managers to stay abreast of the latest developments. Researchers could help managers stay informed on current developments through: (1) periodic informal presentations on ongoing research projects; (2) conducting workshops to explain research results (and their implications) to land managers; (3) developing technical information bulletins for managers; (4) establishing a system for distributing reprints to appropriate agency biologists and planners; and (5) making greater efforts to publish their results in a broad spectrum of journals that includes applied journals. In these busy times, additional steps may be required beyond publication of research results to ensure that those results are incorporated in management decisions. Even small steps may pay big dividends here, because research results often are shared widely among managers once their usefulness is established.

Third, researchers should understand that management of public lands may at times be driven more by external attitudes and policies than by scientific considerations (Sabatier, Loomis, and McCarthy 1996; Grumbine 1997; Newton, this volume). Debates on land-use policy are frequently dominated by special interests using distorted or partially correct information. By communicating their results to the public and to political decision makers, and by clearly explaining the implications of those results, researchers may be able to influence public attitudes and help shape land-use policies. This could be accomplished in a number of ways, including public presentations, formal briefings, and newspaper articles or through environmental education programs. Gains may be slow in this arena, and the task is thankless. In the absence of participation by knowledgeable individuals, however, debates over land-use policy will continue to be dominated by special interests, and decision makers may not have access to unbiased information.

Acknowledgments

G. A. Goodwin, T. G. Grubb, S. J. Hejl, M. N. Kochert, M. Patton-Mallory, and J. M. Marzluff provided helpful reviews of this paper.

Literature Cited

Abrams P.A. (1995) Monotonic or unimodal diversity-productivity gradients: What does competition theory predict? *Ecology* 76:2019–2027.

Adams P.W. and Hairston A.B. (1996) Using science to direct policy. *Journal of Forestry* 94:27–30.

Adar E. (1996) The effects of hydrological and agricultural management on groundwater quality in the Arava Valley. *Ecology & Environment* 3:249–250.

Aebischer N.J., Coulson J.C., and Colebrook J.M. (1990) Parallel long-term trends across four marine trophic levels and weather. *Nature* 347:753–755.

Agee J. (1993) *Fire Ecology of Pacific Northwest Forests.* Island Press, Washington, DC.

Ainley D.G. and Boekelheide R.J. (1990) *Seabirds of the Farallon Islands: Ecology, Dynamics and Structure of an Upwelling-System Community.* Stanford University Press, Stanford, CA.

Ainley D.G. and Lewis T.J. (1974) The history of the Farallon Island marine bird populations, 1854–1972. *Condor* 76:432–446.

Alaska Sea Grant (1996) *Solving Bycatch: Considerations for Today and Tomorrow.* Alaska Sea Grant College Program Report 96–03.

Alberts A.C., Richman A.D., Tran D., Sauvajot R., McCalvin C., and Bolger D.T. (1993) Effects of habitat fragmentation of native and exotic plants in southern California coastal sage scrub. In *Interface between Ecology and Land Development in California.* J.E. Keeley (ed.), pp. 103–110. Southern California Academy of Sciences, Los Angeles, CA.

Allee W.C., Emerson A.D., Park O., Park T., and Schmidt K.P. (1949) *Principles of Animal Ecology.* Saunders, Philadelphia, PA.

Allen D.H., Franzreb K.E., and Escano R.E. (1993) Efficacy of translocation strategies for Red-cockaded Woodpeckers. *Wildlife Society Bulletin* 21:155–159.

Allen D.L. (1991) An insert technique for constructing artificial Red-cockaded Woodpecker cavities. GTR-SE-73. USDA Forest Service, Asheville, NC.

Allen E.B. (1988) Some trajectories of succession in Wyoming sagebrush grassland: Implications for restoration. In *The Reconstruction of Disturbed Arid Lands: An Ecological Approach.* E.B. Allen (ed.), pp. 89–112. Westview Press, Boulder, CO.

Allen E.B. and Knight D.H. (1984) The effects of introduced annuals on secondary succession in sagebrush grassland, Wyoming. *Southwestern Naturalist* 29:407–421.

Allen J.J. (1876) The extinction of the Great Auk at the Funk Islands. *American Naturalist* 10:48.

Allen R.P. and Norton A.H. (1931) *An Inspection of the Colonies of Seabirds on the Coast of Maine.* Unpublished ms., National Audubon Society, New York.

Alverson D.L. (1991) Commercial fisheries and the Steller sea lion (*Eumetopias jubatus*): The conflict arena. *Fish. Res. Inst. Rep.* FRI-UW-9106. Fish Research Institute, Seattle, WA.

Ambrose S. (1994) The evolution of a bird project. *Wingspan* 14:22–23.

Andelman S.J. and Stock A. (1994) *Management, research and monitoring priorities of the conservation of Neotropical migratory landbirds that breed in Oregon.* Washington Department of Natural Resources, Olympia, WA.

Anders A.D. (1996) *Survival and Habitat Selection of Fledgling Wood Thrush in a Forested Landscape.* MS thesis, University of Missouri, Columbia.

Anders A.D., Dearborn D.C., Faaborg J., and Thompson F.R. (1997) Juvenile survival in a population of Neotropical migrant birds. *Conservation Biology* 11:698–707.

Anderson D.E., Rongstad O.J., and Mytton W.R. (1989) Response of nesting Red-tailed Hawks to helicopter overflights. *Condor* 91:296–299.

Anderson D.E., Rongstad O.J., and Mytton W.R. (1990) Home range changes in raptors exposed to increased human activity levels in southeastern Colorado. *Wildlife Society Bulletin* 18:134–142.

Anderson D.R. and Burnham K.P. (1992) Demographic analysis of Northern Spotted Owl populations. In *Draft Recovery Plan for the Northern Spotted Owl,* pp. 319–328. USDA Fish and Wildlife Service, Portland, OR.

Anderson D.W., Volg S., and Keith J.O. (1980) The human influence on seabird nesting success: Conservation implications. *Biological Conservation* 18:339–345.

Anderson J.E. (1991) A conceptual framework for evaluating and quantifying naturalness. *Conservation Biology* 5:347–352.

Anderson K.P. and Ursin E. (1977) A multispecies extension to the Beverton and Holt Theory of fishing, with accounts of phosphorus circulation and primary production. *Meddr. Danm. Fisk.-mog Havunders.* 7:319–435.

Anderson M.C. and Mahoto D. (1995) Demographic models and reserve designs for the California Spotted Owl. *Ecological Applications* 5:639–647.

Anderson R.J. (1985) Bald Eagles and forest management. *Forestry Chronicles* 61: 189–193.

Andrade M.A. (1992) *Aves silvestres do Minas Gerais.* Conselho International para a Preservacão das Aves (CIPA), Secão Panamericana, USA. Belo Horizonte, Minas Gerais Brazil.

Andrewartha H.G. and Birch L.C. (1984) *The Ecological Web: More on the Distribution and Abundance of Animals.* University of Chicago Press, Chicago, IL.

Angermeier P.L. and Karr J.R. (1994) Biological integrity versus biological diversity as policy directives. *BioScience* 44:690–697.

Ankney C.D. (1996) An embarrassment of riches: Too many geese. *Journal of Wildlife Management* 60:217–223.

Annand E.M. and Thompson III F.R. (1997) Forest bird response to regeneration practices in central hardwood forests. *Journal of Wildlife Management* 61:159–171.

Appelquist H., Drabaek I., and Asbirk S. (1985) Variation in mercury content of guillemot feathers over 150 years. *Marine Pollution Bulletin* 16:244–248.

Arcese P., Smith J.N.M., Hockachka W.M., Rogers C.M., and Ludwig D. (1992) Stability, regulation, and the determination of abundance in an insular Song Sparrow population. *Ecology* 73:805–822.

Archibald G.W. (1977) Supplemental feeding and manipulation of feeding ecology of endangered birds: A review. In *Endangered Birds: Management Techniques for Preserving Threatened Species.* S.A. Temple (ed.), pp. 131–134. University of Wisconsin Press, Madison.

Ardern S.L. and Lambert D.M. (1997) Is the Black Robin in genetic peril? *Molecular Ecology* 6:21–28.

Arguedes, N. (1992) *Genetic Variation and Differentiation in the Ovenbird* (Seiurus auro-capillus) *in Central Missouri and Puerto Rico*. MS thesis, University of Missouri, Columbia.

Armstrong D.P. and McLean I.G. (1995) New Zealand translocations: Theory and practice. *Pac. Conservation Biology* 2:39–54.

Aronoff S. (1989) *Geographic Information Systems: A Management Perspective*. WDL Publications, Ottawa, Canada.

Askins R.A. (1993) Population trends in grassland, shrubland, and forest birds in eastern North America. *Current Ornithology* 11:1–34.

Askins R.A. (1995) Hostile landscapes and the decline of migratory songbirds. *Science* 267:1956–1957.

Askins R.A., Lynch J.F., and Greenberg R. (1990) Population declines in migratory birds in eastern North America. *Current Ornithology* 7:1–57.

Askins R.A. and Philbrick M.J. (1987) Effect of changes in regional forest abundance on the decline and recovery of a forest bird community. *Wilson Bulletin* 99:7–21.

Atkinson C.T., Yorinks N., Woods K.L., Dusek R.J., and Iko W.M. (1995) Wildlife disease and conservation in Hawaii: Pathogenicity of avian malaria (*Plasmodium relictum*) in experimentally infected iiwi (*Vestiaria coccinea*). *Parasitology* 111:S59–S69.

Atkinson I.A.E. (1985) The spread of species of Rattus to oceanic islands and their effects on island avifaunas. In *Conservation of Island Birds: Case Studies for the Management of Threatened Species*. P.J. Moor (ed.), pp. 35–81. ICBP Technical Publication 3, Cambridge, England.

Atkinson I.A.E. (1989) Introduced animals and extinctions. In *Conservation for the Twenty-first Century*. D. Western and M.C. Pearl (eds.), pp. 54-75. Oxford University Press, Oxford, England.

Audubon J.J. (1831) *Ornithological Biography*. Adam Black, Edinburgh.

Australian Nature Conservation Agency (1996) *A Directory of Important Wetlands in Australia*. ANCA, Canberra.

Avise J.C. (1994) *Molecular Markers, Natural History and Evolution*. Chapman and Hall, New York.

Avise J.C. (1995) Mitochondrial DNA polymorphism and a connection between genetics and demography of relevance to conservation. *Conservation Biology* 9:686-690.

Awbrey F.T. and Bowles A.E. (1990) *Effects of aircraft noise and sonic booms on raptors: A preliminary model and a synthesis of the literature on disturbance*. Noise and Sonic Boom Impact Technology Technical Operating Report No. 12. Wright-Patterson Air Force Base, Dayton, OH.

Babbitt B. (1995) Science: Opening the next chapter of conservation history. *Science* 267:1954–1955.

Bahat O. (1986) Raptor nesting in the Judean Desert—Past, present and future trends. *Torgos* 6:8–24.

Bahat O. (1989) *Aspects in the Ecology and Biodynamics of the Golden Eagle* (Aquila chrysaetos homeyeri) *in the Desert Regions of Israel*. MS thesis, Tel Aviv University, Tel Aviv.

Bailey E.P. (1993) *Introduction of foxes to Alaskan islands—History, effects on avifauna, and eradication*. Res. Publ. 193. U.S. Fish and Wildlife Service, Washington, DC.

Bailey R.G. (1980) *Descriptions of the ecoregions of the United States*. Misc. Publ. No. 1391. USDA Forest Service, Washington, DC.

Bailey R.S. (1986) *Food consumption by seabirds in the North Sea in relation to the natural mortality of exploited fish stocks*. CM 1986/G:5. International Council for the Exploration of the Sea, Copenhagen.

Bailey R.S., Furness R.W., Gauld J.A., and Kunzlik P.A. (1991) Recent changes in the population of the sandeel (*Ammodytes marinus* Raitt) at Shetland in relation to estimates of seabird predation. *ICES Mar. Sci. Symp.* 193:209–216.

Bailey T. and Black J.M. (1995) Parasites of wild and captive Nene *Branta sandvicensis* in Hawaii. *Wildfowl* 46:59–65.

Baillie S.R. (1990) Integrated population monitoring of breeding birds in Britain and Ireland. *Ibis* 132:151–166.

Baillie S.R. and Peach W.J. (1992) Population limitation in Palearctic-African migrant passerines. *Ibis* 134, suppl. 1:120–132.

Baines D. (1989) The effects of improvement of upland marginal grasslands on the breeding success of Lapwings *Vanellus vanellus* and other waders. *Ibis* 131:497–506.

Baines D. (1990) The roles of predation, food and agricultural practice in determining the feeding success of the Lapwing *Vanellus vanellus* on upland grasslands. *Journal of Animal Ecology* 59:915–929.

Baines D. (1991) Long-term changes in the European Black Grouse population. *Game Conservancy Review of 1990* 22:157–158.

Baines D. (1996) The implications of grazing and predator management on the habitats and breeding success of Black Grouse *Tetrao tetrix*. *Journal of Applied Ecology* 33:54–62.

Baines D. and Hudson P.J. (1995) The decline of Black Grouse in Scotland and northern England. *Bird Study* 42:122–131.

Bairlein F. (1996) Long-term ecological studies on birds. *Verhandlungen der Deutschen Zoologischen Gesellschaft* 89:165–179.

Baker B.W., Cade B.S., Mangus W.L., and McMillen J.L. (1995) Spatial analysis of Sandhill Crane nesting habitat. *Journal of Wildlife Management* 59:752–758.

Baker G.B., Dettmann E.B., Scotney B.T., Hardy L.J., and Drynan D.A.D. (1995) *Report on the Australian Bird and Bat Banding Scheme, 1984–95*. Australian Nature Conservation Agency, Canberra.

Baker R.H. (1951) The avifauna of Micronesia, its origin, evolution, and distribution. *University of Kansas Publications Museum Natural History* 3:1–359.

Baker-Gabb D.J. (1984) The breeding ecology of twelve species of diurnal raptor in north-western Victoria. *Australian Wildlife Research* 11:145–160.

Baker-Gabb D.J., Benshemesh J.S., and Maher P.N. (1990) A revision of the distribution, status and management of the Plains-wanderer *Pedionomus torquatus*. *Emu* 90:161–168.

Balda R.P. and Bateman G.C. (1972) The breeding biology of the Piñon Jay. *Living Bird* 11:5–42.

Baldassare G.A. and Bolen E.G. (1994) *Waterfowl Ecology and Management*. Academic Press, London.

Ballantine W.J. (1997) Design principles for systems of "no-take" marine reserves. In *The Design and Monitoring of Marine Reserves*. T.J. Pitcher (ed.), pp. 4–5. Fisheries Centre Research Report Vol. 5(1). University of British Columbia, Canada.

Ballou J.D. (1983) Calculating inbreeding coefficients from pedigrees. In *Genetics and Conservation*. C.M. Schoenwald-Cox, S.M. Chambers, F. MacBryde, and L. Thomas (eds.), pp. 509–520. Benjamin-Cummings, Menlo Park, CA.

Ballou J.D. (1995) *Genetic Management, Inbreeding Depression and Outbreeding Depression in Captive Populations*. Ph.D. dissertation, University of Maryland, College Park.

Banko P.C. (1992) Constraints on wild Nene productivity. *Wildfowl* 44:99–106.

Banko W.E. (1980) *The Trumpeter Swan*. University of Nebraska Press, Lincoln, NB.

Banko W.E. and Elder W.H. (1990) History of endemic Hawaiian birds. Population histories—Species accounts: Scrub-grassland birds: Nene—Hawaiian Goose. *Avian History Report* 13a. University of Hawaii, Honolulu, Cooperative National Park Resource.

Barinaga M. (1990) Where have all the froggies gone? *Science* 247:1033–1034.

Barr J.F. (1986) Population dynamics of the Common Loon (*Gavia immer*) associated with mercury-contaminated waters in northwestern Ontario. *Canadian Wildlife Service Occasional Paper* 56.

Barrett G., Ford H.A., and Recher H. (1994) Conservation of woodland birds in a fragmented rural landscape. *Pacific Conservation Biology* 1:245–256.

Barrett R.T. and Furness R.W. (1990) The prey and diving depths of seabirds on Hornøy, North Norway after a decrease in Barents Sea capelin stocks. *Ornis Scandinavica* 21:179–186.

Barrowclough, G.F. 1980. Gene flow, effective population sizes and genetic variance in birds. *Evolution* 34:789–798.

Barrowclough G.F. (1992) Systematics, biodiversity, and conservation biology. In *Systematics, Ecology and the Biodiversity Crisis*. N. Eldridge (ed.), pp. 121–143. Columbia University Press, New York.

Barrowclough G.F. and Gutiérrez R.J. (1990) Genetic variation and differentiation in the Spotted Owl (*Strix occidentalis*). *Auk* 107:737–744.

Barrowclough G.F., Johnson N.K., and Zink R.M. 1985. On the nature of genic variation in birds. *Currrent Ornithology* 2:135–154.

Bart J. (1995) Amount of suitable habitat and viability of Northern Spotted Owls. *Conservation Biology* 9:943–946.

Bart J. and Robson D.S. (1992) Methods and recommendations for population monitoring. In: Recovery plan for the Northern Spotted Owl, Appendix A. Draft. USDOI, Washington, DC.

Basili G. and Temple S.A. (1995) A perilous migration. *Natural History* 104:40–46.

Bateson P. (1982) Preferences for cousins in Japanese quail. *Nature* 295:236–237.

Batt B.D.J., Afton, A.D., Anderson, M.I., Arkney, C.D., Johnson, D.H., Kadlek, J.A., and Krapu, G.L. (1992) *Ecology and Management of Breeding Waterfowl*. University of Minnesota Press, Minneapolis.

Baumgartner T.R., Soutar A., and Ferreira–Bartrina V. (1992) Reconstruction of the history of Pacific sardine and Northern anchovy populations over the past two millennia from sediments of the Santa Barbara Basin, California. *California Cooperative Oceanic Fisheries Investigations Report* 33:24–40.

Bayer M. and Porter W.F. (1988) Evaluation of a guild approach to habitat assessment for forest-dwelling birds. *Environmental Management* 12:797–801.

Bean M.J., Fitzgerald S.G., and O'Connell M.A. (1991) *Reconciling Conflicts under the Endangered Species Act: The Habitat Conservation Planning Experience.* World Wildlife Fund, Washington, DC.

Beck B.B., Rapaport L.G., Stanley Price M.R., and Wilson A.C. (1994) Reintroduction of captive-born animals. In *Creative Conservation—The Interface between Captive and Wild Animals.* P.J.S. Olney, G.M. Mace, and A.T.C. Feistner (eds.), pp. 265–286. Chapman and Hall, London.

Bedard J. and Lapointe G. (1984) Banding returns, arrival times, and site fidelity in the Savannah Sparrow. *Wilson Bulletin* 96:196–205.

Beebe W. (1947) Avian migration at Rancho Grande in north-central Venezuela. *Zoologica* 32:153–168.

Beintema A.J. and Muskens G.J.D.M. (1987) Nesting success of birds breeding in Dutch agricultural grasslands. *Journal of Applied Ecology* 24:743–758.

Beissinger S.R. (1990) On the limits and directions of conservation biology. *BioScience* 40:456–457.

Beissinger S.R. (1995) Modeling extinction in periodic environments: Everglades water levels and Snail Kite population viability. *Ecological Applications* 5:618–631.

Beissinger S.R. and Osborne D.R. (1982) Effects of urbanization on avian community organization. *Condor* 84:75–83.

Ben-Gurion D. (1971) *Israel—A Personal History.* American Israel Publishing Company, Tel Aviv.

Bender M. (ed.) (1991) Aleutian Canada Goose reclassified from endangered to threatened. *Endangered Species Technical Bulletin* XVII:10.

Benirschke K., Lasley B., and Ryder O. (1980) The technology of captive propagation. In *Conservation Biology: An Evolutionary-Ecological Perspective.* M.E. Soulé and B.A. Wilcox (eds.), pp. 225–242. Sinauer Associates, Sunderland, MA.

Bennett A., Backhouse G., and Clark T. (1995) *People and Nature Conservation: Perspectives on Private Land Use and Endangered Species Recovery.* Royal Zoological Society of New South Wales, Sydney.

Bennett S., Brereton R., Mansergh I., Berwick S., Sandiford K., and Wellington C. (1991) *The Potential Effect of the Enhanced Greenhouse Climate Change on Selected Victorian Fauna.* Technical Report No. 123. Victorian Department of Conservation and Environment, Melbourne.

Benshemesh J. (1990) Management of Malleefowl—With regard to fire. In *The Mallee Lands: A Conservation Perspective.* J.C. Noble, P.J. Joss, and G.K. Jones (eds.), pp. 206–211. Commonwealth Scientific and Industrial Research Organization, Melbourne.

Bent A.C. (1919) *Life Histories of North American Diving Birds.* Smithsonian Institution United States National Museum Bulletin 107.Washington, DC.

Bernstein N.P., Baker K.B., and Wilmot S.R. (1990) Changes in a prairie bird population from 1940 to 1989. *Journal of the Iowa Academy of Science* 97:115–120.

Berruti A. and Colclough J. (1987) Comparison of the abundance of pilchard in Cape Gannet diet and commercial catches off the Western Cape South Africa. *South African Journal of Marine Science* 51:863–869.

Berry K.H. (1986) Introduction: Development, testing, and application of wildlife-habitat models. In *Wildlife 2000: Modeling Habitat Relationships of Terrestrial Vertebrates*. J. Verner, M.L. Morrison, and C.J. Ralph (eds.), pp. 3–4. University of Wisconsin Press, Madison.

Beser J.H. and von Helden-Sarnowski S. (1982) Zur Ökologie einer Ackerpopulation des Kiebitzes *Vanellus vanellus*. *Charadrius* 18:93–113.

Bibby C.J. (1995) A global view of priorities for bird conservation: A summary. *Ibis* 137:S247–S248.

Bibby C.J., Collar N.J., Crosby M.J., Heath M.F., Imboden C., Johnson T.H., Long A.J., Stattersfield A.J., and Thurgood S.J. (1992) *Putting Biodiversity on the Map: Priority Areas for Global Conservation*. International Council for Bird Preservation, Cambridge, England.

Bierregaard R.O. (1990) Species composition and trophic organization of the understory bird community in a central Amazonian terra firma forest. In *Four Neotropical Forests*. A.H. Gentry (ed.), pp. 217–236. Yale University Press, New Haven, CT.

Bierregaard R.O., Jr. and Lovejoy T.E. (1989) Effects of forest fragmentation on Amazonian understory bird communities. *Acta Amazonica* 19:215–241.

Bierregaard R.O., Jr., Lovejoy T.E., Kapos V., dos Santos A.A., and Hutchings R.W. (1992) The biological dynamics of tropical rainforest fragments. *BioScience* 42:859–866.

Bildstein K. (1993) Shooting galleries. *American Birds* 47:38–43.

Bingham B.B. and Noon B.R. (1997) Mitigation for habitat "take": Application to habitat conservation planning. *Conservation Biology* 11:127–139.

Biodiversity Support Program, Conservation International, The Nature Conservancy, Wildlife Conservation Society, World Resources Institute, and World Wildlife Fund (1995). *A Regional Analysis of Geographic Priorities for Biodiversity Conservation in Latin America and the Caribbean*. Biodiversity Support Program, Washington, DC.

Bird D.M., Varland D.E., and Negro J.J. (1996) *Raptors in Human Landscapes: Adaptations to Built and Cultivated Environments*. Academic Press, London.

Black J.M. (1991) Reintroduction and restocking: Guidelines for bird recovery programmes. *Bird Conservation International* 1:329–334.

Black J.M. (1995a) *Flyway Conservation and Management Plan for Svalbard Barnacle Geese*. Report to Scottish Natural Heritage, Edinburgh, and Directorate for Nature Conservation, Trondheim, Norway.

Black J.M. (1995b) The Nene recovery initiative: Research against extinction. *Ibis* 137:S153–S160.

Black J.M. and Banko P.C. (1994) Is the Hawaiian Goose saved from extinction? In *Creative Conservation—The Interface between Captive and Wild Animals*. P.J.S. Olney, G.M. Mace, and A.T.C. Feistner (eds.), pp. 394–410. Chapman and Hall, London.

Black J.M., Marshall A.P., Gilburn A., Santos N., Hoshide H., Medeiros J., Mello J, Natividad-Hodges K., and Katahira L. (1997) Survival, movements and breeding of Hawaiian Geese: An assessment of the reintroduction program. *Journal of Wildlife Management* 61:1161–1174.

Black J.M., Prop J., Hunter J., Woog F., Marshall A.P., and Bowler J.M. (1994) Foraging behaviour and energetics of the Hawaiian Goose *Branta sandvicensis*. *Wildfowl* 45:65–109.

Blair R.B. (1996) Land use and avian species diversity along an urban gradient. *Ecological Applications* 6:506–519.

Blaisdell J.P. (1949) Competition between sagebrush seedlings and reseeded grasses. *Ecology* 30:512–519.

Blaisdell J.P. (1953) *Ecological effects of planned burning of sagebrush-grass range on the upper Snake River plains.* U.S. Department of Agriculture. Tech. Bull. 1075.

Blake J.G. (1991) Nested subsets and the distribution of birds on isolated woodlots. *Conservation Biology* 5:58–66.

Blake J.G., Niemi G.J., and Hanowski J.M. (1992) Drought and annual variation in bird populations. In *Ecology and Conservation of Neotropical Migrant Landbirds.* J.M. Hagan III and D.W. Johnston (eds.), pp. 443–454. Smithsonian Institution, Washington, DC.

Blakers M., Davies S.J.J.F., and Reilly P.N. (1984) *The Atlas of Australian Birds.* Melbourne University Press, Melbourne.

Blank T.H., Southwood T.R.E., and Cross D.J. (1967) The ecology of the Partridge I. Outline of the population processes with particular reference to chick mortality and nest density. *Journal of Animal Ecology* 36:549–556.

Blaustein A.R. and Wake D.B. (1990) Declining amphibian populations: A new global phenomenon? *Trends in Ecology and Evolution* 5:203–204.

Block W.M. and Brennan L.A. (1993) The habitat concept in ornithology: Theory and applications. *Current Ornithology* 11:35–91.

Block W.M., Brennan L.A., and Gutierrez R.J. (1986) The use of guilds and guild-indicator species for assessing habitat suitability. In *Wildlife 2000: Modeling Habitat Relationships of Terrestrial Vertebrates.* J. Verner, M.L. Morrison, and C.J. Ralph (eds.), pp. 109–113. University of Wisconsin Press, Madison.

Block W.M., Brennan L.A., and Gutierrez R.J. (1987) Evaluation of guild-indicator species for use in resource management. *Environmental Management* 11:265–269.

Block W.M., Morrison M.L., Verner J., and Manley P.N. (1994) Assessing wildlife-habitat relationships models: A case study with California oak woodlands. *Wildlife Society Bulletin* 22:549–561.

Blockstein D.E. (1995) A strategic approach for biodiversity conservation. *Wildlife Society Bulletin* 23:365–369.

Blockstein D.E. and Tordoff H.B. (1985) Gone forever: A contemporary look at the extinction of the Passenger Pigeon. *American Birds* 39:845–851.

Blouin, S.F. and M. Blouin 1988. Inbreeding avoidance behaviors. *Trends in Ecology and Evolution* 3:230–233.

Bock C.E. (1997) The role of ornithology in conservation of the American West. *Condor* 99:1–6.

Bock, C.E., Bock J.H., and Bennett, B.C. (In press) Songbird Abundance in grasslands at a suburban interface on the Colorado High Plains. In *Ecology and Conservation of Grassland Birds in the Western Hemisphere.* P.D. Vickery and J. Herkert (eds.). Studies in Avian Biology. Cooper Ornithological Society, Berkeley, CA.

Bock C.E., Saab V.A., Rich T.D., and Dobkin D.S/ (1993) Effects of livestock grazing on Neotropical migratory landbirds in western North America. In *Status and Management of Neotropical Migratory Birds.* D.M. Finch and P.W. Stangel (eds.), pp. 296–309. RM-GTR-229. USDA Forest Service, Fort Collins, CO.

Boersma P.D. (1986) Ingestion of petroleum by seabirds can serve as a monitor of water quality. *Science* 231:373–376.

Boersma P.D. (1987) Penguin deaths in the South Atlantic. *Nature* 327:96.

Boersma P.D. and Groom M.J. (1993) Conservation of storm-petrels in the North Pacific. In *The Status, Ecology, and Conservation of Marine Birds of the North Pacific.* K. Vermeer, K.T. Briggs, K.H. Morgan, and D. Siegel-Causey (eds.), pp. 112–121. Canadian Wildlife Service Special Publication Ottawa.

Boersma P.D., Stokes D.L., and Yorio P.M. (1990) Reproductive variability and historical change of Magellanic Penguins (*Spheniscus magellanicus*) at Punta Tombo, Argentina. In *Penguin Biology.* L.S. Davis and J.T. Darby (eds.), pp. 15–43. Academic Press, San Francisco, CA.

Bollinger E.K., Bollinger P.B., and Gavin T.A. (1990) Effects of hay-cropping on eastern populations of the Bobolink. *Wildlife Society Bulletin* 18:142–150.

Bollinger E.K. and Gavin T.A. (1989) The effects of site quality on breeding-site fidelity in Bobolinks. *Auk* 106:584–594.

Bollinger E.K. and Gavin T.A. (1992) Eastern Bobolink populations: Ecology and conservation in an agricultural landscape. In *Ecology and Conservation of Neotropical Migrant Landbirds.* J.M. Hagan III and D.W. Johnston (eds.), pp. 497–506. Smithsonian Institute Press, Washington, DC.

Bondrup-Nielsen S. (1978) *Vocalizations, Nesting, and Habitat Preferences of the Boreal Owl* (Aegolius funereus). MS thesis, University of Toronto, Toronto.

Boone R.B. and Hunter M.L. (1996) Using diffusion models to simulate the effects of land use on grizzly bear dispersal in the Rocky Mountains. *Landscape Ecology* 11:51–64.

Boreiko V.E. (1995) *History of Protected Areas in Ukraine.* Kiev Ecologo-Cultural Centre, Kiev.

Boreiko V.E. (1996) *Unknown Facts in Nature Conservation History in the USSR, Russia, Ukraine.* Kiev Ecologo-Cultural Centre, Kiev.

Botkin D. and Keller E. (1995) *Environmental Science: Earth As a Living Planet.* Wiley and Sons, New York.

Botsford L.W. and Brittnacher J.G. (1992) Detection of environmental influences on wildlife: California quail as an example. In *Wildlife 2001: Populations.* D.C. McCullough and R.H. Barrett (eds.), pp. 158–169. Elsevier Applied Science, London.

Botts P.A. Haney A., Holland K., and Packard S. (1994). Midwest oak ecosystems recovery plan. Unpublished Report, The Nature Conservancy, Chicago, IL.

Bourne W.R.P. (1976) Seabirds and pollution. In *Marine Pollution.* R. Johnston (ed.), pp. 403–502. Academic Press, London.

Bowerman W., Kubiak T., Holt J., Evans D., Eckstein R., Sindelar C., Best D., and Kozie K. (1994) Observed abnormalities in mandibles of nestling Bald Eagles *Haliaeetus leucocephalus*. *Bulletin of Environmental Contam. Toxicol.* 53:450–457.

Boyce M.S. (1992) Population viability analysis. *Annual Review of Ecology and Systematics* 23:481–506.

Boyce M.S., Meyer J.S., and Irwin L.L. (1994) Habitat-based PVA for the Northern Spotted Owl. In *Statistics in Ecology and Animal Population Monitoring.* D.J. Fletcher and B.F.J. Manly (eds.), Otago Conference Series 2:63–85. University of Dunedin, N.Z.

Boyd H. (1991) Science and craft in waterfowl management in North America. In *Bird Population Studies: Relevance to Conservation and Management.* C.M. Perrins, J-D. Lebreton, and G.J.M. Hirons (eds.), pp. 526–541. Oxford University Press, London.

Braithwaite L.W., Belbin L., and Austin M.P. (1993) Land use allocation and biological conservation in the Batemans Bay forests of New South Wales. *Australian Forestry* 56:4–21.

Braun C.E., Baker M.F., Eng R.L., Gashwiler J.S., and Schroeder M.H. (1976) Conservation committee report on effects of alteration of sagebrush communities on the associated avifauna. *Wilson Bulletin* 88:165–171.

Braun C.E., Enderson J.H., Fuller M.R., Linhart Y.B., and Marti C.D. (1996) *Northern Goshawk and Forest Management in the Southwestern United States.* Technical Review 96–2. The Wildlife Society, Bethesda, MD.

Brawn J.D. and Balda R.P. (1988) Population biology of cavity nesters in northern Arizona: Do nest sites limit breeding densities? *Condor* 90:61–71.

Brawn J.D., Karr J.R., and Nichols J.D. (1995) Demography of birds in a Neotropical forest: Effects of allometry, taxonomy, and ecology. *Ecology* 76:41–51.

Brawn J.D., Collins T.M., Medina M., and Bermingham E. (1996) Associations between physical isolation and geographical variation within three species of Neotropical birds. *Molecular Ecology* 5:33–46.

Brawn J.D. and Robinson S.K. (1996) Source-sink population dynamics may complicate the interpretation of long-term census data. *Ecology* 77:3–12.

BRD (Biological Resources Division) (1996) *Strategic Science Plan.* U.S. Department of the Interior, Geological Survey, Reston, VA.

Breiman L., Friedman J.H., Ohshen R.A., and Stone C.J. (1984) *Classification and Regression Trees.* Wadsworth, Belmont, CA.

Breininger D.R., Larson V.L., Duncan B.W., Smith R.B., Oddy D.M., and Goodchild M.F. (1995) Landscape patterns of Florida Scrub Jay habitat use and demographic success. *Conservation Biology* 9:1442–1453.

Breininger D.R., Larson V.L., Oddy D.M., Smith R.B., and Barkaszi M.J. (1996) Florida Scrub-Jay demography in different landscapes. *Auk* 113:617–625.

Briggs S.A. and Criswell J.H. (1978) Gradual silencing of spring in Washington: Selective reproduction of species of birds found in three woodland areas over the past 30 years. *Atlantic Naturalist* 32:19–26.

Britten H.B. 1996. Meta-analyses of the association between multilocus heterozygosity and fitness. *Evolution* 50:2158–2164.

Brittingham M.C. (1986) A survey of avian mortality at winter feeders. *Wildlife Society Bulletin* 14:445–450.

Brittingham M.C. and Temple S.A. (1988) Impacts of supplemental feeding on survival rates of Black-capped Chickadees. *Ecology* 69:581–589.

Brittingham M.C. and Temple S.A. (1992) Use of winter bird feeders by Black-capped Chickadees. *Journal of Wildlife Management* 56:103–110.

Brock M.K. and White B.N. (1992) Application of DNA fingerprinting to the recovery program of the endangered Puerto Rican Parrot. *Proceedings of the National Academy of Science, USA* 89:1121–1125.

Brooker L.C. and Brooker M.G. (1994) A model for the effects of fire and fragmentation on the population viability of the Splendid Fairy-wren. *Pacific Conservation Biology* 1:344–358.

Brooker M.G., Rowley I., Adams M., and Baverstock P.R. (1990) Promiscuity: An inbreeding avoidance mechanism in a socially monogamous species? *Behavioral Ecology and Sociobiology* 26:191–199.

Brooks D.R., Mayden R.L., and McLennan D.A. (1992) Phylogeny and biodiversity conserving our evolutionary legacy. *Trends in Ecology and Evolution* 7:55–59.

Brothers N. (1991) Albatross mortality and associated bait loss in the Japanese longline fishery in the Southern Ocean. *Biological Conservation* 55:255–268.

Brown J.H. (1984) On the relationship between abundance and distribution of species. *American Naturalist* 124:255–279.

Brown J.H. (1995) *Macroecology*. University of Chicago Press, Chicago.

Brown J.H., Mehlman D., and Stevens G.C. (1995) Spatial variation in abundance. *Ecology* 76:2028–2043.

Brown J.H., Stevens G.C., and Kaufman D.M. (1996) The geographic range: Size, shape, and internal structure. *Annual Review of Ecology and Systematics* 27:597–623.

Brown L.R. (1993) The world transformed—envisioning an environmentally safe planet. *Futurist* 27:16–21.

Bruce M.D. (1978) The Golden Honeyeater (*Cleptornis marchei*): Notes on behaviour, vocalisations and taxonomic affinities. *Bonner Zoological Beitraege* 29:441–445.

Bucher E.H. and Nores M. (1988) Present status of birds in steppes and savannas of northern and central Argentina. In *Ecology and Conservation of Grassland Birds*. P.D. Goriup (ed.), pp. 71–80. International Council for Bird Preservation Technical Publication No. 7.

Buckley P.A., Foster M.S., Morton E.S., Ridgley R.S., and Buckley F.G. (1985) Neotropical Ornithology. *Ornithological Monographs* 36:1–1041.

Buechner M. and Sauvajot R. (1996) Conservation and zones of human activity: The spread of human disturbance across a protected landscape. In *Biodiversity in Managed Landscapes*. R.C. Szaro and D.W. Johnston (eds.), pp. 605–629. Oxford University Press, New York.

Bunnel F.L., Dunbar D., Koza L., and Ryder G. (1981) Effects of disturbance on the productivity and survivorship of White Pelicans in British Columbia. *Colonial Waterbirds* 4:2–11.

Bunnell F.B. (1995) Forest-dwelling vertebrate fauna and natural fire regimes in British Columbia: Patterns and implications for conservation. *Conservation Biology* 9:636–644.

Bunting S.C. (1985) Fire in sagebrush-grass ecosystems: Successional changes. In *Rangeland Fire Effects—A Symposium*. K. Saunders and J. Durham (eds.), pp. 7–11. USDOI, Bureau of Land Management, Boise, ID.

Burbidge A.A. (1993) Conservation biology in Australia: Where should it be heading, will it be applied? In *Conservation Biology in Australia and Oceania*. C. Moritz and J. Kikkawa (eds.), pp. 27–37. Surrey Beatty, Sydney.

Bureau of Land Management (1982) *Integrated Habitat Inventory and Classification System*. Manual Section 6602. U.S. Department of Interior, Bureau of Land Management, Washington, DC.

Bureau of Land Management (1996) *BLM priority information needs for fiscal year 1997 and fiscal year 1998*. BLM memo 1702(480) dated June 7, 1996, from BLM Director to the Director of National Biological Service. USDOI, Bureau of Land Management, Washington, DC.

Burger A.E. (1993) Estimating the mortality of seabirds following oil spills: Effects of spill volume. *Marine Pollution Bulletin* 26:140–143.

Burger A.E. and Cooper J. (1984) The effects of fisheries on seabirds in South Africa and Namibia. In *Marine Birds: Their Feeding Ecology and Commercial Fisheries Relationships*.

D.N. Nettleship, G.A. Sanger, and P.F. Springer (eds.), pp. 150–160. Canadian Wildlife Service Ottawa.

Burger A.E. and Fry D.M. (1993) Effects of oil pollution on seabirds in the northeast Pacific. In *The Status, Ecology, and Conservation of Marine Birds of the North Pacific*. K. Vermeer, K.T. Briggs, K.H. Morgan, and D. Siegel-Causey (eds.), pp. 254–263. Special Publication of the Canadian Wildlife Service Ottawa.

Burger J. and Gochfeld M. (1990) *The Common Tern: Breeding Behavior and Biology*. Columbia University Press, New York, NY.

Burger J. and Gochfeld M. (1994) Predation and effects of humans on island-nesting seabirds. In *Seabirds on Islands: Threats, Case Studies and Action Plans*. D.N. Nettleship, J. Burger and M. Gochfeld (eds.), pp. 39–67. BirdLife International.

Burger L.D., Burger L.W., Jr., and Faaborg J. (1994) Effects of prairie fragmentation on predation on artificial nests. *Journal of Wildlife Management* 58:249–254.

Burgman M.A., Akçakaya H.R., and Loew S.S. (1988) The use of extinction models for species conservation. *Biol. Conserv.* 43:9–25.

Burgman M.A., Ferson S., and Akçakaya H.R. (1993) *Risk Assessment in Conservation Biology*. Chapman and Hall, London.

Burhans D.E. (1996) *Anti-brood parasite defenses and nest-site selection by forest-edge songbirds in central Missouri*. Ph.D. thesis, University of Missouri, Columbia.

Burke T., Hanotte O., Bruford M.W., and Cairns E. (1991) Multilocus and single locus minisatellite analysis in population biological studies. In *DNA Fingerprinting: Approaches and Applications*. T. Burke, G. Dolf, A. J. Jeffreys, and R. Wolff (eds.), pp. 154–168. Birkhäuser Verlag, Basel, Switzerland.

Burkett E.E. (1995) Marbled Murrelet food habits and prey ecology. In *Ecology and Conservation of the Marbled Murrelet*. C.J. Ralph, G.L. Hunt Jr., M.G. Raphael, and J.F. Piatt (eds.), pp. 223–246. PSW-GTR-152. USDA Forest Service, Albany, CA.

Burnham K.P., Anderson D.R. and White G.C. (1996) Chapter 13. Meta-analysis of vital rates of the Northern Spotted Owl. *Studies in Avian Biology* 17:92–101.

Bury R.B. and Luckenbach R.A. (1983) Vehicular recreation in arid land dunes: Biotic responses and management alternatives. In *Environmental Effect of Off-Road Vehicles, Impact and Management in Arid Regions*. R.H. Webb and H.H. Wilshier (eds.), pp. 207–221. Springer-Verlag, New York.

Bushman E.S. and Therres G.D. (1988) *Habitat management guidelines for forest interior breeding birds of coastal Maryland*. Wildlife Technical Publ. 88–1. Maryland Department of Natural Resources, Forest, Park and Wildlife Service, Annapolis, MD.

Butcher G.S. and Lowe J.D. (1990) *Population trends of twenty species of migratory birds as revealed by Christmas Bird Counts, 1963–87*. Final Report, Cooperative Agreement No. 14–16–0009–88–941, U.S. Fish and Wildlife Service, Office of Migratory Bird Management, Washington, DC.

Butler D. and Merton D. (1992) *The Black Robin*. Oxford University Press, Oxford, England.

Butler P.J. (1992) Parrots, pressures, people, and pride. In *New World Parrots in Crisis: Solutions from Conservation Biology*. S.R. Beissinger and N.F.R. Snyder (eds.), pp. 25–46. Smithsonian Institution Press, Washington, DC.

Byrd G.V. (In press) Current breeding status of the Aleutian Canada Goose, a recovering endangered species. In *Proceedings of the International Canada Goose Symposium*.

Cade T.J. (1986) Reintroduction as a method of conservation. *Raptor Research Report* 5:72–84.

Cade T.J. (1990) Peregrine Falcon recovery. *Endangered Species Update* 8:40–43.

Cade T.J., Enderson J.H., Thelander C.G., and White C.M. (1988) *Peregrine Falcon Populations: Their Management and Recovery*. The Peregrine Fund, Boise, ID.

Cade T.J. and Jones C.G. (1993) Progress in the restoration of the Mauritius Kestrel. *Conservation Biology* 7:169–175.

Cade T.J., Martell M., Redig P., Septon G., and Tordoff H. (1996) Peregrine Falcons in urban North America. In *Raptors in Human Landscapes: Adaptations to Built and Cultivated Environments*. D.M. Bird, D.E. Varland, and J.J. Negro (eds.), pp. 3–13. Academic Press, London.

Cade T.J. and Temple S.A. (1995) Management of threatened bird species: Evaluation of the hands-on approach. *Ibis* 137:S161–S172.

Callaghan D.A. (In press) *Ducks, Geese, Swans and Screamers: An Action Plan for the Conservation of Anseriformes*. International Union for the Conservation of Nature, Gland.

Callaghan D.A. and Green A.J. (1993) Wildfowl at risk, 1993. *Wildfowl* 44:149–169.

Cameron J. and Abouchar J. (1996) The status of the precautionary principle in international law. In *The Precautionary Principle and International Law: The Challenge of Implementation*. D. Freestone and E. Hay (eds.), pp. 29–52. Kluwer Law International, The Hague, The Netherlands.

Campbell E.W., III (1996) *The Effect of Brown Tree Snake* (Boiga irregularis) *Predation on the Island of Guam's Extant Lizard Assemblages*. Ph.D. dissertation Ohio State University, Columbus.

Camphuysen C.J. (1989) *Beached Bird Surveys in the Netherlands 1915–1988*. Netherlands Stookolieslachtoffer-Onderzzoek, Amsterdam.

Canaday C., (1997) Loss of insectivorous birds along a gradient of human impact in Amazonia. *Biological Conservation* 77:63–77.

Capparella A. (1988) Genetic variation in neotropical birds: Implications for the speciation process. *Proceedings of the International Ornithological Congress* 19:1658–1664.

Capparella A. (1992) Neotropical avian diversity and riverine barriers. *Proceedings of the International Ornithological Congress* 20:307–316.

Carey A.B., Horton S.P., and Biswell B.L. (1992) Northern Spotted Owls: Influence of prey base and landscape character. *Ecological Monographs* 62:223–250.

Carey A.B., Reid J.A., and Horton S.P. (1990) Spotted Owl home range and habitat use in southern Oregon Coast ranges. *Journal of Wildlife Management* 54:11–17.

Carpenter S.R., Chisholm S.W., Krebs C.J., Schindler D.W., and Wright R.F. (1995) Ecosystem experiments. *Science* 269:324–327.

Carr E. (1994) Power to the people: The new prize. *The Economist* 27:3–18.

Carrillo E. and Vaughan C. (1994) *La Vida Silvestre de Mesoamerica: Diagnóstico y Estrategia para su Conservación*. Editorial de la Universidad Nacional (UNA). Heredia, Costa Rica.

Carroll R., Augspurger C., Dobson A., Franklin J., Orians G., Reid W., Tracy R., Wilcove D., and Wilson J. (1996) Strengthening the use of science in achieving the goals of the Endangered Species Act: An assessment by the Ecological Society of America. *Ecological Applications* 6:1–11.

Carter H.R., Gilmer D.S., Takekawa J.E., Lowe R.W., and Wilson U.W. (1995) Breeding seabirds in California, Oregon, and Washington. In *Our Living Resources: A Report to the Nation on the Distribution, Abundance, and Health of U.S. Plants, Animals, and Ecosystems*. E.T. Laroe, G.S. Farris, C.E. Puckett, P.D. Doran, and M.J. Mac (eds.), pp. 43–49. U.S. Department of the Interior, National Biological Service, Washington, DC.

Carter H.R., Sowls A.L., Rodway M.S., Wilson U.W., Lowe R.W., Gress F., and Anderson D.W. (1995) Population size, trends, and conservation problems of the Double-crested Cormorant on the Pacific Coast of North America. *Colonial Waterbirds* 18:189–215.

Caswell H. (1989) *Matrix Population Models: Construction, Analysis, and Interpretation.* Sinauer Associates, Sunderland, MA.

Caughley G. (1994) Directions in conservation biology. *Journal of Animal Ecology* 63:215–244.

Caughley G. and Gunn A. (1996) *Conservation Biology in Theory and Practice.* Blackwell Science, Cambridge, MA.

CBSG (Captive Breeding Specialist Group) (1991a) *Disease and Conservation of Threatened Species.* Unpublished ms. Apple Valley, MN.

CBSG (Captive Breeding Specialist Group) (1991b) *Genetic Management Considerations for Threatened Species with a Detailed Analysis of the Florida Panther.* Unpublished ms. Apple Valley, MN.

Chamberlain C.P., Blum J.D., Holmes R.T., Sherry T.W., and Graves G.R. (1997). The use of isotope tracers for identifying populations of migratory birds. *Oecologia* 109:132–141.

Chapdelaine G., Laporte P., and Nettleship D.N. (1987) Population, productivity, and DDT contamination trends of Northern Gannets (*Sula bassanus*) at Bonaventure Island, Quebec, 1967–1984. *Canadian Journal of Zoology* 65:2922–2926.

Cheke A.S. (1987) An ecological history of the Mascarene Islands, with particular reference to the extinctions and introductions of land vertebrates. In *Studies of Mascarene Island Birds.* A. Diamond, A.S. Cheke, and H.F.I. Elliott (eds.), pp. 5–89. Cambridge University Press, Cambridge, England.

Chesser R.T. (1995) Migration in South America: An overview of the austral system. *Bird Conservation International* 4:91–108.

Chivers D.J. (1991) Guidelines for re-introductions: Procedures and problems. *Zoological Society of London Symposia* 62:89–100.

Christensen N.L., Bartuska A.M., Brown J.H., Carpenter S., D'Antonio C., Francis R., Franklin J.F., MacMahon J.A., Noss R.F., Parsons D.J., Peterson C.H., Turner M.G., and Woodmansee R.G. (1996) The report of the Ecological Society of America Committee on the Scientific Basis for Ecosystem Management. *Ecological Applications* 6:665–690.

Christiansen S., Lou O., Muller M., and Wohlmuth H. (1981) The spring migration of raptors in southern Israel and Sinai. *Sandgrouse* 3:1–42.

Christidis L. and Boles W.E. (1994) *The Taxonomy and Species of Birds of Australia and Its Territories.* Royal Australasian Ornithologists Union, Melbourne.

Churcher P.B. and Lawton J.H. (1987) Predation by domestic cats in a English village. *Journal of Zoology* 212:439–455.

Clark R.B. (1984) Impact of oil pollution on seabirds. *Environmental Pollution (Ser. A)* 33:1–22.

Clark T.W. (1995) Learning as a strategy for improving endangered species conservation. *Endangered Species Update* 13:5–6, 22–23.

Clark T.W., Reading R.P., and Clarke A.L. (1994) *Endangered Species Recovery: Finding the Lessons, Improving the Process.* Island Press, Washington, DC.

Cline D.R., Wentworth C., and Barry T.W. (1979) Social and economic values of marine birds. In *Conservation of Marine Birds of Northern North America.* J.C. Bartonek and D.N. Nettleship (eds.), pp. 173–182. U.S. Fish and Wildlife Service. Wildlife Research Report 11.

Clobert J., Lebreton J.D., and Allaine D. (1987) A general approach to survival rate estimation by recaptures or resightings of marked birds. *Ardea* 75:133–142.

Clout M.N. and Craig J.L. (1995) The conservation of critically endangered flightless birds in New Zealand. *Ibis* 137:S181–S191.

Clout M.N., Denyer K., James R.E., and McFadden I.G. (1995) Breeding success of New Zealand Pigeons (*Hemiphaga novaeseelandiae*) in relation to control of introduced mammals. *New Zealand Journal of Ecology* 19:209–212.

Coats S. and Phelps W., Jr. (1985) The Venezuelan Red-siskin: Case history of an endangered species. *Ornithological Monographs* 36:977–985.

Cody M.L. (ed.) (1985) *Habitat Selection in Birds.* Academic Press, Orlando, FL.

Cogger H.G. (1993) Biodiversity conservation in Australia: Development of a national strategy. In *Conservation Biology in Australia and Oceania.* C. Moritz and J. Kikkawa (eds.), pp. 297–304. Surrey Beatty, Sydney.

Cohen J.E. (1995) *How Many People Can the Earth Support?* W.W. Norton and Company, New York, NY.

Colburn T., Dumanoski D., and Myers J.P. (1996) *Our Stolen Future: Are We Threatening Our Fertility, Intelligence and Survival?* Penguin Books, New York.

Colburn T., vom Saal F., and Soto A. (1993) Developmental effects of endocrine disrupting chemicals in wildlife and humans. *Environmental Health Perspectives* 101:378–384.

Cole D.N. (1990) Trampling disturbance and recovery of cryptogamic soil crusts in Grand Canyon National Park. *Great Basin Naturalist* 50:321–325.

Collar N.J. and Andrew P. (1988) *Birds to Watch: The ICBP World Checklist of Threatened Species.* International Council for Bird Preservation, Technical Publication No. 8. Smithsonian Institution Press, Washington, DC.

Collar N.J., Crosby M.J., Stattersfield A.J. (1994) *Birds to Watch 2: The World List of Threatened Birds.* BirdLife International., Cambridge, U.K.

Collar N.J., Gonzaga L.P., Krabbe N., Madroño Nieto A., Naranjo L.G., Parker T.A., III, and Wege D.C. (1992) *Threatened Birds of the Americas.* Third edition, part two. ICBP / IUCN, Smithsonian Institution Press, Washington DC.

Collar N.J. and Stuart S.N. (1985) *Threatened Birds of Africa and Related Islands.* International Union for the Conservation of Nature and Natural Resources and International Council for Bird Preservation, Cambridge, England.

Collar N.J. and Andrew P. (eds.) (1992) *Birds to Watch: the ICBP World Checklist of Threatened Species.* International Council for Bird Preservation, Technical Publication No. 8. Smithsonian Institution Press, Washington, DC.

Commonwealth of Australia (1992) *National Forest Policy Statement: A New Focus for Australia's Forests.* Advance Press, Perth.

Conover M.R. and Chasko G.G. (1985) Nuisance Canada Goose problems in the eastern United States. *Wildlife Society Bulletin* 13:228–233.

Conroy M.J. (1993) The use of models in natural resource management: Prediction, not prescription. *Transactions of the North American Wildlife and Natural Resources Conference* 58:509–519.

Conroy M.J., Cohen Y., James F.C., Matsinos Y.C., and Maurer B.A. (1995) Parameter estimation, reliability, and model improvement for spatially explicit models of animal populations. *Ecological Applications* 5:17–19.

Conroy M.J. and Noon B.R. (1996) Mapping of species richness for conservation of biological diversity: Conceptual and methodological issues. *Ecological Applications* 6:763–773.

Conry P.J. (1988) High nest predation by Brown Tree Snakes on Guam. *Condor* 90:478–482.

Cooke A.S. (1979) Eggshell characteristics of Gannets, Shags, and great Black-backed Gulls exposed to DDE and other environmental pollutants. *Environmental Pollution* 19:47–65.

Cooke A.S., Bell A.A., and Prestt I. (1976) Egg shell characteristics and incidence of shell breakage for Grey Herons (*Ardea cinerea*) exposed to environmental pollutants. *Environmental Pollution* 11:59–84.

Cooper A., Rhymer J., James H., Olson S., McIntosh C., Sorenson M., and Fleischer R. (1996). Ancient DNA and island endemics. *Nature* 381:484.

Cooper J. (1977) Food, breeding and coat colours of feral cats on Dassen Island. *Journal of African Zoology* 12:250–252.

Cooper J., Hockey P.A.R., and Brooke P.K. (1983) Introduced mammals on South and South West Africa islands: History, effects on birds and control. In *Proceedings of the Symposium on Birds and Man*. L.J. Bunning (ed.), pp. 179–203. Johannesburg, South Africa.

Cooper J.A. (1991) Canada Goose management in the Minneapolis–St. Paul International airport. In *Wildlife Conservation in Metropolitan Environments*. L.W. Adams and D.L. Leedy (eds.), pp. 175–183. National Institute of Urban Wildlife, Columbia, MD.

Copeyon C.K. (1990) A technique for constructing cavities for the Red-cockaded Woodpecker. *Wildlife Society Bulletin* 18:303–311.

Copeyon C.K., Walters J.R., and Carter J.H. (1991) Induction of Red-cockaded Woodpecker group formation by artificial cavity construction. *Journal of Wildlife Management* 55:549–556.

Costa R. (1995) Biological opinion on the U.S. Forest Service Environmental Impact Statement for the Management of the Red-cockaded Woodpecker and Its Habitat on National Forests in the Southern Region. In USDA Forest Service, Final Environmental Impact Statement, vol. 2, Management Bulletin R8–MB73, pp. 1–192.

Cott H.B. (1953–54) The exploitation of wild birds for their eggs. *Ibis* 95:409–449, 643–675 and 96:129–149.

Coulson J.C. (1991) The population dynamics of culling Herring Gulls and Lesser Black-backed Gulls. In *Bird Population Studies: Relevance to Conservation and Management*. C.M. Perrins, J.D. Lebreton, and G.J.M. Hirons (eds.), pp. 479–497. Oxford University Press, New York.

Coulson J.C., Deans I.R., Potts G.R., Robinson J., and Crabtree A.N. (1972) Changes in organochloride contamination of the marine environment of eastern Britain monitored by shag eggs. *Nature* 236:454–456.

Court L., Yosef R., Bahat O., and Kaplan D. (In press) Griffon Vulture *(Gyps fulvus)* nest surveillance project at the Gamla Nature Reserve, Israel: 1996 conservation report. *Vulture News.*

Covacevich J. and Archer M.C. (1975) The distribution of the Cane Toad, *Bufo marinus*, in Australia and its effects on indigenous vertebrates. *Memoirs of the Queenslaved Museum* 17:305–310.

Covington W.W. and Moore M.M. (1994) Post settlement changes in natural fire regimes and forest structure: Ecological restoration of old-growth ponderosa pine forests. *Journal of Sustainable Forestry* 2:153–181.

Cox R. (1990) *One of the Wonder Spots of the World: Macquarie Island Nature Reserve.* Department of Parks, Wildlife and Heritage, Tasmania.

Cracknell G., Madsen M., and Fox A.D. (1997) *Western Palearctic Goose Population Review.* Kalo, Denmark.

Craig R.J. (1989) Observations on the foraging ecology and social behavior of the Bridled White-eye. *Condor* 91:187–192.

Craig R.J. (1990) Foraging behavior and microhabitat use of two species of white-eyes (Zosteropidae) on Saipan, Micronesia. *Auk* 107:500–505.

Craig R.J. (1992) Territoriality, habitat use and ecological distinctness of an endangered Pacific Island reed-warbler. *Journal of Field Ornithology* 63:436–444.

Craig R.J. (1994) Regeneration of native Mariana Island forest in disturbed habitats. *Micronesica* 26:97–106.

Craig R.J. (1996) Seasonal population surveys and natural history of a Micronesian bird community. *Wilson Bulletin* 108:246–267.

Craig R.J., Kaipat R., Lussier B.A., and Sabino H. (1992) Foraging differences between small passerines on Aguiguan and Saipan. In *The Aguiguan Expedition.* R.J. Craig (ed.), pp. 16–22. Proceedings of the Marianas Research Symposium 1, Northern Marianas College. Saipan, Northern Mariana Islands.

Craig R.J. and Taisacan E. (1994) Notes on the ecology and population decline of the Rota Bridled White-eye. *Wilson Bulletin* 106:165–169.

Crawford R.J.M., Boonstra H.G.D., Dyer B.M., and Upford L. (1995) Recolonization of Robben Island by African Penguins, 1983–1992. In *The Penguins: Ecology and Management.* P. Dann, I. Norman, and P. Reilly (eds.), pp. 333–363. Surrey Beatty and Sons, Chipping Norton, Australia.

Crawford R.J.M., Shelton P.A., Batchelor A.L., and Clinning D.F. (1980) Observations on the mortality of juvenile Cape Cormorants *Phalacrocorax capensis* during 1975 and 1979. *Fisheries Bulletin of South Africa* 13:69–75.

Crawford R.J.M., Shelton P.A., Cooper J., and Borrke R.K. (1983) Distribution, population size and conservation of the Cape Gannet *Morus capensis. South African Journal of Marine Science* 1:153–174.

Cressie N. (1993) *Statistics for Spatial Data Analysis.* John Wiley, New York.

Crick H.Q.P., Dudley C., Evans A.D., and Smith K.W. (1994) Causes of nest failure among buntings in the UK. *Bird Study* 41:88–94.

Crocker-Bedford C. (1990) *Status of the Queen Charlotte Goshawk.* Unpublished report to the Alaska Region of the Forest Service, Juneau, AK.

Crome F.H.J. and Moore L.A. (1990) Cassowaries in north-eastern Queensland: Report of a survey and a review and assessment of their status and conservation and management needs. *Austrian Wildlife Research* 17:369–386.

Croxall J.P., Evans P.G.H., and Schreiber R.W. (eds.) (1984) *Status and Conservation of the World's Seabirds*. ICBP Technical Publication No. 2. Page Brothers. Norwich, England.

Cruz J.B. and Cruz F. (1987) Conservation of the Dark-rumped Petrel *Pterodroma phaeopygia* in the Galapagos Islands, Ecuador. *Biological Conservation* 42:303–311.

Csuti B.A., Scott J.M., and Estes J. (1987) Looking beyond species-oriented conservation. *Endangered Species Update* 5:4.

Curnutt J.L., Pimm S.L., and Maurer B.A. (1996) Population variability of sparrows in space and time. *Oikos* 76:131–144.

Currie D.J. (1991) Energy and large-scale patterns of animal- and plant-species richness. *American Naturalist* 137:27–49.

da Silva J.M.C., Uhl C., and Murray G. (1996) Plant succession, landscape management, and the ecology of frugivorous birds in abandoned Amazonian pastures. *Conservation Biology* 10:491–503.

Dahlberg M.L. and Day R.H. (1985) Observations of man-made objects on the surface of the North Pacific Ocean. In *Proceedings of the Workshop on the Fate and Impact of Marine Debris, 27–29 November 1984, Honolulu, Hawaii*. R.S. Shomura and H.O. Yoshida (eds.), pp. 198–212. U.S. Department of Commerce National Oceanic and Atmospheric Administration Technical Memorandum, National Marine Fisheries Service. NOAA-TM-NMFS-SWFC-54, La Jolla, CA.

Dale V.H., Pearson S.M., Offerman H.L., and O'Neill R.V. (1994) Relating patterns of land-use change to faunal biodiversity in the Central Amazon. *Conservation Biology* 8:1027–1036.

Danilenko A.K., Boere G.C., and Lebedeva E.L. (1996) Looking for recent breeding grounds of Slender-billed Curlew: Habitat approach. *Wader Study Group Bulletin* 8:17–21.

Danin A. (1992a) Flora and vegetation of Israel and adjacent areas. *Bocconea* 3:18–42.

Danin A. (1992b) Report on Iter Mediterraneum II. *Bocconea* 3:5–17.

Darling F.F. (1938) *Bird Flocks and Breeding Cycle*. Cambridge University Press, Cambridge, England.

Darveau M., Beauchesne P., Belanger L., Huot J., and Larue P. (1995) Riparian forest strips as habitat for breeding birds in boreal forest. *Journal of Wildlife Management* 59:67–78.

Daubenmire R.F. (1970) *Steppe vegetation of Washington*. Technical Bulletin No. 62. Washington Agricultural Experimental Station, Pullman, WA.

Daubenmire R.F. and Daubenmire J.B. (1968) *Forest vegetation of eastern Washington and northern Idaho*. Technical Bulletin No. 60. Washington Agricultural Experimental Station, Pullman, WA.

Daubenmire R.F. (1968) *Plant Communities: A Textbook of Plant Synecology*. Harper and Row, New York.

Davis G.E., Halvorson W.L., and Ehorn W.H. (1988) Science and management in U.S. national parks. *Bulletin of Ecological Society of America* 69:111–114.

Davis J.S., Nelson A.C., and Dueker K.J. (1994) The new 'burbs: The exurbs and their implications for planning policy. *Journal of the American Planning Association* 60:45–59.

Davis J.W., Goodwin G.A., and Ockenfels R.A. (1983) *Snag habitat management.* RM-GTR-99. USDA Forest Service, Ft. Collins, CO.

Davis S.K. (1994) *Cowbird Parasitism, Predation, and Host Selection in Fragmented Grasslands of Southwestern Manitoba.* MS thesis, University of Manitoba, Winnipeg.

de Boer L.E.M. (1994) Development of coordinated genetic and demographic breeding programmes. In *Creative Conservation—The Interface between Captive and Wild Animals.* P.J.S. Olney, G.M. Mace, and A.T.C. Feistner (eds.), pp. 304–311. Chapman and Hall, London.

de Korte J. (1984) Status and conservation of seabird colonies in Indonesia. In *Status and Conservation of the World's Seabirds.* J.P. Croxall, P.G.H. Evans, and R.W. Schreiber (eds.), pp. 527–546. ICBP Technical Publication No. 2. Page Brothers, Norwich, England.

Dedon M.F., Laymon S.A., and Barrett R.H. (1986) Evaluating models of wildlife-habitat relationships of birds in black oak and mixed-conifer habitats. In *Wildlife 2000: Modeling Habitat Relationships of Terrestrial Vertebrates.* J. Verner, M.L. Morrison, and C.J. Ralph (eds.), pp. 115–120. University of Wisconsin Press, Madison.

Degnan S.M. (1993) The perils of single gene-trees—mitochondrial versus single-copy nuclear DNA variation in white-eyes (Aves: Zosteropidae). *Molecular Ecology* 2:219–225.

DeGraaf R.M. and Rudis D.D. (1983) *New England wildlife: Habitat, natural history, and distribution.* NE-GTR-108. USDA Forest Service, Radnor, PA.

DeGraaf R.M., Tilghman N.G., and Anderson S.H. (1985) Foraging guilds of North American birds. *Environmental Management* 9:493–536.

DeGraaf R.M. and Wentworth J.M. (1981) Urban bird communities and habitats in New England. *Transactions of the North American Wildlife and Natural Resources Conference* 46:396–413.

DeGraaf R.M., Witman G.M., Lanier J.W., Hill B.J., and Keniston J.M. (1980) *Forest Habitat for Birds of the Northeast.* USDA Forest Service, Radnor, PA.

DeGraaf R.M., Yamasaki M., Leak W.B., and Lanier J.W. (1992) *New England wildlife: Management of forested habitats.* NE-GTR-144. USDA Forest Service. Radnor, PA.

del Hoyo J., Elliott A., and Sargatal J. (eds.) (1996) *Handbook of the Birds of the World. vol. 3. Hoatzin to Auks.* Lynx Edicions, Barcelona.

Delisle J.M. (1995) *Avian Use of Fields Enrolled in the Conservation Reserve Program in Southeast Nebraska.* MS thesis, University of Nebraska, Lincoln.

Dementiev G.P. (1962) Do we need to kill raptors? *Okhota I Okhotnichie Khozyaistvo* 11:25–26.

den Boer P.J. (1981) The survival of populations in heterogeneous and variable environment. *Oecologia* 50:39–53.

DeSante D.F. (1990) The role of recruitment in the dynamics of a Sierran subalpine bird community. *American Naturalist* 136:429–455.

DeSante D.F. (1992) Monitoring Avian Productivity and Survivorship (MAPS): A sharp, rather than blunt, tool for monitoring and assessing landbird populations. In *Wildlife 2001: Populations.* D.R. McCullough and R.H. Barrett (eds.), pp. 511–521. Elsevier Applied Science, London.

DeSante D.F. (1995) Suggestions for future directions for studies of marked migratory landbirds from the perspective of a practitioner in population management and conservation. *Journal of Applied Statistics* 22:949–965.

DeSante D.F. (1997) *General Evaluation of the Monitoring Avian Productivity and Survivorship (MAPS) Program*. The Institute for Bird Populations, Point Reyes Station, CA. 128 pp.

DeSante D.F. (In press) Management implications of patterns of productivity and survivorship from the MAPS Program. In *Proceedings of the 1995 International Partners in Flight Workshop*. R. Bonney, L. Niles, and D. Pashley (eds.), Cape May, NJ.

DeSante D.F., Burton K.M., and O'Grady D.R. (1996) The Monitoring Avian Productivity and Survivorship (MAPS) program fourth and fifth annual reports (1993 and 1994). *Bird Populations* 3:67–120.

DeSante D.F., Burton K.M., Saracco J.F., and Walker B.L. (1995) Productivity indices and survival rate estimates from MAPS, a continent-wide programme of constant-effort mist netting in North America. *Journal Applied Statistics* 22:935–947.

DeSante D.F. and George T.L. (1994) Population trends in the landbirds of western North America. *Studies in Avian Biology* 15:173–190.

Desrochers A. and Hannon S.J. (In press) Gap-crossing decisions by dispersing forest songbirds. *Conservation Biology*.

Di Silvestro R. (1996) What's killing the Swainson's Hawk? *International Wildlife* 6(3):38–43.

Diamond J.M. (1975) Assembly of species communities. In *Ecology and Evolution of Communities*. M.L. Cody and J.M. Diamond (eds.), pp. 342–444. Belknap Press, Cambridge, MA.

Diamond J.M. (1989) Overview of recent extinctions. In *Conservation for the Twenty-first Century*. D. Western and M. Pearl (eds.), pp. 37–41. Oxford University Press, New York.

Dijack W.D. (1996) *Landscape Characteristics Affecting the Distribution of Mammalian Predators*. MS thesis, University of Missouri, Columbia.

Dinerstein E., Olson D.M., Graham D.J., Webster A.L., Primm S.A., Bookbinder M.P., and Ledec G. (1995) *A Conservation Assessment of the Terrestrial Ecoregions of Latin America*. World Wildlife Fund in association with the World Bank, Washington, DC.

Dinerstein E. and Wikramanayake E.D. (1993) Beyond "hotspots": How to prioritize investments to conserve biodiversity in the Indo-Pacific Region. *Conservation Biology* 7:53–65.

Dmowski K. and Kozakiewicz M. (1990) Influence of a shrub corridor on movements of passerine birds to a lake littoral zone. *Landscape Ecology* 4:99–108.

Dobkin D.S. (1992) *Neotropical migrant landbirds in the Northern Rockies and Great Plains*. R1-93-34, USDA Forest Service, Missoula, MT.

Dobler F.C. (1994) Washington state shrubsteppe ecosystem studies with emphasis on the relationship between nongame birds and shrub and grass cover densities. In *Proceedings—Ecology and Management of Annual Rangelands*. S.B. Monsen and S.G. Kitchen (comps), pp. 149–161. INT-GTR-313. USDA Forest Service, Boise, ID.

Dobson A.P., Rodriguez J.P., Roberts W.M., and Wilcove D.S. (1997) Geographic distribution of endangered species in the United States. *Science* 275:550–553.

Donald P.F. and Forrest C. (1995) The effects of agricultural change on population size of Corn Buntings *Miliaria calandra* on individual farms. *Bird Study* 42:205–215.

Donnelly P. and Tavaré S. (1995) Coalescents and genealogical structure under neutrality. *Annual Review of Genetics* 29:401–421.

Donovan T.M., Jones P.W., Annand E., and Thompson F.R., III (1997) Variation in local-scale edge effects: Mechanisms and landscape context. *Ecology* 78:2064–2075

Donovan T.M., Lamberson R.H., Kimber A., Thompson F.R., III, and Faaborg J. (1995) Modeling the effects of habitat fragmentation on source and sink demography of Neotropical migrant birds. *Conservation Biology* 9:1396–1407.

Donovan T.M., Thompson F.R., III, Faaborg J., and Probst J.R. (1995) Reproductive success of migratory birds in habitat sources and sinks. *Conservation Biology* 9:1380–1395.

Doughty R.W. (1975) *Feather Fashions and Bird Preservation: A Study in Nature Protection.* University of California Press, Berkeley.

Dovrat E. (1980) A summary of autumn migration along the Kfar Qassem route, Autumn 1980. *Torgos* 1:32–47.

Drennan, M.P. (1987). Petit Manan National Wildlife Refuge, 1986: Changes in nesting seabird populations after three years of gull management. U.S. Fish and Wildlife Service report. Milbridge, ME.

Drury W.H. (1965) Gulls vs. terns: Clash of coastal nesters. *Massachusetts Audubon* 1965:207–211.

Drury W.H. (1973) Population changes in New England seabirds. *Bird-Banding* 44:267–313.

Drury W.H. (1974) Population changes in New England seabirds. *Bird-Banding* 45:1–15.

Dryzc A. (1990) Understory bird assemblages in various types of lowland tropical forest in Tambopata Reserve, SE Peru (with faunistic notes). *Acta Zoologica Cracoviensia* 33:215–233.

Duffy D.C. (1983) The foraging ecology of Peruvian seabirds. *Auk* 100:800–810.

Duffy D.C. (1988) Ornithology in Central and South America: Cause for optimism? *Auk* 105:395–396.

Duffy D.C. (1994a) Afterwards: An agenda for managing seabirds and islands. In *Seabirds on Islands: Threats, Case Studies and Action Plans.* D.N. Nettleship, J. Burger, and M. Gochfeld (eds.), pp. 311–318. BirdLife International, Cambridge, England.

Duffy D.C. (1994b) The guano islands of Peru: The once and future management of a renewable resource. In *Seabirds on Islands: Threats, Case Studies and Action Plans.* D.N. Nettleship, J. Burger, and M. Gochfeld (eds.), pp. 68–76. BirdLife International, Cambridge, England.

Duffy D.C. (1994c) Towards a world strategy for seabird sanctuaries. *Colonial Waterbirds* 17:200–206.

Duffy D.C. and Schneider D.C. (1994) Seabird-fishery interactions: A manager's guide. In *Seabirds on Islands: Threats, Case Studies and Action Plans.* D.N. Nettleship, J. Burger, and M. Gochfeld (eds.), pp. 26–38. BirdLife International, Cambridge, England.

Dunning J.B., Jr. and Brown J.H. (1982) Summer rainfall and winter sparrow densities: A test of the food limitation process. *Auk* 99:123–129.

Dunning, J.B. Jr., Stewart D.J., Danielson B.J., Noon B.R., Root T.L., Lamberson R.H., and Stevens E.E. (1995) Spatially explicit population models: Current forms and future uses. *Ecological Applications* 5:3–11.

Dunning J.B., Jr. and Watts B.D. (1990) Regional differences in habitat occupancy by Bachman's Sparrow. *Auk* 107:463–472.

Ecomap (1993) *National Hierarchical Framework of Ecological Units.* USDA Forest Service, Washington, DC.

Edminster C.B. and Olsen W.K. (1996) Thinning as a tool in restoring and maintaining diverse structure in stands of southwestern ponderosa pine. In *Conference on Adaptive Ecosystem Restoration and Management: Restoration of Cordilleran Conifer Landscapes of North America.* W.W. Covington and P.K. Wagner (eds.), pp. 62–68. RM-GTR-278. USDA Forest Service, Fort Collins, CO.

Edwards S.V. (1993) Long-distance gene flow in a cooperative breeder detected in genealogies of mitochondrial DNA sequences. *Proceedings of the Royal Society of London B* 252:177–185.

Edwards S.V., Grahn M., and Potts W.K. 1995. Dynamics of MHC evolution in birds and crocodilians: Amplification of class II genes with degenerate primers. *Molecular Ecology* 4:719–730.

Edwards T.C., Jr., Deshler E.T., Foster D., and Moisen G.G. (1996) Adequacy of wildlife habitat relation models for estimating spatial distributions of terrestrial vertebrates. *Conservation Biology* 10:263–270.

Ehrlich P.R. (1988) The loss of diversity: Causes and consequences. In *Biodiversity.* E.O. Wilson (ed.), pp. 21–27. National Academy Press, Washington, DC.

Ehrlich P.R. (1994) Biodiversity and ecosystem function: Need we know more? In *Biodiversity and Ecosystem Function.* E.D. Schulze and H.A. Mooney (eds.), Springer-Verlag, New York.

Ehrlich P.R. and Ehrlich A.H. (1996) *Betrayal of Science and Reason.* Island Press, Washington, DC.

Eisentraut M. (1950) Das Fehlen endemischer und das Auftreten landfremder Eidechsen auf den beiden Hauptinseln der Balearen, Mallorca und Menorca. *Zoologische Beitraege* 1:3–11.

Eisner T., Lubchenko J., Wilson E.O., Wilcove D.S., and Bean M.J. (1995) Building a scientifically sound policy for protecting endangered species. *Science* 268:1231–1232.

Ellegren H. (1991) DNA typing of museum specimens of birds. *Nature* 354:113.

Ellegren H. (1992) Polymerase-chain-reaction (PCR) analysis of microsatellites—a new approach to studies of genetic relationships in birds. *Auk* 109:886–895.

Elliott J.E. and Noble D.G. (1993) Chlorinated hydrocarbon contaminants in marine birds of the temperate North Pacific. In *The Status, Ecology, and Conservation of Marine Birds of the North Pacific.* K. Vermeer, K.T. Briggs, K.H. Morgan, and D. Siegel-Causey (eds.), pp. 241–253. Special Publication of the Canadian Wildlife Service, Ottawa.

Elliott J.E., Norstrum R.J., and Keith J.H.A. (1988) Organochlorides and eggshell thinning in northern gannets from eastern Canada 1968–1984. *Environmental Pollution* 52:81–102.

Elliot R.D. (1991) The management of the Newfoundland turr hunt. *Canadian Wildlife Service Occasional Paper* 69:29–35.

Ellis D.H., Ellis C.H., and Mindell D.P. (1991) Raptor responses to low-level jet aircraft and sonic booms. *Environmental Pollution* 74:53–83.

Ellis-Joseph S., Green A.J., and Hewston N. (1992) *Global Captive Waterfowl Conservation and Management Action Plan.* Captive Breeding Specialist Group unpublished ms. Apple Valley, MN.

Elyseev N.V. (ed.) (1983) *Red Data Book of Russian Federation. Animals.* Rosselkhozizdat, Moscow.

Emlen J.T. (1974) An urban bird community in Tucson, Arizona: Derivation, structure, regulation. *Condor* 76:184–197.

Emlen J.T., DeJong M.J., Jaeger M.J., Moermond T.C., Rusterholz K.A., and White R.P. (1986) Density trends and range boundary constraints of forest birds along a latitudinal gradient. *Auk* 103:791–803.

Emlen S.T. (1991) Evolution of cooperative breeding in birds and mammals. In *Behavioural Ecology: An Evolutionary Approach,* 3rd ed. J.R. Krebs and N.B. Davies (eds.), pp. 301–337. Blackwell Scientific Publications. Oxford, England

Engbring J. (1983) Forest birds of Guam in critical danger. Endangered Species Technical Bulletin No.8., U.S. Fish and Wildlife Service, Washington, DC.

Engbring J. and Fritts T.H. (1988) Demise of an insular avifauna: The Brown Tree Snake on Guam. *Transactions of the Western Section of the Wildlife Society* 24:31–37.

Engbring J. and Pratt H.D. (1985) Endangered birds in Micronesia: Their history, status, and future prospects. In *Bird Conservation.* S.A. Temple (ed.), pp. 71–105. University of Wisconsin Press, Madison, WI.

Engbring J. and Ramsey F.L. (1984) *Distribution and Abundance of the Forest Birds of Guam: Results of a 1981 Survey.* FWS/OBS-84/20, U.S. Fish and Wildlife Service, Washington, DC.

Engbring J., Ramsey F.L., and Wildman V.J. (1986) Micronesian forest bird survey, 1982: Saipan, Tinian, Aguiguan, and Rota. Unpublished report to U.S. Fish and Wildlife Service, Honolulu, HI.

Engel K.A., Young L.S., Roppe J.A., Wright C.P., and Mulrooney M. (1993) Controlling raven fecal contamination of transmission line-insulators. In *Proceedings of the Avian Interactions with Utility Structures International Workshop.* J.W. Huckabee (ed.), pp. 10-1–10-14. Electrical Power Research Institute, Palo Alto.

Engels T.M. and Sexton C.W. (1994) Negative correlation of Blue Jays and Golden-cheeked Warblers near an urbanizing area. *Conservation Biology* 8:286–290.

Escano R.E.F. (1995) Red-cockaded Woodpecker extinction or recovery: Summary of status and management on our national forests. In *Red-cockaded Woodpecker: Recovery, Ecology and Management.* D.L. Kulhavy, R.G. Hooper, and R. Costa (eds.), pp. 28–35. Stephen F. Austin State University, Nacogdoches, Texas.

Estes J.A. (1995) Top-level carnivores and ecosystem effects: Questions and approaches. In *Linking Species and Ecosystems.* C.G. Jones and J.H. Lawton (eds.), pp. 151–158. Chapman and Hall, New York.

Estes J.A. and Palmisano J.R. (1974) Sea otters: Their role in structuring nearshore communities. *Science* 185:1058–1060.

Evans A.D. and Smith K.W. (1994) Habitat selection of Cirl Buntings *Emberiza cirlus* wintering in Britain. *Bird Study* 41:81–87.

Evans M.I. (1994) *Important bird areas in the Middle East.* BirdLife International (Birdlife Conserv. Series No. 2), Cambridge, England.

Faaborg J. (1979) Qualitative patterns of avian extinction on Neotropical landbridge islands: Lessons for conservation. *Journal of Applied Ecology* 16:99–107.

Faaborg J. and Arendt W.J. (1995) Survival rates of Puerto Rican birds: Are islands really that different? *Auk* 112:503–508.

Faaborg J., Brittingham M., Donovan T., and Blake J. (1995) Habitat fragmentation in the temperate zone. In *Ecology and Management of Neotropical Migratory Birds*. T.E. Martin and D.M. Finch (eds.), pp. 357–380. Oxford University Press, New York.

Facemire C.T., Gross, T., and Guillette L. (1995) Reproductive impairment in the Florida panther: Nature or nurture? *Environmental Health Perspectives Supplement* 103:79–86.

Falconer D.S. (1981) *Introduction to Quantitative Genetics*, 2nd edition. Longman, London.

Fancy S.G., Sugihara R.T., Jeffrey J.J., and Jacob J.D. (1993) Site tenacity of the endangered Palila. *Wilson Bulletin* 105:587–596.

Farman J.C., Gardiner B.G., and Shanklin J.D. (1985) Large losses of total ozone in Antarctica reveal seasonal ClOx/NOx interaction. *Nature* 315:207–210.

Feare C.J. (1976) The exploitation of Sooty Tern eggs in the Seychelles. *Biological Conservation* 10:169–181.

Feare C.J. (1984) Seabirds as a resource: Use and management. In *Biogeography and Ecology of the Seychelles Islands*. D. Stoddart (ed.), pp. 593–606. W. Junk, Amsterdam.

Feare C.J. (1994) Changes in numbers of common starlings and farming practice in Lincolnshire. *British Birds* 87:200–204.

Fearnside P.M. (1990) The rate and extent of deforestation in Brazilian-Amazonia. *Environmental Conservation* 17:213–226.

Fearnside P.M. (1993) Deforestation in Brazilian Amazonia: The effect of population and land tenure. *Ambio* 22:537–545.

Feldman R.A., Freed L.A., and Cann R.L. (1995) A PCR test for avian malaria in Hawaiian birds. *Molecular Ecology* 4:663–674.

FEMAT (Forest Ecosystem Management Assessment Team) (1993) *Forest ecosystem management: an ecological, economic, and social assessment*. U.S. Government Printing Office 794–478. Washington, DC.

Fernandez-Duque E. and Valeggia C. (1995) Meta-analysis: A valuable tool in conservation research. *Conservation Biology* 8:555–561.

Ferraris J.D. and Palumbi S.R. (1996) *Molecular Zoology: Advances, Strategies and Protocols*. Wiley-Liss, New York.

Finch D.M. (1991) Population ecology, habitat requirements, and conservation of Neotropical migratory birds. RM-GTR-205. USDA Forest Service, Ft. Collins, CO.

Finch D.M. and Patton-Mallory M. (1993) Closing the gap between research and management. In *Status and Management of Neotropical Migratory Birds*. D.M. Finch and P.W. Stangel (eds.), pp 12–16. RM-GTR-229. USDA Forest Service, Ft. Collins, CO.

Finch D.M. and Stangel P.W. (eds.) (1993) *Status and Management of Neotropical Migratory Birds*. Gen. Tech. Rep. RM-GTR-229. USDA Forest Service, Ft. Collins, CO.

Finch D.M., Ganey J.L., Yong W., Kimball R., and Sallabanks R. (1997) Effects and interactions of fire, logging, and grazing. In *Songbird Ecology in Southwestern Ponderosa Pine Forests: A Literature Review*. W.M. Block and D.M. Finch (eds.), pp. 103–136. RM-GTR-292. USDA Forest Service, Ft. Collins, CO.

Fish and Wildlife Service (1980) Habitat Evaluation Procedures (HEP). *Ecological Services Manual 102*. U.S. Department of Interior. Washington, DC.

Fish and Wildlife Service (1981) Standards for the development of Suitability Index models. *Ecological Services Manual 103*. U.S. Department of Education, Washington, DC.

Fisher H.I. (1971) Experiments on homing in Laysan Albatrosses (*Diomedea immutabilis*). *Condor* 73:389–400.

Fjeldså J. (1994) Geographical patterns for relict and young species of birds in Africa and South America and implications for conservation priorities. *Biodiversity and Conservation* 3:207–226.

Fjeldså J. (1995) Have ornithologists "slept through the class"? On the response of ornithology to the "Biodiversity crisis" and the "Biodiversity Convention." *Journal of Avian Biology* 26:89–93.

Flannery T.F. (1994) *The Future-Eaters: An Ecological History of the Australasian Lands and People*. Reed, Sydney.

Flather C.H. and Sauer J.R. (1996) Using landscape ecology to test hypotheses about large-scale abundance patterns in migratory birds. *Ecology* 77:28–35.

Fleischer R.C. (1983) A comparison of theoretical and electrophoretic assessments of genetic structure in populations of the House Sparrow (*Passer domesticus*). *Evolution* 37:1001–1009.

Fleischer R.C. (1996) Application of molecular methods to the assessment of genetic mating systems in vertebrates. In *Molecular Zoology: Advances, Strategies and Protocols*. J.D. Ferraris and S.R. Palumbi (eds.), pp. 133–161. Wiley-Liss, New York.

Fleischer R.C., Conant S.C., and Morin M. (1991) Population bottlenecks and genetic variation in native and introduced populations of the Laysan Finch (*Telespiza cantans*). *Heredity* 66:125–130.

Fleischer R.C., Fuller G., and Ledig D. (1995) Genetic structure of endangered Clapper Rail (*Rallus longirostris*) populations in southern California. *Conservation Biology* 9:1234–1243.

Fleischer R.C. and Rothstein S.I. (1988) Known secondary contact and rapid gene flow among subspecies and dialects in the brown-headed cowbird. *Evolution* 42:1146–1158.

Fleischer R.C., Tarr C.L., and Pratt T.K. (1994) Genetic structure in the Palila, an endangered Hawaiian Honeycreeper, as assessed by DNA fingerprinting. *Molecular Ecology* 3:383–392.

Fleishner T.L. (1994) Ecological costs of livestock grazing in western North America. *Conservation Biology* 8:919–921.

Flesness N.R. and Foose T.J. (1990) The role of captive breeding in conservation of species. In *1990 International Union for the Conservation of Nature Red List of Threatened Animals*. World Conservation Monitoring Centre (eds.), pp. xi–xv. IUCN, Gland.

Flint V.E. (1995) *The Catalogue of the Land Vertebrate Animals of Russia*. All Russian Institute for Nature Protection, Moscow.

Foose T.J. and Ballou J.D. (1988) Population management: Theory and practice. *International Zoo Yearbook* 27:26–41.

Ford H.A., Barrett G., and Howe R.W. (1995) Effect of habitat fragmentation and degradation on bird communities in Australian eucalypt woodland. In *Functioning and Dynamics of Natural and Perturbed Ecosystems*. D. Bellan-Santini, G. Bonin, and C. Emig (eds.), pp. 99–116. Lavoisier, Paris.

Ford H.G., Barrett G., and Recher H. 1995. Birds in a degraded landscape-safety nets for capturing regional biodiversity. Pages 43–50 in *Nature Conservation 4: The Role of Networks*. D.A. Saunders, J.L. Craig and E.M. Mattiske (eds.). Surrey Beatty & Sons, Chipping Norton, New South Wales, Australia

Forsman E.D. (1976) *A Preliminary Investigation of the Spotted Owl in Oregon*. MS thesis, Oregon State University, Corvallis.

Forsman E.D., Meslow E.C., and Wight H.M. (1984) Distribution and biology of the Spotted Owl in Oregon. *Wildlife Monographs* 87:1–64.

Forsman E.D., S. DeStephano, M.G. Raphael, and R.J. Guitiérrez. 1996. Demography of the Northern Spotted Owl. *Studies in Avian Biology* 17:1–122.

Fortin M.-J. (1994) Edge detection algorithms for two-dimensional ecological data. *Ecology* 75:956–965.

Foster M.S. (1993) Research, conservation, and collaboration: The role of visiting scientists in developing countries. *Auk* 110:414–417.

Fralish J.S., Anderson R.C., Ebinger J.E., and Szafoni R. (1994) *Proceedings of the North American Conference on Barrens and Savannas.* Illinois State University, Bloomington.

Frankham R. (1995a) Conservation genetics. *Annual Review of Genetics* 29:305–327.

Frankham R. (1995b) Inbreeding and extinction: A threshold effect. *Conservation Biology* 9:792–799.

Franklin A.B., Anderson D.R., Forsman E.D., Burnham K.P., and Wagner F.W. (1996) Methods for collecting and analyzing demographic data on the Northern Spotted Owl. *Studies in Avian Biology* 17:12–20.

Franklin D.C., Menkhorst P.W., and Robinson J.L. (1989) Ecology of the Regent Honeyeater *Xanthomza phrygia. Emu* 89:140–154.

Franklin J.F. (1988) Structural and functional diversity in temperate forests. In *Biodiversity.* E.O. Wilson (ed.), pp. 166–175. National Academy Press, Washington, DC.

Franklin J.F. (1993a) Lessons from old-growth. *Journal of Forestry* 9112:11–13.

Franklin J.F. (1993b) Preserving biodiversity: Species, ecosystems, or landscapes? *Ecological Applications* 3:202–205.

Franklin J.F. (1994) Response to Tracy and Brussard. *Ecological Applications* 4:208–209.

Franklin J.F. and Dyrness C.T. (1973) *Natural Vegetation of Oregon and Washington.* PNW-GTR-8. USDA Forest Service, Portland, OR.

Franklin K. (1987) Endangered species: Where to from here? *American Forests* 93:57–58, 60, 74–76.

Frazer N. (1992) Sea turtle conservation and halfway technologies. *Conservation Biology* 6:179–184.

Freemark K.E. and Collins B. (1992) Landscape ecology of birds breeding in temperate forest fragments. In *Ecology and Conservation of Neotropical Migrant Landbirds.* J.M. Hagen III and D.W. Johnston (eds.), pp. 443–454. Smithsonian Institution Press, Washington, DC.

Freemark K.E., Dunning J.B., Hejl S., and Probst J.R. (1995) A landscape ecology perspective for research, conservation, and management. In *Ecology and Management of Neotropical Migratory Birds: A Synthesis and Review of Critical Issues.* T.E. Martin and D.M. Finch (eds.), pp. 381–427. Oxford University Press, New York.

Frémont J.C. (1845) *Report of the exploring expedition to the Rocky Mountains in the year 1842, and to Oregon and Northern California in the years 1843–44.* Gales and Seaton. Washington, DC.

French H.F. (1997) Learning from the ozone experience. In *State of the World 1997.* L.L. Brown (project director), pp. 151–171, 212–218. A Worldwatch Institute Report on Progress toward a Sustainable Society. W.W. Norton, New York.

Fretwell S.D. (1972) *Populations in a Seasonal Environment.* Princeton University Press, Princeton, New Jersey.

Fretwell S.D. (1986) Distribution and abundance of the Dickcissel. *Current Ornithology.* 4:211–242.

Friedel M.H., Foran B.D., and Stafford Smith D.M. (1990) Where the creeks run dry or ten feet high: Pastoral management in arid Australia. *Proceedings of the Ecological Society of Australia* 16:185–194.

Friesen L.E., Eagles P.F.J., and MacKay R.J. (1995) Effects of residential development on forest-dwelling Neotropical migrant songbirds. *Conservation Biology* 9:1408–1414.

Frissell C.A. (1993) Topology of extinction and endangerment of native fishes in the Pacific Northwest and California (U.S.A.). *Conservation Biology* 7:342–354.

Frissell, C.A., Nawa R.K., and Noss R. (1992) Is there any conservation biology in "New Perspectives?": a response to Salwasser. *Conservation Biology* 6:461–464.

Fritts T.H. (1988) *The brown tree snake,* Boiga irregularis, *a threat to Pacific Islands.* Biol. Rep. 88(31). U.S. Fish and Wildlife Service, Washington, DC.

Fritts T.H. and McCoid M.J. (1991) Predation by the Brown Tree Snake on poultry and other domesticated animals in Guam. *The Snake* 23:75–80.

Fritts T.H., McCoid M.J., and Haddock R.L. (1994) Symptoms and circumstances associated with bites by the Brown Tree Snake (Colubridae: *Boiga irregularis*) on Guam. *Journal of Herpetology* 28:27–33.

Fritts T.H. and Rodda G.H. (1995) Invasions of the Brown Tree Snake. In *Our Living Resources: A Report to the Nation on the Distribution, Abundance, and Health of U.S. Plants, Animals, and Ecosystems.* E. LaRoe, G. Farris, C. Puckett, P. Doran, and M. Mac (eds.), pp. 454–456. U.S. National Biological Service, Washington, DC.

Fritts T.H., Scott N.J., Jr., and Savidge J.A. (1987) Activity of the arboreal Brown Tree Snake (*Boiga irregularis*) on Guam as determined by electrical power outages. *The Snake* 19:51–58.

Fritzell E.K. (1975) Effects of agricultural burning on nesting waterfowl. *Canadian Field Naturalist* 89:21–27.

Frost P.G.H., Siegfried W.R. and Cooper J. (1976) Conservation of the Jackass Penguin (*Spheniscus demersus* [L]). *Biological Conservation* 9:79–99.

Frumkin R. (1986) The status of breeding raptors in the Israeli deserts, 1980–1985. *Sandgrouse* 8:42–57.

Frumkin R. (1989) First Sparrowhawk (*Accipiter nisus*) nesting records in Israel. *Torgos* 8:48–49.

Frumkin R. and Adar M. (1989) First breeding records of Sparrowhawk in Israel. *Bulliten of the Ornithological Society of the Middle East* 23:20–22.

Frumkin R. and Man S. (1984) Raptor nesting in the desert areas of Israel, 1980–84. *Torgos* 4:11–60.

Fry D., Peard R., Speich S., and Toone C. (1987) Sex ratio skew and breeding patterns of gulls: Demographic and toxicological considerations. *Studies in Avian Biology* 10:26–43.

Fry D. and Toone M. (1981) DDT-induced feminization of gull embryos. *Science* 213:922–24.

Fry D.M., Fefer S.I., and Sileo L. (1987) Ingestion of plastic debris by Laysan Albatrosses and Wedge-tailed Shearwaters in the Hawaiian Islands. *Marine Pollution Bulletin* 18:339–343.

Fry M.E., Risser R.J., Stubbs H.A., and Leighton J.P. (1986) Species selection for habitat-evaluation procedures. In *Wildlife 2000: Modeling Habitat Relationships of Terrestrial Vertebrates.* J. Verner, M.L. Morrison, and C.J. Ralph (eds.), pp. 105–108. University of Wisconsin Press, Madison.

Fuller R.J., Gregory R.D., Gibbons D.W., Marchant J.H., Wilson J.D., Baillie S.R., and Carter N. (1995) Population declines and range contractions among lowland farmland birds in Britain. *Conservation Biology* 9:1425–1441.

Furness R.W. (1982) Competition between fisheries and seabirds. *Advances in Marine Biology* 20:225–307.

Furness R.W. (1993) Birds as monitors of pollutants. In *Birds As Monitors of Environmental Change.* R.W. Furness and J.J.D. Greenwood (eds.), pp. 86–143. Chapman and Hall, London.

Furness R.W. and Cooper J. (1982) Interactions between breeding seabird and pelagic fish populations in the southern Benguela regions. *Marine Ecology Progress Series* 8:243–250.

Furness R.W., Ensor K., and Hudson A.V. (1992) The use of fishery waste by gull populations around the British Isles. *Ardea* 80:105–113.

Furness R.W. and Monaghan P. (1987) *Seabird Ecology.* Chapman and Hall, New York.

Gaines D. (1988) *Birds of Yosemite and the East Slope.* Artemisia Press, Lee Vining, CA.

Galbraith H. (1988) The effects of territorial behaviour on lapwing populations. *Ornis Scandinavica* 19:134–138.

Galeotti P., Morimando F., and Violani C. (1991) Feeding ecology of the Tawny Owls (*Strix aluco*) in urban habitats (northern Italy). *Bollettino Di Zoologia* 58:143–150.

Galushin V.M. (1963) Consideration of raptors in particular conditions. *Okhota i Okhotnichie Khozyaistvo* 6:24–27.

Galushin V.M. (1982) Role of raptors in ecosystems. *Scientific Reviews, USSR Academy of Sciences* 11:158–238.

Galushin V.M. (1994) Long-term changes in birds of prey populations within European Russia and neighbouring countries. In *Bird Numbers 1992.* E.J.M. Hagemeijer and T.J. Verstrael (eds.), pp. 139–141. International Conference of Bird Census Committee, Voorburg/Meerlen, Netherlands.

Galushin V.M. (1996) Recent status and population trends of raptors in European Russia. *2nd International Conference on Raptors, Abstracts.* Urbino, Italy, p. 80.

Gandini P., Boersma P.D., Frere E., Gandini M., Holik T., and Lichtschein V. (1994) Magellanic penguins (*Spheniscus magellanicus*) affected by chronic petroleum pollution along coast of Chubut, Argentina. *Auk* 111:20–27.

Garner K.M. (1978) Management of blackbird and starling winter roost problems in Kentucky and Tennessee. In *Proceedings Eighth Vertebrate Pest Conference.* W.E. Howard and R.E. Marsh (eds.), pp. 54–59. University of California, Davis.

Garnett S. (1992a) *The Action Plan for Australian Birds.* Australian National Parks and Wildlife Service, Canberra.

Garnett S. (1992b) *Threatened and Extinct Birds of Australia.* Royal Australasian Ornithologists Union and Australian National Parks and Wildlife Service, Canberra.

Gates J.E. and Gysel L.W. (1978) Avian nest dispersion and fledging success in field-forest ecotones. *Ecology* 59:871–883.

Gauthreaux S.A. Jr. (1992) The use of weather radar to monitor long-term patterns of trans-Gulf migration in spring. In *Ecology and Conservation of Neotropical migrant land-*

birds. J.M. Hagan III and D.W. Johnston (eds.), pp. 96–199. Smithsonian Institution Press, Washington, DC.

Gavin T.A. and Bollinger E.K. (1988) Reproductive correlates of breeding site fidelity in Bobolinks (*Dolichonyx oryzivorus*). *Ecology* 69:96–103.

Gavrilo M.V., Tertitsky G.M., Pokrovskaya I.V., and Golovkin A.N. (1994) The archipelago ornithofauna. In *Environment and Ecosystems of the Frantz Josef Land (Archipelago and Shelf)*. G.G. Matishov (ed.), pp. 71–85. Russian Academy of Sciences, Apatity.

Gehlbach F.R. (1988) Population and environmental features that promote adaptation to urban ecosystems: The Case of Eastern Screech Owls (*Otus asio*) in Texas. In *Symposium on Avian Population Responses to Man-Made Environments*. L. Tomialojc and F.R. Gehlbach (eds.), pp. 1809–1813. Acta XIX Congressus Internationalis. Ornithological Ottawa, Canada.

Gehlbach F.R. (1994) *The Eastern Screech Owl: Life History, Ecology, and Behavior in the Suburbs and Countryside*. Texas A & M University Press, College Station.

Gehlbach F.R. (1996) Eastern Screech Owls in Suburbia: A model of raptor urbanization. In *Raptors in Human Landscapes: Adaptations to Built and Cultivated Environments*. D.M. Bird, D.E. Varland, and J.J. Negro (eds.), pp. 69–74. Academic Press, London.

Geissler P.H. and Noon B.R. (1981) Estimates of avian population trends from the North American Breeding Bird Survey. *Studies in Avian Biology* 6:42–51.

Geissler P.H. and Sauer J.R. (1990) Topics in route-regression analysis. In *Survey designs and statistical methods for the estimation of avian population trends*. J.R. Samora and S. Droge (eds.), USDOI-FWS Biological Report 90(1), Washington, DC.

Geisy J., Ludwig J., and Tillit D. (1994) Deformities in birds of the Great Lakes region: Assigning causality. *Environmental Science and Technology* 28:128–135.

George T.L., Fowler A.C., Knight R.L., and McEwen L.C. (1992) Impacts of a severe drought on grassland birds in western North Dakota. *Ecological Applications* 2:275–284.

Georges M., Dietz A.B., Mishra A., Nielsen D., Sargeant L.S., Sorensen A., Steele M.R., Zhao X.Y., Leipold H., Womack J.E., and Lathrop M. (1993) Microsatellite mapping of the gene causing weaver disease in cattle will allow the study of an associated quantitative trait locus. *Proceedings National Academy of Science* 90:1058–1062.

Gersh J. (1996) Subdivide and Conquer: concrete, condos, and the second conquest of the American West. *Amicus Journal* 18:14–20.

Ghosh, S. and Collins F.S. (1996) The geneticist's approach to complex disease. *Annual Review of Medicine* 47:333–353.

Giannecchini J. (1993) Ecotourism: New partners, new relationships. *Conservation Biology* 7:429–432.

Gibbons D.W., Reid J.B., and Chapman R.A. (1993) *The New Atlas of Breeding Birds in Britain and Ireland: 1988–1991*. Poyser, London.

Gibbons P. and Lindenmayer D.B. (1996) Issues associated with the retention of hollow-bearing trees within eucalypt forests managed for wood production. *Forest Ecology and Management* 83:245–279.

Gibbs J.P. (1991) Avian nest predation in tropical wet forest: An experimental study. *Oikos* 60:155–161.

Gibbs J.P. and Faaborg J. (1990) Estimating the viability of Ovenbird and Kentucky Warbler populations in forest fragments. *Conservation Biology* 4:193–196.

Gibbs W.W. (1995) Lost science in the third world. *Scientific American* 273:92–99.

Gilpin M. (1996) Metapopulations and wildlife conservation: Approaches to Modeling spatial structure. In *Metapopulations and Wildlife Conservation.* D.R. McCullough (ed.), pp. 11–27. Island Press, Washington, DC.

Gilpin M.E. and Soulé M.E. (1986) Minimum viable populations: Process of species extinction. In *Conservation Biology, the Science of Scarcity and Diversity.* M. E. Soulé (ed.), pp. 19–34. Sinauer Associates, Sunderland, MA

Gipps J.H.W. (1991) *Beyond Captive Breeding: Reintroducing Endangered Mammals to the Wild.* Clarendon Press, Oxford.

Glenn T.C. (1997) *Genetic Bottlenecks in Long-Lived Vertebrates: Mitochondrial and Microsatellite DNA Variation in American Alligators and Whooping Cranes.* Ph.D. dissertation, University of Maryland, College Park.

Gochfeld M. (1980) Mechanisms and adaptive value of reproductive synchrony in colonial seabirds. In *Behavior of Marine Animals.* J. Burger, B.L. Olla, and H.E. Winn (eds.), pp. 207–265. Plenum Press, New York.

Goldsmith F.B. (ed.) (1991) *Monitoring for Conservation and Ecology.* Chapman and Hall, New York.

Gotelli N.J. and Graves G.R. (1996) *Null Models in Ecology.* Smithsonian Institution Press, Washington, DC.

Götmark F. (1992) Naturalness as an evaluation criterion in nature conservation: A response to Anderson. *Conservation Biology* 6:455–458.

Goudswaard R. (1991) The search for the Campbell Island Flightless Teal *Anas aucklandica nesiotis. Wildfowl* 42:145–148.

Gould G.I., Jr. (1977) Distribution of the Spotted Owl in California. *Western Birds* 8:131–146.

Gradus Y. (1996) Changing approaches in urban and regional development in Beer Sheva and its surrounding towns. *Ecology and Environment* 3:255–260.

Graham F., Jr. (1990) *The Audubon Ark.* Alfred A. Knopf, New York.

Grant B.R. and Grant P.R. (1989) *Evolutionary Dynamics of a Natural Population.* University of Chicago Press, Chicago.

Graul W.D. and Miller G.C. (1984) Strengthening ecosystem management approaches. *Wildlife Society Bulletin* 12:282–289.

Graul W.D., Torres J., and Denney R. (1976) A species-ecosystem approach for nongame programs. *Wildlife Society Bulletin* 4:79–80.

Green A.J. (1992a) The status and conservation of the White-winged Wood Duck *Cairina scutulata. IWRB Special Publication* 17:1–115.

Green A.J. (1992b) Wildfowl at risk, 1992. *Wildfowl* 43:160–184.

Green A.J. (1996) Analysis of globally threatened Anatidae in relation to threats, distribution, migration patterns, and habitat use. *Conservation Biology* 10:1435–1445.

Green A.J., Black J.M., and Ellis-Joseph S. (1993) Conservation planing for globally threatened Anseriforms. In *Waterfowl and Wetland Conservation in the 1990s —A Global Perspective. IWRB Special Publication* 26:128–133.

Green A.J. and Ellis S. (1994) Wildfowl conservation: Implications of the Anseriform conservation assessment and management plan. *International Zoo Yearbook* 33:114–118.

Green R.E. (1988) Stone-curlew conservation. *Royal Society for the Protection of Birds Conservation Review* 2:30–33.

Green R.E. (1995) Diagnosing the causes of bird population declines. *Ibis* 137:S47–S45.

Green R.E. and Hirons G.J.M. (1991) The relevance of population studies to the conservation of threatened birds. In *Bird Population Studies—Relevance to Conservation and Management*. C.M. Perrins, J.-D. Lebreton, and G.J.M. Hirons (eds.), pp. 594–633. Oxford University Press, New York.

Greenberg R. (1996) Managed forest patches and the diversity of birds in Southern Mexico. In *Forest Patches in Tropical Landscapes*. J. Schelas and R. Greenberg (eds.), pp. 59–90. Island Press, Washington, DC.

Greenwood J.J.D., Baillie S.R., Crick H.P.Q., Marchant J.H., and Peach W.J. (1993) Integrated population monitoring: Detecting the effects of diverse changes. In *Birds As Monitors of Environmental Change*. R.W. Furness and J.J.D. Greenwood (eds.), pp. 267–342. Chapman and Hall, New York.

Greenwood J.J.D., Baillie S.R., Gregory R.D., Peach W.J., and Fuller R.J. (1995) Some new approaches to conservation monitoring of British breeding birds. *Ibis* 137:S16–S28.

Greenwood P.J. (1980) Mating systems, philopatry, and dispersal in birds and mammals. *Animal Behaviour* 28:1140–1162.

Greenwood P.J. and Harvey P.H. (1982) The natal and breeding dispersal of birds. *Annual Review of Ecology and Systematics* 13:1–21.

Greenwood R.J., Sargeant A.B., Johnson D.H., Cowardin L.M., and Shaffer T.L. (1987) Mallard nest success and recruitment in prairie Canada. *Transactions of the North American Wildlife Natural Resources Conference* 52:298–309.

Greenwood R.J., Sargeant A.B., Johnson D.H., Cowardin L.M., and Shaffer T.L. (1995) Factors associated with duck nest success in the prairie pothole region of Canada. *Wildlife Monographs* 128:1–57.

Gretton A. (1991) *The ecology and conservation of the Slender-billed Curlew* (Numenius tenuirostris). Technical Publication Number 9. International Council for Bird Preservation, Cambridge, England.

Griffith B., Scott J.M., Carpenter J.W., and Reed C. (1989) Translocation as a species conservation tool: Status and strategy. *Science* 245:477–480.

Grimmett R.F.A. and Jones T.A. (1989) *Important Bird Areas in Europe*. International Council for Bird Preservation, Cambridge, England.

Grossberg R.K., Levitan D.R., and Cameron B.B. (1996) Characterization of genetic structure and genealogies using RAPD-PCR markers: A random primer for the novice and nervous. In *Molecular Zoology: Advances, Strategies and Protocols*. J.D. Ferraris and S.R. Palumbi (eds.), pp. 67–100. Wiley-Liss, New York.

Grout D.J., Lusk M., and Fancy S.C. (1996) *Results of the 1995 Mariana Crow Survey on Rota*. Unpublished report to U.S. Fish and Wildlife Service, Honolulu, HI.

Grumbine R.E. (1994) What is ecosystem management? *Conservation Biology* 8:27–38.

Grumbine R.E. (1997) Reflections on "What is ecosystem management?" *Conservation Biology* 11:41–47.

Grzybowski J.A. (1982) Population structure in grassland bird communities during winter. *Condor* 84:137–152.

Gutiérrez R.J. (1994) Conservation planning: Lessons from the Spotted Owl. In *Sustainable Ecological Systems: Implementing an Ecological Approach to Land Management*. W.W. Covington and L.F. DeBano (eds.), pp. 51–58. RM-GTR-247. USDA Forest Service, Ft. Collins, CO.

Gutiérrez R.J. and Harrison S. (1996) Applying metapopulation theory to Spotted Owl management: A history and critique. In *Metapopulations and Wildlife Conservation*. D.R. McCullough (ed.), pp.167–186. Island Press, Washington, DC.

Gutzwiller K.J. (1995) Recreational disturbance and wildlife communities. In *Wildlife and Recreationists: Coexistence through Management and Research*. R.L. Knight and K.J. Gutzwiller (eds.), pp. 169–181. Island Press, Washington, DC.

Guyot I. (1988) Relationships between shag feeding areas and human fishing activities in Corsica (Mediterranean Sea). In *Proceedings 3rd International Conference of the Seabird Group*. M.L. Tasker (ed.), pp. 22–23. Seabird Group, Glasgow.

Hagan J.M. (1995) Environmentalism and the science of conservation biology. *Conservation Biology* 9:975–977.

Hagan J.M. and Johnston D.W. (eds.) (1992) *Ecology and Conservation of Neotropical Migratory Landbirds*. Smithsonian Institution Press, Washington, DC.

Hagan J.M., Vander Haegen W.M., and McKinley P.S. (1996) The early development of forest fragmentation effects on birds. *Conservation Biology* 10:188–202.

Haig S.M. and Avise J.C. (1996) Avian conservation genetics. In *Conservation Genetics, Case Histories from Nature*. J.C. Avise and J.L. Hamrick (eds.), pp. 160–189. Chapman and Hall, New York.

Haig S.M., Ballou J.D., and Derrickson S.R. (1990) Management options for preserving genetic diversity: Reintroduction of Guam rails to the wild. *Conservation Biology* 4:290–300.

Haig S.M., Belthoff J.R., and Allen D.H. (1993) Population viability analysis for a small population of Red-cockaded Woodpeckers and an evaluation of enhancement strategies. *Conservation Biology* 7:289–300.

Haig S.M., Gratto-Trevor C.L., Mullins T.D., and Colwell M.A. (1997) Population identification of western hemisphere shorebirds throughout the annual cycle. *Molecular Ecology* 6:413–427.

Haig S.M. and Oring L.W. (1988) Genetic differentiation of Piping Plovers across North America. *Auk* 105:260–267.

Haig S.M., Rhymer J.M., and Heckel D.G. (1994) Population differentiation in randomly amplified polymorphic DNA of Red-cockaded Woodpeckers *Picoides borealis*. *Molecular Ecology* 3:581–595.

Haila Y (1986) North European land birds in forest fragments: Evidence for area effects? In *Wildlife 2000: Modeling Habitat Relationships of Terrestrial Vertebrates*. J. Verner, M.L. Morrison, and C.J. Ralph (eds.), pp. 315–319. University of Wisconsin Press, Madison.

Haila Y. and Hanski I.K., (1993) Birds breeding on small British islands and extinction risks. *American Naturalist* 142:1025–1029.

Haila Y., Hanski I.K., and Raivio S. (1993) Turnover of breeding birds in small forest fragments: The "sampling" hypothesis corroborated. *Ecology* 74:714–725.

Hale P. and Lamb D. (eds.) (In press) *Conservation outside nature reserves*. Surrey Beatty, Sydney.

Hamel P.B. (1992) *Land manager's guide to the birds of the South*. The Nature Conservancy, Southeastern Region, Chapel Hill, NC.

Hamel P.B., Smith W.P., Twedt D.T., Woehr J.R., Morris E., Hamilton R.B., and Cooper R.J. (1996) *A land manager's guide to point counts of birds in the Southeast*. SO-GTR-120. USDA Forest Service, Asheville, NC.

Hamilton M. (1994) *Ex situ* conservation of wild plant species: Time to reassess the genetic assumptions and implications of seed banks. *Conservation Biology* 8:39–49.

Hanotte O., Burke T., Armour J.A.L., and Jeffreys A.J. (1991) Hypervariable minisatellite DNA sequences in the Indian Peafowl *Pavo cristatus*. *Genomics* 9:587–597.

Hanotte O., Zanon C., Pugh A., Greig C., Dixon A., and Burke T. (1994) Isolation and characterization of microsatellite loci in a passerine bird: The Reed Bunting *Emberiza schoeniclus*. *Molecular Ecology* 3:529–530.

Hansen A.J., Garman S.L., Marks B., and Urban D.L. (1993) An approach for managing vertebrate diversity across multiple-use landscapes. *Ecological Applications* 3:481–496.

Hansen A.J., McComb W.C., Raphael M.G., and Hunter M.L. (1995) Bird habitat relationships in natural and managed forests in the west Cascades of Oregon. *Ecological Applications* 5:555–569.

Hansen A.J. and Rotella J.R. (In press) Abiotic factors and biodiversity. In *Managing Forests for Biodiversity*. M.L. Hunter (ed.). Cambridge University Press, London.

Hanski I. (1982) Dynamics of regional distribution: The core and satellite species hypothesis. *Oikos* 38:210–221.

Hanski I. (1989) Metapopulation dynamics: Does it help to have more of the same? *Trends in Ecology and Evolution* 4:113–114.

Hanski I. (1994) A practical model of metapopulation dynamics. *Journal of Animal Ecology* 63:151–162.

Hanski I.K., Fenske T.J., and Niemi G.J. (1996) Lack of edge effect in nesting success of breeding birds in managed forest landscapes. *Auk* 113:578–585.

Hanski I. and Gilpin M.E. (1996) *Metapopulation Biology: Ecology, Genetics and Evolution*. Academic Press, San Diego.

Hanson H.C. (1965) *The Giant Canada Goose*. Southern Illinois University Press, Carbondale.

Hanson H.C. and Jones R.L. (1976) *The Biogeochemistry of Blue, Snow, and Ross' Geese*. Special Publication #1, Illinois Natural History Survey. Southern Illinois University Press, Carbondale.

Harper K.T. and Marble J.R. (1988) A role for nonvascular plants in management of arid and semiarid rangelands. In *Vegetation Science Applications for Rangeland Analysis and Management*. P.T. Tueller (ed.), pp. 135–169. Kluwer Academic Press, Boston.

Harper S. (1987) The persistence of ant-following birds in small Amazonian forest fragments. *Acta Amazonica* 19:249–263.

Harrington G.N. and Sanderson K.D. (1994) Recent contraction of wet sclerophyll forest in the wet tropics of Queensland due to invasion by rainforest. *Pacific Conservation Biology* 1:319–327.

Harris M.P. (1970) The biology of an endangered species, the Dark-rumped Petrel (*Pterodroma phaeopygia*) in the Galapagos Islands. *Condor* 72:76–84.

Harrison C.S., Naughton M.B., and Fefer S.I. (1984) The status and conservation of seabirds in the Hawaiian Archipelago and Johnston Atoll. In *Status and Conservation of the World's Seabirds*. J.P. Croxall, P.G.H. Evans, and R.W. Schreiber (eds.), pp. 513–526. ICBP International Council for Bird Preservation Technical Publication No. 2. Page Brothers, Norwich, England.

Harrison S. (1991) Local extinction in a metapopulation context: An empirical evaluation. *Biological Journal of the Linnean Society* 42:73–88.

Hart J. (1996) *Storm over Mono*. University of California Press, Los Angeles.

Hatch J.J. (1970) Predation and piracy by gulls at a ternery in Maine. *Auk* 87:244–254.

Hatch S.A. and Sanger G.A. (1992) Puffins as predators on juvenile pollack and other forage fish in the Gulf of Alaska. *Marine Ecology Program Series* 80:1–14.

Haufler J.B. (1994) An ecological framework for forest planning for forest health. *Journal of Sustainable Forestry* 2:307–316.

Haufler J.B. (1995) Forest industry partnerships for ecosystem management. *Transactions of the North American Wildlife Natural Resources Conference* 60:422–432.

Haufler J.B. and Irwin L.L. (1994) An ecological basis for forest planning for biodiversity and resource use. *Proceedings of the International Union of Game Biologists* 21:73–81.

Haufler J.B., Mehl C.A., and Roloff G.J. (1996) Using a coarse-filter approach with species assessment for ecosystem management. *Wildlife Society Bulletin* 24:200–208.

Hayes F.N. and Dumbell G.S. (1989) Progress in Brown Teal *Anas a. chlorotis* conservation. *Wildfowl* 40:137–140.

Haynes A.H. (1987) Human exploitation of seabirds in Jamaica. *Biological Conservation* 41:99–124.

Hayward G.D. and Escaño R.E. (1989) Goshawk nest-site characteristics in western Montana and northern Idaho. *Condor* 91:476–479.

Hayward G.D. (In press) Forest management and conservation of Boreal Owls in North America. *Journal of Raptor Research*.

Hayward G.D. and Hayward P.H. (1993) Boreal Owl. In *The Birds of North America*. A. Poole and F. Gill (eds.), pp. 1–20. Academy of Natural Science Philadelphia, PA; and American Ornithologists' Union, Washington, DC.

Hayward G.D., Hayward P.H., and Garton E.O. (1993) Ecology of Boreal Owls in the northern Rocky Mountains, USA. *Wildlife Monographs* 124:1–59.

Hayward G.D., Hayward P.H., Garton E.O., and Escaño R.E. (1987) Revised breeding distribution of the Boreal Owl in the northern Rocky Mountains. *Condor* 89:431–432.

Hayward G.D., Steinhorst R.K., and Hayward P.H. (1992) Monitoring Boreal Owl populations with nest boxes: Sample size and cost. *Journal of Wildlife Management* 56:776–784.

Hayward G.D. and Verner J. (1994) *Flammulated, Boreal, and Great Gray Owls in the United States: A technical conservation assessment.* RM-GTR-253. USDA Forest Service, Fort Collins, CO.

Hazevoet C.J. (1996) Conservation and species list: Taxonomic neglect promotes the extinction of endemic birds, as exemplified by taxa from eastern Atlantic Islands. *Bird Conservation International* 6:181–196.

Hedrick P.W., Lacy R.C., Allendorf F.W., and Soulé M.E. (1996) Directions in conservation biology: Comments on Caughley. *Conservation Biology* 10:1312–1320.

Hedrick P.W. and Miller P.S. (1992) Conservation genetics: Techniques and fundamentals. *Ecological Applications* 2:30–46.

Heinz G.H. (1979) Methylmercury: Reproductive and behavioral effects on three generations of Mallard Ducks. *Journal of Wildlife Management* 43:394–401.

Hejl S.J., Hutto R.L., Preston C.R., and Finch D.M. (1995) The effects of silvicultural treatments in the Rocky Mountains. In *Ecology and Management of Neotropical Migratory Birds: A Synthesis and Review of Critical Issues.* T.E. Martin and D.M. Finch (eds.), pp. 220–244. Oxford University Press, New York.

Hejl S.J. and Paige L.C. (1994) A preliminary assessment of birds in continuous and frag- mented forests of western redcedar/western hemlock in northern Idaho. In *Interior Cedar–Hemlock–White Pine Forests: Ecology and Management*. D.M. Baumgartner, J.E. Lotan, and J.R. Tonn (eds.), pp. 189–197. Washington State University Cooperative Extension, Pullman, WA.

Hendricks P. and Reinking D.L. (1994) Investigator visitation and predation rates on bird nests in burned and unburned tallgrass prairie in Oklahoma: An experimental study. *Southwestern Naturalist* 39:196–200.

Hengeveld R. (1990) *Dynamic biogeography*. Cambridge University Press, Cambridge.

Henny C.J., Galushin V.M., Dudin P.I., Khrustov A.V., Moseikin V.N., Sarychev V.S., and Turchin V.G. (1996) Organochlorine pesticides and polychlorinated biphenyls in hawk, falcon, eagle, and owl eggs from the Lipetsk, Voronezh, and Saratov regions, Russia, 1992–1993. *2nd International Conference on Raptors, Abstracts*. Urbino, Italy, p. 100.

Henny C.J., Galushin V.M., and Kuznetsov A.V. (1995) Organochlorine pesticides and PCBs in Osprey eggs from the upper Volga river, Russia, 1992. *Holarctic Birds of Prey International Conference*. Badajoz, Spain, p. 108.

Henny C.J., Ganusevich S.A., Ward P.F., and Schwartz T.R. (1994) Organochlorine pesticides, chlorinated dioxins and furans, and PCBs in Peregrine Falcon *Falco pere- grinus* eggs from the Kola peninsula, Russia. In *Raptor Conservation Today*. B.U. Mey- burg and R.D. Chancellor (eds.), pp. 739–749. World Working Group on Birds of Prey and Owls, Pica Press, London.

Henny C.J., Overton W.S., and Wight H.M. (1970) Determining parameters for popu- lations by using structural models. *Journal of Wildlife Management* 34:1133–1141.

Henson L. (1993) Backlash to "lashing back"—Feedback from a regional forester. *Inner Voice* September:12–13.

Heppell S.S., Walters J.R., Crowder, L.B. (1994) Evaluating management alternatives for Red-cockaded Woodpeckers: A management approach. *Journal of Wildlife Manage- ment* 58:479–487.

Heredia B., Rose L., and Painter M. (1996) *Globally Threatened Birds in Europe: Action Plans*. BirdLife International, Strasbourg, Austria.

Herkert J.R. (1994a) Breeding bird communities of midwestern prairie fragments: The effects of prescribed burning and habitat-area. *Natural Areas Journal* 14:128–135.

Herkert J.R. (1994b) The effects of habitat fragmentation on midwestern grassland bird communities. *Ecological Applications* 4:461–471.

Herkert J.R. (1994c) Status and habitat selection of the Henslow's Sparrow in Illinois. *Wilson Bulletin* 106:35–45.

Herkert J.R. (1997) Bobolink *Dolichonyx oryzivorus* population decline in agricultural landscapes in the midwestern USA. *Biological Conservation* 80:107–112.

Herkert J.R., Sample D.W., and Warner R.E. (1996) Management of grassland land- scapes for the conservation of migratory birds. In *Managing Midwest Landscapes for the Conservation of Neotropical Migratory Birds*. F.R. Thompson (ed.), pp. 89–116. NC- GTR-187. USDA Forest Service, St. Paul, MN.

Hernández-Baños B.E., Peterson A.T., Navarro-Sigüenza A.G., and Escalante-Pliego P.B. (1995) Bird faunas of the humid montane forests of Mesoamerica: Biogeographic patterns and priorities for conservation. *Bird Conservation International* 5:251–278.

Heywood V.H. (1995) *Global Biodiversity Assessment*. Cambridge University Press, Cambridge.

Higgins K.F. (1986) A comparison of burn season effects on nesting birds in North Dakota mixed-grass prairie. *Prairie Naturalist* 18:219–228.

Hilborn R. and Walters C.J. (1981) Pitfalls of environmental baseline and process studies. *Environmental Impact Assessment Review* 2:265–278.

Hill A.V.S. (1996) Genetic susceptibility to malaria and other infectious diseases: From the MHC to the whole genome. S75–S84. Symposium of the British Society of Parasitology 33: 575–584.

Hillis D.M., Moritz C., and Mable B.K. (1996) *Molecular Systematics*, 2nd edition. Sinauer Associates, Sunderland, MA.

Hinsley S.A., Pakeman R., Bellamy P.E., and Newton I. (1996) Influences of habitat fragmentation on bird species distributions and regional population sizes. *Proceedings of the Royal Society of London B* 263:307–313.

Hites R.A. (1990) Environmental behaviour of chlorinated dioxins and furans. *Accounts of Chemical Research* 23:194–201.

Hitt S. (1992) The triumph of politics over science: Goshawk management in the Southwest. *Inner Voice* 4:1, 11, 12.

Hobbs R.J. (1992) The role of corridors in conservation: Solution or bandwagon. *Trends in Ecology and Evolution* 77:389–392.

Hobson K.A. and Clark R.G. (1992) Assessing avian diets using stable isotopes II: Factors influencing diet-tissue fractionation. *Condor* 94:181–188.

Hobson K.A. and Wassenaar L.I. (1997) Linking breeding and wintering grounds of Neotropical migrant songbirds using stable hydrogen isotopic analysis of feathers. *Oecologia* 109:142–148.

Hodges M.F., Jr. and Krementz D.G. (1996) Neotropical migratory breeding bird communities in riparian forests of different widths along the Altamaha River, Georgia. *Wilson Bulletin* 108:496–506.

Hoelzel A.R. (1992) *Molecular Genetic Analysis of Populations*. IRL Press, Oxford University Press, Oxford, England.

Hollander A.D., Davis F.W., and Stoms D.M. (1994) Hierarchical representations of species distributions using maps, images and sighting data. In *Mapping the Diversity of Nature*. R.I. Miller (ed.), pp. 71–88. Chapman and Hall, London.

Holmes R.T., Marra P.P., and Sherry T.W. (1996) Habitat-specific demography of breeding Black-throated Blue Warblers (*D. caerulescens*): Implications for population dynamics. *Journal of Animal Ecology* 65:183–195.

Holmes R.T. and Sherry T.W. (1988) Assessing population trends of New Hampshire forest birds: Local vs. regional patterns. *Auk* 105:756–768.

Holmes R.T. and Sherry T.W. (1992) Site fidelity of migratory warblers in temperate breeding and Neotropical wintering areas: Implications for population dynamics, habitat selection and conservation. In *Ecology and Conservation of Neotropical Migrant Landbirds*. J.M. Hagan III and D.W. Johnston (eds.), pp. 563–575. Smithsonian Institution Press, Washington, DC.

Holthausen R.S., Raphael M.G., McKelvey K.S., Forsman E.D., Starkey E.E., and Seaman D.E. (1995) *The contribution of federal and nonfederal habitat to persistence of the Northern Spotted Owl on the Olympic Peninsula, Washington: Report of the reanalysis team.* PNW-GTR- 352. USDA Forest Service, Portland, OR.

Hoover J.P. and Brittingham M.C. (1993) Regional variation in cowbird parasitism of Wood Thrushes. *Wilson Bulletin* 105:228–238.

Hoover R.L. and Wills D.L. (eds.) (1984) *Managing Forested Lands for Wildlife*. Colorado Division of Wildlife, Denver.

Horiuchi S. (1992) Stagnation in the decline of the world population growth rate during the 1980s. *Science* 257:761–765.

Hoshide H.M., Price A.J., and Katahira L. (1990) A progress report on Nene (Branta sandvicensis) in Hawaii Volcanoes National Park from 1974–89. *Wildfowl* 41:152–155.

Howe M.A., Geissler P.H., and Harrington B.A. (1989) Population trends of North American shorebirds based on the International Shorebird Survey. *Biological Conservation* 49:185–199.

Howe R.W., Davis G.J., and Mosca V. (1991) The demographic influence of "sink" populations. *Biological Conservation* 57:239–255.

Howell T.R. (1985) Eugene Eisenmann and the study of Neotropical birds. *Ornithological Monographs* 36:1–4.

Howell T.R. and Cade T.J. (1954) The birds of Guadalupe Island in 1953. *Condor* 56:283–291.

Hsu W. and Melville D.S. (1994) Seabirds of China and adjacent seas: Status and conservation. In *Seabirds on Islands: Threats, Case Studies and Action Plans*. D.N. Nettleship, J. Burger, and M. Gochfeld (eds.), pp. 10–218. BirdLife International, Cambridge, England.

Hudson A.V. and Furness R.W. (1988) The behaviour of seabirds foraging at fishing boats around Shetland. *Ibis* 131:225–237.

Hudson R.R. (1990) Gene genealogies and the coalescent process. *Oxford Surveys of Evolutionary Biology* 7:1–44.

Huenneke L.F. (1994) Involving academic scientists in conservation research: Perspectives of a plant ecologist. *Ecological Applications* 5:209–214.

Hughes A.L. (1991) MHC polymorphism and the design of captive breeding programs. *Conservation Biology* 5:249–251.

Humphries C.J., Margules C.R., Pressey R.L., and Vane-Wright R.I. (1996) *Priority Areas Analysis: Systematic Methods of Conserving Biodiversity*. Oxford University Press, Oxford, England.

Hunt G.L., Jr. (1972) Influence of food distribution and human disturbance on the reproductive success of Herring Gulls. *Ecology* 53:1051–1061.

Hunter M.L. (1996) *Fundamentals in Conservation Biology*. Blackwell Science, Cambridge, MA.

Hunter W.C. (1990) *Handbook for Nongame Bird Management and Monitoring in the Southeast Region*. U.S. Fish and Wildlife Service, Atlanta, GA.

Hurley J.F., Salwasser H., and Shimamoto K. (1982) Fish and wildlife capability models and special habitat criteria. *Cal-Nevada Wildlife Transactions* 1982:40–48.

Huston M.A. (1994) *Biological diversity—the Coexistence of Species on Changing Landscapes*. Cambridge University Press, Cambridge, England.

Hutchins M., Paul E., and Bahner B. (1996) *AZA Micronesian Kingfisher: Species Survival Plan Action Plan*. American Association of Zoological Parks and Aquariums, Bethesda, MD.

Hutto R.L. (1988) Is tropical deforestation responsible for the reported declines in Neotropical migrant populations? *American Birds* 42:375–379.

Hutto R.L. (1995) Composition of bird communities following stand-replacement fires in northern Rocky Mountain (U.S.A.) conifer forests. *Conservation Biology* 9:1041–1058.

Hutto R.L. (in press) *Distribution and habitat relationships of landbirds in the USFS Northern Region*. RM-GTR-xxx. USDA Forest Service, Ft. Collins, CO.

Hutto R.L., Pleschet S.M., and Hendricks P. (1986) A fixed-radius point count method for nonbreeding and breeding season use. *Auk* 103:593–602

Hutto R.L., Hejl S.J., Preston C.R., and Finch D.M. (1993) Effects of silvicultural treatments on forest birds in the Rocky Mountains: Implications and management recommendations. In *Status and Management of Neotropical Migratory Birds*. D.M. Finch and P.W. Stangel (eds.), pp. 386–391. RM-GTR-229. USDA Forest Service, Fort Collins, CO.

Hutto R.L. and Hoffland J. (1996) USDA Forest Service Northern Region Landbird Monitoring Project: Field Methods. in-house report, Wildlife Biology, USDA Northern Region, Missoula, MT.

Hutto R.L., Reel S., and Landres P.B. (1987) A critical evaluation of the species approach to biological conservation. *Endangered Species Update* 4:1–4.

ICBP (1992) Putting Biodiversity on the Map: Priority Areas for Global Conservation. International Council for Bird Preservation. Cambridge, U.K.

Igl L.D. and Johnson D.H. (1996) Dramatic increase of Le Conte's Sparrow in Conservation Reserve Program fields in the northern Great Plains. *Prairie Naturalist* 27:89–94.

Ilani G. (1981) Summary of Lappet-face Vulture count during 1955–1981. *Torgos* 1:35–40.

Ilyichev V.D. and Flint V.E. (eds.) (1982) Birds of the USSR. Nauka, Moscow.

Innes J., Warburton B., Williams D., Speed H., and Bradfield P. (1995) Large-scale poisoning of ship rats (*Rattus rattus*) in indigenous forest of the North Island, New Zealand. *New Zealand Journal of Ecology* 19:5–17.

Irwin L.L. (1994) Improving wildlife habitat models for assessing forest ecosystem health. *Journal of Sustainable Forestry* 2:293–306.

Irwin L.L. and Wigley T.B. (1993) Toward an experimental basis for protecting forest wildlife. *Ecological Applications* 3:213–217.

IUCN, UNEP, WWF (1991) *Caring for the Earth: A strategy for sustainable living*. Gland, Switzerland.

IUCN/SSC Reintroduction Specialist Group (1997) IUCN/SSC Guidelines for Reintroduction. http://www.rbgkew.org.uk/conservation/RSGguidelines.html.

Iverson C., Hayward G.D., Titus K., Degayner G., Lowell R., Crocker-Bedford C., Schempf P., and Lindell J. (1996) *Conservation Assessment for Northern Goshawk in Southeast Alaska*. PNW-GTR-387. USDA Forest Service, Portland, OR.

Jaffe M. (1994) *And No Birds Sing*. Simon and Schuster, New York.

James F.C. (1983) Environmental component of morphological differentiation in birds. *Science* 221:184–186.

James F.C., Hess C.A., and Kufrin D. (1997) Species-centered environmental analysis: Indirect effects of fire history on Red-cockaded Woodpeckers. *Ecological Applications* 7:118–129.

James F.C. and McCulloch C.E. (1995) The strength of inferences about causes of trends in populations. In *Ecology and Management of Neotropical Migratory Birds*. T.E. Martin and D.M. Finch (eds.), pp. 40–51. Oxford University Press, New York.

James F.C., McCulloch C.E., and Wiedenfeld D.A. (1996) New approaches to the analysis of population trends in land birds. *Ecology* 77:13–27.

James F.C., Wiedenfeld D.A., and McCulloch C.E. (1992) Trends in breeding populations of warblers: Declines in the southern highlands and increases in the lowlands. In *Ecology and Conservation of Neotropical Migrant Landbirds.* J.M. Hagen III and D.W. Johnston (eds.), pp. 43–56. Smithsonian Institution Press, Washington, DC.

James P.C. (1987) Ornithology in Central and South America. *Auk* 104:348–349.

Janzen D.H. (1986) The eternal external threat. In *Conservation Biology: The Science of Scarcity and Diversity.* M.E. Soulé (ed.), pp. 286–303. Sinauer Associates, Sunderland, MA.

Jarvi S.I., Gee G.F., Miller M.M., and Briles W.E. (1995) A complex alloantigen system in Florida Sandhill Cranes *Grus canadensis pratensis*: Evidence for the major histocompatibility (B) system. *Journal of Heredity* 86:348–353.

Jasanoff S. (1990) *The Fifth Branch: Science Advisers As Policymakers.* Harvard University Press, Cambridge, MA.

Jeffreys A.J., Royle N.J., Wilson V., and Wong Z. (1988) Spontaneous mutation rates to new length alleles at tandem-repetitive hypervariable loci in human DNA. *Nature* 332:278–281.

Jehl J.R. (1972) On the cold trail of an extinct petrel. *Pacific Discovery* 25:24–29.

Jehl J.R., Jr. and Johnson N.K., (eds.) (1993) A century of change in western North America. *Studies in Avian Biology* No. 15.

Jenkinson M.A. (1993) The American Ornithologists' Union's support of Latin American ornithology. *Auk* 110:659–660.

Jennings M.D. (1993) *Natural terrestrial cover classification: Assumptions and definitions.* Gap analysis Tech. Bull. 2. U.S. Fish and Wildlife Service, Idaho Cooperative Fish and Wildlife Research Unit, Moscow, ID.

Jenny M. (1990) Nahrungsökologie der Feldlerche *Alauda arvensis* in einer intensiv genutzten Agrarlandschaft des schweizeruschen Mittelandes. *Ornithologische Beobachter* 87:31–53.

Jimenez J.A., Hughes K.A., Alaks G., Graham L., and Lacy R.C. (1994). An experimental study of inbreeding depression in a natural habitat. *Science* 265: 271–273

Johns A.D. (1991) Response of Amazonian rain forest birds to habitat modification. *Journal of Tropical Ecology* 7:417–437.

Johnson A.R., Milne B.T., and Wiens J.A. (1992) Diffusion in fractal landscapes: Simulations and experimental studies of tenebrionid beetle movements. *Ecology* 73:1968–1983.

Johnson D.H. (1996) Management of northern prairies and wetlands for the conservation of Neotropical migratory birds. In *Managing Midwest Landscapes for the Conservation of Neotropical Migratory Birds.* F.R. Thompson (ed.), pp. 53–67. NC-GTR-187. USDA Forest Service, St. Paul, MN.

Johnson D.H. (1997) Effects of fire on bird populations in mixed-grass prairie. In *Ecology and Conservation of Great Plains Vertebrates.* F.L. Knopf and F.B. Samson (eds.), pp. 181–206. Springer-Verlag, New York.

Johnson D.H., Nichols J.D. and Schwartz M.D. (1992) Population dynamics of breeding waterfowl. In *Ecology and Management of Breeding Waterfowl.* B.D.J. Batt, A.D. Afton, M.G. Anderson, C.D. Ankney, D.H. Johnson, J.A. Kadlec, and G.L. Krapu (eds.), pp. 446–485.

Johnson H.F. and Stattersfield A.J. (1990) A global review of island endemic birds. *Ibis* 132:167–180.

Johnson K.N., Crim S., and Barber K. (1993) *Sustainable harvest levels and short-term timber sales for options considered in the report of the Forest Ecosystem Management Assessment Team: Methods, results, and interpretations.* USDA Forest Service, Portland, OR.

Johnson R.G. and Temple S.A. (1986) Assessing habitat quality for birds nesting in fragmented tallgrass prairies. In *Wildlife 2000.* J.A. Verner, M.I. Morrison, and C.J. Ralph (eds.), pp. 245–250. University of Wisconsin Press, Madison.

Johnson R.G. and Temple S.A. (1990) Nest predation and brood parasitism of tallgrass prairie birds. *Journal of Wildlife Management* 54:106–111.

Jones C.G., Heck W., Leis R.E., Mungroo Y., Slade G., and Cade T. (1995) The restoration of the Mauritius Kestrel *Falco punctatus* population. *Ibis* 137:S173–S180.

Jones E. 1977. Ecology of the ferel cat, *Felis catus* (L.) (Carnivora: Felidae) on Macquarie Island. *Australian Wildlife Research* 4:249–62.

Jones L.L. and DeGange A.R. (1988) Interactions between seabirds and fisheries in the North Pacific Ocean. In *Seabirds and Other Marine Vertebrates: Competition, Predation and Other Interactions.* J. Burger (ed.), pp. 269–291. Columbia University Press, New York.

Jones R.D., Jr. and Byrd G.V. (1979) Interrelations between seabirds and introduced animals. In *Conservation of Marine Birds of Northern North America.* J.C. Bartonek and D.N. Nettleship (eds.), pp. 221–226. Wildlife Research Report 11, U.S. Fish and Wildlife Service, Washington, DC.

Joseph L. and Moritz C. (1994) Mitochondrial DNA phylogeography of birds in eastern Australian rainforests: First fragments. *Australian Journal of Zoology* 42:385–403.

Kadlec J.A. and Drury W.H. (1968) Structures of the New England Herring Gull population. *Ecology* 49:644–676.

Kaiser A. and Berthold P. (1994) Population trends of resting migratory passerines at the Mettnau Peninsula, Germany: First annual report on the MRI-Program (1992 and 1993). *Bird Populations* 2:127–135.

Kaiser G.W. & Forbes L.S. (1992) Climatic and Oceanographic influences on island use in four burrow-nesting alcids. *Ornis Scandinavica* 23: 1–6

Kaiser J. (1997) When a habitat is not a home. *Science* 276:1636–1638.

Kalliola R., Salo J., Puhakka M., and Rajasitta M. (1991) New site formation and colonizing vegetation in primary succession on the western Amazon floodplains. *Journal of Ecology* 79:877–901.

Kamada M. and Nakagoshi N. (1993) Pine Forest Structure in a Human-Dominated Landscape System in Korea. *Ecological Research* 8:35–46.

Kampp K., Nettleship D.N., and Evans P.G.H. (1994) Thick-billed Murres of Greenland: Status and prospects. In *Seabirds on Islands: Threats, Case Studies and Action Plans.* D.N. Nettleship, J. Burger, and M. Gochfeld (eds.), pp. 133–154. BirdLife International, Cambridge, England.

Kantrud H.A. and Kologiski R.L. (1982) *Effects of soils and grazing on breeding birds of uncultivated upland grasslands of the northern great plains.* Wildlife Research Report 15. USDOI-FWS, Washington, DC.

Kanyamibwa S., Bairlein F., and Schierer A. (1993) Comparison of survival rates between populations of the White Stork *Ciconia ciconia* in central Europe. *Ornis Scandinavica* 24:297–302.

Kareiva P. (1990) Population dynamics in spatially complex environments: Theory and data. *Philisophical Transactions of the Society of London (B)* 330:175–190.

Kareiva P. and Anderson M. (1989) Spatial aspects of species interactions: The wedding of models and experiments. In *Community Ecology*. A. Hastings (ed.), pp. 35–50. Springer-Verlag, New York.

Karr J.R. (1982) Avian extinctions on Barro Colorado Island, Panama: A reassessment. *American Naturalist* 119:220–239.

Karr J.R. (1987) Biological monitoring and environmental assessment: A conceptual framework. *Environmental Management* 11:249–256.

Karr J.R. (1990) The avifauna of Barro Colorado Island and the Pipeline Road, Panama. In *Four Neotropical Forests*. A.H. Gentry (ed.), pp. 183–198. Yale University Press, New Haven, CT.

Karr J.R. (1991) Biological integrity: A long-neglected aspect of water resource management. *Ecological Applications* 1:66–84.

Karr J.R. and Brawn J.D. (1990) Food resources of understory birds in Central Panama: Quantification and effects on avian populations. *Studies in Avian Biology* 13:58–64.

Karr J.R. and Freemark K.E. (1983) Habitat selection and environmental gradients: Dynamics in the "stable" tropics. *Ecology* 64:1481–1494.

Karr J.R., Nichols J.D., Klimkiewicz M.K., and Brawn J.D. (1990) Survival rates of tropical and temperate forest birds: Will the dogma survive? *American Naturalist* 136:277–291.

Karr J.R., Robinson S.K., Blake J.G., and Bierregaard R.O. (1990) Birds of four Neotropical forests. In *Four Neotropical Rainforests*. A.H. Gentry (ed.), pp. 237–272. Yale University Press, New Haven, CT.

Kattan G.H., Alvarez-Lopez H., and Giraldo M. (1994) Forest fragmentation and bird extinctions: San Antonio eighty years later. *Conservation Biology* 8:138–146.

Kear J. (1975) Breeding endangered wildfowl as an aid to their survival. In *Breeding Endangered Species in Captivity*. R.D. Martin (ed.), pp. 49–60. Academic Press, London.

Kear J. (1977) Captive propagation of waterfowl. In *Endangered Birds: Management Techniques for Preserving Threatened Species*. S.A. Temple (ed.), pp. 243–250. University of Wisconsin Press, Madison.

Kear J. (1979) Wildfowl at risk. *Wildfowl* 30:159–161.

Kear J. and Berger A. (1980) *The Hawaiian Goose: An Experiment in Conservation*. Poyser, Calton, England.

Kear J. and Williams G. (1978) Wildfowl at risk. *Wildfowl* 29:5–21.

Keast A. (1995) Habitat and species loss: The birds of Sydney fifty years ago and now. *Australian Zoology* 30:3–25.

Keith J.O. (1966) Insecticide contaminations in wetland habitats and their effects on fish-eating birds. *Journal of Applied Ecology* 3:71–85.

Kennedy E.T., Costa R., and Smathers W.M., Jr. (1996) New directions for Red-cockaded Woodpecker habitat conservation. *Journal of Forestry* 94:22–26.

Kennedy P.L., Ward J.M., Rinker A., and Gessaman J.A. (1994) Post-fledging areas in Northern Goshawk home ranges. *Studies in Avian Biology* 16:75–82.

Kenward R.E., Pfeffer R.H., Bragin E.A., and Levin A.S. (1996) Preliminary modelling of Saker populations: Implications for sustainability. *2nd International Conference on Raptors, Abstracts.* Urbino, Italy, p. 32.

Kenward R.E., Pfeffer R.H., Bragin E.A., Levin A.S., and Kovshar A.F. (1995) The status of Saker Falcons in Kazakhstan. In *Proceedings of the Specialist Workshop, Abu Dhabi, United Arab Emirates.* Middle East Falcon Research Group, pp. 131–142.

Kepler C.B. (1978) Captive propagation of Whooping Cranes: A behavioral approach. In *Endangered Birds: Management Techniques for Preserving Threatened Species.* S.A. Temple (ed.), pp. 231–241. University of Wisconsin Press, Madison.

Kershaw M., Mace G.M., and Williams P.H. (1995) Threatened status, rarity and diversity as alternative selection measures for protected areas: A test using Afrotropical antelopes. *Conservation Biology* 9:324–334.

Keuhler C., Kuhn M., McIlraith B., and Campbell G. (1994) Artificial incubation and hand-rearing of Alala (*Corvus hawaiiensis*) eggs removed from the wild. *Zoo Biology* 13:257–266.

Kincaid A.L. and Stoskopf M.K. (1987) Passerine dietary iron overload syndrome. *Zoo Biology* 6:79–88.

King M. (1995) Sustainable use—A hunter's concept. In *Conservation through Sustainable Use of Wildlife.* G.C. Grigg, P.T. Hale, and D. Lunney (eds.), pp. 282–287. Surrey Beatty, Sydney.

King, W.B. (ed.) 1974. *Pelagic studies of seabirds in the central and eastern Pacific Oceans.* Smithsonian Contributions in Zoology 158. Washington, DC.

King W.B. (1978–79) *Red Data Book, 2. Aves,* 2nd edition. International Union for the Conservation of Nature, Morges, Switzerland.

Kirsch L.M. and Kruse A.D. (1972) Prairie fires and wildlife. *Proceedings of the Tall Timbers Fire Ecology Conference* 12:289–303.

Kleiman D.G., Stanley Price M.R., and Beck B.B. (1994) Criteria for reintroductions. In *Creative Conservation: The Interface between Captive and Wild Populations.* P.J.S. Olney, G.M. Mace, and A.T.C. Feistner (eds.), pp. 287–303. Chapman and Hall, London.

Knick S.T. and Rotenberry J.T. (1995) Landscape characteristics of fragmented shrubsteppe habitats and breeding passerine birds. *Conservation Biology* 9:1059–1071.

Knick S.T. and Rotenberry J.T. (In press) Landscape characteristics of disturbed shrubsteppe habitats in southwestern Idaho. *Landscape Ecology.*

Knick S.T., Rotenberry J.T., and Zarriello T.J. (1997) Supervised classification of Landsat thematic mapper imagery in a semiarid rangeland by nonparametric discriminant analysis. *Photogrammetric Engineering and Remote Sensing* 63:79–86.

Knight R.L. (1990) Ecological principles applicable to the management of urban ecosystems. In *Perspectives in Urban Ecology.* E.A. Webb and S.Q. Foster (eds.), pp. 24–34. Denver Museum of Natural History and Thorne Ecological Institute, Denver.

Knight R.L., Andersen D.E., Bechard M.J., and Marr N.V. (1989) Geographic variation in nest-defense behaviour of the Red-tailed Hawk, *Buteo jamaicensis. Ibis* 131:22–26.

Knight R.L. and Fitzner R.E. (1985) Human disturbance and nest site placement in Black-billed Magpies. *Journal of Field Ornithology* 56:153–157.

Knight R.L., Grout D.J., and Temple S.A. (1987) Nest defense behavior of the American Crow in urban and rural areas. *Condor* 89:175–177.

Knight R.L. and Gutzweiller K.J. (eds.) (1995) *Wildlife and Recreationists—Coexistence through Management and Research.* Island Press, Washington, DC.

Knight R.L. and Knight S.K. (1984) Responses of wintering Bald Eagles to boating activity. *Journal of Wildlife Management* 48:999–1004.

Knight R.L. and Temple S.A. (1986) Why does intensity of avian nest defense increase during the nesting cycle? *Auk* 103:318–327.

Knight R.L., Wallace G.N., and Riebsame W.E. (1995) Ranching the view: Subdivisions versus agriculture. *Conservation Biology* 9:459–461.

Knopf F.L. (1994) Avian assemblages on altered grasslands. *Studies in Aviation Biology* 15:247–257.

Knopf F.L. (1996a) Perspectives on grazing nongame bird habitats. In *Rangeland and Wildlife*, P.R. Krausman (ed.), pp. 51–58. Society for Range Management, Denver, CO.

Knopf F.L. (1996b) Prairie legacies—birds. In *Prairie Conservation: Preserving North America's Most Endangered Ecosystem*. F.B. Samson and F.L. Knopf (eds.), pp. 135–148. Island Press, Washington, DC.

Knopf F.L. and Rupert J.R. (1995) Habitat and habits of Mountain Plovers in California. *Condor* 97:743–751.

Knopf F.L. and Rupert J.R. (1996) Reproduction and movements of Mountain Plovers breeding in Colorado. *Wilson Bulletin* 108:28–35.

Knopf F.L. and Rupert J.R. (In press) Use of cropfields by breeding Mountain Plovers. *Studies in Aviation Biology*.

Knopf F.L. and Samson F.B. (1997) Conservation of grassland vertebrates. In *Ecology and Conservation of Great Plains Vertebrates*. F.L. Knopf and F.B. Samson (eds.), pp. 273–289. Springer-Verlag, New York.

Kochert M.N. and Pellant M. (1986) Multiple use in the Snake River Birds of Prey Area. *Rangelands* 8:217–220.

Költringer C., Dodeikat G., and Curio E. (1993) Anti-predator behavior of Black Grouse *Tetrao tetrix* chicks as influenced by hen-rearing vs. hand-rearing. In *Proceedings of the 6th International Symposium on Grouse*. D. Jenkins (ed.), pp. 81–83. World Pheasant Association, Cambridge, England.

Kooiker G. von (1984) Brutökologische Untersuchungen an einer Population des Kiebitzes *Vanellus vanellus*. *Vogelwelt* 105:121–137.

Korschgen CE (1979) *Coastal Waterbird Colonies: Maine*. Biological Service Program FWS/OBS-79/09. U.S. Fish and Wildlife Service, Washington, DC.

Krajewski C. (1994) Phylogenetic measures of biodiversity: A comparison and critique. *Biological Conservation* 69:33–39.

Krasnov Y.V. (1995) *Seabirds. Environment and Ecosystems of Novaya Zemlya: Archipelago and Shelf.* Kola Science Centre, Russian Academy of Sciences, Apatity, pp. 138–147.

Krebs J.R. (1970) Regulation of numbers in the Great Tit (Aves: Passeriformes). *Journal of Zoology* 162:317–333.

Kremen C. (1992) Assessing the indicator properties of species assemblages for natural areas monitoring. *Ecological Applications* 2:203–217.

Kremen C., Merenlender A.M., and Murphy D.D. (1994) Ecological monitoring: A vital need for integrated conservation and development programs in the tropics. *Conservation Biology* 8:388–397.

Kress S.W. (1983) The use of decoys, sound recordings, and gull control for re-establishing a tern colony in Maine. *Colonial Waterbirds* 6:185–196.

Kress S.W. and Nettleship D.N. (1988) Re-establishment of Atlantic Puffins (*Fratercula arctica*) at a former breeding site in the Gulf of Maine. *Journal of Field Ornithology* 59(2):161–170.

Kress S.W., Weinstein E.H., and Nisbet I.C.T. (eds.) (1983) The status of tern populations in northeastern United States and adjacent Canada. *Colonial Waterbirds* 6:84–106.

Krever V., Dinerstein E., Olson D., and Williams L. (eds.) (1994) *Conserving Russia's Biological Diversity: An Analytical Framework and Initial Investment Portfolio.* Corporate Press. Landover, MD.

Krivenko V.G. (1991) *Waterfowls and their conservation.* Agropromizdat, Moscow.

Kubiak T.J., Harris H.J., Smith L.M., Schwarz T.R., Stalling D.L., Tirck J.A., Sileo L., Docherty D.E., and Erdman T.C. (1989) Microcontaminants and reproductive impairment of the Forster's Tern on Green Bay, Lake Michigan—1983. *Archives of Environmental Contamination and Toxicology* 18:706–727.

Kuchler A.W. (1964) *Manual to accompany the map, potential natural vegetation of the conterminous United States.* Special Publication No. 36. American Geographic Society, NY.

Lacy R.C. (1989) Analysis of founder representation in pedigrees: Founder equivalents and founder genome equivalents. *Zoo Biology* 8:111–123.

Lacy R.C., Alaks G., and Walsh A. (1996) Hierarchical analysis of inbreeding depression in *Peromyscus polionotus. Evolution* 50:2187–2200.

Lacy R.C., Petric A.M., and Warneke M. (1993) Inbreeding and outbreeding depression in captive populations of wild animal species. In *The Natural History of Inbreeding and Outbreeding.* N.W. Thornhill (ed.), pp. 352–374. University of Chicago Press, Chicago.

LaHaye W.S., Gutiérrez R.J., and Ackakaya H.R. (1994) Spotted Owl metapopulation dynamics in southern California. *Journal of Animal Ecology* 63:775–785.

Laist D.W. (1987) Overview of the biological effects of lost and discarded plastic debris in the marine environment. *Marine Pollution Bulletin* 18:319–326.

Lancia R.A., Braun C.E., Collopy M.W., Dueser R.D., Kie J.G., Martinka C.J., Nichols J.D., Nudds T.D., Porath W.R., and Tilghman N.G. (1996) ARM! For the future: Adaptive resource management in the wildlife profession. *Wildlife Society Bulletin* 24:436–442.

Lande R. (1988) Genetics and demography in biological conservation. *Science* 241:1455–1459.

Lande R. (1994) Risk of population extinction from fixation of new deleterious mutations. *Evolution* 48:1460–1469.

Lande R. and Barrowclough G.F. (1987) Effective population size, genetic variation, and their use in population management. In *Viable Populations for Conservation.* M.E. Soulé (ed.), pp. 87–124. Cambridge University Press, Cambridge.

Landres P.B. (1983) Use of the guild concept in environmental impact assessment. *Environmental Management* 7:393–398.

Landres P.B., Verner J., and Thomas J.W. (1988) Ecological uses of vertebrate indicator species: A critique. *Conservation Biology* 2:316–328.

Landsberg H.E. (1981) *The Urban Climate.* Academic Press, London.

Lapushkin V.A., Demidova M.I., Shepel A.I., and Fisher S.V. (1995) Nesting of the Pallid Harrier in the Perm region. *Data on bird distribution in Ural, Preduralye and Western Siberia.* Ural Branch. Russian Academy of Sciences, Ekaterinburg.

LaRoe E.T. (1993) Implementation of an ecosystem approach to endangered species conservation. *Endangered Species Update* 10:3–6.

Latta S.C., Wunderle J.M., Jr., Terranova E., and Pagan M. (1995) An experimental study of nest predation in a subtropical wet forest following hurricane disturbance. *Wilson Bulletin* 107:590–602.

Lautenschlager R.A. and Bowyer R.T. (1985) Wildlife management by referendum: When professionals fail to communicate. *Wildlife Society Bulletin* 13:564–570.

Lawrence N. and Murphy D. (1992) New perspectives or old priorities? *Conservation Biology* 6:465–468.

Lawton J.H. (1993) Range, population abundance and conservation. *Trends in Ecology and Evolution* 8:409–413.

Laymon S.A. and Reid J.A. (1986) Effects of grid-cell size on tests of a Spotted Owl HSI model. In *Wildlife 2000: Modeling Habitat Relationships of Terrestrial Vertebrates*. J. Verner, M.L. Morrison, and C.J. Ralph (eds.), pp. 93–97. University of Wisconsin Press, Madison.

Lebedeva Y.A. (1996) In the Russian Bird Conservation Union. *Towards a Sustainable Russia* 3:27–28.

Lebreton J.D. and Clobert J. (1991) Bird population dynamics, management, and conservation: The role of mathematical modelling. In *Bird Population Studies: Relevance to Conservation and Management*. C.M. Perrins, J.D. Lebreton, and G.J.M. Hirons (eds.), pp. 105–125. Oxford University Press, New York.

Lebreton J.D., Burnham K.P., Clobert, J. and Anderson D.R. (1992) Modeling survival and testing biological hypotheses using marked animals: A unified approach with case studies. *Ecological Monographs* 62:67–118.

Leck C.F. (1979) Avian extinctions in an isolated tropical wet-forest preserve, Ecuador. *Auk* 96:343–352.

Lee D.P., Honda K., Tatsukawa R., and Won P.O. (1989) Distribution and residue levels of mercury, cadmium, and lead in Korean birds. *Bulletin of Environmental Contamination and Toxicology* 43:550–555.

Lee F.B., Schroeder C.H., Kuck T.L., Schoonover L.J., Johnson M.A., Nelson H.K., and Beauduy C.A. (1984) *Rearing and Restoring Giant Canada Geese in the Dakotas*. North Dakota Game and Fish Department, Bismarck.

Leedy D.L. (1987) Role of research. *Wildlife Society Bulletin* 15:115–126.

Leeton P., Christidis L., and Westerman M. (1993) Feathers from museum bird skins— A good source of DNA for phylogenetic studies. *Condor* 95:465–466.

Legendre L. and Legendre P. (1983) *Numerical Ecology*. Elsevier, New York.

Lehmkuhl J.F. and Raphael M.G. (1993) Habitat pattern around Northern Spotted Owl locations on the Olympic Peninsula, Washington. *Journal of Wildlife Management* 57:302–315.

Lélé S. and Norgaard R.B. (1996) Sustainability and the scientist's burden. *Conservation Biology* 10:354–365.

Lenssen N.K. (1992) *Empowering Development: The New Energy Equation*. World Watch Institute, Washington, DC.

Leopold A. (1933) *Game Management*. Charles Scribner's Sons, New York.

Leshem Y. (1983) The status of the breeding raptors of Israel. *Torgos* 6:15–16.

Leshem Y. and Yom-Tov Y. (1996a) The magnitude and timing of migration by soaring raptors, pelicans and storks over Israel. *Ibis* 138:188–203.

Leshem Y. and Yom-Tov Y. (1996b) The use of thermals by soaring raptors. *Ibis* 138:667–674.

Lesica P. and Allendorf F.W. (1995) When are peripheral populations valuable for conservation? *Conservation Biology* 9:753–760.

Levey D.J. and Stiles F.G. (1992) Evolutionary precursors of long-distance migration: Resource availability and movement patterns in Neotropical landbird. *American Naturalist* 140:447–476.

Levins R. (1969) Some demographic and genetic consequences of environmental heterogeneity for biological control. *Bulletin of the Entomological Society of America* 15:237–240.

Levins R. (1970) Extinction. In *Some Mathematical Questions in Biology. Lectures in Mathematics in the Life Sciences, Volume 2.* M. Gerstenhaber (ed.), pp. 77–107. American Mathematics Society, Providence, RI.

Lewis J.C. (1990) Captive propagation in the recovery of Whooping Cranes. *Endgangered Species Update* 8:46–48.

Lindenmayer D.B., Burgman M.A., Akçakaya H.R., Lacy R.C., and Possingham H.P. (1995) A review of the generic computer programs ALEX, RAMAS/space and VORTEX for modelling the viability of wildlife populations. *Biological Modelling* 82:161–174.

Lindenmayer D.B. and Lacy R.C. (1995) Metapopulation viability of Leadbeater's possum, *Gymnobelideus leadleateri*, in fragmented old-growth forests. *Ecological Applications* 5:164–182.

Lindsey G.D., Fancy S.G., Reynolds M.H., Pratt T.K., Wison K.A., Banko P.C., and Jacobi J.D. (1995) Population structure and survival of the Palila. *Condor* 97:528–535.

Line L. (1996) Lethal migration. *Audubon* September–October:50–56, 93–94.

Liu J., Dunning J.B., and Pulliam H.R. (1995) Potential effects of a forest management plan on Bachman's Sparrows (*Aimophila aestivalis*): Linking a spatially explicit model with GIS. *Conservation Biology* 9:62–75.

Livingstone D.R., Donkin P., and Walker C.H. (1992) Pollutants in marine ecosystems: An overview. In *Persistent Pollutants in Marine Ecosystems.* C.H. Walker and D.R. Livingstone (eds.), pp. 235–263. Pergamon Press, New York.

Lockley R.M. (1953) *Puffins.* Devin-Adair, New York.

Loiselle B.A. and Blake J.G. (1991) Temporal variation in birds and fruits along an elevational gradient in Costa Rica. *Ecology* 72:180–193.

Loiselle B.A. and Blake J.G. (1992) Population variation in a tropical bird community. *BioScience* 42:838–845.

Loiselle B.A. and Hoppes W.G. (1983) Nest predation in insular and mainland lowland rainforest in Panama. *Condor* 85:93–95.

Lombard A.T. (1995) The problems with multi-species conservation: Do hotspots, ideal reserves coincide? *South African Journal of Zoology* 30:145–163.

Lomolino M.V. and Channell R. (1995) Splendid isolation: Patterns of geographic range collapse in endangered mammals. *Journal of Mammology* 76:335–347.

Long J.L. (1981) *Introduced Birds of the World.* Universe, New York.

Lovejoy T.E., Rankin J.M., Bierregaard R.O., Brown K.S., Jr, Emmons L.H., and VandeVoort M.E. (1984) Ecosystem decay of Amazon forest remnants. In *Extinction.* M.E. Nitecki (ed.), pp. 295–325. University of Chicago Press, Chicago.

Loyn R.H. (1980) Bird populations in successional forests of Mountain Ash *Eucalyptus regnans* in central Victoria. *Emu* 85:213–230.

Loyn R.H. (1985) Birds in fragmented forests in Gippsland, Victoria. In *Birds of Eucalypt Forests and Woodlands: Ecology, Conservation, Management*. A. Keast, H.F. Recher, H. Ford, and D. Saunders (eds.), pp. 323–331. Surrey Beatty, Sydney.

Loyn R.H., Runnalls R.G., Forward G.Y., and Tyers J. (1983) Territorial Bell Miners and other birds affecting populations of insect prey. *Science* 221:1411–1413.

Lubchenco J., Allison G.W., Bohnsack J., Lubomudrov L., Munro G.R., Palumbi S.R., and Parrish J.K. (In prep) Marine protected areas: New recipes for salty preserves.

Ludwig D., Hilborn R. and Walters C. (1993) Uncertainty, resource exploitation, and conservation: Lessons from history. *Science* 260:17–36.

Lymn N. and Temple S.A. (1991) Land-use changes in the Gulf Coast Region: Links to declines in midwestern Loggerhead Shrike populations. *Passenger Pigeon* 53:315–325.

Lynch C.B. (1977) Inbreeding effects upon animals derived from a wild population of *Mus musculus*. *Evolution* 31:526–537.

Lynch J.F. (ed.) (1980) Bird populations—a litmus test of the environment. Proceedings of the Mid-Atlantic Natural History Symposium. *Atlantic Naturalist* 33:1–48.

Lynch J.F. and Whigham D.F. (1984) Effects of forest fragmentation on breeding bird communities in Maryland, U.S.A. *Biological Conservation* 28:287–324.

Lynch M. (1991) Analysis of population genetic structure by DNA fingerprinting. In *DNA Fingerprinting: Approaches and Applications*. T. Burke, G. Dolf, A.J. Jeffreys, and R. Wolff (eds.), pp. 113–126. Birkhäuser Verlag, Basel, Switzerland.

Lynch M. (1996) A quantitative-genetic perspective on conservation issues. In *Conservation Genetics, Case Histories from Nature*. J.C. Avise and J.L. Hamrick (eds.), pp. 471–501. Chapman and Hall, New York.

Lynch M., Conery J., and Burger R. (1995) Mutation accumulation and the extinction of small populations. *American Naturalist* 146:489–518.

Lynch M. and Milligan B.G. (1994) Analysis of population genetic structure with RAPD markers. *Molecular Ecology* 3:91–99.

MacArthur R.H. and Wilson E.O. (1967) *The Theory of Island Biogeography*. Princeton University Press, Princeton, NJ.

MacCluer J.W., VandeBerg J.L., Read B., and Ryder O.A. (1986) Pedigree analysis by computer simulation. *Zoo Biology* 5:147–160.

Machtans C.S., Villard M.-A., and Hannon S.J. (1996) Use of riparian buffer strips as movement corridors by forest birds. *Conservation Biology* 10:1366–1379.

Mack R.N. (1981) Invasion of *Bromus tectorum* L. into western North America: An ecological chronicle. *Agro-Ecosystems* 7:145–165.

Mack R.N. and Thompson J.N. (1982) Evolution in steppe with few large, hoofed mammals. *American Naturalist* 119:757–773

MacLaren P.A. (1992) Bridging the gap between biologists and managers—Resource management audits. In *Science and the Management of Protected Areas*. J.H. Willison, S. Bondrup-Nielsen, S. Drysdale, C. Herman, T.B. Munro, and T.L. Pollock (eds.), pp. 181–184. Reprinted in *Developments in Landscape Management and Urban Planning* 7.

MaGuire L.A., Wilhere G.F., and Dong Q. (1995) Population viability analysis for Red-cockaded Woodpeckers in the Georgia piedmont. *Journal of Wildlife Management* 59:533–542.

Male B. (1995) Recovery action for threatened species—an Australian perspective. *Ibis* 137:S204–S208.

Maloney R.F. and McLean I.G. (1995) Historical and experimental learned predator recognition in free-living New Zealand Robins. *Animal Behavior* 50:1193–1201.

Mangel M., Talbot L.M., Meffe G.K., Agardy M.T., Alverson D.L., Barlow J., Botkin D.B., Budowski G., Clark T., Cooke J., Crozier R.H., Dayton P.K., Elder D.L., Fowler C.W., Funtowicz S., Giske J., Hofman R.J., Holt S.J., Kellert S.R., Kimball L.A., Ludwig D., Magnusson K., Malayang B.S., Mann C., Norse E.A., Northridge S.P., Perrin W.F., Perrings C., Peterman R.M., Rabb G.B., Regier H.A., Reynolds J.E., Sherman K., Sissenwine M.P., Smith T.D., Starfield A., Taylor R.J., Tillman M.F., Toft C., Twiss J.R.J., Wilen J., and Young T.P. (1996) Principles for the conservation of wild living resources. *Ecological Applications* 6:338–362.

Manly B.F.J., McDonald L.L., and Thomas D.L. (1993) *Resource Selection by Animals*. Chapman and Hall, London.

Mann C.M. (1995) Filling in Florida's gaps: Species protection done right? *Science* 269:318–320.

Mannan R.W., Morrison M.L., and Meslow E.C. (1984) Comment: The use of guilds in forest bird management. *Wildlife Society Bulletin* 12:426–430.

Manuwal D.A. and Ettl G. (1990) *Nuisance Waterfowl at Public Waterfront Parks in the Seattle Metropolitan Area*. College of Forest Resources, University of Washington, Seattle.

Marchant J.H., Hudson R., Carter S.P., and Whittington P. (1990) *Population Trends in British Breeding Birds*. British Trust for Ornithology, Tring, England.

Mares M.A. (1986) Conservation in South America: Problems, consequences, and solutions. *Science* 233:734–739.

Marini M.A., Robinson S.K., and Heske E.J. (1995) Effects of edges on nest predation and nest predators. *Biological Conservation* 74:203–213.

Marquiss M. and Newton I. (1982) The Goshawk in Britain. *British Birds* 75:243–260.

Marshall A. and Black J.M. (1992) The effect of rearing experience on subsequent behaviour traits in captive-reared Hawaiian geese: Implications for the re-introduction programme. *Bird Conservation International* 2:131–147.

Marshall J.T., Jr. (1949) The endemic avifauna of Saipan, Tinian, Guam and Palau. *Condor* 51:200–221.

Martin B.P. (1990) *The Glorious Grouse: A Natural and Unnatural History*. David and Charles, London.

Martin K., Stacey P.B., and Braun C.E. (1998) Demographic rescue and maintenance of population stability in grouse—Beyond metapopulations. White-tailed Ptarmigan populations following site and regional level failure. *Wildlife Biology* 3:295–296.

Martin T.E. (1992) Breeding productivity considerations: What are the appropriate habitat features for management? In *Ecology and Conservation of Neotropical Migrant Landbirds* J.M. Magan III and D.W. Johnston (eds.), pp. 455–473. Smithsonian Intitution Press, Washington, DC.

Martin T.E. (1993) Nest predation among vegetation layers and habitat types: Revising the dogmas. *American Naturalist* 141:897–913.

Martin T.E. (1995) Avian life history evolution in relation to nest sites, nest predation, and food. *Ecological Monographs* 65:101–127.

Martin T.E. (1996) Life history evolution in tropical and south temperate birds: What do we really know? *Journal of Avian Biology* 27:263–272.

Martin T.E., Donovan T.M., Thompson III F.R., George T.L., Vickery P., and Gauthreaux S.A. (Unpublished report) *Priority research needs for conservation of neotropical migrants.* Partners in Flight National Research Working Subcommittee, Missoula, MT.

Martin T.E. and Finch D.M. (eds.) (1995) *Ecology and Management of Neotropical Migratory Birds.* Oxford University Press, New York.

Martin T.E. and Geupel G.R. (1993) Nest-monitoring plots: Methods for locating nests and monitoring success. *Journal of Field Ornithology* 64:507–519.

Martin T.E. and Karr J.R. (1986) Temporal dynamics of Neotropical birds with special reference to frugivores in second-growth woods. *Wilson Bulletin* 98:38–60.

Martin T.E., Paine C., Conway C.J., Hochachka W., Allen P., and Jenkins W. (1997) *BBIRD (Breeding biology research and monitoring database) field protocol.* Available from T.E. Martin. Montana Cooperative Wildlife Research Unit, University of Montana, Missoula, MT 59812.

Marzluff J.M. (1988) Do Pinyon Jays alter nest placement based on prior experience? *Animal Behaviour* 36:1–10.

Marzluff J.M. (1997) Effects of urbanization and recreation on songbirds. In *Songbird Ecology in Southwestern Ponderosa Pine Forests: A Literature Review.* W.M. Block and D.M. Finch (eds.), pp. 89–102. RM-GTR-292. USDA Forest Service, Fort Collins, CO.

Marzluff J.M. and Balda R.P. (1989) Causes and consequences of female-biased dispersal in a flock-living birds, the Pinyon Jay. *Ecology* 70:316–328.

Marzluff J.M. and Balda R.P. (1992) *The Pinyon Jay.* T. and A. D. Poyser, London.

Marzluff J.M., Boone R.B., and Cox G.W. (1994) Historical changes in populations and perceptions of native pest bird species in the West. *Studies in Avian Biology* 15:202–220.

Marzluff J.M., McFadzen S.M. (1996) Do Standardized brood counts accurately measure productivity? *Wilson Bulletin* 108:151–153.

Marzluff J.M., Valutis L.L., and Witmore K.D. (1995) *Captive Propagation and Reintroduction of Social Birds.* Sustainable Ecosystems Institute, Meridian, ID.

Masatomi H. (1991) Wintering population of the Tancho *Grus japonensis* in Hokkaido 1989–1990 and 1990–1991. *Journal of Environmental Science Laboratory, Senshu University* 2:171–179.

Mason D.J. (1996) Responses of understory birds to selective logging, enrichment strips, and vine cutting. *Biotropica* 28:296–309

Matter H. (1982) Einfluss intensiver Fieldbewirtschaftung auf der Bruterfolg des Kiebitzes *Vanellus vanellus* in Mittel europa. *Ornithologishe Beobachter* 79:1–24.

Matthysen E., Adrianensen F., and Dhondt A.A. (1995) Dispersal distances of Nuthatches, *Sitta europaea,* in a highly fragmented forest landscape. *Oikos* 72:375–381.

Maurer B.A. (1994) *Geographical Population Analysis.* Blackwell Scientific, Oxford.

Maurer B.A. and Villard M.-A. (1994) Population density: Geographic variation in abundance of North American birds. *Research and Exploration* 10:306–317.

Maurer B.A. and Villard M.-A. (1996) Continental scale ecology and Neotropical migratory birds: How to detect declines amid the noise. *Ecology* 77:1–2.

May R.M. (1994) The effects of spatial scale on ecological questions and answers. In *Large-Scale Ecology and Conservation Biology.* P.J. Edwards, R.M. May, and N.R. Webb (eds.), pp. 1–17. Blackwell Science, Oxford.

May R.M., Beddington J.R., Clark C.W., Holt S.J., and Laws R.M. (1979) Management of multispecies fisheries. *Science* 250:267–275.

May R.M., Lawton J.H., and Stork N.E. (1995) Assessing extinction rates. In *Extinction Rates*. J.H. Lawton and R.M. May (eds.), pp. 1–24. Oxford University Press, Oxford.

Mayfield H.F. (1993) Kirkland's Warblers benefit from large forest tracts. *Wilson Bulletin* 105:351–353.

Mazzeo R. (1953) Homing of the Manx Shearwater. *Auk* 70:200–201.

McCleery R.H. and Perrins C.M. (1985) Territory size, reproductive success and population dynamics in the Great Tit, *Parus major*. In *Behavioural Ecology: Ecological Consequences of Adaptive Behaviour*. R.M. Sibly and R.H. Smith (eds.), pp. 353–373. Blackwell Scientific Publications, Oxford.

McDonald D.B. and Caswell H. (1993) Matrix methods for avian demography. In *Current Ornithology*. D. Power (ed.), pp. 139–185. Plenum Press, New York.

McFadzen M.E. and Marzluff J.M. (1996) Mortality of Prairie Falcons during the fledgling-dependence period. *Condor* 98:791–800.

McFarland D.C. (1988) The composition, microhabitat use and response to fire of the avifauna of subtropical heathlands in Coloola National Park, Queensland. *Emu* 88:249–257.

McGarigal K. and McComb W.C. (1995) Relationships between landscape structure and breeding birds in the Oregon Coast Range. *Ecological Monographs* 65:235–260.

McIntyre S., Barrett G.W., Kitching R.L., and Recher H.F. (1992) Species triage—seeing beyond wounded rhinos. *Conservation Biology* 6:604–606.

McIvor D.E. and Conover M.R. (1994) Perceptions of farmers and non-farmers toward management of problem wildlife. *Wildlife Society Bulletin* 22:212–219.

McKelvey K., Noon B.R., and Lamberson R.H. (1993) Conservation planning for species occupying fragmented landscapes: The case of the Northern Spotted Owl. In *Biotic Interactions and Global Change*. P.M. Kareiva, J.G. Kingsolver, and R.B. Huey (eds.), pp. 424–450. Sinauer, Boston, MA.

McKnight D.E. and Knoder D.E. (1979) Resource development along coasts and on the ocean floor: Potential conflicts with marine bird conservation. In *Conservation of Marine Birds of Northern North America*. J.C. Bartonek and D.N. Nettleship (eds.), pp. 183–194. Wildlife Research Report 11. U.S. Fish and Wildlife Service, Washington, DC.

McLean I.G., Lundie-Jenkins G., and Jarman P.J. (1994) Training captive Rufous Hare-wallabies to recognize predators. In *Reintroduction Biology of Australian and New Zealand Fauna*. M. Serena (ed.), pp. 177–182. Surrey Beatty and Sons, Chipping Norton, New South Wales, Australia.

McNab W.H. and Avers P.E. (comp.) (1994) *Ecological subregions of the United States: Section descriptions*. USDA Forest Service, Administrative Publication WO-WSA-5. Washington, DC.

McNeeley J.A. (1992) The sinking ark: Pollution and the world-wide loss of biodiversity. *Biodiversity Conservation* 1:2–18.

McNeeley J.A. (1994) Protected areas for the 21st century: Working to provide benefits to society. *Biodiversity Conservation* 3:390–405.

McOrist S. (1989) Some diseases of free-living Australian birds. In *Disease and Threatened Birds*. J.E. Cooper (ed.), pp. 63–68. International Council for Bird Preservation, Cambridge, England.

McShea W.J., McDonald V, Morton E.S., Meier R., and Rappole J.H. (1995) Long-term trends in habitat selection by Kentucky Warblers. *Auk* 112:375–381.

Medin D.E. (1990) Birds of a Shadscale *Atriplex confertifolia* habitat in east central Nevada, USA. *Great Basin Naturalist* 50:295–298.

Meffe G.K. (1992) Techno-arrogance and halfway technologies: Salmon hatcheries on the Pacific Coast of North America. *Conservation Biology* 6:350–354.

Meffe G.K. and Carroll R.C (1994) *Principles of Conservation Biology*. Sinauer Associates, Sunderland, MA..

Melvin E.F. and Conquest L. (1996) *Reduction of seabird bycatch in salmon drift gillnet fisheries: 1995 sockeye/pink salmon fishery final report*. Project number A/FP-2(a). WSG AS 96–01. Washington Sea Grant Program. Olympia, WA.

Mena P.A. and Suárez L. (eds.) (1993) *La Investigación para la Conservación de la Diversidad Biológica en el Ecuador*. EcoCiencia, Quito, Ecuador.

Mendelssohn H. (1972) The impact of pesticides on bird life in Israel. International Council for Bird Preservation *ICBP Bulletin* 11:75–104.

Mendelssohn H. and Leshem Y (1983) The status and conservation of vultures in Israel. In *Vulture Biology and Management*. S.R. Wilbur and J.A. Jackson (eds.), pp. 86–98. University Califorina Press, Berkely, CA.

Merriam G. (1984) Connectivity: A fundamental characteristic of landscape pattern. In *Proceedings of the First International Seminar on Methodology in Landscape Ecological Research and Planning*. J. Brandt and P. Agger (eds.), pp. 5–15. International Association for Landscape Ecology, Roskilde, Denmark.

Merriam G. (1988) Landscape dynamics in farmland. *Trends in Ecology and Evolution* 3:16–20.

Merriam G. (1991) Corridors and connectivity: Animal populations in heterogeneous environments. In *Nature Conservation 2: The Role of Corridors*. D.A. Saunders and R.J. Hobbs (eds), pp. 133–142. Surrey Beatty, Chipping-Norton, New South Wales.

Meyer de Schauensee R. (1966) *The Species of Birds of South America and Their Distribution*. Livingston Publishing, Narbeth, PA.

Meyer de Schauensee R. and Phelps W.H., Jr. (1978) *A Guide to the Birds of Venezuela*. Princeton University Press, Princeton, NJ.

Milberg P. and Tyrberg T. (1993) Naive birds and noble savages—A review of man-caused extinctions of island birds. *Ecography* 16:229–250.

Miller B. and Mullette K.J. (1985) Rehabilitation of an endangered Australian bird: The Lord Howe Island Woodhen *Tricholimnas sylvestris* (Schlater). *Biological Conservation* 34:55–95.

Miller C. (1995) *100 Birds of Belize*. American Bird Conservancy, Gallon Jug, Belize.

Miller G.H., Magee J.W., and Jull A.J.T. (1997) Low-latitude glacial cooling in the Southern Hemisphere from amino-acid racemization in Emu eggshells. *Nature* 385:241–244.

Miller M.L. (1993) The rise of coastal and marine tourism. *Ocean Coastal Management* 20:181–199.

Miller P.S. and Hedrick P.W. (1991) MHC polymorphism and the design of captive breeding programs: Simple solutions are not the answer. *Conservation Biology* 5:556–558.

Mills G.S., Dunning J.B.J., and Bates J.M. (1989) Effects of urbanization on breeding bird community structure in southwestern desert habitats. *Condor* 91:416–428.

Mills L.S. and Smouse P.E. (1994) Demographic consequences of inbreeding in small remnant populations. *American Naturalist* 144:412–431.

Mills L.S., Hayes S.G., Baldwin C., Wisdom M.J., Citta J., Mattson D.J., and Murphy K. (1996) Factors leading to different viability predictions for a Grizzly Bear data set. *Conservation Biology* 3:863–873.

Mills L.S., Soulé M.E., and Doak D.F. (1993) The keystone-species concept in ecology and conservation. *BioScience* 43:219–224.

Milne B.T., Johnson A.R., Keitt T.H., Hatfield C.A., David J., and Hraber P.T. (1996) Detection of critical densities associated with piñon-juniper woodland ecotones. *Ecology* 77:805–821.

Mineau P., Fox G.A., Norstrom R.J., Weseloh D.V., Hallett D.J., and Ellenton J.A. (1984) Using the Herring Gull to monitor levels and effects of organochloride contamination in the Canadian Great Lakes. In *Toxic Contaminants in the Great Lakes*. J.O. Nriagu and M.S. Simmons (eds.), pp. 425–452. Wiley, New York.

Mittermeier R.A. (1988) Primate diversity and the tropical forest: Case studies from Brazil and Madagascar and the importance of megadiversity countries. In *Biodiversity*. E.O. Wilson (ed.), pp. 145–154. National Academy Press, Washington, DC.

Mitton, J.B. (1994) Molecular approaches to population biology. *Annual Review of Ecology and Systematics* 25: 45–69.

Mizutani H., Fukuda M., and Kabaya Y. (1990) Carbon isotope ratio of feathers reveals feeding behavior of cormorants. *Auk* 107:400–437.

Moir W.H., Geils B., Benoit M.A., and Scurlock D. (1997) Ecology of southwestern ponderosa pine forests. In *Songbird Ecology in Southwestern Ponderosa Pine Forests: A Literature Review*. W.M. Block and D.M. Finch (eds.), pp. 3–27. RM-GTR-292. USDA Forest Service, Fort Collins, CO.

Monaghan P. and Coulson J.C. (1977) Status of large gulls nesting on buildings. *Bird Study* 24:89–104.

Monsen S.B. and Kitchen S.G. (eds.) (1994) *Proceedings—Ecology and Management of Annual Rangelands*. INT-GTR-313. USDA Forest Service, Boise, ID.

Montana Bird Distribution Committee (1996) *P.D. Skaar's Montana Bird Distribution*, 5th edition. Special Publication No. 3. Montana Natural Heritage Program, Helena, MT.

Montevecchi W.A. (1991) Incidence and types of plastic in gannets' nests in the northwest Atlantic. *Canadian Journal of Zoology* 69:295–297.

Montevecchi W.A. (1993) Birds as indicators of change in marine prey stocks. In *Birds As Monitors of Environmental Change*. R.W. Furness and J.J.D. Greenwood (eds.), pp. 17–266. Chapman and Hall, London.

Montevecchi W.A. and Myers R.A. (1992) Monitoring fluctuations in pelagic fish availability with seabirds. *Canadian Atlantic Fisheries Science Advisory Committee Research Documents* 92/94:1–22.

Montevecchi W.A. and Tuck L.M. (1987) *Newfoundland Birds: Exploitation, Study, Conservation*, Publ. 21. Nuttall Ornithology Club, Cambridge, England

Montrey H. (1991) Forest Service research in the southwest: Reflections and projections. In *A Southwestern Mosaic: Proceedings of the Southwestern Region New Perspectives University Colloquium*. D.C. Hayes, J.S. Bumstead, and M.T. Richards (eds.), pp. 34–36. RM-GTR- 216. USDA Forest Service, Ft. Collins, CO.

Moore W.S. (1995) Inferring phylogenies from mtDNA variation: Mitochondrial-gene trees versus nucelar-gene trees. *Evolution* 49:718–726.

Moorehouse R.J. and Powlesland R.G. (1991) Aspects of the ecology of the Kakapo *Strigops habroptilus* liberated on Little Barrier Island (Huaturu), New Zealand. *Biological Conservation* 56:349–365.

Moors P.J. and Atkinson I.A.E. (1984) Predation on seabirds by introduced animals, and factors affecting its severity. In *Status and Conservation of the World's Seabirds*. J.P. Croxall, P.G.H. Evans, and R.W. Schreiber (eds.) pp. 667–690. International Council for Bird Preservation Technical Publication No. 2. Page Brothers, Norwich, England.

Morales G., Novo I., Bigio D., Luy A., and Rojas-Suárez F. (eds.) (1994) *Biología y Conservación de los Psitácidos de Venezuela*. Editorial Giavimar, Caracas, Venezuela.

Morgan P., Aplet G.H., Haufler J.B., Humphries H.C., Moore M.M., and Wilson W.D. (1994) Historical range of variability: A useful tool for evaluating ecosystem change. *Journal of Sustainable Forestry* 2:87–112.

Mori S.A., Boom B.M., and Prance G.T. (1981) Distribution patterns and conservation of eastern Brazilian coastal forest tree species. *Brittonia* 33:233–245.

Morison S.E. (1953) *New Guinea and the Marianas, March 1944–Aug 1944*. Little, Brown, Boston, MA.

Moritz C. (1994a) Applications of mitochondrial DNA analysis in conservation: A critical review. *Molecular Ecology* 3:401–411.

Moritz C. (1994b) Defining "Evolutionarily Significant Units" for conservation. *Trends in Ecology and Evolution* 9:373–375.

Morozov V.V. (1995) Status, distribution and trends of the Lesser White-fronted Goose (*Anser erythropus*) population in Russia. *Bulletin of Geese Study Group of Eastern Europe and Northern Asia. Menzbir Ornithological Society*. No. 1.

Morris R.J. (1980) Floating plastic debris in the Mediterranean. *Marine Pollution Bulletin* 11:125.

Morrison M.L. and Marcot B.G. (1995) An evaluation of resource inventory and monitoring program used in National Forest planning. *Environmental Management* 19:147–156.

Morrison M.L., Marcot B.G., and Mannan R.W. (1992) *Wildlife-Habitat Relationships: Concepts and Applications*. University of Wisconsin Press, Madison.

Morse S. (1996) *Source-Sink Dynamics of Kentucky Warblers in a Complex Forested Landscape*. MSc thesis, University of Illinois at Urbana-Champaign.

Mosconi S.L. and Hutto R.L. (1982) The effect of grazing on the land birds of a western Montana riparian habitat. In *Proceedings of the Wildlife-Livestock Relationships Symposium*. L. Nelson and J.M. Peek (co-chairmen), pp. 221–233. Forest, Wildlife and Range Experiment Station, University of Idaho, Moscow, ID.

Moser M., Bibby C., Newton I., Pienkowski M., Sutherland W.J., Ulfstrand S., and Wynne G. (1995) Bird conservation: The science and the action. *Ibis* 137:S1–S8.

Moser M., Prentice R.C., and van Vessem J. (1993) *Waterfowl and Wetland Conservation in the 1990s—a Global Review*. Proceedings of the International Waterfowl and Wetlands Research Bureau Symposium, St. Peterburg, Florida.

Moulton M.P. (1993) The all-or-none pattern in introduced Hawaiian passeriforms: The role of competition sustained. *American Naturalist* 141:105–119.

Moyle P.B. and Williams J.E. (1990) Biodiversity loss in the temperate zone: Decline of the native fish fauna of California. *Conservation Biology* 4:275–284.

Mundy NI., Winchell C.S., and Woodruff D.S. (1997) Genetic differences between the endangered San Clemente Island Loggerhead Shrike *Lanius ludovicianus mearnsi* and two neighbouring subspecies demonstrated by mtDNA control region and cytochrome b sequence variation. *Molecular Ecology* 6:29–37.

Murawski S.A. (1991) Can we manage our multispecies fisheries? *Fisheries* 16:5–13.

Murawski S.A. (1996) Meeting the challenges of bycatch: New rules and new tools. In *Solving Bycatch: Considerations for Today and Tomorrow*. Alaska Sea Grant College Program, Anchorage, AK. Report 96–03.

Murcia C. (1995) Edge effects in fragmented forests: Implications for conservation. *Trends in Ecology and Evolution* 10:58–62.

Murdoch W.W. (1993) Individual-based models for predicting effects of global change. In *Biotic Interactions and Global Change*. P.M. Kareiva, J.G. Kingsolver, and R.B. Huey (eds.), pp. 147–162. Sinauer Associates, Sunderland, MA.

Murdoch W.W. (1994) Population regulation in theory and practice. *Ecology* 73:271–287.

Murie O.J. (1959) Fauna of the Aleutian Islands and Alaska Peninsula. *U.S. Fish and Wildlife Service, North America Fauna* 61:1–364.

Murphy D., Wilcove D., Noss R., Harte J., Safina C., Lubchenko J., Root T., Sher V., Kaufman L., Bean M. and Pimm S. (1994) On reauthorization of the Endangered Species Act. *Conservation Biology* 8:1–3.

Murphy D.D. and Noon B.D. (1991) Coping with uncertainty in wildlife biology. *Journal of Wildlife Management* 55:773–782.

Murphy R.C. (1936) *Oceanic Birds of South America*. American Museum of Natural History, New York.

Myers N. (1988) Threatened biotas: "Hotspots" in tropical forests. *The Environmentalist* 8:1–20.

Myers N. (1990) The biodiversity challenge: Expanded hot-spots analysis. *The Environmentalist* 10:243–256.

Myers N. (1996) The rich diversity of biodiversity issues. In *Biodiversity II—Understanding and Protecting Our Biological Resources*. M.L. Reaka-Kudla, D.E. Wislon, and E.O. Wilson (eds.), pp. 125–138. Joseph Henry Press, Washington, DC.

Myers R.H. (1990) *Classical and Modern Regression with Applications*. PWS-KENT, Boston.

Nathan R. (1996) Criteria assignment in designing the most important sites for the conservation of breeding passerines in Israel. *Ecology and Environment* 3:229–238.

Nathan R., Safriel U.N., and Shirihai H. (1996) Extinction and vulnerability to extinction at distribution peripheries: An analysis of the Israeli breeding fauna. *Israel Journal of Zoology* 42:361–383.

National Rangeland Management Working Group (1996) *Draft National Strategy for Rangeland Management*. Department of the Environment, Sport and Territories, Canberra.

Nechaev V.A (1991) *Birds of Sakhalin Island*. Far East Science Centre, Vladivostok.

Nei M. (1987) *Molecular Evolutionary Genetics*. Columbia University Press, New York.

Nei M., Maruyama T., and Chakraborty R. (1975) The bottleneck effect and genetic variability in populations. *Evolution* 29:1–10.

Nei M. and Takahata N. (1993) Effective population size, genetic diversity, and coalescence time in subdivided populations. *Journal of Molecular Evolution* 37:240–244.

Neigel J.E., Ball R.M., and Avise J.C. (1991) Estimation of single generation migration distances from geographic variation in animal mitochondrial DNA. *Evolution* 45:423–432.

Nelson, J.B., (1979) Seabirds: Their biology and ecology. A and W Publishers, New York

Nelson R.D. and Salwasser H. (1982) The Forest Service wildlife and fish habitat relationship program. *Transactions of the North American Wildlife and Natural Resources Conference* 47:174–183.

Nettleship D.N. (1972) Breeding success of the Common Puffin (*Fratercula arctica*) on different habitats at Great Island, Newfoundland. *Ecological Monographs* 42:239–268.

Nettleship D.N. and Birkhead T.R. (1985) *The Atlantic Alcidae: The Evolution, Distribution and Biology of the Auks Inhabiting the Atlantic Ocean and Adjacent Water Areas.* Academic Press, Orlando.

Nettleship D.N. and Chapdelaine G. (1988) Population size and status of the Northern Gannet *Sula bassanus* in North America, 1984. *Journal of Field Ornithology* 59:120–127.

Nettleship D.N. and Evans P.G.H. (1985) Distribution and status of the Atlantic Alcidae. In *The Atlantic Alcidae: The Evolution, Distribution and Biology of the Auks Inhabiting the Atlantic Ocean and Adjacent Water Areas.* D.N. Nettleship and T.R. Birkhead (eds.), pp. 53–154. Academic Press, Orlando.

Nettleship D.N. and Peakall D.B. (1987) Organochloride residue levels in three high Arctic species of colonially breeding seabirds from Prince Leopold Island. *Marine Pollution Bulletin* 18:434–438.

Nettleship D.N., Sanger G.A., and Springer P.F. (1984) *Marine Birds: Their Feeding Ecology and Commercial Fisheries Relationships. Proceedings of the Pacific Seabird Group Symposium.* Seattle, Washington, 6–8 January 1982.

Newmark W.D. (1995) Extinction of mammal populations in western North American national parks. *Conservation Biology* 9:512–526.

Newton I. (1986) *The Sparrowhawk.* Poyser, Calton, England.

Newton I. (1989) *Lifetime Reproduction in Birds.* Academic Press, London.

Newton I. (1991) Habitat variation and population regulation in Sparrowhawks. *Ibis* 133 (suppl.):76–88.

Newton I. (1995) The contribution of recent research on birds to ecological understanding. *Journal of Animal Ecology* 64:675–696.

Newton I. and Marquiss M. (1986) Population regulation in Sparrowhawks. *Journal of Animal Ecology* 55:463–480.

Nichols J.D. (In press) Monitoring is not enough: On the need for a model-based approach to migratory bird management. In *Proceedings of the 1995 International Partners in Flight Workshop.* R. Bonney, L. Niles, and D. Pashley (eds.). Cape May, NJ.

Nichols J.D., Noon B.R., Stokes S.L., and Hines J.E. (1981) Remarks on the use of mark-recapture methodology in estimating avian population size. *Studies in Avian Biology* 6:121–136.

Nisbet I.C.T. (1973) Terns in Massachusetts: Present numbers and historical changes. *Bird Banding* 44:27–55.

Nisbet I.C.T. (1975) Selective effects of predation in a tern colony. *Condor* 77:221–226.

Nisbet I.C.T. (1994) Effects of pollution on marine birds. *Birdlife Conservation Series* 1:8–25.

Nishimoto M. (1996) *The status of Black-footed and Laysan albatrosses at Sand Island, Midway Atoll during the 1993–94 breeding season.* Unpublished report, U.S. Fish and Wildlife Service, Kealia Pond National Wildlife Refuge. Kihei, Hawaii.

Noon B.R. (1992) Thoughts on the monitoring of terrestrial vertebrates. In *Needs Assessment: Monitoring Neotropical Migratory Birds.* G.S. Butcher and S. Droege (eds.), pp. 49–50. Cornell Laboratory of Ornithology, Ithaca, NY.

Noon B.R. (1993) Book review: Wildlife habitat relationships, concepts and applications. *Journal of Wildlife Management* 57:934–936.

Noon B.R. and McKelvey K.S. (1996) Management of the Spotted Owl: A case history in conservation biology. *Annual Review of Ecology and Systematics* 27:135–162.

Noon B.R. and Sauer J.R. (1992) Population models for passerine birds: Structure, parameterization, and analysis. In *Wildlife 2001: Populations*. D.C. McCullough and R.H. Barrett (eds.), pp. 441–464. Elsevier Applied Science, London.

Nørrevang A. (1986) Tradition of seabird fowling in the Faroes: An ecological basis for sustained fowling. *Ornis Scandinavica* 17:275–281.

Norris C.A. (1947) Report on the distribution and status of the Corncrake. *British Birds* 40:226–244.

North M. (1993) *Stand structure and truffle abundance associated with Northern Spotted Owl habitat.* Ph.D. Dissertation, University of Washington, Seattle.

Norton A.H. (1923) Notes on the birds of the Knox County region. *Maine Naturalist* 3:1–4.

Norton A.H. (1924a) Notes on birds of the Knox County region. *Maine Field Naturalist* 4:35–39, 95–100.

Norton A.H. (1924b) Notes on birds of the Knox County region. *Maine Field Naturalist* 5:1–5.

Norton T.W. (1997) Conservation and management of eucalypt ecosystems. In *Eucalypt Ecology: Individuals to Ecosystems*. J.E. Williams and J.C.Z. Woinarski (eds.), pp. 373–401. Cambridge University Press, Cambridge, England.

Norton T.W. and May S.A. (1994) Towards sustainable forestry in Australian temperate eucalypt forests: Ecological impacts and priorities for conservation, research and management. In *Ecology and Sustainability of Southern Temperate Ecosystems*. T.W. Norton and S.R. Dovers (eds.), pp. 10–30. Commonwealth Scientific and Industrial Research Organisation, Melbourne.

Noss R.F. (1987) From plant communities to landscapes in conservation inventories: A look at The Nature Conservancy. *Biological Conservation* 41:11–37.

Noss R.F. (1990) Indicators for monitoring biodiversity: A hierarchical approach. *Conservation Biology* 4:355–364.

Noss R.F. (1995) Foreword. In *Mosaic Landscapes and Ecological Processes*. L. Hansson, L. Fahrig, and G. Merriam (eds.), pp. xiii–xv. Chapman and Hall. London.

Noss R.F. (1996) Ecosystems as conservation targets. *Trends in Ecology and Evolution* 11:351.

Noss R.F. and Cooperrider A. (1994) *Saving Nature's Legacy: Protecting and Restoring Biodiversity*. Island Press, Washington, DC.

Noss R.F., LaRoe E.T., III, and Scott J.M. (1995) Endangered ecosystems of the United States: A preliminary assessment of loss and degradation. Biological Report 28, U.S. Department of the Interior, National Biological Service, Washington, DC.

Noss R.F., O'Connell M.A., and Murphy D.D. (1997) *The Science of Conservation Planning*. Island Press, Washington, DC.

Nott M.P., Rogers E., and Pimm S. (1995) Modern extinctions in the kilo-death range. *Current Biology* 5:14–17.

NRC (National Research Council) (1992) *The Scientific Bases for the Preservation of the Hawaiian Crow*. National Academy Press, Washington, DC.

NRC (National Research Council) (1993) *A Biological Survey for the Nation*. National Academy Press, Washington, DC.

NRC (National Research Council) (1995a) *Review of EPA's Environmental Monitoring and Assessment Program—Overall Evaluation*. National Academy Press, Washington, DC.

NRC (National Research Council) (1995b) *Science and the Endangered Species Act.* National Academy Press, Washington, DC.

Nusser J.A., Goto R., Ledig D., Fleischer R.C., and Miller M.M. (1996) Genetic diversity of populations of the endangered Light-footed Clapper Rail revealed by RAPD analysis. *Molecular Ecology* 5:463–472.

Nuzzo V.A. (1986) Extent and status of Midwest oak savanna: Presettlement and 1985. *Natural Areas Journal* 6:6–36.

O'Brien S.J. (1994) Genetic and phylogenetic analyses of endangered species. *Annual Review of Genetics* 28:467–489.

O'Brien S.J. and Mayr E. (1991) Bureaucratic mischief: Recognizing endangered species and subspecies. *Science* 251:1187–1188.

O'Conner R.J. (1992) Population variation in relation to migratory status in some North American birds. In *Ecology and Conservation of Neotropical Migrant Landbirds.* J.M. Hagan III and D.W. Johnson (eds.), pp. 64–74. Smithsonian Institution Press, Washington, DC.

O'Connor R.J., Jones M.T., White D., Hunsaker C., Loveland T., Jones B., and Preston E. (1996) Spatial partitioning of environmental correlates of avian biodiversity in the conterminous United States. *Biodiversity Letters* 3:97–110.

O'Connor R.J. and Shrubb M. (1986) *Farming and Birds.* Cambridge University Press, Cambridge, England.

Office of Technology Assessment (1987) *Technologies to Maintain Biological Diversity.* U.S. Government Printing Office, Washington, DC.

Ogi H., Yatsu, Y.C., Hatanaka H., and Nitta A. (1993) The mortality of seabirds by driftnet fisheries in the North Pacific. In *International North Pacific Fisheries Commission Bulletin No. 53: III. Catch and Fishery Impact (All Species).* J. Ito, W. Shaw, and R.L. Burgner (eds.), pp. 499–518.

Ohlendorf H.M. (1993) Marine birds and trace elements in the temperate North Pacific. In *The Status, Ecology, and Conservation of Marine Birds of the North Pacific.* K. Vermeer, K.T. Briggs, K.H. Morgan, and D. Siegel-Causey (eds.), pp. 232–240. Special Publication of the Canadian Wildlife Service Ottawa.

Ohlendorf H.M. and Fleming W.J. (1988) Birds and environmental contaminants in San Francisco and Chesapeake bays. *Marine Pollution Bulletin* 19:487–495.

Olney P.J.S., Mace G.M., and Feistner A.T.C. (1994) *Creative Conservation: The Interface between Captive and Wild Populations.* Chapman and Hall, London.

Olson S.L. (1977) Additional notes on subfossil bird remains from Ascension Island. *Ibis* 119:37–43.

Olsson M. and Reutergårdh L. (1986) DDT and PCB pollution trends in the Swedish aquatic environment. *Ambio* 15:103–109.

O'Neil L.J. and Carey A.B. (1986) Introduction: When habitats fail as predictors. In *Wildlife 2000: Modeling Habitat Relationships of Terrestrial Vertebrates.* J. Verner, M.L. Morrison, and C.J. Ralph (eds.), pp. 207–208. University of Wisconsin Press, Madison.

O'Neil T.A., Steidl R.J., Edge W.D., and Csuti B. (1996) Using wildlife communities to improve vegetation classification for conserving biodiversity. *Conservation Biology* 9:1482–1491.

Oniki Y. (1979) Is nesting success of birds low in the tropics? *Biotropica* 11:60–69.

Opdam P. (1991) Metapopulation theory and habitat fragmentation: A review of holarctic breeding bird studies. *Landscape Ecolology* 5:93–106.

Orians G.H. (1993) Endangered at what level? *Ecological Applications* 3:206–208.

Orians G.H. (1997) Basic and applied ecology: A false dichotomy. *Society of Conservation Biology Newsletter* 4:1–2.

Ormerod S.J., O'Halloran J.O., Gribbon S.D., and Tyler S.J. (1991) The ecology of Dippers *Cinclus cinclus* in relation to stream acidity in upland Wales: Breeding performance, calcium physiology, and nestling growth. *Journal of Applied Ecology* 28:419–433.

Osborne D. (1957) Archaeological occurrence of pronghorn antelope, bison, and horse in the Columbia Plateau. *Science Monthly* 77:260–269.

Ounsted M.L. (1991) Re-introducing birds: Lessons to be learned for mammals. *Symposium of the Zoological Society of London* 62:75–85.

Owen M. and Black J.M. (1990) *Waterfowl Ecology*. Blackie Publications Glasgow.

Owen M. and Norderhaug M. (1977) Population dynamics of Barnacle Geese *Branta leucopsis* breeding in Svalbard, 1948–1976. *Ornis Scandinavica* 8:161–174.

Owen R.B. and Galbraith W.J. (1989) Earthworm biomass in relation to forest types, soil, and land use: Implications for woodcock management. *Wildlife Society Bulletin* 17:130–136.

Pacific Seabird Group (1994) *Methods of Surveying for Marbled Murrelets in Forests: A Protocol for Land Management and Research*. Pacific Seabird Group, Seattle, WA.

Page G.W., Quinn P.L., and Warriner J.C. (1989) Comparison of the breeding of hand-reared and wild-reared Snowy Plovers. *Conservation Biology* 3:198–201.

Paige L.C. (1990) *Population Trends of Songbirds in Western North America*. MS thesis, University of Montana, Missoula.

Paine R.T. (1966) Food web complexity and species diversity. *American Naturalist* 100:65–75.

Paine R.T. (1969) A note on trophic complexity and community stability. *American Naturalist* 103:91–93.

Paine R.T., Wootton J.T., and Boersma P.D. (1990) Direct and indirect effects of Peregrine Falcon predation on seabird abundance. *Auk* 112:390–401.

Palmer D.A. (1986) *Habitat Selection, Movements and Activity of Boreal and Saw-whet Owls*. MS thesis, Colorado State University, Boulder.

Palmer, R.S. (1949) *Maine Birds*. Bulletin of the Museum of Comparative Zoology, Cambridge, MA.

Parker T.A. and Bailey B. (1991) *A Biological Assessment of the Alto Madidi Region and Adjacent Areas of Northwest Bolivia*. Conservation International Rapid Assessment Program Working Papers. University of Chicago Press, Chicago.

Parker T.A., Parker S.A., and Plenge M.A. (1982) *An Annotated Checklist of Peruvian Birds*. Buteo Books, Vermilion, SD.

Parker T.A., III, Schulenberg T.S., Kessler M., and Wust W.H. (1995) Natural history and conservation of the endemic avifauna in north-west Peru. *Bird Conservation International* 5:201–232.

Parks and Wildlife Commission of the Northern Territory (1995) *A Trial Management Program for the Red-tailed Black Cockatoo Calyptorhynchus banksii in the Northern Territory of Australia*. Parks and Wildlife Commission of the Northern Territory, Darwin.

Parrish J.K. (1995) Influence of group size and habitat type on reproductive success in Common Murres (*Uria aalge*). *Auk* 112:390–401.

Parrish J.K. and Boersma P.D. (1995) Muddy waters. *American Scientist* 83:112–115.

Parrish J.K. and Paine R.T. (1996) Ecological interactions and habitat modification in nesting Common Murres, *Uria aalge*. *Bird Conservation International* 6:261–269.

Parslow J.L.F., Jeffries D.J., and Hanson J.M. (1973) Gannet mortality incidents in 1972. *Marine Pollution Bulletin* 4:41–44.

Partners in Flight (1992) *Needs assessment: Monitoring Neotropical migratory birds*. Procedures of Monitoring Working Group Partners in Flight meeting. Arlington, VA, September 1991.

Pashley D.N. and Barrow W.C. (1992) Effects of land use practices on Neotropical migratory birds in bottomland hardwood forests. In *Ecology and Conservation of Neotropical Migrant Landbirds*. J.M. Hagan III and D.W. Johnston (eds.), pp. 315–320. Smithsonian Institution Press, Washington, DC.

Paton P.W.C. (1994) The effect of edge on avian nest success: How strong is the evidence? *Conservation Biology* 8:17–26.

Patterson J.H. (1995) The North American Waterfowl Management Plan and Wetlands for the Americas programmes: A summary. *Ibis* 137:S215–S218.

Patton D.R. (1992) *Wildlife Habitat Relationships in Forested Ecosystems*. Timber Press, Portland.

Paxinos E., McIntosh C., Ralls K., and Fleischer R. (1997) Methods for detecting and studying endangered kit foxes using DNA markers from dung. *Molecular Ecology* 6:483–486.

Paz U. (1987) *The Birds of Israel*. Ministry of Defense Publications, Tel Aviv.

Peach W., Baillie S., and Underhill L. (1991) Survival of British Sedge Warblers *Acrocephalus schoenobaenus* in relation to west African rainfall. *Ibis* 133:300–305.

Peach W.J., Buckland S.T., and Baillie S.R. (1990) Estimating survival rates using mark-recapture data from multiple ringing sites. *The Ring* 13:87–102.

Peach W.J., Buckland S.T., and Baillie S.R. (1996) The use of constant effort mist-netting to measure between-year changes in the abundance and productivity of common passerines. *Bird Study* 43:142–156.

Peach W.J., Thompson P.S., and Coulson J.C. (1994) Annual and long-term variation in the survival rates of British Lapwings *Vanellus vanellus*. *Journal of Animal Ecology* 63:60–70.

Pearl, M. (1992) Conservation of Asian primates: Aspects of genetics and behavioral ecology that predict vulnerability. In *Conservation Biology: The Theory and Practice of Nature Conservation, Preservation and Management*. P.L. Fiedler, S.K. Jain (eds.), pp. 297–320. Chapman and Hall, New York.

Pease C.M. and Grzybowski J.A. (1995) Assessing the consequences of brood parasitism and nest predation on seasonal fecundity in passerine birds. *Auk* 112:343–363.

Pedler L. and Burbidge A.H. (1995) The range and status of the Nullabor Quail-thrush. *South Australian Ornithologist* 32:45–52.

Pellant M. (1990) The cheatgrass-wildfire cycle—Are there any solutions? In *Proceedings of a Symposium on Cheatgrass Invasion, Shrub Die-off, and Other Aspects of Shrub Biology and Management*. E.D. McArthur, E.M. Romney, S.D. Smith, and P.T. Tueller (comp.), pp. 11–17. INT-GTR-276. USDA Forest Service, Boise, ID.

Pemberton J.M., Slate J., Bancroft D.R., and Barrett J.A. (1995) Nonamplifying alleles at microsatellite loci: A caution for parentage and population studies. *Molecular Ecology* 4:249–252.

Penland, S.T. 1984. Avian responses to a gradient of urbanization in Seattle, Washington. Ph.D. dissertation. University of Washington.

Peres C.A., and Terborgh J.W. (1995) Amazonian nature preserves: An analysis of the defensibility status of existing conservation units and design criteria for the future. *Conservation Biology* 9:34–46.

Perrins C.M., Lebreton J.-D., and Hirons G.J.M. (1991) *Bird Population Studies: Relevance to Conservation and Management.* Oxford University Press, New York.

Peterjohn B.G., Sauer J.R., and Link W.A. (1996) The 1994 and 1995 summary of the North American Breeding Bird Survey. *Bird Populations* 3:48–66.

Peterjohn B.G., Sauer J.R., and Robbins C.S. (1995) Population trends from the North American Breeding Bird Survey. In *Ecology and Management of Neotropical Migratory Birds.* T.E. Martin and D.M. Finch (eds.), pp. 3–39. Oxford University Press, New York.

Peters R.L. (1996) Hope for the Red-cockaded? *Defenders* 71:27–32.

Petersen K.L. and Best L.B. (1987) Effects of prescribed burning on nongame birds in a sagebrush community. *Wildlife Society Bulletin* 15:317–325.

Petty S.J., Shaw G., and Anderson D.I.K. (1994) Value of nest boxes for population studies and conservation of owls in coniferous forest in Britain. *Journal of Raptor Research* 28:134–142.

Phillips K. (1990) Where have all the frogs and toads gone? *BioScience* 40:422–424.

Piatt J.F. and Lensink C.J. (1989) *Exxon Valdez* bird toll. *Nature* 342:865–866.

Piatt J.F., Lensink C.J., Butler W., Kendziorek M., and Nysewandser D.K. (1990) Immediate impact of the "Exxon Valdez" oil spill on marine birds. *Auk* 107:387–397.

Piatt J.F. and Nettleship D.N. (1987) Incidental catch of marine birds and mammals in fishing nets off Newfoundland, Canada. *Marine Pollution Bulletin* 18:344–349.

Pimm S.L. (1987) The snake that ate Guam. *Trends in Ecology and Evolution* 2:293–295.

Pimm S.L. and Askins R.A. (1995) Forest losses predict bird extinction in eastern North America. *Proceedings of the National Academy of Science* 92:9343–9347.

Pimm S.L., Moulton M.P., and Justice L.J. (1994) Bird extinction in the Pacific. *Philosophical Transactions of the Royal Society of London, Series B* 344:29–33.

Pimm S.L., Russell G.J., Gittleman J.L., and Brooks T.M. (1995) The future of biodiversity. *Science* 269:347–350.

Pitelka F.A. (1942) High population of breeding birds within an artificial habitat. *Condor* 44:172–174.

Plesnik J. (1990) Long-term study of some urban and extra-urban populations of the Kestrel (*Falco tinnunculus* L.). In *Bird Census and Atlas Studies.* K. Stastiny and V. Bejcek (eds.), pp. 453–458. XIth International Conference on Bird Census and Atlas Work, Prague.

Podolsky R.H. and Kress S.W. (1989) Plastic debris incorporated into Double-Crested Cormorant nests in the Gulf of Maine. *Journal of Field Ornithology* 60:248–250.

Pokrovskaya I.V. and Tertitsky G.M. (1993) Recent hunting avifauna at Novaya Zemlya. *Novaya Zemlya* 2:91–97. Moscow.

Pollock K.H., Nichols J.D., Brownie C., and Hines J.E. (1990) *Statistical inference for capture-recapture experiments.* Wildlife Monographs 107.

Poole A.F. (1989) *Ospreys: A Natural and Unnatural History*. Cambridge University Press, Cambridge.

Pope J.G. (1991) The ICES (International Council for the Exploration of the Sea) Multispecies Assessment Working Group: Evolution, insights and future problems. *ICES Marine Science Symposium* 193:22–33.

Porneluzi P. (1996) *Demography and viability of Ovenbird populations in fragmented versus unfragmented landscapes*. Ph.D. thesis, University of Missouri, Columbia.

Porneluzi P., Bednarz J.C., Goodrich L.J., Zawada N., and Hoover J. (1993) Reproductive performance of territorial Ovenbirds occupying forest fragments and a contiguous forest in Pennsylvania. *Conservation Biology* 7:618–622.

Porter R.D. and Wiemeyer S.N. (1969) Dieldrin and DDT: Eggshells and reproduction. *Science* 165:199–200.

Potts G.R. (1986) *The Partridge: Pesticides, Predation and Conservation*. Collins, London.

Potts G.R. and Aebischer N.J. (1991) Modelling the population dynamics of the Grey Partridge: Conservation and management. In *Bird Population Studies*. C.M. Perrins, J.-D. Lebreton, and G.J.M. Hirons (eds.), pp. 373–390. Oxford University Press, Oxford.

Power M.E., Tilman D., Estes J.A., Menge B.A., Bond W.J., Mills L.S., Daily G., Castilla J.C., Lubchenco J., and Paine R.T. (1996) Challenges in the quest for Keystones. *BioScience* 46:609–620.

Pradel R., Hines J.E., Lebreton J.D., Nichols J.D., and Viallefont A. (1997) Estimating survival probabilities and proportions of "transients" using capture-recapture data. *Biometrics* 53:60–72.

Pratt H.D., Bruner P.L., and Berrett D.G. (1987) *A Field Guide to the Birds of Hawaii and the Tropical Pacific*. Princeton University Press, Princeton, NJ.

Prendergast J., Quinn R.M., Lawton J.H., Eversham B.C, and Gibbons D.W. (1993) Rare species, the coincidence of diversity hotspots and conservation strategies. *Nature* 365:335–337.

Pressey R.L. (1994) *Ad hoc* reservations: Forward and backward steps in developing representative reserve systems? *Conservation Biology* 8:662–668.

Priddel D. (1990) Conservation of the Malleefowl in New South Wales: An experimental management study. In *The Mallee Lands: A Conservation Perspective*. J.C. Noble, P.J. Joss, and G.K. Jones (eds.), pp. 71–74. Commonwealth Scientific and Industrial Research Organisation, Melbourne.

Priddel D. and Wheeler R. (1994) Mortality of captive-raised Malleefowl, *Leipoa ocellata*, released into a mallee remnant within the wheat-belt of New South Wales. *Wildlife Research* 21:543–552.

Probst J.R. and Thompson F.R., III (1996) A multi-scale assessment of the geographic and ecological distribution of midwestern Neotropical migratory birds. In *Management of Midwestern Landscapes for the Conservation of Migrant Landbirds*. NC-GTR-187. USDA Forest Service, St. Paul, MN.

Pulliam H.R. (1988) Sources, sinks, and population regulation. *American Naturalist* 132:652–661.

Pulliam H.R. and Babbitt B. (1997) Science and the protection of endangered species. *Science* 275:499–501.

Pulliam H.R. and Dunning J.B., Jr. (1987) The influence of food supply on local density and diversity of sparrows. *Ecology* 68:1009–1014.

Pulliam H.R., Dunning J.B., and Liu J. (1992) Population dynamics in complex landscapes: A case study. *Ecological Applications* 2:165–177.

Pulliam H.R., Liu J., Dunning J.B, Stewart D.J., and Bishop T.D. (1995) Modelling animal populations in changing landscapes. *Ibis* 137:S120–S126.

Quigley T.M., Haynes R.W., and Russell R.T. (eds.) (1996) *Integrated scientific assessment for ecosystem management in the interior Columbia basin and portions of the Klamath and Great Basins.* PNW-GTR-382. USDA Forest Service, Portland, OR.

Quinn T. (1995) Using public sighting information to investigate coyote use of urban habitat. *Journal of Wildlife Management* 59:238–245.

Quinn T.W. (1992) The genetic legacy of mother goose–phylogeographic patterns of Lesser Snow Goose *Chen caerulescens caerulescens* maternal lineages. *Molecular Ecology* 1:105–117.

Radovich J. (1981) The collapse of the California sardine fishery: What have we learned? In *Resource Management and Environmental Uncertainty: Lessons from Coastal Upwelling Fisheries.* M.H. Glantz and J.D. Thompson (eds.), pp. 107–136. Wiley and Sons, New York.

Raish C., Yong W., and Marzluff J.M. (1997) Contemporary human uses of southwestern ponderosa pine forests. In *Songbird Ecology in Southwestern Ponderosa Pine Forests: A Literature Review.* W.M. Block and D.M. Finch (eds.), pp. 28–42. RM-GTR-292. USDA Forest Service, Fort Collins, CO.

Ralls K., Ballou J.D., and Templeton A. (1988) Estimates of lethal equivalents and the cost of inbreeding in mammals. *Conservation Biology* 2:185–193.

Ralls K., Harvey P.H., and Lyles A.M. (1986) Inbreeding in natural populations of birds and mammals. In *Conservation Biology: The Science of Scarcity and Diversity.* M.E. Soulé (ed.), pp. 35–56. Sinauer Associates, Sunderland, MA.

Ralph C.J. and Fancy S.G. (1994) Demography and movements of the endangered Akepa and Hawaii creeper. *Wilson Bulletin* 106:615–628.

Ralph C.J., Geupel G.R., Pyle P., Martin T.E., and DeSante D.F. (1993) *Handbook of field methods for monitoring landbirds.* PSW-GTR-144. USDA Forest Service, Albany, CA.

Ralph C.J., Hunt G.L., Raphael M.G., and Piatt J.F. (tech. eds.) (1995) *Ecology and conservation of the Marbled Murrelet.* PSW-GTR-152. USDA Forest Service, Albany, CA.

Ramsey F.L. and Harrod L.A. (1995) Results from avian surveys of Rota and Tinian Islands, Northern Marianas, 1982 and 1994. Unpublished report to U.S. Fish and Wildlife Service, Honolulu, HI.

Rand D.M. (1996) Neutrality tests of molecular markers and the connection between DNA polymorphism, demography and conservation biology. *Conservation Biology* 10:665–671.

Rands M.R.W. (1985) Pesticide use on cereals and the survival of partridge chicks: A field experiment. *Journal of Applied Ecology* 22:49–54.

Raphael M.G., Young J.A., McKelvey K., Galleher B.M., and Peeler K.C. (1994) A simulation analysis of population dynamics of the Northern Spotted Owl in relation to forest management alternatives. *Final Environmental Impact Statement on Management of*

Habitat for Late-Successional and Old-Growth Forest Related Species within the Range of the Northern Spotted Owl. USDA Forest Service, Portland, OR.

Rappole J.H., Powell G.V.N., and Sader S.A. (1994) Remote-sensing assessment of tropical habitat availability for a Nearctic migrant: The Wood Thrush. In *Mapping the Diversity of Nature.* R.I. Miller (ed.), pp. 91–103. Chapman and Hall, New York.

Rasanen M.E., Salo J.S., and Kalliola R.J. (1987) Fluvial perturbance in the western Amazon basin: Regulation by long-term sub-Andean tectonics. *Science* 238:1398–1401.

Ratcliffe D.A. (1993) *The Peregrine Falcon,* 2nd edition. T. and A.D. Poyser. London, England.

Ratti J.T. and Garton E.O. (1996) Research and experimental design. In *Research and Management Techniques for Wildlife and Habitats.* T.A. Bookout (ed.), p. 1–23. Wildlife Society, Bethesda, MD.

Rave E.H. (1995) Genetic analysis of wild populations of Hawaiian Geese using DNA fingerprinting. *Condor* 97:82–90.

Rave E.H., Fleischer R.C., Duvall F., and Black J.M. (1994) Genetic analyses through DNA fingerprinting of captive populations of Nene. *Conservation Biology* 8:744–751.

Rave E.H., Fleischer R.C., Duvall F., and Black J.M. (Submitted) Effects of inbreeding on reproductive success in captive populations of Hawaiian Geese.

Ravkin J.S. (1984) *Space structure of bird populations within the forest zone (Western and Central Siberia).* Nauka, Novosibirsk.

Reader R. (1988) Using the guild concept in the assessment of tree harvesting effects on understory herbs: A cautionary note. *Environmental Management* 12:803–808.

Recher H.F. (1996) Conservation and management of eucalypt forest vertebrates. In *Conservation of Faunal Diversity in Forested Landscapes.* R. DeGraff and I. Miller (eds.), pp. 339–388. Chapman and Hall, London.

Recher H.F. and Lim L. (1990) A review of current ideas of the extinction, conservation and management of Australia's terrestrial vertebrate fauna. *Proceedings of the ecological Society of Australia* 16:287–301.

Recher H.F., Shields J., Kavanagh R.P., and Webb G. (1987) Retaining remnant mature forest for nature conservation at Eden, New South Wales: A review of theory and practice. In *Nature Conservation: The Role of Remnants of Native Vegetation.* D.A. Saunders, G.W. Arnold, A.A. Burbidge, and A.J.M. Hopkins (eds.), pp. 177–194. Surrey Beatty, Sydney.

Redford K.H. and Robinson J.G. (1987) The game of choice: Patterns of Indian and colonist hunting in the Neotropics. *American Anthropology* 89:650–667.

Reichel J.D. (1991) The status and conservation of seabirds in the Mariana Islands. In *Seabird Status and Conservation: A Supplement.* J.P. Croxall (ed.), pp. 248–262. International Council for Bird Preservation Technical Publication 11.

Reichel J.D. and Glass P.O. (1991) Checklist of the birds of the Mariana islands. *'Elepaio* 51:3–10.

Reichel J.D. and Lemke T.O. (1994) Ecology and extinction of the Mariana Mallard. *Journal of Wildlife Management* 58:199–205.

Reid J. and Fleming M. (1992) The conservation status of birds in arid Australia. *Rangelands Journal* 14:65–91.

Reid V.W. (1992) How many species will there be? In *Tropical Deforestation and Species Extinction*. T.C. Whitmore and J.A. Sayer (eds.), pp. 55–73. Chapman and Hall, London.

Remsen J.V., Jr. (1995) The importance of continued collecting of bird specimens to ornithology and bird conservation. *Bird Conservation International* 5:145–180.

Remsen J.V., Jr. and Parker T.A. (1983) Contribution of river-created habitats to bird species richness in Amazonia. *Biotropica* 15:223–231.

Remsen J.V., Jr. and Parker T.A. (1995) Bolivia has the opportunity to create the planet's richest park for terrestrial biota. *Bird Conservation International* 5:181–199.

Rensselaerville Roundtable (1995) *Navigating into the Future. Rensselaerville Roundtable: Integrating Science and Policymaking*. 386–111/00505. U.S. Government Printing Office, Washington, DC.

Reynolds R.T. (1983) *Management of western coniferous forest habitat for nesting accipiter hawks*. RM-GTR-102. USDA Forest Service, Fort Collins, CO.

Reynolds R.T., Graham R.T., Reiser M.H., Bassett R.L., Kennedy P.L., Boyce D.A., Jr., Goodwin G., Smith R., and Fisher E.L. (1992) *Management recommendations for the Northern Goshawk in the southwestern United States*. RM-GTR-217. USDA Forest Service, Fort Collins, CO.

Rhymer J.M. and Simberloff D. (1996) Extinction by hybridization and introgression. *Annual Review of Ecology and Systematics* 27:83–109.

Rice J.C. (1992) Multispecies interactions in marine ecosystems: Current approaches and implications for study of seabird populations. In *Wildlife 2001: Populations*. D.R. McCullough and R.H. Barrett (eds.), pp. 586–601. Elsevier, London.

Rice S.M., Guthery F.S., Spears G.S., DeMaso S.J., and Koerth B.H. (1993) A precipitation-habitat model for Northern Bobwhites on semiarid rangeland. *Journal of Wildlife Management* 57:92–102.

Rich P.V. and Baird R.F. (1986) History of the Australian avifauna. *Current Ornithology* 4:97–139.

Rich T.D. (1978) Cowbird parasitism of Sage and Brewer's Sparrow. *Condor* 80:348.

Rickard W.H., Hedlund J.D., and Fitzner R.E. (1977) Elk in the shrub-steppe region of Washington: An authentic record. *Science* 196:1009–1010.

Ridgely R.S. (1976) *A Guide to the Birds of Panama*. Princeton University Press, Princeton, NJ.

Ridgely R.S. and Gwynne J.A., Jr. (1989) *Birds of Panama*, 2nd edition. Princeton University Press, Princeton, NJ.

Ridgely R.S. and Tudor G. (1994) *Birds of South America Vol 1: The Oscine Passerines*. University of Texas Press, Austin, Texas.

Riebsame W.E., Gosnell H., and Theobald D.M. (1996) Land use and landscape change in the Colorado Mountains I: Theory, scale, and pattern. *Mountain Research and Development* 16:395–405.

Riffell S.K., Gutzwiller K.J., and Anderson S.H. (1996) Does repeated human intrusion cause cumulative declines in avian richness and abundance? *Ecological Applications* 6:492–505.

Ringold P.L., Alegria J., Czaplewski R.L., Mulder B.S., Tolle T., and Burnett K. (1996) Adaptive monitoring design for ecosystem management. *Ecological Applications* 6:745–747.

Risser P.G. (1993) Making ecological information practical for resource managers. *Ecological Applications* 3:37–38.

Robbins C.S. (1979) Effects of forest fragmentation on bird populations. In *Proceedings of the Workshop on Management of North-Central and Northeastern Forests for Nongame Birds.* R.M. Degraaf and N. Tilghman (eds.), pp. 198–212. NC-GTR-51. USDA Forest Service, St. Paul, MN.

Robbins C.S., Bystrak D. and Geissler P.H. (1986) *The Breeding Bird Survey: Its First Fifteen Years, 1965–1979.* Resource Publication 157, U.S. Fish and Wildlife Service, Washington, DC.

Robbins C.S., Sauer J.R., Greenberg R.S., and Droege S. (1989) Population declines in North American birds that migrate to the Neotropics. *Proceedings of the National Academy of Science* 86:7658–7662.

Robbins C.S., Sauer J.R., and Peterjohn B.G. (1993) Population trends and management opportunities for neotropical migrants. In *Status and Management of Neotropical Migratory Birds.* D.M. Finch and P.W. Stangel (eds.), pp. 17–23. RM-GTR-229. USDA Forest Service, Ft. Collins, CO.

Roberts T.H. (1987) Construction of guilds for habitat assessment. *Environmental Management* 11:473–477.

Roberts T.H. and O'Neil L.J. (1985) Species selection for habitat assessments. *Transactions of the North American Wildlife Natural Resources Conference* 50:352–362.

Robinson D. (1991) Threatened birds in Victoria: Their distributions, ecology and future. *Victoria Nature* 108:67–77.

Robinson D. and Traill B.J. (1996) Conserving woodland birds in the wheat and sheep belts of southern Australia. *Royal Australian Ornithology Union Conservation Statement* 10:1–16.

Robinson J.G. (1992) The limits to caring: Sustainable living and the loss of biodiversity. *Conservation Biology* 7:20–28.

Robinson J.G. and Redford K.H. (1991) *Neotropical Wildlife Use and Conservation.* University of Chicago Press, Chicago.

Robinson S.K. (1993) Conservation problems of Neotropical migrant land birds. *Transactions North American Wildlife Natural Resources Conference* 58:379–89.

Robinson S.K. (1995) Habitat selection and foraging ecology of raptors in Amazonian Peru. *Biotropica* 26:443–458.

Robinson S.K. (1996) Threats to breeding Neotropical migratory birds in the Midwest. In *Management of Midwestern Landscapes for the Conservation of Migrant Landbirds*, pp. 1–21. NC- GTR-187. USDA Forest Service, St. Paul, MN.

Robinson S.K. (1997) The case of the missing songbirds. *Consequences* 3:3–15.

Robinson S.K. (In press) *Birds of a Peruvian oxbow lake: Populations, resources, predation, and social behavior.* Ornithological Monographs 48.

Robinson S.K., Grzybowski J.A., Rothestein S.I., Brittingham M.C., Petit L.J., and Thompson F.R. (1993) Management implications of cowbird parasitism on Neotropical migrant songbirds. *Status and Management of Neotropical Migratory Birds.* D.W. Finch and P.W. Stangel (eds.), pp. 93–102. GTR-RM-229. USDA Forest Service, Ft. Collins, CO.

Robinson S.K. and Terborgh J. (1997) Bird community dynamics along primary successional gradients of an Amazonian whitewater river. *Ornithological Monographs* 641–672.

Robinson S.K., Terborgh J., and Munn C.A. (1990) Lowland tropical forest bird communities of a site in Western Amazonia. In *Biogeography and Ecology of Forest Bird Communities*. A. Keast (ed.), pp. 229–258. Academic Press, The Hague, the Netherlands.

Robinson S.K., Thompson F.R., III, Donovan T.M., Whitehead D.R., and Faaborg J. (1995) Regional forest fragmentation and the nesting success of migratory birds. *Science* 267:1987–1990.

Robinson S.K. and Wilcove D.S. (1994) Forest fragmentation in the temperate zone and its effects on migratory songbirds. *Bird Conservation International* 4:233–249.

Roca R.L. (1994) *Oilbirds of Venezuela: Ecology and Conservation*. No. 24. Nuttall Ornithological Club, Cambridge, MA

Rockwell R.F. and Barrowclough G.F. (1987) Gene flow and the genetic structure of populations. In *Avian Genetics*. F. Cooke (ed.), pp. 223–255. Academic Press, London.

Rodda G.H. and Fritts T.H. (1992) The impact of the introduction of the Brown Tree Snake, *Boiga irregularis*, on Guam's lizards. *Journal of Herpetology* 26:166–174.

Rodda G.H., Fritts T.H., and Conry P.J. (1992) Origin and population growth of the brown tree snake, *Boiga irregularis*, on Guam. *Pacific Science* 46:46–57.

Rodda G.H., Fritts T.H., McCoid M.J. and Campbell E.W., III (1998) An overview of the biology of the Brown Treesnake, *Boiga irregularis*, a costly introduced pest on Pacific Islands. In *Problem Snake Management: Habu and Brown Treesnake Examples*. G.H. Rodda, Y. Sawani, D. Chiszar and H. Tanaka (eds.), Cornell University Press, Ithaca, NY

Rodenhouse N.L. and Best L.B. (1983) Breeding ecology of Vesper Sparrows in corn and soybean fields. *American Midland Naturalist* 110:265–275.

Rodenhouse N.L., Best L.B., O'Connor R.J., and Bollinger E.K. (1995) Effects of agricultural practices and farmland structures. In *Ecology and Management of Neotropical Migratory Birds*. T.E. Martin and D.M. Finch (eds.), pp. 269–293. Oxford University Press, New York.

Rodriguez J.P. and Rojas-Suárez F. (1995) *Libro Rojo de la Fauna Venezolana*. Provita, Caracas, Venezuela.

Rodway M.S., Lemon M.J.F., and Summers K.S. (1990) *British Columbia seabird colony inventory: Report #4—Scott Islands. Census results from 1982 to 1989 with reference to the Nestucca oil spill*. Technical Report Series No. 86. Candian Wildlife Service Pacific and Yukon Region, British Columbia.

Roemmich D. (1992) Ocean warming and sea level rise along the southwest U.S. coast. *Science* 257:373–375.

Roemmich D. and McGowan J.A. (1995) Climatic warming and the decline of zooplankton in the California current. *Science* 267:1324–1326.

Rojas M. (1992) The species problem and conservation: What are we protecting? *Conservation Biology* 6:170–178.

Rojek N. and Conant S. (1996) Food preferences and response to novel native berries by captive reared Nene goslings. *Wildfowl* 48: 8:26–39.

Romesburg H.C. (1981) Wildlife science: Gaining reliable knowledge. *Journal of Wildlife Management* 45:293–313.

Root T. (1988a) Energy constraints on avian distributions and abundances. *Ecology* 69:330–339.

Root T. (1988b) Environmental factors associated with avian distributional boundaries. *Journal of Biogeography* 15:489–505.

Rose P.M. and Scott D.A. (1994) Waterfowl Population Estimated. *IWRB Publication 29*. Slimbridge, England.

Rosemarin A. (1988) Ecotoxicology on the upswing—But where are the ecologists? *Ambio* 17:359–?.

Rosenberg D.K. (1997) *Evaluation of the Monitoring Avian Productivity and Survivorship (MAPS) Program*. Institute for Bird Populations, Point Reyes Station, CA. 83 pp.

Rosenberg D.K., DeSante D.F. and Hines J.E. (In press) Monitoring survival rates of landbirds at varying spatial scales: An application of the MAPS Program. In *Proceedings of the 1995 Partners in Flight International Workshop*. R. Bonney, L. Niles, and D. Pashley (eds.). Cape May, NJ.

Rosenberg G.H. (1990) Habitat selection and foraging behavior by birds of Amazonian river islands in northeastern Peru. *Condor* 92:427–443.

Rosenberg K.V., Terrill S.B., and Rosenberg G.H. (1987) Value of suburban habitats to desert riparian birds. *Wilson Bulletin* 99:642–654.

Rosenberg K.V. and Wiedenfeld D.A. (1993) *Directory of Neotropical Ornithology*. American Ornithologists' Union, Washington, DC.

Rosenfield R.N., Bielefeldt J., Affeldt J.L., and Beckmann D.J. (1996) Urban nesting biology of Cooper's Hawks in Wisconsin. In *Raptors in Human Landscapes: Adaptations to Built and Cultivated Environments*. D.M. Bird, D.E. Varland, and J.J. Negro (eds.), pp. 41–44. Academic Press, London.

Rosenzweig M.L. (1985) Some theoretical aspects of habitat selection. In *Habitat Selection in Birds*. M.L. Cody (ed.), pp. 517–540. Academic Press, Orlando, FL.

Rosenzweig M.L. (1991) Habitat selection and population interactions: The search for mechanism. *American Naturalist* 137:S5–S28.

Rosenzweig M.L. (1992) Species diversity gradients: We know more and less than we thought. *Journal of Mammalogy* 73:715–730.

Rossi R.E., Mulla D.J., Journel A.G., and Franz E.H. (1992) Geostatistical tools for modeling and interpreting ecological spatial dependence. *Ecological Monographs* 62:277–314.

Rotenberry J.T. (1985) The role of habitat in avian community composition: Physiognomy or floristics? *Oecologia* 67:213–217.

Rotenberry J.T. (1986) Habitat relationships of shrubsteppe birds: Even "good" models cannot predict the future. In *Wildlife 2000: Modeling Habitat Relationships of Terrestrial Vertebrates*. J. Verner, M.L .Morrison, and C.J. Ralph (eds.), pp. 217–221. University of Wisconsin Press, Madison.

Rotenberry J.T. and Wiens J.A. (1980) Temporal variation in habitat structure and shrubsteppe bird dynamics. *Oecologia* 47:1–9.

Rotenberry J.T. and Wiens J.A. (1989) Reproductive biology of shrubsteppe passerine birds: Geographical and temporal variation in clutch size, brood size, and fledging success. *Condor* 91:1–14.

Rothstein S.I. (1994) The cowbird's invasion of the Far West: History, causes and consequences experienced by host populations. *Studies in Avian Biology* 15:301–315.

Rowley I.C.R. and Russell E. (1991) Demography of passerines in the temperate Southern Hemisphere. In *Bird Population Studies*. C.M. Perrins, J.D. Leberton, and G.J.M. Hirons (eds.), pp. 22–44. Oxford University Press, Oxford.

Rudzitis G. and Johansen H. (1989) Migration into western wilderness counties: Causes and consequences. *Western Wildlands* Spring:19–23.

Ruggiero L.F., Hayward G.D., and Squires J.R. (1994) Viability analysis in biological evaluations: Concepts of population viability analysis, biological population, and ecological scale. *Conservation Biology* 8:364–372.

Ruiz Muller M. (1996) Implementing CDB provisions on access to genetic resources: Recent developments in Latin America. *Biodiversity Bulletin* 1:24.

Ryan M.R. (1986) Nongame management in grassland and agricultural ecosystems. In *Management of Nongame Wildlife in the Midwest: A Developing Art*. J.B. Hale, L.B. Best, and R.L. Clawson (eds.), pp. 117–136. Bookcrafters, Chelsea, MI.

Ryan P.G. (1987a) The effects of ingested plastics on seabirds: Correlations between plastic load and body condition. *Environmental Pollution* 46:119–125.

Ryan P.G. (1987b) The incidence and characteristics of plastic particles ingested by seabirds. *Marine Environment Research* 23:175–206.

Ryder O.A. (1986) Species conservation and systematics: The dilemma of subspecies. *Trends in Ecology and Evolution* 1:9–10.

Ryser F.A. (1985) *Birds of the Great Basin: A natural history*. University of Nevada Press, Reno.

Ryti R. (1992) Effect of the focal taxon on the selection of nature reserves. *Ecological Applications* 2:404–410.

Saab V.A., Bock C.E., Rich T.D., and Dobkin D.S. (1995) Livestock grazing effects in western North America. In *Ecology and Management of Neotropical Migratory Birds*. T.E. Martin and D.M. Finch (eds.), pp. 311–353. Oxford University Press, New York.

Saavedra C.J. and Freese C. (1986) Prioridades biológicas de conservación en los Andes Tropicales. *Parques* 11:8–11.

Sabatier P., Loomis J., and McCarthy C. (1996) Attitudes and decisions within the Forest Service: Is there a connection? *Journal of Forestry* 94(1):42–46.

Safriel U. (1968) Bird migration at Eilat, Israel. *Ibis* 110:283–320.

Sallabanks R. (1996) *Forest birds of the ecosystem diversity matrix for the Interior Columbia River Basin*. Sustainable Ecosystems Institute, Meridian, ID.

Salo J., Kalliola R., Hakkinen I., Makinen Y., Niemela P., Puhakka M., and Coley P.D. (1986) River dynamics and the diversity of Amazonian lowland forest. *Nature* 322:254–258.

Salomonsen F. (1970) Birds useful to man in Greenland. In *Productivity and Conservation in Northern Circumpolar Lands*. International Union for Conservation of Nature and Natural Resources, Morges, Switzerland.

Salwasser H. (1991) New perspectives for sustaining diversity in U.S. national forest ecosystems. *Conservation Biology* 5:567–569.

Salwasser H. (1992) From new perspectives to ecosystem management: Response to Frissell et al. and Lawrence and Murphy. *Conservation Biology* 6:469–472.

Sample D.W. (1989) *Grassland Birds in Southern Wisconsin: Habitat Preference, Population Trends, and Response to Land Use Changes*. MS thesis, University of Wisconsin, Madison.

Samson F.B. and Knopf F.L. (1994) Prairie conservation in North America. *BioScience* 44:418–421.

Samson F.B. and Knopf F.L. (1996) *Prairie Conservation: Preserving North America's Most Endangered Ecosystem*. Island Press, Washington, DC.

Sauer J.R. and Droege S. (1992) Geographic patterns in population trends of Neotropical migrants in North America. In *Ecology and Conservation of Neotropical Migrant Land-birds*. J.M. Hagan III and D.W. Johnston (eds.), pp. 26–42. Smithsonian Institution Press, Washington, DC.

Sauer J.R. and Droege S. (eds.) (1990) *Survey designs and statistical methods for the estimation of avian population trends*. Biology Report 90(1). U.S. Fish and Wildlife Service, Washington, DC.

Sauer J.R., Schwartz S., Peterjohn B.G., and Hines J.E. (1996) The North American Breeding Bird Survey home page, Version 94.3. Patuxant Wildlife Research Center, Laurel, MD.

Saunders D.A. (1989) Changes in the avifauna of a region, district and remnant as a result of fragmentation of native vegetation: The Wheatbelt of Western Australia. A case study. *Biological Conservation* 50:99–135.

Saunders D.A., Arnold G.W., Burbidge A.A., and Hopkins A.J.M. (1987) *Nature Conservation: The Role of Remnants of Native Vegetation*. Surrey Beatty, Sydney.

Saunders D.A. and Curry P.J. (1990) The impact of agricultural and pastoral industries on birds in the southern half of Western Australia: Past, present and future. *Proceedings of the Ecological Society of Australia* 16:303–321.

Saunders D.A. and Hobbs R.J. (eds.) (1991) *Nature Conservation 2: The Role of Corridors*. Surrey Beatty, Sydney.

Saunders D.A., Hobbs R.J., and Ehrlich P.R. (1993) *Nature Conservation 3: Reconstruction of Fragmented Ecosystems*. Surrey Beatty, Sydney.

Saunders D.A., Hobbs R.J., and Margules C.R. (1991) Biological consequences of ecosystem fragmentation: A review. *Conservation Biology* 5:18–32.

Savidge J.A. (1987) Extinction of an island forest avifauna by an introduced snake. *Ecology* 68:660–668.

Schafale M.P. and Weakley A.S. (1990) *Classification of the Natural Communities of North Carolina: Third Approximation*. North Carolina Natural Heritage Program, Raleigh, NC.

Schaffner F.C. (1986) Trends in Elegant Tern and Northern anchovy populations in California. *Condor* 88:347–354.

Schamberger M. and Farmer A. (1978) The habitat evaluations procedures: Their application in project planning and impact evaluation. *Transactions of the North American Wildlife and Natural Resources Conference* 43:274–283.

Schelhas J. and Greenberg R. (1996) *Forest Patches in Tropical Landscapes*. Island Press, Washington, DC.

Schena M., Shalon D., Heller R., Chai A., Brown P.O., and Davis R.W. (1996) Parallel human genome analysis: Microarray-based expression monitoring of 1000 genes. *Proceedings of the National Academy of Science* 93:10614–10619.

Schieck J. (1997) Biased detection of bird vocalizations affects comparisons of bird abundance among forested habitats. *Condor* 99:179–190.

Schieck J., Lertzman K., Nyberg B., and Page R. (1995) Effects of patch size on birds in old-growth montane forests. *Conservation Biology* 9:1072–1084.

Schläpfer A. (1988) Populätionsökologie der Feldlerche *Alauda arvensis* in der intensiv genutzten Agrarlandschaft. *Ornithologische Beobachter* 85:309–371.

Schmidt K.F. (1997) "No-take" zones spark fisheries debate. *Science* 277:489–491.

Schmitz P.E. (1912) Die Tagraubvogel Palastinas. *Das Helige Land* 56:224–230.

Schodde R. (1990) The bird fauna of the Mallee—Its biogeography and future. In *The Mallee Lands: A Conservation Perspective.* J.C. Noble, P.J. Joss, and G.K. Jones (eds.), pp. 61–70. Commonwealth Scientific and Industrial Research Organization, Melbourne.

Schoener T.E. and Spiller D.A. (1987) High population persistence in a system with high turnover. *Nature* 330:474–477.

Schueck L.S. and Marzluff J.M. (1995) Influence of weather on conclusions about effects of human activities on raptors. *Journal of Wildlife Management* 59:674–682.

Schulz T.T. and Joyce J.A. (1992) A spatial application of a marten habitat model. *Wildlife Society Bulletin* 20:74–83.

Scott J.M. (1994) Preserving and restoring avian diversity: A search for solutions. *Studies in Avian Biology* 14:340–348.

Scott J.M. and Carpenter J.W. (1987) Release of captive-reared or translocated endangered birds: What do we need to know? *Auk* 104:544–545.

Scott J.M., Csuti B., Jacobi J.D., and Estes J.E. (1987) Species richness: A geographical approach to protection of biological diversity. *BioScience* 39:782–788.

Scott J.M., Davis F., Csuti B., Noss R., Butterfield B., Caicco S., Groves C., Edwards T.C., Jr., Ulliman J., Anderson H., D'Erchia F., and Wright R.G. (1993) *Gap Analysis: A Geographic Approach to Protection of Biological Diversity.* Wildlife Monographs No. 123.

Scott J.M., Temple S.A., Harlow D., Shaffer M. (1994) Restoration and management of endangered species. In *Research and Management Techniques for Wildlife and Habitats.* A. Bookhout (ed.), pp. 531–539. Wildlife Society, Bethesda, MD.

Scotts D.J. (1991) Old-growth forests: Their ecological characteristics and value to forest-dependent vertebrate fauna of south-east Australia. In *Conservation of Australia's Forest Fauna.* D. Lunney (ed.), pp. 147–159. Royal Zoological Society of New South Wales, Sydney.

Seal U.S. (1991) Life after extinction. *Zoological Society London Symposium* 62:39–56.

Seal U.S., Foose T.J., and Ellis S. (1994) Conservation assessment and management plans (CAMPs) and global captive action plans (GCAPs). In *Creative Conservation: The Interface between Captive and Wild Populations.* P.J.S. Olney, G.M. Mace, and A.T.C. Feistner (eds.), pp. 312–328. Chapman and Hall, London.

Sealy S.G. (1996) Evolution of host defenses against brood parasitism: Implications of puncture-ejection by a small passerine. *Auk* 113:346–355.

Sealy S.G. and Carter H.R. (1984) At-sea distribution and nesting habitat of the Marbled Murrelet in British Columbia: Problems in the conservation of a solitary nesting seabird. In *Status and Conservation of the World's Seabirds.* J.P. Croxall, P.G.H. Evans, and R.W. Schreiber (eds.), pp. 737–756. ICBP Technical Publication No. 2. Page Brothers, Norwich, England.

Seba D.B. and Corcoran E.F. (1969) Surface slicks as concentrators of pesticides in the marine environment. *Pesticide Monitoring Journal* 3:190–193.

Seddon P.J. and Davis L.S. (1989) Nest-site selection by Yellow-eyed Penguins. *Condor* 91:653–659.

Serena M. (1982) The status and distribution of the Willow Flycatcher in selected portions of the Sierra Nevada, 1982. California Department of Fish and Game Administrative Report 82–5, Sacramento, CA.

Serena M. (ed.) (1994) *Reintroduction Biology of Australian and New Zealand Fauna.* Surrey Beatty, Sydney.

Serventy D.L. (1967) Aspects of the population ecology of the Short-tailed Shearwater (*Puffinus tenuirostris*). *Proceedings of the International Ornithological Congress* 14:165–190.

Serventy D.L., Serventy U., and Warham J. (1971) Seabird conservation problems in Australia. In *The Handbook of Australian Birds*. D.L. Serventy, U. Serventy, and J. Warham (eds.), pp. 40–44. A.H. and A.W. Reed, Sydney.

Severinghaus W.D. (1981) Guild theory developed as a mechanism for assessing environmental impact. *Environmental Management* 5:187–190.

Shaffer M.L. (1987) Minimum population sizes for species conservation. *BioScience* 31:131–134.

Shaffer M.L. (1990) Population viability analysis. *Conservation Biology* 4:39–40.

Sharrock J.T.R. (1976) *The Atlas of Breeding Birds in Britain and Ireland*. T. and A.D. Poyser, Berkhamstead, England.

Sherry T.W. (1984) Comparative dietary ecology of sympatric insectivorous Neotropical flycatchers (Tyrannidae). *Ecological Monographs* 54:313–338.

Sherry T.W. and Holmes R.T. (1992) Population fluctuations in a long-distance Neotropical migrant: Demographic evidence for the importance of breeding season events in the Redstart. In *Ecology and Conservation of Neotropical Migrant Landbirds*. J.M. Hagan III and D.W. Johnston (eds.), pp. 431–442. Smithsonian Institution Press. Washington, DC.

Sherry T.W. and Holmes R.T. (1995) Summer versus winter limitation of populations: What are the issues and what is the evidence? In *Ecology and Management of Neotropical Migratory Birds: A Synthesis and Review of Critical Issues*. T.E. Martin and D.M. Finch (eds.), pp. 85–120. Oxford University Press, New York.

Shirihai H. (1996) *The Birds of Israel*. Academic Press, London, England.

Shirihai H. and Christie D. (1992) Raptor migration at Eilat. *British Birds* 85:141–186.

Short H.L. and Hestbeck J.B. (1995) National biotic resource inventories and GAP analysis. *BioScience* 45:535–539.

Short L.L. (1984) Priorities in ornithology: The urgent need for tropical research and researchers. *Auk* 101:892–893.

Shrader-Frechette K.S. and McCoy E.D. (1994) Biodiversity, biological uncertainty, and setting conservation priorities. *Biology and Philosophy* 9:167–195.

Shrubb M. and Lack P.C. (1991) The numbers and distribution of Lapwings *Vanellus vanellus* nesting in England and Wales in 1987. *Bird Study* 38:20–37.

Sierra Nevada Ecosystem Project (1996) *Final report to Congress, vol. II, Assessments and scientific basis for management options*. Centers for Water and Wildland Resources, University of California, Davis.

Sieving K.E. (1992) Nest predation and differential insular extinction among selected forest birds of central Panama. *Ecology* 73:2310–2328.

Sieving K.E. and Karr J.R. (In press) Avian extinction and persistence mechanisms in lowland Panama. In *Tropical Forest Remnants: Ecology, Management, and Conservation of Fragmented Communities*. W.F. Laurance, R.O. Bierregard, and C. Moritz (eds.). University of Chicago Press, Chicago.

Silva J.L. and Strahl S.D. (1991) Human Impact on Populations of Chachalacas, Guans, Curassows (Galliformes: Cracidae) in Venezuela. In *Neotropical Wildlife Use and Conservation*. J.G. Robinson and K.H. Redford (eds.), University of Chicago Press, Chicago, IL.

Silveira C.E. (1995) The Black-eared Miner. *Australian Bird Watcher* 16:96–109.

Simberloff D. (1986) Are we on the verge of a mass extinction in tropical rain forest? In *Dynamics of Extinction* D.K. Elliot (ed.), pp. 165–180. Wiley, New York.

Simberloff D. (1988) The contribution of population and community biology to conservation science. *Annual Review of Ecology and Systematics* 19:473–511.

Simberloff D. (1995) Habitat fragmentation and population extinction of birds. *Ibis* 137:S105–S111.

Simberloff D. and Cox J. (1987) Consequences and costs of conservation corridors. *Conservation Biology* 1:63–71.

Simpson D., Sedjo R., and Reid J. (1996) Valuating biodiversity: An application to genetic prospecting. *Journal of Political Economy* 104:163–185.

Sinderman C.J. (1982) *Winning the Games Scientists Play.* Plenum Press, New York.

Sisk T.D., Llauner A.E., Switky K.R., and Ehrlich P.R. (1994) Identifying extinction threats: Global analyses of the distribution of biodiversity and the expansion of the human enterprise. *BioScience* 44:592–604.

Sissenwine M.P. and Daan N. (1991) An overview of multispecies models relevant to management of living resources. *International Council for the Exploration of the Sea Marine Science Symposium* 193:6–11.

Skellam J.G. (1951) Random dispersal in theoretical populations. *Biometrika* 38:196–218.

Skira I. (1996) Aboriginal people and muttonbirding in Tasmania. In *Sustainable Use of Wildlife by Aboriginal Peoples and Torres Strait Islanders.* M. Bomford and J. Caughley (eds.), pp. 167–175. Australian Government Publishing Service, Canberra.

Sklyarenko S.L. (1995) The illegal capture of Saker Falcons in Kazakhstan. *World Working Group on Birds of Prey and Owls Newsletter* 21/22:14–15.

Skole D. and Tucker C. (1993) Tropical deforestation and habitat fragmentation in the Amazon: Satellite data from 1978–1988. *Science* 260:1905–1909.

Skutch A.F. (1985) Clutch size, nesting success, and predation on nests of Neotropical birds, reviewed. *Ornithological Monographs* 36:575–594.

Slatkin M. and Barton N.H. (1989) A comparison of three indirect methods for estimating average levels of gene flow. *Evolution* 43:1349–1368.

Slatkin M. and Maddison W.P. (1989) A cladistic measure of gene flow inferred from the phylogenies of alleles. *Genetics* 123:603–613.

Smales I., Menkhorst P., and Horrocks G. (1995) The Helmeted Honeyeater recovery program. In *People and Nature Conservation: Perspectives on Private Land Use and Endangered Species Recovery.* A. Bennett, G. Backhouse, and T. Clark (eds.), pp. 35–44. Royal Zoological Society of New South Wales, Sydney.

Small M.F. and Hunter M.L. (1988) Forest fragmentation and avian nest predation in forested landscapes. *Oecologia* 76:62–64.

Smith G.T. (1985) Population and habitat selection of the Noisy Scrub-bird, *Atrichornis clamosus*, 1962–83. *Australian Wildlife Research* 12:479–485.

Smith J.N.M., Taitt M.J., Rogers C.M., Arcese P., Keller L.F., Cassidy L.E.V., and Hochachka W.M. (1996) A metapopulation approach to the population biology of the Song Sparrow *Melospiza melodia. Ibis* 138:120–128.

Smith J.P., Griswell R.E., and Hayes J.P. (1997) *A research problem analysis in support of the Cooperative Forest Ecosystem Research (CFER) Program.* Final Report. U.S. Geological Survey–Biological Resources Division. Corvallis, OR.

Smith P. and Smith J. (1994) Historical change in the bird fauna of western New South Wales: Ecological patterns and conservation implications. In *Future of the Fauna of*

Western New South Wales. D. Lunney, S. Hand, P. Reed, and D. Butcher (eds.), pp. 123–147. Royal Zoological Society of New South Wales, Sydney.

Smith P.A. (1994) Autocorrelation in logistic regression modelling of species' distributions. *Global Ecology and Biogeography Letters* 4:47–61.

Smith S.M. (1978) The "underworld" in a territorial sparrow: Adaptive strategy for floaters. *American Naturalist* 112:571–582.

Smyth A. and Young J. (1996) Observations on the endangered Black-breasted Button-quail *Turnix melanogaster* breeding in the wild. *Emu* 96:202–206.

Snow D.W. (1958) The breeding of the Blackbird *Turdus merula* at Oxford. *Ibis* 100:1–30.

Snyder N.F.R., Derrickson S.R., Beissinger S.R., Wiley J.W., Smith T.B., Toone W.D., and Miller B. (1996) Limitations of captive breeding in endangered species policy. *Conservation Biology* 10:338–348.

Snyder N.F.R., Koenig S.E., Koschmann J., Snyder H.A., and Johnson T.B. (1994) Thick-billed Parrot releases in Arizona. *Condor* 96:845–862.

Snyder N.F.R. and Snyder H.A. (1989) Biology and conservation of the California Condor. *Current Ornithology* 6:175–263.

Soderquist T.R. (1994) The importance of hypothesis testing in reintroduction biology: Examples from the reintroduction of the carnivorous marsupial *Phascogale tapoatafa*. In *Reintroduction Biology of Australian and New Zealand Fauna*. M. Serena (ed.), pp. 159–164. Surrey Beatty and Sons, Chipping Norton, New South Wales, Australia.

Sorenson M.D. and Fleischer R.C. (1996) Multiple independent transpositions of mito-chondrial DNA control region sequences to the nucleus. *Proceedings of the National Academy of Science* 93:15239–15243.

Sotherton N.W. (1991) Conservation headlands: A practical combination of intensive cereal farming and conservation. In *The Ecology of Temperate Cereal Fields*. L.G. Firbank, N. Carter, J.F. Darbyshire, and G.R. Potts (eds.), pp. 373–397. Blackwell, Oxford.

Soulé M.E. (1980) Thresholds for survival: Maintaining fitness and evolutionary potential. In *Conservation Biology: An Evolutionary–Ecological Perspective*. M.E. Soulé and B.A. Wilcox (eds.), pp. 151–169. Sinauer Associates, Sunderland, MA.

Soulé M.E. (1985) What is conservation biology? *BioScience* 35:727–734.

Soulé M.E. (ed.) (1986) *Conservation Biology: The Science of Scarcity and Diversity*. Sinauer Associates, Sunderland, MA.

Soulé M.E. (ed.) (1987) *Viable Populations for Conservation*. Cambridge University Press, Cambridge.

Soulé M.E. (1991) Conservation: Tactics for a constant crisis. *Science* 253:744–750.

Soulé M.E., Bolger D.T., Alberts A.C., Wright J., Sorice M., and Hill S. (1988) Reconstructed dynamics of rapid extinctions of chaparral-requiring birds in urban habitat islands. *Conservation Biology* 2:75–92.

Soulé M.E. and Kohm K.A. (eds.) (1989) *Research Priorities for Conservation Biology*. Island Press, Washington, DC.

Soulé M.E. and Simberloff D. (1986) What do genetics and ecology tell us about the design of nature reserves? *Biological Conservation* 35:19–40.

Soulé M.E. and Wilcox B.A. (eds.) (1980) *Conservation Biology*. Sinauer Associates, Sunderland, MA.

Soutar A. and Issacs J.D. (1974) Abundance of pelagic fish during the 19th and 20th centuries as recorded in anaerobic sediment off the Californias. *Fisheries Bulletin* 72:257–273.

Southern W.E. (1974) Copulatory wing-flagging: A synchronizing stimulus for nesting Ring-billed Gulls. *Bird-Banding* 45:210–216.

Southwood T.R.E. (1978) *Ecological Methods—With Particular Reference to the Study of Insect Populations,* 2nd edition. Chapman and Hall, London.

Sowls L.W. and Bartonek J.C. (1974) Seabirds—Alaska's most neglected resource. *Transactions of the North American Wildlife Natural Resources Conference* 39:117–126.

Sparrow H.R., Sisk T.D., Ehrlich P.R., and Murphy D.D. (1994) Techniques and guidelines for monitoring Neotropical butterflies. *Conservation Biology* 8:800–809.

Sparrowe R.D. (1995) Wildlife managers—Don't forget to dance with the one that brung you. *Wildlife Society Bulletin* 23:556–563.

Speiser R. and Bosakowski T. (1987) Nest site selection by Northern Goshawks in northern New Jersey and southeastern New York. *Condor* 89:387–394.

Springer A.M. (1992) A review: Walleye pollock in the North Pacific—how much difference do they really make? *Fisheries and Oceanography* 1:80–96.

Squires J.R. and Ruggiero L.F. (1996) Nest-site preference of Northern Goshawks in southcentral Wyoming. *Journal of Wildlife Management* 60:170–177.

Stacey P.B., Johnson V.A., and Taper M.L. (1997) Migration within metapopulations: The impact upon local population dynamics. In *Metapopulation Biology: Ecology, Genetics and Evolution.* I. Hanski and M. Gilpin (eds.), pp. 267–291. Academic Press, New York.

Stacey P.B. and Taper M. (1992) Environmental variation and the persistence of small populations. *Ecological Applications* 2:18–29.

Stalmaster M.V. and Gessaman J.A. (1984) Ecological energetics and foraging behavior of overwintering Bald Eagles. *Ecological Monographs* 54:407–428.

Stanley Price M.R. (1989) *Animal Reintroductions: The Arabian Oryx in Oman.* Cambridge University Press, Cambridge.

Stanley Price M.R. (1991) A review of mammal re-introductions, and the role of the Re-introduction Specialist Group of IUCN/SSC. *Zoological Society of London Symposium* 62:9–26.

Stanners D. and Bordeau P. (1994) *Europe's Environment: The Dobrís Assessment.* European Environment Agency, Copenhagen.

Steadman D.W. (1992) Extinct and extirpated birds from Rota, Mariana Islands. *Micronesica* 25:71–84.

Steadman D.W. (1995a) Determining the natural distribution of resident birds in the Mariana Islands, phase I. Preliminary report (22 April 1995) on the results of field work conducted during June–July 1994. Unpublished report on file with U.S. Fish and Wildlife Service, Honolulu, HI.

Steadman D.W. (1995b) Prehistoric extinction of Pacific island birds. *Science* 267:1123–1131.

Steadman D.W. (1996) Human-caused extinction of birds. In *Biodiversity II—Understanding and Protecting Our Biological Resources.* M.L. Reaka-Kudla, D.E. Wilson, and E.O. Wilson (eds.), pp.139–161. Joseph Henry Press, Washington, DC.

Steele J.H. (1996) Regime shifts in fisheries management. *Fisheries Research* 25:19–23.

Steele R., Pfister R.D., Ryker R.A., and Kittams J.A. (1981) *Forest habitat types of central Idaho.* INT-GTR-114. USDA Forest Service, Boise, ID.

Steenhof K. (1982) Use of an automated geographic information system by the Snake River Birds of Prey Research Project. *Computers Environment and Urban Systems* 7:245–251.

Steenhof K. (ed.) (1994) Technology transfer. In *Snake River Birds of Prey National Conservation Area 1994 Annual Report,* pp. 2–10. Raptor Research and Technical Assistance Center, U.S. Department of the Interior, National Biological Survey. Boise, ID.

Steidl R.J., Hayes J.P., and Schauber E. (1997) Statistical power analysis in wildlife research. *Journal of Wildlife Management* 61:270–279.

Stephens S. and Maxwell S. (eds.) (1996) *Back from the Brink: Refining the Threatened Species Recovery Process.* Surrey Beatty, Sydney.

Stewart G. and Hull A.C. (1949) Cheatgrass in southern Idaho. *Ecology* 30:58–74.

Stiles F.G. (1988) Altitudinal movements of birds on the Caribbean slope of Costa Rica: Implications for conservation. *Memoirs of the California Academy of Science* 12:243–258.

Stinson C.M. and Stinson D.W. (1994) Nest sites, clutch size and incubation behavior in the Golden White-eye. *Journal of Field Ornithology.* 65:65–69.

Stinson D.W. and Glass P.O. (1992) The Micronesian Megapode (*Megapodius laperouse*): Conservation and research needs. *Zoologische Verhandelingen (Leiden)* 278:53–55.

Stinson D.W., Ritter M.W., and Reichel J.D. (1991) The Mariana Common Moorhen: Decline of an island endemic. *Condor* 93:38–43.

Stith B.M., Fitzpatrick J.W., Woolfenden G.E., and Pranty B. (1996) Classification and conservation of metapopulations: A case study of the Florida Scrub Jay. In *Metapopulations and Wildlife Conservation.* D.R. McCullough (ed.), pp. 187–215. Island Press, Washington, DC.

Stone C.P., Walker R.L., Scott J.M., and Banko P.C. (1983) Hawaiian Goose management and research: Where do we go from here? '*Elepaio* 44:11–15.

Stophlet J.J. (1946) Birds of Guam. *Auk* 63:534–540.

Storch I. (1995) Annual home ranges and spacing patterns of Capercaillie in central Europe. *Journal of Wildlife Management* 59:392–400.

Stotz D.F., Fitzpatrick J.W., Parker T.A., III, and Moskovits D.K. (1996) *Neotropical Birds: Ecology and Conservation.* University of Chicago Press, Chicago.

Stouffer P. and Bierregaard R.O., Jr. (1995a) Effects of forest fragmentation on understory hummingbirds in Amazonian Brazil. *Conservation Biology* 9:1085–1094.

Stouffer P. and Bierregaard R.O., Jr. (1995b) Use of Amazonian forest fragments by understory insectivorous birds. *Ecology* 76:2429–2445.

Strahl S.D. (1992) Furthering avian conservation in Latin America. *Auk* 109:680–682.

Strahl S.D. and Grajal A. (1991) Conservation of large avian frugivores and the management of Neotropical protected areas. *Oryx* 25:50–55.

Stribley J.M. (1993) *Factors Influencing Cowbird Distributions in Forested Landscapes of Northern Michigan.* MS thesis, Michigan State University, East Lansing.

Stroud D. and Glue D. (1991) *Britain's Birds in 1989–90.* British Trust for Ornithology/Nature Conservancy Council, Thetford, UK.

Stuart S.N. (1991) Re-introductions: To what extent are they needed? *Zoological Society of London Symposium* 62:27–38.

Suárez A.V., Pfenning K.S., and Robinson S.K. (In press) Edges are not all equal: Nesting success of a disturbance-dependent songbird. *Conservation Biology.*

Swartzman G.L. and Haar R.T. (1983) Interactions between fur seal populations and fisheries in the Bering Sea. *Fisheries Bulletin* 81:121–132.

Szaro R.C. (1986) Guild management: An evaluation of avian guilds as a predictive tool. *Environmental Management* 10:681–688.

Szaro R.C. and Balda R.P. (1982) *Selection and Monitoring of Avian Indicator Species: An Example from a Ponderosa Pine Forest in the Southwest*. USDA Forest Service, Fort Collins, CO.

Szymczak M.R. (1975) *Canada Goose Restoration along the Foothills of Colorado*. Colorado State Wildlife Technical Publication 31:1–64.

Tajima F. (1989) The effect of change in population size on DNA polymorphism. *Genetics* 123:597–601.

Takekawa J.E., Carter H.R., and Harvey T.E. (1990) Decline of the Common Murre in central California, 1980–1986. *Studies in Avian Biology* 14:149–163.

Tarr C.L. and Fleischer R.C. (1993) Mitochondrial DNA variation and evolutionary relationships in the Amakihi complex. *Auk* 110:825–831.

Tarr C.L. and Fleischer R.C. (1995) A molecular assessment of genetic variability and population differentiation in the endangered Mariana Crow (*Corvus kubaryi*). *Report to U.S. Fish and Wildlife Service*, Honolulu, HI.

Tarvin K.A. and K.G. Smith (1995) Microhabitat factors influencing predation and success of suburban Bluejay (*Cyanocitta cristata*) nests. *Journal of Avian Biology* 26:296–304.

Tasker M.L., Furness R.W., Harris M.P., and Bailey R.S. (1989) *Food consumption of seabirds in the North Sea (abstract)*. International Council for the Exploration of the Sea Symposium 66. Multispecies Models to Management of Living Resources. International Council for the Exploration of the Sea, Copenhagen.

Tatner P. (1982) The breeding biology of magpies *Pica pica* in an urban environment. *Journal of Zoology, London* 197:559–581.

Tautz D. (1989) Hypervariability of simple sequences as a general source for polymorphic DNA markers. *Nucleic Acids Research* 17:6463–6471.

Taylor A.L. and Forsman E.D. (1976) Recent range extensions of the Barred Owl in western North America, including the first record for Oregon. *Condor* 78:560–561.

Taylor P.D., Fahrig L., Henein K., and Merriam G. (1993) Connectivity is a vital element of landscape structure. *Oikos* 68:571–573.

Taylor R.H. (1979) How the Macquarie Island Parakeet became extinct. *New Zealand Journal of Ecology* 2:42–45.

Tear T.H., Scott J.M., Hayward P.H., and Griffith B. (1993) Status and prospects for success of the Endangered Species Act: A look at recovery plans. *Science* 262:976–977.

Tear T.H., Scott J.M., Hayward P.H., and Griffith B. (1995) Recovery plans and the endangered species act: Are criticisms supported by data? *Conservation Biology* 9:182–195.

Temple S.A. (1977) The concept of managing endangered birds. In *Endangered Birds: Management Techniques for Preserving Threatened Species*. S.A. Temple (ed.), pp. 3–7. University of Wisconsin Press, Madison.

Temple S.A. (1986) The problem of avian extinctions. *Current Ornithology* 3:453–485.

Temple S.A. (ed.) (1978) *Endangered Birds: Management Techniques for Preserving Threatened Species*. University of Wisconsin Press, Madison.

Temple S.A. and Cary J.R. (1988) Modeling dynamics of habitat-interior bird populations in fragmented landscapes. *Conservation Biology* 2:340–347.

Temple S.A. and Wiens J.A. (1989) Bird populations and environmental changes: Can birds be bio-indicators? *American Birds* 43:260–270.

Temple S.A. and Wilcox B.A. (1986) Introduction: Predicting effects of habitat patchiness and fragmentation. In *Wildlife 2000*. J. Verner, M.L. Morrison, and C.J. Ralph (eds.), pp. 261–262. University of Wisconsin Press, Madison.

Terborgh J. (1975) Faunal equilibria and the design of wildlife preserves. In *Tropical Ecological Systems: Trends in Terrestrial and Aquatic Research*. F.B. Golley and E. Medina (eds.), pp. 369–380. Springer-Verlag, Berlin.

Terborgh J. (1983) *Five New World Primates*. Princeton University Press, Princeton, NJ.

Terborgh J. (1989) *Where Have All the Birds Gone?* Princeton University Press, Princeton, NJ.

Terborgh J. and Petren K. (1991) Development of habitat structure through succession in an Amazonian floodplain forest. In *Habitat Structure: The Physical Arrangement of Objects in Space*. S.S. Bell, E.D. McCoy, and H.R. Mushinsky (eds.), pp. 28–46. Chapman and Hall, New York.

Terborgh J. and Weske J.S. (1969) Colonization of secondary habitats by Peruvian birds *Ecology* 50:765–782.

Terborgh J. and Winter B. (1983) A method for siting parks and reserves with special reference to Colombia and Ecuador. *Biological Conservation* 33:95–117.

Terborgh J., Robinson S.K., Parker T.A., III, Munn C.A., and Pierpont N. (1990) Structure and organization of an Amazonian forest bird community. *Ecological Monographs* 60:213–238.

Theobald D.M., Gosnell H., and Riebsame W.E. (1996) Land use and landscape change in the Colorado Mountains II: A case study of the East River Valley. *Mountain Research and Development* 16:407–418.

Thibodeau F.R. (1983) Endangered species: Deciding which species to save. *Environmental Management* 7:101–107.

Thiollay J.-M. (1996) Distributional patterns of raptors along altitudinal gradients in the northern Andes and effects of forest fragmentation. *Journal of Tropical Ecology* 12:535–560.

Thomas J.W. (1988) Effectiveness—the hallmark of the natural resource management professional. *Transactions North American Wildlife and Natural Resources Conference* 51:27–38.

Thomas J.W. (ed.) (1979) *Wildlife habitats in managed forests: The Blue Mountains of Oregon and Washington*. U.S. Department of Agriculture, Forest Service. Agriculture Handbook Number 553, Washington, DC.

Thomas J.W., Forsman E.D., Lint J.B., Meslow E.C., Noon B.R., and Verner J. (1990) *A conservation strategy for the Northern Spotted Owl*. 791–171/20026. U.S. Government Printing Office, Washington, DC.

Thompson D.R., Furness R.W., and Walsh P.M. (1992) Historical changes in mercury concentrations in the marine ecosystem of the north and northeast Atlantic Ocean as indicated by seabird feathers. *Journal of Applied Ecology* 29:79–84.

Thompson D.R., Hamer K.C., and Furness R.W. (1991) Mercury accumulation in Great Skuas *Catharacta skua* of known age and sex, and its effects on breeding and survival. *Journal of Applied Ecology* 28:672–684.

Thompson F.R., III (1993) Simulated responses of a forest interior migrant bird population to forest management options in central hardwood forests of the United States. *Conservation Biology* 7:325–333.

Thompson F.R., III (ed.) (1996) *Management of midwestern landscapes for the conservation of migrant landbirds*. NC-GTR-187. USDA Forest Service, St. Paul, MN.

Thompson F.R., III, Probst J.R., and Raphael M.G. (1993) Silvicultural options for

Neotropical migratory birds. In *Status and Management of Neotropical Migratory Birds*. RM-GTR-229. USDA Forest Service, Ft. Collins, CO.

Thompson F.R., III, Robinson S.K., Donovan T.M., Faaborg J., and Whitehead D.R. (In press) Biogeographic, landscape, and local factors affecting cowbird abundance and host parasitism levels. In *The Ecology and Management of Cowbirds*. T. Cook, S.K. Robinson, S.I. Rothstein, and S.G. Sealy (eds.). University of Texas Press, Austin.

Thompson F.R., III, Robinson S.K., Whitehead D.R., and Brawn J.D. (1996) Management of central hardwood landscapes for the conservation of Neotropical migratory birds. In *Management of Midwestern Landscapes for the Conservation of Migrant Landbirds*. NC-GTR-187. USDA Forest Service, St. Paul, MN.

Thomsen J.B. and Mulliken T.A. (1992) Trade in Neotropical Psittacines and its conservation implications. In *New World Parrots in Crisis: Solutions from Conservation Biology*. S.R. Beissinger and N.F.R. Snyder (eds.), pp. 221–239. Smithsonian Institution Press, Washington, DC.

Thorn C.R., McAda D.P., and Kernodle J.M. (1993) *Geohydrologic framework and hydrologic conditions in the Albuquerque Basin, central New Mexico*. Water Resources Investigations Report 93–4149. U.S. Geological Survey, Washington, DC.

Tiainen J., Hanski I.K., Pakkala T., Piiruoinen J., and Yrjölä R. (1989) Clutch-size, nestling growth and nestling mortality of the Starling *Sturnus vulgaris* in south Finnish agro-environments. *Ornis Fennica* 66:41–48.

Tidemann S.C., McOrist S., Woinarski J.C.Z., and Freeland W.J. (1992) Parasitism of wild Gouldian Finches *Erythrura gouldiae* by the air-sac mite *Sternostoma tracheacolum*. *Journal of Wildlife Diseases* 28:80–84.

Tilman D. (1989) Ecological experimentation: Strengths and conceptual problems. In *Long-term studies in ecology*. G.E. Likens (ed.), pp. 136–157. Springer-Verlag, New York.

Tinbergen J.M., van Balen J.H., and van Eck H.M. (1985) Density dependent survival in an isolated Great Tit population (*Parus major*): Kluyver's data reanalysed. *Ardea* 73:38–48.

Toland B.R. (1986) Nesting ecology of Northern Harriers in southwest Missouri. *Transactions of the Missouri Academy of Sciences* 20:49–57.

Tomback D.F. and Taylor C.L. (1986) Tourist impact on Clark's Nutcracker foraging activities in Rocky Mountain National Park. In *Proceedings of the Fourth Triennial Conference on Research in National Parks and Equivalent Reserves*. F.J. Singer (ed.), pp. 158–172. Colorado State University, Fort Collins, CO.

Tomialojc L. (1979) The impact of predation on urban and rural Woodpigeon (*Columba palumbus* [L.]) populations. *Polish Ecological Studies* 5:141–220.

Tomialojc L. and Gehlbach F.R. (1988) Introduction and concluding remarks. In *Symposium on Avian Population Responses to Man-Made Environments*. L. Tomialojc and F.R. Gehlbach (eds.), pp. 1777 and 1824–1825. Acta XIX Congressus Internationalis Ornithologici, Ottawa, Canada.

Tomich P.Q., Wilson N., and Lamoureaux C.H. (1968) Ecological factors on Manana Island, Hawaii. *Pac. Sci.* 12:352–368.

Tomlenson C., Mace G., Black J.M., and Hewston N. (1991) Management of an inbred species: The White-Winged Wood Duck. *Wildfowl* 42:123–133.

Toth E.F. and Baglien J.W. (1986) Summary: When habitats fail as predictors—the manager's viewpoint. In *Wildlife 2000: Modeling Habitat Relationships of Terrestrial Vertebrates*. J. Verner, M.L. Morrison, and C.J. Ralph (eds.), pp. 255–256. University of Wisconsin Press, Madison.

Tracy C.R. and Brussard P.F. (1994) Preserving biodiversity: Species in landscapes. *Ecological Applications* 4:205–207.

Triggs S.J., Williams M.J., Marshall S.J., and Chambers G.K. (1992) Genetic structure of Blue Duck (*Hymenolaimus malacorhynchos*) populations revealed by DNA fingerprinting. *Auk* 109:80–89.

Trine C.L. (1996) *Mechanisms underlying population dynamics of a migratory songbird in a fragmented forest.* Ph.D. Thesis, University of Illinois at Champaign-Urbana.

Tristram H.B. (1885) *The Fauna and Flora of Palestine.* Palestine Exploration Fund, London.

Tucker G.M. (1992) Effects of agricultural practice on field use by invertebrate-feeding birds in winter. *Journal of Applied Ecology* 29:779–790.

Tucker G.M. and Heath M.F. (1994) *Birds in Europe: Their Conservation Status.* BirdLife International, Cambridge, England.

Turchin P. (1991) Translating foraging movements in heterogeneous environments into the spatial distribution of foragers. *Ecology* 72:1253–1266.

Turner M.G., Arthaud G.J., Engstrom R.T., Hejl S.J., Liu J., Loeb S., and McKelvey K. (1995) Usefulness of spatially explicit models in land management. *Ecological Applications* 5:12–16.

Turner M.G., Wu Y., Romme W.H., and Wallace L.L. (1993) A landscape simulation model of winter foraging by large ungulates. *Ecological Modelling* 69:163–184.

Uhl C. (1988) Restoration of degraded lands in the Amazon Basin. In *Biodiversity.* E.O.Wilson and F.M. Peter (eds.), pp. 326–332. National Academy Press, Washington, DC.

Uhl C., Clark K., Clark H., and Murphy P. (1981) Early plant succession after cutting and burning in the upper Rio Negro region of the Amazon Basis. *Journal of Ecology* 69:631–649.

Underwood A.J. (1995) Ecological research and (and research into) environmental management. *Ecological Applications* 5:232–247.

U.S. Antarctic Marine Living Resources Program (1995) *AMLR 1994/95 Field Season Report—Objectives, Accomplishments and Tentative Conclusions.* J. Rosenberg (ed.). Administrative Report LJ-95-13. NOAA, LaJolla, CA.

U.S. Department of Commerce (1981) *Final environmental impact statement on the incidental take of Dall Porpoise in the Japanese salmon fishery.* National Marine Fisheries Service, National Oceanic and Atmospheric Administration, Washington, DC.

USDA (U.S. Department of Agriculture), Forest Service (1992) *Final environmental impact statement on management for the Northern Spotted Owl in the national forests.* USDA Forest Service, Portland, OR.

USDA (U.S. Department of Agriculture), Forest Service (1995a) *Record of decision, final environmental impact statement for the management of the Red-cockaded Woodpecker and its habitat on national forests in the southern region, vol. 1.* Management Bulletin R8-MB-73. USDA Forest Service, Atlanta, GA.

USDA (U.S. Department of Agriculture), Forest Service (1995b) *Wildlife, fish and sensitive plant habitat management.* USDA Forest Service, Washington, DC.

USDA (U.S. Department of Agriculture), Forest Service (1997a) *Draft environmental impact statement for the revised land and resource management plan, national forests in Florida.* Management Bulletin R8-MB-77B, USDA Forest Service, Atlanta, GA.

USDA (U.S. Department of Agriculture), Forest Service (1997b) *Tongass Land and Resource Managment Plan.* Management Bulletin R10-MB-338DD, USDA Forest Service, Juneau, AK.

USDA and USDOI (U.S. Department of Agriculture and U.S. Department of the Interior) (1994) *Final supplemental environmental impact statement on management of habitat for late-successional and old-growth forest related species within the range of the Northern Spotted Owl.* USDA and USDOI, Portland, OR.

USDOI (U.S. Department of the Interior) (1979a) *Snake River Birds of Prey environmental impact statement.* Bureau of Land Management, Boise District, Boise, ID.

USDOI (U.S. Department of the Interior) (1979b) *Snake River Birds of Prey special research report to the secretary of the interior.* Bureau of Land Management, Boise District, Boise, ID.

USDOI (U.S. Department of the Interior) (1985) *Snake River Birds of Prey area management plan.* Bureau of Land Management, Boise District, Boise, ID.

USDOI (U.S. Department of the Interior) (1988) *Special status species management.* U.S. Department of the Interior. Bureau of Land Management, Washington, DC.

USDOI (U.S. Department of the Interior) (1992) *Final draft recovery plan for the Northern Spotted Owl.* USDOI, Washington, DC.

USDOI (U.S. Department of the Interior) (1995) *Snake River Birds of Prey National Conservation Area management plan.* Bureau of Land Management, Lower Snake River District Office, Boise, ID.

USDOI (U.S. Department of the Interior) (1996a) *Effects of military training and fire in Snake River Birds of Prey National Conservation Area.* BLM/IDARNG Research Project Final Report. U.S. Geology Service, Biological Resources Division, Snake River Field Station, Boise, ID.

USDOI (U.S. Department of the Interior) (1996b) *Wildlife Research Report for Navy-Leased Lands on the Island of Tinian, Commonwealth of the Northern Mariana Islands.* Unpublished report to U.S. Navy, Pacific Division of the Naval Facilities Engineering Command, Pearl Harbor, HI.

USDOI (U.S. Department of the Interior), Fish and Wildlife Service (1996) *Endangered Spp. Bull.* 21:1–5.

USGS (U.S. Geological Survey) (1996) *Strategic scientific plan.* U.S. Geological Survey, Biological Resources Division, Washington, DC.

Vader W., Barrett R.T., Erikstad K.E., and Strann K.B. (1990) Differential responses of Common and Thick-billed Murres to a crash in the capelin stock in the southern Barents Sea. *Studies in Avian Biology* 14:175–180.

Valdes A.M., Slatkin M., and Friemer N.B. (1993) Allele frequencies at microsatellite loci: The stepwise mutation model revisited. *Genetics* 133:737–749.

Vale T.R. (1975) Presettlement vegetation in sagebrush-grass areas of the intermountain west. *Journal of Range Management* 28:32–36.

Valiela I., Parsons D.J. and Johnson A.E. (1989) Additional views. In *Long-term Studies in Ecology: Approaches and Alternatives.* G.E. Likens (ed.), pp. 158–169. Springer-Verlag, New York.

van Aarde R.J. (1979) Distribution and density of feral cat *Felis catus* on Marion Island. *South African Journal of Antarctic Research* 9:14–19.

van Aarde R.J. (1980) The diet and feeding behaviour of feral cats, *Felis catus* at Marion Island. *South African Journal of Antarctic Research* 10:123–128.

Van den Bosch F., Hengeveld R., and Metz J.A.J. (1992) Analyzing the velocity of animal range expansion. *Journal of Biogeography* 19:135–150.

Vane-Wright, R.I., Humphries C.J., and Williams P.H. (1991) What to protect—systematics and the agony of choice. *Biological Conservation* 55:235-254.

van Halewyn R. and Norton R.L. (1984) The status and conservation of seabirds in the Caribbean. In *Status and Conservation of the World's Seabirds*. J.P. Croxall, P.G.H. Evans, and R.W. Schreiber (eds.), pp. 169–222. ICBP Technical Publication No. 2. Page Brothers, Norwich, England.

Van Horn M.A., Gentry R.M., and Faaborg J. (1995) Patterns of Ovenbird (*Seiurus aurocapillus*) pairing success in Missouri forest tracks. *Auk* 112:98–106.

Van Horne B. (1983) Density as a misleading indicator of habitat quality. *Journal of Wildlife Management* 47:893–901.

van Riper C., III, van Riper S.G., Goff M.L., and Laird M. (1986) The epizootiology and ecological significance of malaria in Hawaiian land birds. *Ecological Monographs* 56:327–344.

VanDruff L.W., Bolen E.G., and San Julian G.J. (1994) Management of urban wildlife. In *Research and Management Techniques for Wildlife and Habitat,* 5th edition. T.A. Bookhout (ed.), pp. 507–530. Wildlife Society, Bethesda, MD.

Vardi J., Benvenisti R., and Seroussi S. (eds.) (1995) *Development options for cooperation: The Middle East/ East Mediterranean region.* Version IV, Government of Israel, Tel Aviv.

Vardi J., Benvenisti R., Seroussi S., and Dilevsky J. (eds.) (1996) *Programs for regional cooperation—1997.* Government of Israel, Tel Aviv.

Varley G.C. and Gradwell G.R. (1960) Key factors in population studies. *Journal of Animal Ecology* 29:399–401.

Varvio S.-L., Chakraborty R., and Nei M. (1986) Genetic variation in subdivided populations and conservation genetics. *Heredity* 57:189–198.

Veit R.R. and Lewis M.A. (1996) Dispersal, population growth, and the Allee effect: Dynamics of the House Finch invasion of eastern North America. *American Naturalist* 148:255–274.

Veit R.R., McGowan J.A., Ainley D.G., Wahl T.R., and Pyle P. (1997) Apex marine predator declines ninety percent in association with changing oceanic climate. *Global Change Biology* 3:23–28.

Veitch C.R. (1994) Habitat repair: A necessary prerequisite to translocation of threatened birds. In *Reintroduction Biology of Australian and New Zealand Fauna.* M. Serena (ed.), pp. 97–104. Surrey Beatty and Sons, Chipping Norton, New South Wales, Australia.

Venn D.R. and Fisher J. (1993) Red-tailed Black-cockatoo *Calyptorhynchus banksii graptogyne.* Action Statement no. 37. Victorian Department of Conservation and Natural Resources, Melbourne.

Verboom J., Schotman A., Opdam P., and Metz J.A.J. (1991) European Nuthatch metapopulations in a fragmented agricultural landscape. *Oikos* 61:149–156.

Vermeer K. (1963) The breeding ecology of the Glaucous-winged Gull (*Larus glaucescens*) on Mandarte Island, B.C. *Occasional Papers Serial British Columbia Provincial Museum* 13:1–104.

Verner J. (1984) The guild concept applied to management of bird populations. *Environmental Management* 8:1–14.

Verner J., McKelvey K.S., Noon B.R., Gutierrez R.J., Gould G.I., Jr, and Beck T.W. (tech. coords.) (1992) *The California Spotted Owl: A technical assessment of its current status.* PSW- GTR-133. USDA Forest Service, Berkeley, CA.

Vickery P.D., Herkert J.R., Knopf F.L., Ruth J., and Keller C.E. (In press) Grassland birds: An overview of threats and recommended management strategies. In *Proceedings of the 1995 Partners in Flight Conference,* Cape May, NJ.

Vickery P.D., Hunter M.L., Jr., and Melvin S.M. (1994) Effects of habitat-area on the distribution of grassland birds in Maine. *Conservation Biology* 8:1087–1097.

Vickery P.D., Hunter M.L., Jr., and Wells J.V. (1992a) Evidence of incidental nest predation and its effects on nests of threatened grassland birds. *Oikos* 63:281–288.

Vickery P.D., Hunter M.L., Jr., and Wells J.V. (1992b). Is density an indicator of breeding success? *Auk* 109:706–710.

Vickery P.D., Hunter M.L., Jr., and Wells J.V. (1992c) Use of a new reproductive index to evaluate relationship between habitat quality and breeding success. *Auk* 109:697–705.

Viella F.J. and Arnizaut A.B. (1994) Making the best of Mother Nature: Managing the Puerto Rican Parrot after Hurricane Hugo. *Endangered Species Technical Bulletin* 19:10–11.

Villard M.-A., Freemark K.E., and Merriam G. (1992) Metapopulation theory and Neotropical migrant birds in temperate forests: An empirical investigation. In *Ecology and Conservation of Neotropical Migrant Landbirds*. J.M. Hagan and D.W. Johnston (eds.), pp. 474–482. Smithsonian Institution Press, Washington, DC.

Villard M.-A., Martin P.R., and Drummond C.G. (1993) Habitat fragmentation and pairing success in the Ovenbird (*Seiurus aurocapillus*). *Auk* 110:759–768.

Villard M.-A. and Maurer B.A. (1996) Geostatistics as a tool for examining the hypothesized declines in migratory songbirds. *Ecology* 77:59–68.

Villard M.-A., Merriam G., and Maurer B.A. (1995) Dynamics in subdivided populations of Neotropical migratory birds in a fragmented temperate forest. *Ecology* 76:27–40.

Villard M.-A. and Taylor P.D. (1994) Tolerance to habitat fragmentation influences the colonization of new habitat by forest birds. *Oecologia* 98:393–401.

Vincent J. (1966–71) *Red Data Book, Aves*. International Union for the Conservation of Nature, Morges, Switzerland.

Vitousek, P.M. (1994) Beyond global warming: Ecology and global change. *Ecology* 75:1861–1876.

Vitousek P.M., D'Antonio C.M., Loope L.L., and Westbrooks R. (1996) Biological invasions as global environmental change. *American Scientist* 84:468–478.

Vogler A.P. and DeSalle R. (1994) Diagnosing units of conservation management. *Conservation Biology* 8:354–363.

Volkov A.E. (ed.) (1996) *Strict Nature Reserves (Zapovedniks) of Cultural Revolution in Soviet Union*. Indiana University Press, Bloomington, IN.

Wagner F.J., Meslow E.C., Bennett G.M., Larson C.J., Small S.M., and DeStefano S. (1996) Demography of Northern Spotted Owls in the southern Cascades and Siskiyou Mountains, Oregon. *Studies in Avian Biology* 17:67–76.

Wakeley J. and Hey J. (1997) Estimating ancestral population parameters. *Genetics* 145:847–855.

Walker B.W. (1995) Conserving biological diversity through ecosystem resilience. *Biological Conservation* 9:747–752.

Walker C.H. (1990) Persistent pollutants in fisheating seabirds—Bioaccumulation, metabolism and effects. *Aquatic Toxicology* 17:293–324.

Walker C.H. (1992) The ecotoxicology of persistent pollutants in marine fish-eating birds. In *Persistent Pollutants in Marine Ecosystems*. C.H. Walker and D.R. Livingstone (eds.), pp. 211–232. Pergamon Press, New York.

Walters C.J. (1986) *Adaptive Management of Natural Resources*. Macmillan, New York.

Walters C.J. and Holling C.S. (1990) Large-scale management experiments and learning by doing. *Ecology* 71:2060–2068.

Walters J.R. (1991) Application of ecological principles to the management of endangered species: The case of the Red-cockaded Woodpecker. *Annual Review of Ecology and Systematics* 22:505–523.

Walters J.R., Carter J.H., III, Doerr P.D., and Copeyon C.K. (1995) Response to drilled artificial cavities by Red-cockaded Woodpeckers in the North Carolina Sandhills: 4-year assessment. In *Red-cockaded Woodpecker: Recovery, Ecology and Management.* D.L. Kulhavy, R.G. Hooper, and R. Costa (eds.), pp. 380–384. Stephen F. Austin State University, Nacogdoches, TX.

Warham J. (1990) *The Petrels—Their Ecology and Breeding Systems.* Academic Press, New York.

Warner R.E. (1963) Recent history and ecology of the Laysan Duck. *Condor* 65:2–23.

Warner R.E. (1994) Agricultural land use and grassland habitat in Illinois: Future shock for Midwestern birds? *Conservation Biology* 8:147–156.

Watkins D. (1995) East Asian–Australasian Shorebird Reserve Network proposal. *Stilt* 27:7–10.

Watson J.C., Carlson D.L., Taylor W.E., and Milling T.E. (1995) Restoration of the Red-cockaded Woodpecker population on the Francis Marion National Forest: Three years post Hugo. In *Red-cockaded Woodpecker: Recovery, Ecology and Management.* D.L. Kulhavy, R.G. Hooper, and R. Costa (eds.), pp. 172–182. Stephen F. Austin State University, Nacogdoches, TX.

WCMC (World Conservation Monitoring Center) (1992) *Global biodiversity: Status of the world's living resources.* Chapman and Hall, London.

Weatherhead P.J. (1986) How unusual are unusual events? *American Naturalist* 128:150–154.

Weatherhead P.J. and Forbes M.R.L. (1994) Natal philopatry in passerine birds: Genetic or ecological influences? *Behavioral Ecology* 5:426–433.

Weber P. (1995) Protecting oceanic fisheries and jobs. In *State of the World 1995—A Worldwatch Institute Report on Progress Toward a Sustainable Society.* L. Brown (ed.), pp. 21–37. W.W. Norton. New York.

Wege D.C. and Long A.J. (1995) *Key Areas for Threatened Birds in the Neotropics.* Bird Life Conservation Series 5. Bird Life International, Cambridge, England.

Weigensberg I. and Roff D.A. (1996) Natural heritabilities: Can they be reliably estimated in the laboratory? *Evolution* 50:2149–2157.

Weller M. (1988) *Wildfowl in Winter.* University of Minnesota Press, Minneapolis.

Wennergren U., Ruckelshaus M., and Kareiva P. (1995) The promise and limitations of spatial models in conservation biology. *Oikos* 74:349–356.

Wenny D.G., Clawson R.L., Sheriff S.L., and Faaborg J. (1993) Population variation, habitat selection, and minimum area requirements of three forest interior warblers in central Missouri. *Condor* 95:968–979.

Weyerhaeuser Company (1992) *Doak Mountain management plan for forest health and eagle habitat.* Weyerhaeuser Company. Tacoma, WA.

Wheeler A.G. and Calver M.C. (1995) Resource partitioning in an island community of insectivorous birds during winter. *Emu* 96:23–31.

Whisenant S.G. (1990) Changing fire frequencies on Idaho's Snake River plains: Ecological and management implications. In *Proceedings: Symposium on Cheatgrass Invasion,*

Shrub Die-Off, and Other Aspects of Shrub Biology and Management. E.D. McArthur, E.M. Romney, S.D. Smith, and P.T. Tueller (eds.), pp. 4–10. INT-GTR-276. USDA Forest Service, Boise, ID.

Whitcomb R.F., Robbins C.S., Lynch J.F., Whitcomb B.L., Klimkiewicz M.K., and Bystrak D. (1981) Effects of forest fragmentation on avifauna of the eastern deciduous forest. In *Forest Island Dynamics in Man-Dominated Landscapes.* R.L. Burgess and B.M. Sharpe (eds.), pp. 125–206. Springer-Verlag, New York.

White G.C. and Garrott R.A. (1990) *Analysis of Wildlife Radio-Tracking Data.* Academic Press, San Diego.

White J. (1994) How the terms savanna, barrens, and oak openings were used in early Illinois. In *Proceedings of the North American Conference on Barrens and Savannas.* Fralish et al. (eds.), pp. 25–64. Illinois State University, Bloomington.

White J.R., Hofmann P.S., Urquhart K.A.F., Hammond D., and Baumgartner S. (1989) *Selenium Verification Study, 1987–1988: A Report to the California State Water Resources Control Board.* California Department of Fish and Game, Sacramento.

White R.R. (1988) Winter grounds and migration patterns of the Upland Sandpiper. *American Birds* 42:1247–1253

Whitehead P.J., and Tschirner K. (1991) Lead shot ingestion and lead poisoning of Magpie Geese (*Anseranas semipalmata*) foraging in a northern Australian hunting reserve. *Biological Conservation* 58: 99–118

Whitehead P.J., Wilson B.A., and Saalfeld K. (1992) Managing the Magpie Goose in the Northern Territory: Approaches to conservation of mobile fauna in a patchy environment. In *Conservation and Development Issues in North Australia.* I. Moffatt and A. Webb (eds.), pp. 90–104. North Australian Research Unit, Darwin.

Whitney B.M. and Pacheco J.P. (1995) Distribution and conservation status of four *Myrmotherula* antwrens (Formicariidae) in the Atlantic Forest of Brazil. *Bird Conservation International* 5:421–439.

Wiens J.A. (1976) Population responses to patchy environments. *Annual Review of Ecology and Systematics* 7:81–120.

Wiens J.A. (1983) Avian community ecology: An iconoclastic view. In *Perspectives in Ornithology.* A.H. Brush and G.A. Clark, Jr. (eds.), pp. 355–403. Cambridge University Press, Cambridge.

Wiens J.A. (1989) Spatial scaling in ecology. *Functional Ecology* 3:385–397.

Wiens J.A. (1994) Habitat fragmentation: Island vs. landscape perspectives on bird conservation. *Ibis* 137:S97–S104.

Wiens J.A. (1995) Landscape mosaics and ecological theory. In *Mosaic Landscapes and Ecological Processes.* L. Hansson, L. Fahrig, and G. Merriam (eds.), pp. 1–26. Chapman and Hall, London.

Wiens J.A., Crist T.O., With K.A., and Milne B.T. (1995) Fractal patterns of insect movement in microlandscape mosaics. *Ecology* 76:663–666.

Wiens J.A. and Milne B.T. (1989) Scaling of "landscapes" in landscape ecology, or landscape ecology from a beetle's perspective. *Landscape Ecology* 3:87–96.

Wiens J.A. and Rotenberry J.T. (1980) Bird community structure in cold shrub deserts: Competition or chaos? *Acta XVII Congressus Internationalis Ornithologici*:1063–1070.

Wiens J.A. and Rotenberry J.T. (1981) Habitat associations and community structure of birds in shrubsteppe environments. *Ecological Monographs* 51:21–41.

Wiens J.A. and Rotenberry J.T. (1985) Response of breeding passerine birds to range-land alteration in a North American shrubsteppe locality. *Journal of Applied Ecology* 22:655–668.

Wiens J.A., Rotenberry J.T., and Van Horne B. (1985) Territory size variation in shrub-steppe birds. *Auk* 102:500–505.

Wiens J.A., Rotenberry J.T., and Van Horne B. (1986) A lesson in the limitations of field experiments: Shrubsteppe birds and habitat alteration. *Ecology* 67:365–376.

Wilcove D.S. (1985) Nest predation in forest tracts and the decline of migratory song-birds. *Ecology* 66:1211–1214.

Wilcove D.S. (1994a) Response: Tracy and Bussard (1994). *Ecological Applications* 4:207–208.

Wilcove D.S. (1994b) Turning conservation goals into tangible results: The Case of the Spotted Owl and old-growth forests. In *Large Scale Ecology and Conservation Biology*. P.J. Edwards, R.M. May, and N.R. Webb (eds.), pp. 313–329. Blackwell, London.

Wilcove D.S., McClellan C.H., and Dobson A.P. (1986) Habitat fragmentation in the temperate zone. In *Conservation Biology: The Science of Scarcity and Diversity*. M.E. Soulé (ed.), pp. 237–256. Sinauer Associates, Sunderland, MA.

Wilcove D.S., McMillan M., and Winston K.C. (1993) What exactly is an endangered species? An analysis of the U.S. endangered species list: 1985–1991. *Conservation Biology* 7:87–93.

Wilcove D.S. and Terborgh J.W. (1984) Patterns of population decline in birds. *American Birds* 38:10–13.

Wiles G.J. (1987) Current research and future management of Marianas Fruit Bats (Chiroptera:Pteropodidae) on Guam. *Australian Mammalogy* 10:93–95.

Wiles G.J., Aguon C.F., Davis G.W., and Grout D.J. (1995) The status and distribution of endangered animals and plants in northern Guam. *Micronesica* 28:31–49.

Wiles G.J., Beck R.E., and Amerson A.B. (1987) The Micronesian Megapode on Tinian, Mariana Islands. *'Elepaio* 47:1–3.

Wiley J.W. and Wunderle J.M., Jr. (1993) The effects of hurricanes on birds, with special reference to Caribbean islands. *Bird Conservation International* 3:319–349.

Williams B.L. and Marcot B.G. (1991) Use of biodiversity indicators for analyzing and managing forest landscapes. *Transactions of the North American Wildlife and Natural Resources Conference* 56:613–627.

Williams G.L., Russell K.R., and Seitz W.K. (1978) Pattern recognition as a tool in the ecological analysis of habitat. In *Classification, inventory, and analysis of fish and wildlife habitat*. A. Marmelstein (ed.), pp. 521–531. FWS/OBS-78/76. U.S. Department of the Interior, Fish and Wildlife Service, Washington, DC.

Williams M. (1991) Social and demographic characteristics of Blue Duck *Hymenolaimus malacorhynchos*. *Wildfowl* 42:65–86.

Williams M. and Robertson C.J.R. (1997) The Campbell Island Teal *Anas auklandica nesiotis*: History and review. *Wildfowl* 47:134–165.

Williams N. (1995) Slow start for Europe's habitat protection plan. *Science* 269:320–322.

Williams P.H., Gibbons D., Margules C., Rebelo A., Humphries C., and Pressey R. (1996) A comparison of richness hotspots, rarity hotspots, and complementary areas for conserving diversity of British birds. *Conservation Biology* 10:155–174.

Williams T.D. (1995) *The Penguins: Spheniscidae*. Oxford University Press, New York.

Willis E.O. (1974) Populations and local extinctions of birds on Barro Colorado Island Panama. *Ecological Monographs* 44:153–169.

Willis E.O. (1979) The composition of avian communities in remanescent woodlots in southern Brazil. *Papeis Avulsos de Zoologia* 33:1–25.

Willis E.O. and Eisenmann E. (1979) A revised list of the birds of Barro Colorado Island, Panama. *Smithsonian Contributions in Zoology* 291:1–31.

Willson M.F. (1995) Biodiversity and ecological processes. In *Biodiversity in Managed Landscapes: Theory and practice*. R. Szaro and D. Johnson (eds.). Oxford University Press, New York.

Willson M.F., de Santo T., Sabag C., and Armesto J.J. (1994) Avian communities of fragmented south-temperate rainforests in Chile. *Conservation Biology* 8:508–520.

Wilme L. (1994) Status, distribution and conservation of two Madagascar birds species endemic to Lake Alaotra: Delacour's Grebe *Tachybaptus rufolavatus* and Madagascar Pochard *Aythya innota*. *Biological Conservation* 69:15–21.

Wilmut I., Schnieke A.E., McWhir J., Kind A.J., and Campbell K.H.S. (1997) Viable offspring derived from fetal and adult mammalian cells. *Nature* 385:810–813.

Wilson A. (1812) *American Ornithology*. Bradford and Inskeep, Philadelphia.

Wilson A.C. and Stanley Price M.R. (1994) Reintroduction as a reason for captive breeding. In *Creative Conservation: The Interface between Captive and Wild Populations*. P.J.S. Olney, G.M. Mace, and A.T.C. Feistner (eds.), pp. 243–264. Chapman and Hall, London.

Wilson E.O. (1992) *The Diversity of Life*. Belknap Press, Cambridge, MA.

Wingate D.B. (1977) Excluding competitors from Bermuda Petrel nesting burrows. In *Endangered Birds: Management Techniques for Preserving Threatened Species*. S. A. Temple (ed.), pp. 93–102. University of Wisconsin Press, Madison.

Wingate D.B. (1985) The restoration of Nonsuch Island as a living museum of Bermuda's pre-colonial biome. *ICBP Technical Publication* 3:225–238.

With K.A. (1994) Using fractal analysis to assess how species perceive landscape structure. *Landscape Ecology* 9:25–36.

With K.A. and Crist T.O. (1995) Critical thresholds in species responses to landscape structure. *Ecology* 76:2446–2459.

Witteman G.J. and Beck R.E., Jr. (1991) Decline and conservation of the Guam Rail. In *Wildlife Conservation: Present Trends and Perspective for the 21st Century*. N. Maruyama et al. (eds.), pp. 173–177. Proceedings of the International Symposium on Wildlife Conservation, Tokyo.

Wittenberger J.F. (1978) The breeding biology of an isolated Bobolink population in Oregon. *Condor* 80:355–371.

Woinarski J.C.Z. (1993) Australian tropical savannas, their avifauna, conservation status and threats. In *Birds and Their Habitats: Current Knowledge and Conservation Priorities in Queensland*. C. Catterall, P. Driscoll, K. Hulsman, and A. Taplin (eds.), pp. 45–63. Queensland Ornithological Society, Brisbane.

Woinarski J.C.Z. and Recher H.F. (In press) Impact and response: A review of the effects of fire on the Australian avifauna. *Pacific Conservation Biology*.

Woinarski J.C.Z., Whitehead P.J., Bowman D.M.J.S., and Russell-Smith J. (1992) Conservation of mobile species in a variable environment: The problem of reserve design in the Northern Territory, Australia. *Global Ecology Biogeography Letters* 2:1–10.

Woldhek S. (1980) *Bird Killing in the Mediterranean*. European Committee for the Prevention of Mass Destruction of Migratory Birds. Zeist, The Netherlands.

Woodford M.H. and Rossiter P.B. (1994) Disease risks associated with wildlife translocation projects. In *Creative Conservation: The Interface between Captive and Wild Populations.* P.J.S. Olney, G.M. Mace, and A.T.C. Feistner (eds.), pp. 178–200. Chapman and Hall, London.

Woodward P.W. (1972) The natural history of Kure Atoll, northwestern Hawaiian Islands. *Atoll Research Bulletin* 164:1–318.

Woolfenden G.E. and Fitzpatrick J.W. (1991) Florida Scrub Jay ecology and conservation. In *Bird Population Studies—Relevance to Conservation and Management.* C.M. Perrins, J.D. Lebreton, and G.J.M. Hirons (eds.), pp. 542–565. Oxford University Press, New York.

Wooton J.T. and Bell D.A. (1992) A metapopulation model of the Peregrine Falcon in California: Viability and management strategies. *Ecological Applications* 2:307–321.

World Bank (1994) *World Development Report 1994. Infrastructure for Development. World Development Indicators.* Published for the World Bank by Oxford University Press, New York.

World Resources Institute (1992) *World Resources 1992–1993.* Oxford University Press, New York.

World Resources Institute (1994) *World Resources 1994–95.* Oxford University Press, New York.

Wray T., III, Strait K.A., and Whitmore R.C. (1982) Reproductive success of grassland sparrows on a reclaimed surface mine in West Virginia. *Auk* 99:157–164.

Wright H.A., Neuenschwander L.F., and Britton C.M. (1979) *The role and use of fire in sagebrush-grass and pinyon-juniper plant communities.* GTR-INT-58. USDA Forest Service, Boise, ID.

Wunderle J.M. and Latta S.C. (1996) Avian abundance in sun and shade coffee plantations and remnant pine forest in the Cordillera Central, Dominican Republic. *Ornitologia Neotropical* 7:19–34.

Wyllie I. and Newton I. (1991) Demography of an increasing population of Sparrowhawks. *Journal of Animal Ecology* 60:749–766.

Yaffee S.L. (1994) *The Wisdom of the Spotted Owl: Policy Lessons for a New Century.* Island Press, Washington, DC.

Yaffee S.L. (1996) Ecosystem management in practice: The importance of human institutions. *Ecological Applications* 6:724–727.

Yaffee S.L., Phillips A.F., Frentz I.C., Hardy P.W., Maleki S.M., and Thorpe B.E. (1996) Bitterroot Ecosystem Management Research Project. In *Ecosystem Management in the United States: An Assessment of Current Experience,* pp. 93–94. Island Press, Washington, DC.

Yauk C.L. and Quinn J.S. (1996) Multilocus DNA fingerprinting reveals high rate of heritable genetic mutation in Herring Gulls nesting in an industrialized urban site. *Proceedings of the National Academy of Science* 93:12137–12141.

Yensen D.L. (1982) *A grazing history of southwestern Idaho with emphasis on the Birds of Prey Study Area.* U.S. Department of the Interior, Bureau of Land Management, Boise District, Boise, ID.

Yom-Tov Y. (1987) The reproductive rates of Australian passerines. *Australian Wildlife Research* 14:319–330.

Yom-Tov Y. and Mendelssohn H. (1988) Changes in the distribution and abundance of vertebrates in Israel during the 20th Century. In *The Zoogeography of Israel.* Y. Yom-Tov and E. Tchernov (eds.), pp. 515–547. W. Junk, Dordrecht, Holland.

Yom-Tov Y. and Tchernov E. (eds.) (1988) *The Zoogeography of Israel.* W. Junk, Dordrecht, Holland.

Yorio P. and Boersma P.D. (1992) The effects of human disturbance on Magellanic Penguin *(Spheniscus magellanicus)* behavior and breeding success. *Bird Conservation International* 2:161–173.

Yosef R. (1997) Physical distances among individuals in flocks of Greater Flamingoes *(Phoenicopterus ruber)* are affected by human disturbance. *Israel Journal of Zoology* 43:79–85.

Young H.G. and Smith J.G. (1989) The search for the Madagascar Pochard *Aythya innota*: Survey of Lac Alaotra, Madagascar, October–November, 1989. *Dodo, Journal of the Wildlife Preservation Trust* 26:17–34.

Young J.A. and Evans R.A. (1978) Population dynamics after wildfires in sagebrush grasslands. *Journal of Range Management* 31:283–289.

Young M.K. (1995) *Conservation Assessment for Inland Cutthroat Trout.* RM-GTR-256. USDA Forest Service, Fort Collins, CO.

Zahavi A. (1955) Observations of birds in the Hula swamp. *Salit* 1:2–14, 22–26.

Zahavi A. (1957) The breeding birds of Hula swamp and lake. *Ibis* 99:600–607.

Zimmerman J.L. (1988) Breeding season habitat selection by the Henslow's Sparrow *(Ammodramus henslowii)* in Kansas. *Wilson Bulletin* 100:17–24.

Zimmerman J.L. (1992) Density-independent factors affecting the avian diversity of the tallgrass prairie community. *Wilson Bulletin* 104:85–94.

Zimmerman J.L. (1993) *The Birds of Konza: The Avian Ecology of the Tallgrass Prairie.* University Press of Kansas, Lawrence.

Zimmerman J.L. (1997) Avian community responses to fire, grazing, and drought in the tallgrass prairie. In *Ecology and Conservation of Great Plains Vertebrates.* F.L. Knopf and F.B. Samson (eds.), pp. 167–180. Springer-Verlag, New York.

Zink R.M. and Kale H.W. (1995) Conservation genetics of the extinct Dusky Seaside Sparrow. *Biological Conservation* 74:69–71.

Contributing Authors

Edward B. Arnett, Weyerhaeuser Company, P.O.Box 275, Springfield, OR 97477

Steven R. Beissinger, Division of Ecosystem Science, University of California, Berkeley, CA 94720–3110

Jeffrey M. Black, The Wildfowl & Wetlands Trust, Slimbridge, Gloucestershire, GL2 7BT, United Kingdom (Currently: Department of Wildlife, Humboldt State University, Arcata, CA 95526)

P. Dee Boersma, Zoology Department, Box 351800, University of Washington, Seattle, WA 98195

Jeffrey D. Brawn, Illinois Natural History Survey, 609 E. Peabody Dr., Champaign, IL 61068

Earl W. Campbell III, Ohio Cooperative Fish and Wildlife Research Unit, Ohio State University, 1735 Neil Ave., Columbus OH 43210

Michael W. Collopy, U.S.G.S., Biological Resources Division, Forest and Rangeland Ecosystem Science Center, Corvallis, OR 97331

Cecelia M. Dargan, U.S.D.A. Forest Service, Coconino National Forest, Flagstaff, AZ 86001

Scott R. Derrickson, Smithsonian Institution, Conservation and Research Center, Front Royal, VA 22630

David F. DeSante, The Institute for Bird Populations, P.O. Box 1346, Point Reyes Station, CA 94956

Therese M. Donovan, U.S.D.A. Forest Service, North Central Forest Experiment Station, 1–26 Agriculture Building, Columbia, MO 65211

John Faaborg, Division of Biological Sciences, University of Missouri, Columbia, MO 65211

Robert C. Fleischer, National Zoological Park, Smithsonian Institution, Washington, DC 20008

Beth M. Galleher, U.S.D.A. Forest Service, Pacific Northwest Research Station, 3625 93rd Ave. SW, Olympia, WA 98512

Vladimir M. Galushin, Russian Bird Conservation Union, Kibalchicha St. #6, Building 5, Office 110, Moscow 129278, Russia

Joseph L. Ganey, U.S.D.A. Forest Service, Rocky Mountain Forest and Range Experiment Station, Flagstaff AZ 86001

Frederick R. Gehlbach, Department of Biology, Baylor University, P.O. Box 97338, Waco, TX 76798

James F. Gore, U.S.D.A. Forest Service, 324 25th Street, Ogden, UT 84401

Alejandro Grajal, Wildlife Conservation Society, 185st and Southern Boulevard, Bronx, NY 10460

Kathleen Milne Granillo, U.S.D.I., Fish and Wildlife Service, P.O. Box 1306, Albuquerque, NM 87103

Jonathan B. Haufler, Boise Cascade Corporation, P.O. Box 50, Boise, ID 83728

Gregory D. Hayward, Department of Zoology and Physiology, University of Wyoming, Laramie, WY 82071

Sallie J. Hejl, U.S.D.A. Forest Service, Intermountain Research Station, P.O. Box 8089, Missoula, MT 59807

James R. Herkert, Illinois Endangered Species Protection Board, 524 South Second Street, Springfield, IL 62701

Richard L. Hutto, Division of Biological Sciences, University of Montana, Missoula, MT 59812

Larry L. Irwin, National Council of the Paper Industry for Air and Stream Improvement, P.O. Box 68, Stevensville, MT 59870

Frances C. James, Department of Biological Science, Florida State University, Tallahassee, FL 32306–1100

Fritz L. Knopf, U.S.G.S. Biological Resources Division, 4512 McMurry Ave., Fort Collins, CO 80525

Michael N. Kochert, U.S.G.S., Biological Resources Division, Forest and Rangeland Ecosystem Science Center, Snake River Field Station, 970 Lusk, Boise, ID 83706

Stephen W. Kress, Seabird Restoration Program, National Audubon Society, 159 Sapsucker Woods Rd., Ithaca, NY 14850

Rony Malka, Israel Nature Reserves Authority, 78 Yermiyahu St., Jerusalem 94467, Israel

David A. Manuwal, Wildlife Science Group, College of Forest Resources, University of Washington, Seattle, WA 98195

John M. Marzluff, Wildlife Science Group, College of Forest Resources, University of Washington, Seattle, WA 98195

Brian A. Maurer, Department of Zoology, Brigham Young University, Provo, UT 84602

Kevin S. McKelvey, U.S.D.A. Forest Service, Pacific Southwest Research Station, 1700 Bayview Drive, Arcata, CA 97331

Ian Newton, Institute of Terrestrial Ecology, Monks Wood, Abbots Ripton, Huntingdon, Cambs PE17 2LS, United Kingdom

Julia K. Parrish, Zoology Department, Box 351800, University of Washington, Seattle, WA 98195

Martin G. Raphael, U.S.D.A. Forest Service, Pacific Northwest Research Station, 3625 93rd Ave. SW, Olympia, WA 98512

Scott K. Robinson, Illinois Natural History Survey, 609 E. Peabody Dr., Champaign, IL 61068

W. Douglas Robinson, Department of Ecology, Ethology, and Evolution, University of Illinois, 606 E. Healey, Champaign, IL 61820

Gordon H. Rodda, U.S.G.S. Biological Resources Division, 4512 McMurry Ave., Fort Collins, CO 80525

Daniel K. Rosenberg, Department of Fisheries and Wildlife, Oregon State University, Corvallis, OR 97331

John T. Rotenberry, Natural Reserve System and Department of Biology, University of California, Riverside, CA 92521

Rex Sallabanks, Sustainable Ecosystems Institute, 30 E. Franklin Rd., Suite 50, Meridian, ID 836642

Elise V. Schmidt, Department of Zoology, Brigham Young University, Provo, UT 84602

Noel F. R. Snyder, Wildlife Preservation Trust International, P.O. Box 426, Portal, AZ 85632

John R. Squires, U.S.D.A. Forest Service, Intermountain Research Station, P.O. Box 8089, Missoula, MT 59807

Susan Stenquist, Sustainable Development and Conservation Biology Program, University of Maryland, College Park, MD 20742

Douglas F. Stotz, Field Museum of Natural History, Environment and Conservation Programs, Roosevelt and Lakeshore Drive, Chicago, IL 60605

Frank R. Thompson III, U.S.D.A. Forest Service, North Central Forest Experiment Station, 1–26 Agriculture Building, Columbia, MO 65211

Daniel E. Varland, Rayonier, 3033 Ingram St., Hoquiam, WA 98550

Marc-André Villard, Departement de biologie, Université de Moncton, Moncton, NB E1A3E9, Canada

Jeffrey R. Walters, Department of Biology, Virginia Polytechnic Institute and State University, Blacksburg, VA 24061–0406

Donald R. Whitehead, Department of Biology, Indiana University, Bloomington, IN 47405

John C. Z. Woinarski, Parks and Wildlife Commission of the Northern Territory, P.O. Box 496, Palmerston, Northern Territory 0831, Australia

Reuven Yosef, International Birding Center, P.O. Box 7743, Elat 88106, Israel

Leonard S. Young, Department of Natural Resources, Forest Land Management Research Center, P.O. Box 47014, Olympia, WA 98504

Victor A. Zubakin, Russian Bird Conservation Union, Kibalchicha St. #6, Building 5, Office 110, Moscow 129278, Russia

Index